Medical Images:
Formation, Handling and Evaluation

NATO ASI Series

Advanced Science Institutes Series

A series presenting the results of activities sponsored by the NATO Science Committee, which aims at the dissemination of advanced scientific and technological knowledge, with a view to strengthening links between scientific communities.

The Series is published by an international board of publishers in conjunction with the NATO Scientific Affairs Division

A Life Sciences B Physics	Plenum Publishing Corporation London and New York
C Mathematical and Physical Sciences D Behavioural and Social Sciences E Applied Sciences	Kluwer Academic Publishers Dordrecht, Boston and London
F Computer and Systems Sciences G Ecological Sciences H Cell Biology I Global Environmental Change	Springer-Verlag Berlin Heidelberg New York London Paris Tokyo Hong Kong Barcelona Budapest

NATO-PCO DATABASE

The electronic index to the NATO ASI Series provides full bibliographical references (with keywords and/or abstracts) to more than 30000 contributions from international scientists published in all sections of the NATO ASI Series. Access to the NATO-PCO DATABASE compiled by the NATO Publication Coordination Office is possible in two ways:

- via online FILE 128 (NATO-PCO DATABASE) hosted by ESRIN, Via Galileo Galilei, I-00044 Frascati, Italy.
- via CD-ROM "NATO-PCO DATABASE" with user-friendly retrieval software in English, French and German (© WTV GmbH and DATAWARE Technologies Inc. 1989).

The CD-ROM can be ordered through any member of the Board of Publishers or through NATO-PCO, Overijse, Belgium.

Series F: Computer and Systems Sciences Vol. 98

Medical Images: Formation, Handling and Evaluation

Edited by

Andrew E. Todd-Pokropek

University College London, Department of Medical Physics
Gower Street, London WC1E 6BT, UK

Max A. Viergever

Utrecht University, Computer Vision Research Group
University Hospital Utrecht E02.222
Heidelberglaan 100, 3584 CX Utrecht, The Netherlands

Springer-Verlag
Berlin Heidelberg New York London Paris Tokyo
Hong Kong Barcelona Budapest
Published in cooperation with NATO Scientific Affairs Division

Proceedings of the NATO Advanced Study Institute on The Formation, Handling and Evaluation of Medical Images, held at Povoa de Varzim, Portugal, September 12–23, 1988

Directors

Andrew E. Todd-Pokropek, London, UK
Max A. Viergever, Utrecht, The Netherlands

Scientific Committee

H. H. Barrett, Univ. of Arizona, Tucson, AZ, USA
A. E. Burgess, Univ. of British Columbia, Vancouver, BC, Canada
G. T. Herman, Univ. of Pennsylvania, Philadelphia, PA, USA
J. J. Koenderink, Univ. of Utrecht, Utrecht, The Netherlands
S. M. Pizer, Univ. of North Carolina, Chapel Hill, NC, USA

Sponsor

NATO Scientific Affairs Division

Co-Sponsors

University College London, UK
Utrecht University, The Netherlands
National Science Foundation, USA
ADAC Inc., USA

CR Subject Classification (1991): J.3, I.4

ISBN-13: 978-3-642-77890-2 e-ISBN-13: 978-3-642-77888-9
DOI: 10.1007/ 978-3-642-77888-9

Softcover reprint of the hardcover 1st edition 1992

Typesetting: Camera ready by authors
45/3140 - 5 4 3 2 1 0 - Printed on acid-free paper

Preface

Medical imaging is not only a very important area in diagnostic (and increasingly therapeutic) medicine, but also a very exciting field. Many new techniques are in the process of being developed or extended, most of which are strongly dependent on the availability of digital methods. Thus while conventional x-radiographs still comprise the bulk of the medical images that are acquired in a hospital (although these are also starting to be handled digitally) the so-called digital methods, such as Computerized Tomography (CT) Magnetic Resonance Imaging (MRI), etc., are now often claimed to have a more significant clinical impact. This book is concerned with three stages in the 'life cycle' related to such digital images: their formation or how the data can be acquired; their handling, or how such data may be manipulated to increase their clinical value; and not least, their evaluation, or how their impact and value may be assessed and hopefully improved.

This book, which is a successor to a previous publication in this series entitled "Mathematics and Computer Science in Medical Imaging" (Vol F39) is divided into three parts. The first part comprises a series of reviews in the general subject area, written by authorities in the field, which will provide tutorial material to assist readers in this field understand the topics of general interest. The second and third parts comprise a series of selected papers, on a variety of subjects and application areas, contributed by participants at the NATO Advanced Study Institute at Povoa in Portugal from which this book resulted. However, since the original meeting, a number of the papers as presented here have been revised and upgraded to reflect progress in the field.

The book is intended to be of value and read by scientists and informed clinicians in the area of medical imaging and the handling of digital images in medicine. It is written in general with the idea that the readership has a reasonable understanding of mathematics and of the physical principles involved in medical imaging, and is therefore able to go into a number of important issues at a depth not often found in associated textbooks. Thus, while this book could be used as part of an undergraduate or graduate course in the area of medical physics or medical informatics, it is probably more appropriate as a contribution to further education in this area.

The editors would like in particular to thank all the contributors to this book, the participants of the Advanced Study Institute itself, the scientific committee and, of course, our sponsors. We hope that this book will provide useful tutorial and reference material to its readers and that it will serve to indicate how exciting we believe this rapidly evolving field to be.

London March 1992

A. Todd-Pokropek
M.A. Viergever

Table of Contents

Part 1

An Introduction to and Overview of the Field

IMAGE RECONSTRUCTION AND THE SOLUTION OF INVERSE PROBLEMS IN MEDICAL IMAGING

Harrison H. Barrett

University of Arizona
Tucson, Az, USA

ABSTRACT

An overview of reconstruction methods applicable to medical tomography is given. Analytic methods based on the continuous form of the inverse Radon transform and matrix inversion and pseudoinversion methods based on a discrete formalism are presented. Statistical principles such as maximum likelihood and Bayesian estimation are also discussed.

1. INTRODUCTION

Many different medical imaging modalities are based on reconstruction of an object from a finite set of its projections. Computed tomography (CT), positron emission tomography (PET) and single-photon emission computed tomography (SPECT) all fit into this category, and all are established clinical tools of proven value in diagnosis. Accurate diagnoses therefore depend on good performance of the reconstruction algorithms, and that in turn depends on a good understanding of the basic physical and mathematical principles underlying reconstruction from projections. In this chapter we present an overview of these principles.

In Section 2 we take a brief look at the imaging modalities listed above, explain how the data are collected in each case, and show that under suitable approximations a common mathematical description, namely line-integral projections, can be used for all of the modalities.

Section 3 then takes a more formal look at reconstructions and develops the theory of the Radon transform; the popular convolution or filtered-backprojection algorithm follows from this theory and is derived in Section 4. Many important algorithms, however, follow from a matrix formulation of the problem as presented in Section 5. An overview of various approaches to inverting this matrix is given in Section 6.

One theme that will run through this discussion is solution of inverse problems. In the context of radiological imaging, a *direct* problem is the calculation of the data set that would be produced by some specified imaging system viewing a specified object. An *inverse* problem is the converse: given the data, determine what object produced it, again with some specified imaging system. Inverse problems are almost always *ill-posed*; they may not have a solution at all, the solution may not be unique, or the solution may be very sensitive to noise in the data.

Though couched in the language of medical imaging, this chapter should also serve as an introduction to ill-posed inverse problems in general.

2. PROJECTION IMAGING SYSTEMS

2.1 Projection radiography

The concept of a projection, as we are using the term here, is easily illustrated by the familiar chest radiograph. As shown in Fig. 1, the three-dimensional (3D) structure of the patient's chest is projected down to a two-dimensional (2D) film, so the film record is in some sense a projection of the patient. To make this statement more precise, we need an accurate mathematical description of the developed x-ray film, but this is actually rather difficult. The x rays emerge more or less omnidirectionally from a source with a small but finite spatial extent. As they pass through the patient's body they are attenuated through photoelectric and Compton scattering processes, both of which depend on the spatially inhomogeneous distribution of bone and soft tissue within the patient's body. The x rays then expose the film, which is developed to yield a nonlinear record of the x-ray exposure.

A number of idealizations are often made in order to obtain a tractable mathematical description of these processes. The x-ray source is treated as a point source of monoenergetic x rays, and Compton scatter in the body is neglected. With these approximations, the x rays are attenuated exponentially as they pass through the body, and the process is described by a total attenuation coefficient $\mu(\mathbf{r})$, where $\mathbf{r}$ is a vector denoting position in the body. Then the x-ray intensity at a point D on the detector plane (see Fig. 1) is given by

$$I(D,S) = I_0 \exp\left[- \int_S^D \mu(\mathbf{r})\, d\ell \right], \tag{1}$$

where the integral is along the line from the source point S to the detector point D, and I_0 is the intensity that would arrive at D in the absence of the attenuating body. It is frequently useful to take the logarithm of this equation in order to obtain an expression linear in the unknown quantity $\mu(\mathbf{r})$. We therefore define $\lambda(D,S)$, which is a function of position D on the film plane, by

$$\lambda(D,S) \equiv - \ln [\, I(D,S)/I_0 \,] = \int_S^D \mu(\mathbf{r})\, d\ell \,. \tag{2}$$

This function is referred to as the 2D line-integral projection of the 3D function $\mu(\mathbf{r})$.

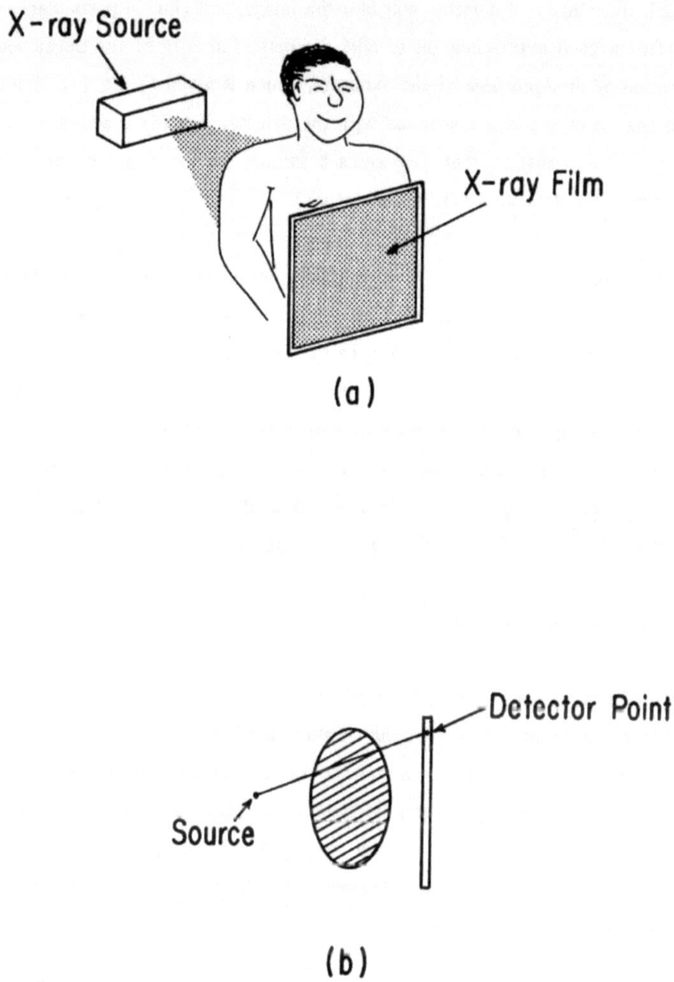

Fig. 1. Conventional chest radiography system. If we neglect scatter and blur due to the detector and the finite source size, the exposure at each point on the detector depends on the attenuation coefficient along the line SD connecting source point and detector location.

A further level of complexity arises if one considers the properties of the image detector in detail, since any real detector will blur the image, and film in particular is highly nonlinear. It is often a good approximation to take the *optical density* of the developed film as a linear function of the *logarithm* of the exposure, which is given by Eq. (1). With this approximation and neglect of the blur associated with the detector, Eq. (2) is adequate to describe the x-ray image. This equation then represents a solution to the direct problem; it enables us to determine the data set given the object $\mu(\mathbf{r})$.

On the other hand, the data set is definitely not adequate to solve the inverse problem, i.e. to determine the full 3D structure of the body as given by $\mu(\mathbf{r})$. One way to see this is to note that there is no way to tell where a particular attenuating object point is located along the line SD; only the line integral matters. Another way to put it is that the single x-ray film represents a superposition of many planes in the 3D object, and there is no way to separate these planes. With either viewpoint, it is easily seen that many different object distributions $\mu(\mathbf{r})$ can yield the same x-ray intensity data at all detector points D, so the data are seriously incomplete for the solution of the inverse problem.

2.2 Longitudinal tomography

The incompleteness of the information in a single projection was recognized already in the 1930s, and a variety of *tomographic* or slice-imaging techniques were developed in an effort to get more accurate information about 3D objects. By moving the source and/or detector during the exposure, the early tomographic systems succeeded in rendering a sharp image of a selected plane while blurring other planes in some way. Since the sharply imaged plane was often parallel to the long axis of the patient's body, these systems were referred to as longitudinal tomographic systems.

Further refinements in longitudinal tomography led to the development of multiplane systems which produced several images, each with a different plane in focus, but none of these systems succeeded in completely eliminating the undesired out-of-focus planes, and therefore none represented an adequate solution to the inverse problem of determining a 3D structure from 2D projections.

2.3 Computed tomography

A breakthrough of immense practical significance occurred in the early 1970s when E.M.I. Ltd. introduced its computed tomography brain scanner (see Fig. 2). In this system, an x-ray source is collimated to a thin pencil beam that passes through the patient's head and is detected by a collimated detector aligned with the pencil beam. Because of the collimation, there is a very low probability of scattered radiation reaching the detector, and the detector reading is well

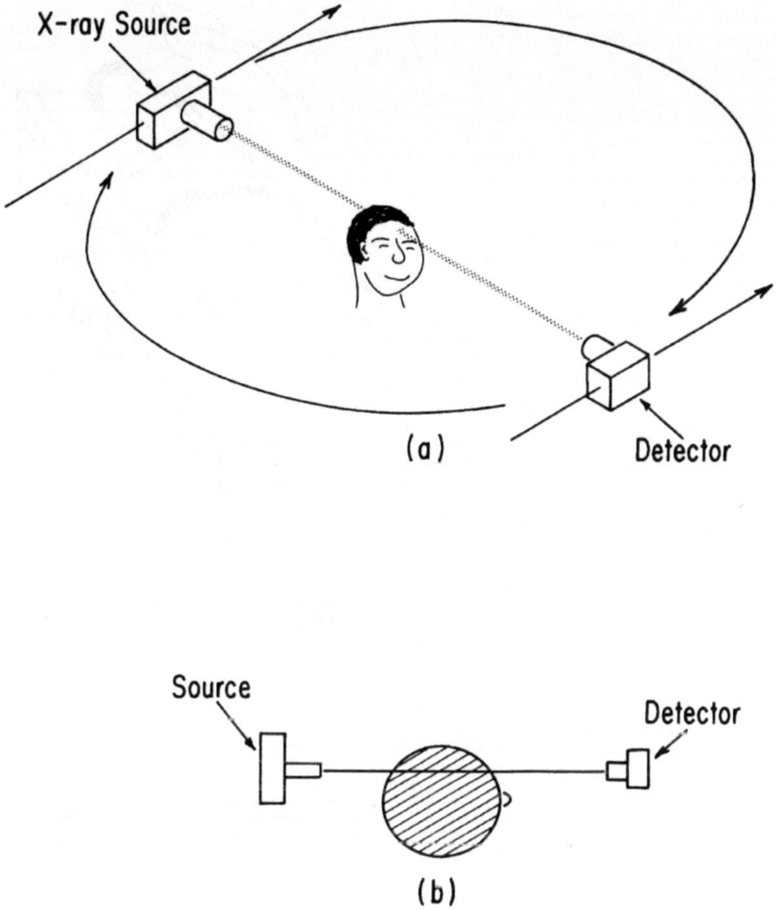

Fig. 2. Schematic representation of a first-generation CT head scanner. A thin pencil beam of x rays passes through the head and is measured with a collimated detector. The source-detector assembly is translated laterally to measure a single 1D projection, $\lambda_\phi(p)$ for fixed ϕ, and then the assembly is rotated slightly to measure projections at other angles.

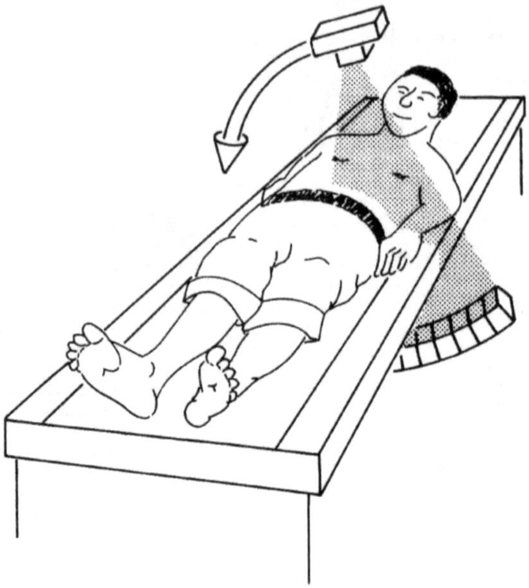

(a)

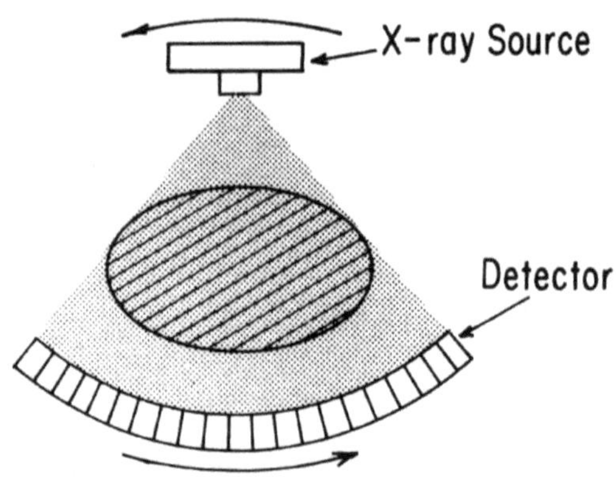

(b)

Fig. 3. Schematic of a modern fan-beam CT body scanner in which an array of detectors is used to measure many line integrals simultaneously. No detector translation is required. After one rotation around the body, sufficient data are available to reconstruct the attenuation distribution in a thin slice of the body.

described by Eq. 1. In this case, however, the detector is not an imaging detector like a piece of film, so it yields only a single number for a fixed position of the source-detector assembly. To acquire a complete data set, the assembly is translated so that the line SD remains in the same orientation but moves laterally. After one translation, the source-detector assembly is indexed slightly in angle (typically by 1^{0}) and another translation is performed. This process is repeated until a full 180^{0} has been covered. In more modern CT systems, a fan beam of x rays and a linear array of detectors are used (Fig. 3), so that many line integrals are measured simultaneously and no translation is necessary. In either case, however, only one thin slice of the head is irradiated, so clearly only this slice can contribute to the data or the reconstruction.

As before it is useful to calculate the logarithm of the intensity data. This quantity is a one-dimensional (1D) function of the lateral displacement, and it is referred to as a 1D projection of the 2D slice being imaged. To the extent that the pencil beam of x rays can be approximated by an ideal line, this projection is given by Eq. (2). The data set obtained in CT consists of many different 1D projections, each at a different projection angle. The inverse problem is to recover $\mu(\mathbf{r})$ from this data set.

The vital difference between CT and longitudinal tomography is the full angular scan used in the former. As we shall see in Section 3, this yields a "complete" data set from which an accurate representation of $\mu(\mathbf{r})$ in the slice can be reconstructed, something that is really impossible in principle with any limited-angle system. In other words, whether or not an adequate solution of the inverse problem can be found depends critically on the data-taking geometry.

2.4 SPECT

Single-photon emission computed tomography or SPECT is a technique for mapping the distribution of a radioactive tracer in nuclear medicine. Each radioactive disintegration results in the emission of a single gamma-ray photon, hence the name. The qualifier *single-photon* is necessary to distinguish the method from two-photon imaging or PET, which we shall discuss in the next section.

Historically, the first imaging system based on reconstruction from projections to be used in clinical medicine was the SPECT system developed by Kuhl and coworkers. Because of poor spatial resolution and inaccurate reconstruction algorithms, this approach did not receive widespread acceptance until after the advent of the E.M.I. scanner. More recent systems, using algorithms very similar to those of CT, are an important adjunct to conventional, nontomographic nuclear imaging.

Modern SPECT systems are usually based on an Anger scintillation camera that is rotated around the body as shown in Fig. 4, though some special-purpose units not based on rotating

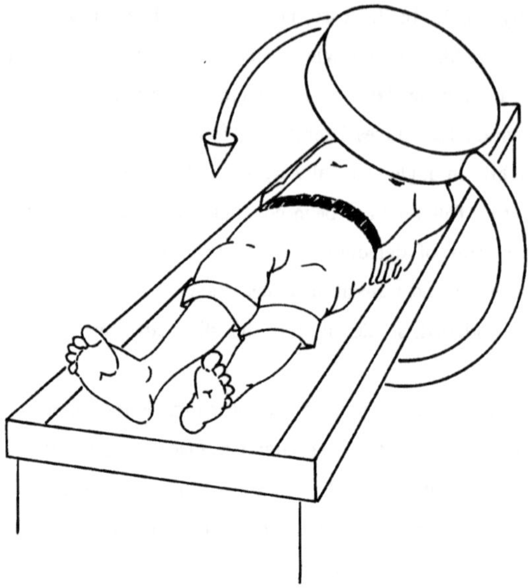

(a)

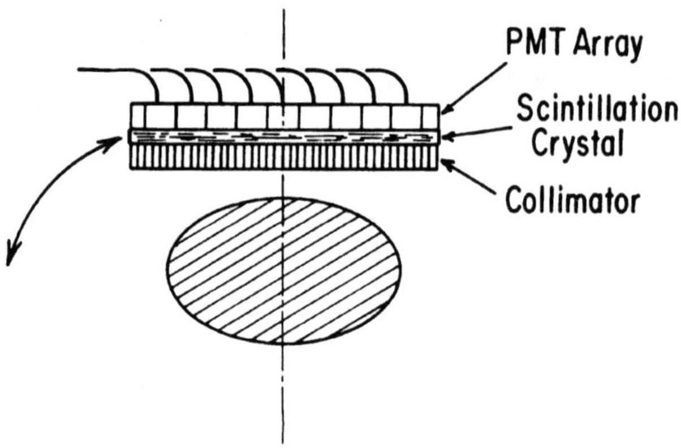

(b)

Fig. 4. Single-photon emission computed tomography or SPECT system for nuclear medicine. The detector is an Anger scintillation camera and parallel-hole collimator that records a full 2D projection simultaneously. Rotation around the body yields sufficient data to reconstruct the activity distribution in a 3D volume.

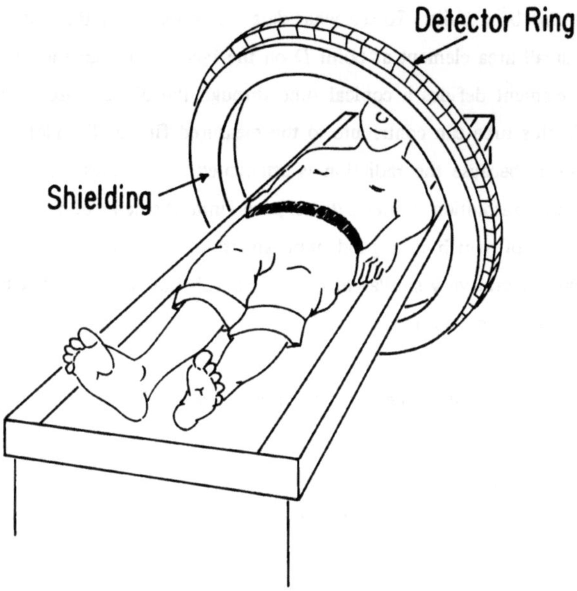

(a)

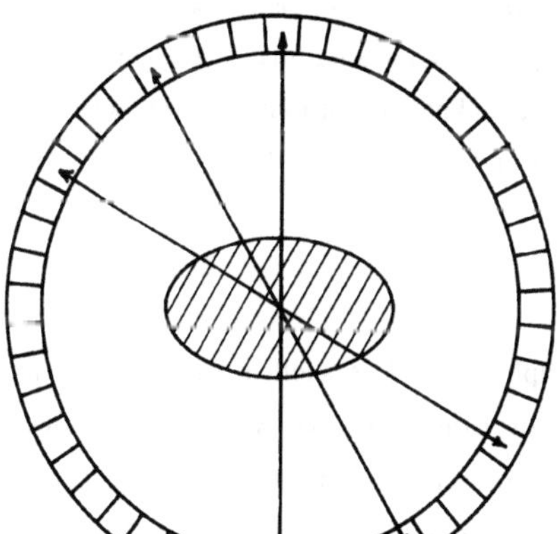

(b)

Fig. 5. Positron-emission tomography or PET system. A full ring of detectors is used without any motion to image a single slice. The rate of coincidence counts in any pair of detectors is related to the integral of the activity along the line connecting the detectors (see text).

cameras have also been built. To see what data are collected by the system depicted in Fig. 4, consider one small area element at point D on the face of the camera. The collimator bore in front of this element defines a conical tube through the object space such that radioactive sources within this tube can contribute to the measured flux at D. Not all sources contribute equally, however, because the radiation is attenuated as it passes through the body. If we neglect the spatial resolution of the detector, the finite diameter of the collection tube and the attenuation (none of which is a good approximation in practice), then the detector at D measures a line integral very similar to that in Eq. (2), but with the activity distribution f(**r**) in place of the attenuation coefficient $\mu(\mathbf{r})$.

If the attenuation is included, the detector measures a weighted line integral of the form

$$g(D,\phi) = \text{const x} \int_{-\infty}^{D} f(\mathbf{r})\, A(\mathbf{r},D,\phi)\, d\ell \,, \tag{3}$$

where the integral runs through the body along a line defined by the detector point D and the collimator, ϕ specifies the orientation of the camera, and the factor $A(\mathbf{r},D,\phi)$ represents the attenuation of gamma rays from the source point **r** to D.

The inverse problem in emission tomography is to recover the activity distribution f(**r**), given a set of measurements {g(D)}. A complication that is evident from Eq. (3) is that the attenuation factor depends on a second unknown distribution, $\mu(\mathbf{r})$, which must be either neglected, approximated, measured separately or estimated from the emission data.

2.5 PET

Positron emission tomography or PET is another method for mapping the distribution of a radioactive tracer, but in this case the radioactive nuclei emit energetic positrons (antielectrons) rather than gamma rays. However, the positrons are rapidly brought to rest within a millimeter or so of their emission point, and they then annihilate by combining with an electron to produce two gamma rays. Conservation of energy requires that the sum of the energies of the two gamma rays equal the sum of the rest-mass energies of the electron and positron (their kinetic energies being negligible). Furthermore, the electron and positron, being essentially at rest before the interaction, have almost no momentum, so the resulting photons cannot have any net momentum either. The only way that energy and momentum can be conserved is for the two photons to have the same energy, namely the rest-mass energy of the electron or positron (511 keV), and to be emitted in opposite directions.

A practical PET system is illustrated in Fig. 5. When an annihilation event occurs in the plane of the detector ring, there is some probability that the two photons will be emitted in the plane also, in which case they will be recorded almost simultaneously by two of the detectors, say D and D′, This simultaneous recording is referred to as a coincidence event. When it occurs, we know that the corresponding annihilation event occurred somewhere along the tube connecting D and D′. If the tube is approximated as a line, the coincidence count rate at these two detectors is given by

$$g(D,D') = \text{const x} \int_{D}^{D'} f(\mathbf{r})\, A(D,D')\, d\ell\,, \tag{4}$$

where the line integral runs along the line connecting D and D′, and A(D,D′) is the appropriate attenuation factor. Note that in this case a line integral has been defined without use of a collimator, so PET systems are much more efficient in collecting photons than are SPECT systems.

The attenuation factor is less of a problem here than in SPECT for two reasons. First, the gamma rays have higher energies than is typical for SPECT, so they are not attenuated as strongly. Second, the attenuation factor depends on only the total line integral of $\mu(\mathbf{r})$ along DD′ and not on the unknown position $\mathbf{r}$ of the annihiliation event; if the annihilation event moves along DD′ towards, say, D, then one gamma ray is more likely to reach D without attenuation, but the other gamma ray is *less* likely to reach D′, and the coincidence rate is in fact unchanged. Nevertheless, A(D,D′) is generally unknown and therefore causes difficulties in solving the inverse problem, which we may here state as recovering the activity distribution $f(\mathbf{r})$ from the coincidence count rates from all *pairs* of detectors D and D′.

3. PROBLEM FORMULATION IN A CONTINUOUS SPACE

In Section 2 we showed that there are several radiographic imaging systems that measure, at least approximately, line integrals of an object. The basic theory of reconstruction of an object from its line integrals was derived long before the advent of x-ray CT, by an Austrian mathematician named Johann Radon (1917). Modern CT may be regarded as an implementation of the inverse Radon transform, which we derive in Section 4, though the inventors of CT were not aware of the work of Radon at the time.

In this section we look more closely at the mathematics of line-integral projection. In Section 3.1 we present a general definition of a line integral, basically stating the Radon transform. In Section 3.2 we look at the information content of a single 1D projection of a 2D object and derive an important theorem that we shall need in Section 4. Section 3.3 discusses sampling requirements, but a full discrete formulation is postponed to Section 5.

Much of the material in Sections 3 and 4 is a condensed version of the author's presentation at an earlier NATO Advanced Study Institute (Barrett, 1988), to which the reader is referred for more detail. In particular, only the case of reconstruction of a 2D object from 1D projections is discussed here; higher dimensions are discussed in the earlier work. Other useful references on the Radon transform are Deans (1983), Helgason (1980) and Barrett (1984).

3.1. Definitions and geometry

Let us begin by considering the 1D projection of a 2D function f(x,y), with the projection being defined by line integrals along a series of lines parallel to the y axis. The projection along the line x=p, denoted $\lambda_0(p)$, is given by

$$\lambda_0(p) = \int_{-\infty}^{\infty} f(p,y)\, dy = \int_{-\infty}^{\infty} dy \int_{-\infty}^{\infty} dx\, f(x,y)\, \delta(p-x) \,. \tag{5}$$

A more general projection is along the line $p = \mathbf{r} \cdot \mathbf{n}$, which is perpendicular to the unit vector $\mathbf{n}$ and a distance p from the origin. In this case the projection is given by

$$\lambda_\phi(p) = \int_{\infty} d^2r\, f(\mathbf{r})\, \delta(p-\mathbf{r}\cdot\mathbf{n}) \,, \tag{6}$$

where the integral runs over the infinite plane, and ϕ is the angle between the x axis and the unit vector $\mathbf{n}$. Note that the Dirac delta functions in Eqs. (5) and (6) are one-dimensional and can be used to perform only one of the two integrals in each equation.

The projection $\lambda_\phi(p)$ can be regarded either as a 1D function of p, parameterized by ϕ, or as a function in a 2D space with polar coordinates p and ϕ. In the latter viewpoint, it is convenient to use a slightly different notation $\lambda(\mathbf{p})$, where $\mathbf{p} = p\mathbf{n}$ is a 2D vector in *Radon space.* The scalar p is thus the magnitude of the vector $\mathbf{p}$, and ϕ is its polar angle relative to the x axis. An operator notation is also useful; we may write

$$\lambda_\phi(p) = \mathcal{R}_2\{f(\mathbf{r})\} \tag{7a}$$

or simply

$$\lambda = \mathcal{R}_2\{f\} \,, \tag{7b}$$

with $\mathcal{R}_2$ being the integral operator implied by Eq. (6).

3.2 The central-slice theorem

To understand what information about the unknown 2D function $f(\mathbf{r})$ is contained in the projection $\lambda_\phi(p)$, let us take the 1D Fourier transform of the projection with respect to p at fixed ϕ:

$$\Lambda_\phi(\nu) = \mathcal{F}_1\{\lambda_\phi(p)\} = \int_{-\infty}^{\infty} dp\ \lambda_\phi(p)\ e^{-2\pi i\nu p} = \int_\infty d^2r \int_{-\infty}^{\infty} dp\ f(\mathbf{r})\ \delta(p-\mathbf{r}\cdot\mathbf{n})\ e^{-2\pi i\nu p}$$

$$= \int_\infty d^2r\ f(\mathbf{r})\ e^{-2\pi i \mathbf{r}\cdot\mathbf{n}\nu} = F(\mathbf{n}\nu) = \mathcal{F}_2\{f(\mathbf{r})\}_{\boldsymbol{\rho}=\mathbf{n}\nu}\ , \qquad (8)$$

where $\mathcal{F}_2\{f(\mathbf{r})\}$ or $F(\boldsymbol{\rho})$ denotes the 2D Fourier transform of $f(\mathbf{r})$, with $\boldsymbol{\rho}$ being the general 2D spatial frequency vector with Cartesian components (ξ,η). Equation (8) is the very important *central-slice theorem.* In words, it says that the 1D Fourier transform of the projection yields one line through the 2D transform of the original object $f(\mathbf{r})$. The line, defined by $\boldsymbol{\rho} = \mathbf{n}\nu$, passes through the origin of the 2D Fourier plane (hence, *central* slice) and makes an angle of ϕ to the ξ axis.

A compact operator expression of Eq. (8) is

$$\mathcal{F}_2 = \mathcal{F}_1\, \mathcal{R}_2\ . \qquad (8')$$

In other words, if we apply the operator $\mathcal{R}_2$ to any function $f(\mathbf{r})$ and then apply $\mathcal{F}_1$ to the result $\lambda_\phi(p)$, it is equivalent to applying $\mathcal{F}_2$ directly to $f(\mathbf{r})$.

If we can obtain $F(\boldsymbol{\rho})$ for all points in the 2D spatial frequency plane, we can reconstruct $f(\mathbf{r})$ by a 2D inverse Fourier transform. The data set is thus said to be complete if all points in the frequency plane are sampled, which will be the case either if the values of p and ϕ satisfy

$$-\infty < p < \infty \quad \text{and} \quad 0 < \phi \le \pi \qquad (9a)$$

or if

$$0 \le p < \infty \quad \text{and} \quad 0 < \phi \le 2\pi\ . \qquad (9b)$$

Note that

$$\lambda_\phi(p) = \lambda_{\phi+\pi}(-p) , \tag{10}$$

since replacing ϕ with $\phi+\pi$ is the same as replacing $\mathbf{n}$ with $-\mathbf{n}$, and the vector $\mathbf{p}$ in Radon space can be written either as $p\mathbf{n}$ or $(-p)(-\mathbf{n})$.

3.3 Sampling

Of course, an actual measured data set cannot satisfy the stringent conditions given above since measurements will be made for a discrete set of p and ϕ values, and it is therefore not possible to determine $F(\boldsymbol{\rho})$ for all spatial frequencies $\boldsymbol{\rho}$. However, if we make some reasonable assumptions about the object $f(\mathbf{r})$, then we can show that its transform $F(\boldsymbol{\rho})$ is adequately determined by a finite set of samples.

If the object Fourier transform has negligible value for frequencies above some maximum, call it ρ_{max}, then we say the object is *essentially band-limited*, and we therefore do not need to know $F(\boldsymbol{\rho})$ outside a circle of radius ρ_{max} in the Fourier plane. We call this region the circle of support of $F(\boldsymbol{\rho})$. Furthermore, if the object $f(\mathbf{r})$ also has a finite size in the space domain, say fitting within a circle of support of diameter D, then sampling theory tells us that we do not need all spatial frequencies, in a continuous sense, to reconstruct the object. Rather, according to the Nyquist condition applied in Fourier space, a finite set with spacing 1/D will suffice. The total number N_s of samples of $F(\boldsymbol{\rho})$ that is required is therefore the area of the circle of support in Fourier space divided by the square of the spacing between samples, or

$$N_s = \pi \rho_{max}^2 D^2 . \tag{11}$$

This number is frequently referred to as the *space-bandwidth product* or the number of *degrees of freedom* of the object.

The above discussion must be modified slightly to apply to computed tomography since the projection data do not yield uniformly spaced samples of $F(\boldsymbol{\rho})$. Let us assume that we measure a 1D projection of $f(\mathbf{r})$ at N_p equally spaced values of p, ranging from $-D/2$ to $D/2$. Since $F(\boldsymbol{\rho})$ is assumed to be negligible for $\rho > \rho_{max}$, it follows from the central-slice theorem that the 1D projection has the same band limit, and the Nyquist sampling condition therefore requires that the spacing between samples Δp satisfy $\Delta p < 1/(2\rho_{max})$. The total number of samples in each projection must then satisfy $N_p = D/\Delta p \geq 2\rho_{max}D$.

If we perform a discrete Fourier transform on the projection, we obtain N_p values of the *1D* transform of the projection, with uniform spacing 1/D between samples. By the central-slice theorem, this set of samples is equivalent to samples of the *2D* transform $F(\boldsymbol{\rho})$ along a radial line in Fourier space.

If projections are recorded at a finite set of angles, with spacing $\Delta\phi = \pi/N_\phi$, then $F(\boldsymbol{\rho})$ is known at sample points on N_ϕ radial lines. The maximum azimuthal spacing between these sample points is at the maximum spatial frequency ρ_{max}, where the spacing is $\rho_{max}\Delta\phi$. Since the radial spacing is $1/D$, it is reasonable to take $\rho_{max}\Delta\phi = 1/D$ also. Though this strategy results in oversampling near the origin of Fourier space, i.e. taking samples more densely than is required by the Nyquist theorem, it at least guarantees that the sampling is adequate over the entire space. It is straightforward to show that the minimum total number of measurements is given by

$$N_{tot} = N_\phi\, N_p = 2\pi\, \rho^2_{max}\, D^2 . \tag{12}$$

This number is just twice the space-bandwidth product of the object as given by Eq. (11); the penalty of a factor of two is the result of oversampling $F(\boldsymbol{\rho})$ near the origin.

4. ANALYTIC INVERSION FORMULAS

In Section 3 we adopted line-integral projections as a reasonable approximate model for CT, SPECT and PET data and formally defined the complete set of line integrals as the Radon transform of the unknown object $f(\mathbf{r})$. In this section we look at one approach to solving the inverse problem of determining $f(\mathbf{r})$ from its projections, namely the inverse Radon transform. Throughout this section we assume that we know the projections $\lambda_\phi(p)$ exactly for all p and ϕ. In other words, we are ignoring the complications due to sampling and noise in the data, but this deficiency will be remedied in later sections.

4.1 The inverse Radon transform

The easiest way to find the inverse Radon transform is to take the inverse 2D Fourier transform of $F(\mathbf{n}\nu)$, as expressed by the central-slice theorem. This is most easily accomplished by letting $\boldsymbol{\rho}$ have polar coordinates (ρ,θ_ρ) and, somewhat unconventionally, taking $-\infty<\rho<\infty$ and $0<\theta_\rho\leq\pi$. With this choice for the range of the coordinates, the area element is given by $d^2\rho = |\rho|d\rho d\theta_\rho$, and the general 2D inverse Fourier transform is given by

$$f(\mathbf{r}) = \mathcal{F}_2^{-1}\{F(\boldsymbol{\rho})\} = \int_{-\infty}^{\infty} |\rho|d\rho \int_0^{\pi} d\theta_\rho\, F(\boldsymbol{\rho})\, e^{2\pi i\boldsymbol{\rho}\cdot\mathbf{r}}, \tag{13}$$

where $\boldsymbol{\rho}\cdot\mathbf{r} = \xi x + \eta y$.

We now specifically invoke the central-slice theorem, eq. (8), which requires that we substitute $\boldsymbol{\rho}=\mathbf{n}\nu$ in $F(\boldsymbol{\rho})$ to obtain $\Lambda\ (\nu)$. This implies that $\rho=\nu$ and $\theta\ =\phi$, so that

$$f(\mathbf{r}) = \int_{-\infty}^{\infty} |\nu| d\nu \int_{0}^{\pi} d\phi \; \Lambda_{\phi}(\nu) \; e^{2\pi i \nu \mathbf{r}\cdot\mathbf{n}} \; . \tag{14}$$

This is one form of the 2D inverse Radon transform, which may be expressed in a handy operator notation as

$$f(\mathbf{r}) = \mathscr{R}_2^{-1}\{\lambda_{\phi}(p)\}, \tag{15a}$$

or simply

$$f = \mathscr{R}_2^{-1}\{\lambda\} \; . \tag{15b}$$

A useful alternative form of the inverse Radon transform may be found by regrouping terms to yield

$$f(\mathbf{r}) = \int_{0}^{\pi} d\phi \left[\int_{-\infty}^{\infty} d\nu \; K_{\phi}(\nu) \; e^{2\pi i \nu p} \right]_{p=\mathbf{r}\cdot\mathbf{n}} \; , \tag{16}$$

where

$$K_{\phi}(\nu) = |\nu| \Lambda_{\phi}(\nu) \tag{17}$$

represents a filtered projection in the 1D frequency domain, with the filter function given by $|\nu|$. We may now rewrite $K_{\phi}(\nu)$ as the 1D Fourier transform of a convolution of two terms:

$$f(\mathbf{r}) = \int_{0}^{\pi} d\phi \; [\lambda_{\phi}(p) * \mathscr{F}_1^{-1}\{|\nu|\}]_{p=\mathbf{r}\cdot\mathbf{n}}, \tag{18}$$

where the asterisk denotes 1D convolution. Equation (18) is still another form of the inverse Radon transform.

4.2 Filtered back-projection

Equation (18) shows that we may reconstruct $f(\mathbf{r})$ in three steps:

1. filter each projection by convolution with $\mathscr{F}_1^{-1}\{|\nu|\}$ (the functional form of which will be discussed later);
2. substitute $\mathbf{r}\cdot\mathbf{n}$ for p in the filtered projections;
3. integrate (sum) over all angles.

Let us discuss steps 2 and 3 first. The substitution of $\mathbf{r}\cdot\mathbf{n}$ for p is the operation of "back-projection", which creates a 2D function of $\mathbf{r}$ from the 1D function of p by smearing the 1D function back in the original direction of projection. Step 3 simply adds up all of the back-projected 2D functions for different projection angles ϕ. In practice, of course, only a finite set of angles is used, and the continuous integral is replaced with a discrete sum.

4.3 The convolution kernel

The remaining problem is to determine the functional form of $\mathscr{F}_1^{-1}\{|\nu|\}$. Note that

$$|\nu| = \nu\ \mathrm{sgn}(\nu) = 2\pi i\nu\ \frac{\mathrm{sgn}(\nu)}{2\pi i}\ , \tag{19}$$

where

$$\mathrm{sgn}(\nu) \equiv \begin{cases} +1 & \text{if } \nu > 0 \\ 0 & \text{if } \nu = 0 \\ -1 & \text{if } \nu < 0 \end{cases} \quad . \tag{20}$$

From tables of Fourier transforms or a simple contour integration, we find that

$$\mathscr{F}_1^{-1}\left\{\frac{\mathrm{sgn}(\nu)}{2\pi i}\right\} = \left(\frac{1}{2\pi^2}\right)\mathscr{P}\{1/p\}\ , \tag{21}$$

where $\mathscr{P}\{1/p\}$ denotes the Cauchy principal value of $1/p$, defined by

$$\int_{-\infty}^{\infty} g(p)\ \mathscr{P}\{1/p\}\ dp \equiv \mathscr{P}\int_{-\infty}^{\infty} \frac{g(p)dp}{p} = \lim_{\epsilon\to 0}\left[\int_{-\infty}^{-\epsilon}\frac{g(p)}{p}\,dp + \int_{\epsilon}^{\infty}\frac{g(p)}{p}\,dp\right]\ , \tag{22}$$

where g(p) is any well-behaved test function. In other words, the effect of the principal value is simply to delete a symmetric infinitesimal neighborhood around the singularity of $1/p$.

The factor $2\pi i\nu$ in eq. (19) may be handled by the Fourier derivative theorem:

$$\mathscr{F}_1^{-1}\{2\pi i\nu G(\nu)\} = dg(p)/dp\ , \tag{23}$$

where $G(\nu)$ is the 1D Fourier transform of g(p). We may also rewrite this theorem by using the fact that the (inverse) Fourier transform of a product is the convolution of the transforms of the individual factors, so that

$$\mathcal{F}_1^{-1}\{2\pi i\nu G(\nu)\} = \mathcal{F}_1^{-1}\{2\pi i\nu\} * g(p). \tag{24}$$

Furthermore, the derivative of a function is obtained by convolving it with the derivative of a delta function:

$$dg(p)/dp = g(p) * \delta'(p). \tag{25}$$

By comparing eqs. (23) and (24) we see that

$$\mathcal{F}_1^{-1}\{2\pi i\nu\} = \delta'(p). \tag{26}$$

We may now put all the pieces together and write

$$\mathcal{F}_1^{-1}\{|\nu|\} = \frac{1}{2\pi^2}\mathcal{P}\left[\frac{1}{p}\right] * \delta'(p) = \frac{1}{2\pi^2}\frac{d}{dp}\mathcal{P}\left[\frac{1}{p}\right]. \tag{27}$$

This equation involves a number of questionable mathematical operations, including the derivative of a function with an infinite discontinuity. The pure mathematician resolves these problems via the theory of generalized functions (Lighthill, 1962), the applied mathematician approaches them through a limiting process, and the engineer uses a technique called "apodization".

4.4 Generalized functions

In fact, we have already introduced the concept of a generalized function in eq. (22) where the function $\mathcal{P}(1/p)$ was defined through its action within an integral. Similarly, the function $(d/dp)\mathcal{P}(1/p)$ is defined by

$$\int_{-\infty}^{\infty} dp\; g(p)\, \frac{d}{dp}\mathcal{P}\left[\frac{1}{p}\right] = -\mathcal{P}\int_{-\infty}^{\infty} dp\; \frac{g'(p)}{p}, \tag{28}$$

where g(p) is a well-behaved test function, and g'(p) is its derivative. The right-hand side of this equation may be formally regarded as resulting from an integration by parts of the left-hand side. With this interpretation in mind, we may adopt the shorthand

$$\frac{d}{dp}\mathcal{P}\left[\frac{1}{p}\right] = -\frac{1}{p^2}. \tag{29}$$

This generalized function behaves just as the ordinary function $-1/p^2$ for $p \neq 0$, but there is an extreme positive singularity, which may be loosely described as an infinite-weight delta function, at p=0. In terms of this generalized function, we have, finally,

$$\mathcal{F}_1^{-1}\{|\nu|\} = -\frac{1}{2\pi^2 p^2} \,. \tag{30}$$

With the shorthand of eq. (29), the inverse Radon transform becomes

$$f(\mathbf{r}) = -\frac{1}{2\pi^2}\int_0^{\pi} d\phi\, [\lambda_\phi(p) * \frac{1}{p^2}]_{p=\mathbf{r}\cdot\mathbf{n}} \,. \tag{31}$$

Since $\mathbf{r}\cdot\mathbf{n} = r[\cos(\theta-\phi)]$, with (r,θ) being the polar coordinates of $\mathbf{r}$, we may also write

$$f(\mathbf{r}) = -\frac{1}{2\pi^2}\int_0^{\pi} d\phi \int_{-\infty}^{\infty} dp\, \frac{\lambda_\phi(p)}{[p - r\cos(\theta-\phi)]^2} \,. \tag{32}$$

This is a very common form for the inverse radon transform, but care must be exercised in interpreting it because of the implied singularity in the integrand. Without this singularity, the result would be nonsensical. Consider what happens in the common case where $f(\mathbf{r})$ is non-negative. Then its projection in any direction is also non-negative, the entire integrand in eq. (32) must be non-negative (except for the singularity), and there is an overall minus sign in front of the integral. The right-hand side is therefore negative, yet it must represent the non-negative quantity on the left; the paradox is resolved through the singularity.

4.5 Apodization

Let us now turn to apodization as another (and more practical) way of dealing with the singularities in the inverse Radon transform. The term apodization comes from the radar literature. It literally means a "cutting off of feet", and in radar it refers to the process of weighting a waveform with some smooth function to eliminate sidelobes ("feet") in the response after signal processing. Here we use it in a somewhat more general sense to refer to any beneficial use of smooth weighting functions. For example, most of the mathematical problems discussed above arose because we were trying to take the Fourier transform of a function, $|\nu|$, that is not absolutely integrable and hence does not possess a Fourier transform in the usual sense. Instead of defining a generalized function to represent the Fourier transform of $|\nu|$, we could instead modify the function to be transformed:

$$|\nu| \rightarrow |\nu|A(\nu) \,, \tag{33}$$

where $A(\nu)$ is some broad, slowly varying function. For example, a suitable choice for $A(\nu)$

$$A(\nu) = \exp(-\pi\alpha^2\nu^2) \ . \tag{34}$$

If α approaches zero, $|\nu|A(\nu)$ approaches $|\nu|$, so we may always recover the original function, but in practice the "roll-off" parameter α is a very useful way to control the response of the system at high spatial frequencies and hence the noise properties of the reconstruction.

We shall now consider the modified form of the inverse Radon transform when the substitution (33) is made. Since we are departing from the mathematically exact formalism at this point, the resulting inverse transform will not represent f(**r**) but rather some approximation to it which we shall denote by $\hat{f}(\mathbf{r})$. The space-domain counterpart of (33) is

$$-\frac{1}{2\pi^2 p^2} \rightarrow -\frac{1}{2\pi^2 p^2} * a(p) \ , \tag{35}$$

where a(p) is the 1D Fourier transform of $A(\nu)$. It is convenient to define

$$h(p) = -\frac{1}{2\pi^2 p^2} * a(p) \ , \tag{36}$$

in terms of which the approximate (apodized) form of eq. (31) becomes

$$\hat{f}(\mathbf{r}) = \int_0^{\pi} d\phi \int_{-\infty}^{\infty} dp \ [\lambda_\phi(p) * h(p)]_{p=r\cos(\theta-\phi)} \ . \tag{37}$$

The steps in the reconstruction are now:

1. filter each projection by convolution with h(p);
2. back-project (substitute $\mathbf{r}\cdot\mathbf{n}$ or $r\cos(\theta-\phi)$ for p in the filtered projections);
3. integrate (sum) over all angles.

This procedure is often referred to as the filtered-backprojection algorithm or as the convolution algorithm.

4.6 MTF and PSF in the convolution algorithm

Since convolution is associative, the substitution (35) is equivalent to replacing $\lambda_\phi(p)$ with $\lambda_\phi(p) * a(p)$ in eq. (31). One easy way to see the effect of this substitution is to use the central-slice theorem to write

$$\mathcal{F}_1\{\lambda_\phi(p) * a(p)\} = \Lambda_\phi(\nu)A(\nu) = F(\mathbf{n}\nu)A(\nu) \equiv \hat{F}(\mathbf{n}\nu) \ , \tag{38}$$

where $\hat{F}(\boldsymbol{\rho})$ is the 2D Fourier transform of $\hat{f}(\mathbf{r})$. In other words, the effect of the apodization is to yield a reconstructed estimate of F( ) that is equal to the true value multiplied by the apodizing function $A(\rho)$, where ρ is the magnitude of the vector $\boldsymbol{\rho}$. Thus, even though $A(\nu)$ was introduced as a 1D weighting function, it now plays the role of the 2D modulation transfer function (MTF) of the reconstruction system. As a result of the central-slice theorem, weighting the projection in the 1D frequency domain with a function $A(\nu)$ is equivalent to weighting the original function $F(\boldsymbol{\rho})$ *with the same apodizing function*, but now regarded as a 2D, rotationally symmetric function.

To gain further insight into the behavior of this important algorithm, let us trace the fate of a point object. In other words, let us calculate the point spread function (PSF) of the system. Since all objects can be described as a superposition of points and since the algorithm is linear, the PSF is a complete description of the system.

A point object in 2D is represented by a 2D delta function $\delta(\mathbf{r}-\mathbf{r}_0)$, where $\mathbf{r}_0$ is the location of the point. The projection of this object is a 1D delta function given by

$$\mathcal{R}_2\{\delta(\mathbf{r}-\mathbf{r}_0\} = \lambda_\phi^\delta(p) = \int_\infty d^2r\, \delta(\mathbf{r}-\mathbf{r}_0)\, \delta(p-\mathbf{r}\cdot\mathbf{n})$$

$$= \delta(p-\mathbf{r}_0\cdot\mathbf{n})\,, \tag{39}$$

where the superscript δ denotes the special case of a point object. If we use this projection in eq. (37), we find easily that

$$\text{PSF} = \hat{f}^\delta(\mathbf{r}) = \int_0^\pi d\phi \int_{-\infty}^{\infty} dp\, [\delta(p-\mathbf{r}_0\cdot\mathbf{n}) * h(p)]_{p=\mathbf{r}\cdot\mathbf{n}}$$

$$= \int_0^\pi d\phi\, h[(\mathbf{r}-\mathbf{r}_0)\cdot\mathbf{n}]\,. \tag{40}$$

Thus the PSF for the combined operations of projection, filtering, back-projection, and summation over projection angles is simply the back-projection of the filter function averaged over ϕ. We see also that these operations taken together constitute a linear shift-invariant system since the PSF is a function only of $\mathbf{r}-\mathbf{r}_0$. The filter function $h(p)$ is of the form of eq. (36), where $a(p)$ is the inverse 1D Fourier transform of some apodizing function $A(\nu)$; if $A(\nu)$ is broad and slowly varying, then the PSF is sharply peaked, approximating a 2D delta function.

Of course, the PSF is also the 2D Fourier transform of the MTF, and we have already seen that $A(\rho)$ serves as the MTF, so it must also be true that

$$\int_0^{\pi} d\phi\ h(\mathbf{r}\cdot\mathbf{n}) = \mathcal{F}_2^{-1}\{A(\rho)\}\ , \tag{41}$$

but we shall leave the explicit proof of this result as an exercise for the student.

4.7 2D filtering

The steps in the convolution algorithm, listed below eq. (37), may also be executed in a different order. We may:

1. back-project;
2. sum over projection angles;
3. filter the resulting 2D back-projected image.

From eq. (38), it is straightforward to show that the reconstruction obtained with this sequence is the same as that obtained by the earlier algorithm, in which filtering is done on the 1D projections before back-projection, provided the 1D and 2D filter functions have the same functional form in their respective frequency domains. However, once again it is instructive to trace the fate of a delta function. We define a new function $b(\mathbf{r})$ that is the result of back-projection and summation without filtering, i.e.

$$b(\mathbf{r}) = \int_0^{\pi} d\phi\ \lambda_\phi(\mathbf{r}\cdot\mathbf{n})\ . \tag{42}$$

In the specific case of a point object located at $\mathbf{r}_0$, we have

$$b^\delta(\mathbf{r}) = \int_0^{\pi} d\phi\ \delta[(\mathbf{r}-\mathbf{r}_0)\cdot\mathbf{n}]\ . \tag{43}$$

To evaluate this expression, we define $\mathbf{R} = \mathbf{r}-\mathbf{r}_0$ and let (R,θ_R) be its polar coordinates. Then

$$b^\delta(\mathbf{r}) = \int_0^{\pi} d\phi\ \delta[R\cos(\theta_R-\phi)]\ . \tag{44}$$

To proceed, we make use of an identity involving delta functions:

$$\delta[q(x)] = \frac{\delta(x-x_0)}{q'(x_0)}\ , \tag{45}$$

integral this identity is to be used in, so that $q(x_0) = 0$, and the derivative $q'(x_0)$ does not vanish at this point. These conditions are satisfied in the present problem where ϕ plays the role of x and $R\cos(\theta_R - \phi)$ plays the role of q(x). The zero of this function occurs at $\phi = \phi_0$, where $\theta_R - \phi_0 = \pi/2$, and the derivative of the function at this point is just R. Therefore,

$$b^\delta(\mathbf{r}) = \int_0^\pi d\phi \, \frac{\delta(\phi-\phi_0)}{R} = \frac{1}{R} = \frac{1}{|\mathbf{r}-\mathbf{r}_0|} \,. \tag{46}$$

For a more general object, we find

$$b(\mathbf{r}) = f(\mathbf{r}) ** \frac{1}{r} \,, \tag{47}$$

where the double asterisk denotes 2D convolution, and r is the magnitude of **r**.

The PSF for back-projection and summation, without any filtering, is thus the cusp-like function 1/r. Although it does have a sharp central peak, 1/r is actually a terrible function for a PSF because of its long wings. When the PSF is convolved with some broad, extended function, the wings build up and cause severe quantitative errors and loss of contrast in the reconstruction. It is for this reason that a further filtering operation is needed. To discover the required form for this 2D filter, let us take the 2D Fourier transform of eq. (47); since it is well known that

$$\mathcal{F}_2\{1/r\} = 1/\rho \,, \tag{48}$$

we find that

$$B(\boldsymbol{\rho}) = \frac{F(\boldsymbol{\rho})}{\rho} \,, \tag{49}$$

where B and F denote the transforms of b and f as usual, and of course ρ is the magnitude of $\boldsymbol{\rho}$. The desired expression for f(**r**) follows by an inverse 2D Fourier transform:

$$f(\mathbf{r}) = \mathcal{F}_2^{-1}\{\rho B(\boldsymbol{\rho})\} \,. \tag{50}$$

Although this is an exact expression, and in fact yet another form of the 2D inverse Radon transform, in practice it is necessary to use an apodizing function to control noise at high spatial frequencies. Thus we obtain an estimate of f(**r**) given by

$$\hat{f}(\mathbf{r}) = \mathcal{F}_2^{-1}\{\rho A(\rho) B(\boldsymbol{\rho})\} \,. \tag{51}$$

To reiterate, the steps in this algorithm are:

1. back-project each projection;
2. sum over projection angles;
3. filter the resulting 2D image b(**r**).

Note that step 3 may be carried out either by multiplying B($\boldsymbol{\rho}$) in the 2D frequency domain by $\rho A(\rho)$, or by convolving b(**r**) with the 2D inverse Fourier transform of that function. If the transforms are performed with the fast Fourier transform algorithm (FFT), the frequency domain is much faster in practice.

5. DISCRETE FORMULATION

Section 4 was based on several unrealistic premises: (1) measured data are exactly described by line integrals along infinitesimal lines; (2) there is no measurement noise; and (3) projections are available for a continuous range of p and ϕ values. In this section we present a more realistic model that will form the basis for discrete reconstruction algorithms to be discussed in Section 6.

5.1 Object representations

The object to be imaged in any imaging system is, of course, defined for all points **r** in some 3D region and therefore, in principle, requires a continuous function f(**r**) for its proper representation. Digital reconstruction algorithms, on the other hand, produce a finite, discrete set of numbers. It is therefore necessary to describe the object by discrete numbers in order to fit into the digital framework.

A popular way to "discretize" a function f(**r**) is to average it over small cubic volume elements or "voxels"; if N voxels are required to cover the region where f(**r**) is nonzero, then this process reduces the continuous function f(**r**) to a set of N numbers f_j , j = 1,2,...,N. These numbers can be arranged into a column vector **f**, the jth component of which is the average of f(**r**) over the jth voxel. Clearly, **f** is not an exact representation of f(**r**) since fine details smaller than the voxel size are lost in the averaging process, but if these details are not of interest or would be lost by the imaging system anyway, the vector **f** may be adequate for our purposes.

There are alternatives to voxels for representing f(**r**). More generally, we can expand f(**r**) in some orthonormal basis:

$$f(\mathbf{r}) \simeq \sum_{j=1}^{N} f_j \, \Phi_j(\mathbf{r}) \; . \tag{52}$$

Orthonormality requires that

$$\int_\infty d^3r\ \Phi_j(\mathbf{r})\ \Phi_k(\mathbf{r}) = \delta_{jk}\ , \tag{53}$$

where the integral runs over all space, and δ_{jk} is the Kronecker symbol (unity if j=k, zero otherwise). Note that nonoverlapping voxels satisfy Eq. (53) if they are suitably normalized, but any other orthogonal functions may also be chosen. An important alternative choice for $\{\Phi_j\}$ is the set of complex exponentials of the form $\exp(2\pi i \boldsymbol{\rho}_j \cdot \mathbf{r})$, which is just the basis set for a Fourier representation. Another important possibility will be given in Section 5.3.

Whatever the choice for $\{\Phi_j\}$, the expansion coefficients f_j in Eq. (52) may be found by

$$f_j = \int_\infty d^3r\ \Phi_j(\mathbf{r})\ f(\mathbf{r})\ , \tag{54}$$

which follows readily from Eqs. (52) and (53).

It is important to note that we are *not* requiring that $\{\Phi_j\}$ form a *complete* set, and indeed it is not possible that a finite set would form a complete basis for expanding the continuous function f(**r**). Therefore the expansion in Eq. (52) is only approximate.

5.2 System representations

A digital imaging system produces a finite number of measurements g_i, i=1,2,....,M. In contrast to the stringent assumptions about the measurements that we made in Section 4 (noise-free line integrals), here we assume only that the mean value of each measurement depends linearly on the unknown object f(**r**). Let $h_i(\mathbf{r})$ be the factor relating an object element $f(\mathbf{r})d^3r$ to the measurement g_i. For example, in CT, $h_i(\mathbf{r})$ represents a thin beam of finite width, while in SPECT it represents both the collimator response and the attenuation factor. With the representation of Eq. (52), we can write, quite generally,

$$g_i = \int_\infty h_i(\mathbf{r})\ f(\mathbf{r})\ d^3r\ + n_i = \sum_{j=1}^{N} H_{ij}\ f_j + n_i\ , \tag{55}$$

where n_i is the noise in the ith measurement, and

$$H_{ij} = \int_{\infty} h_i(\mathbf{r})\, \Phi_j(\mathbf{r})\, d^3\mathbf{r}\ . \tag{56}$$

In matrix form, Eq. (55) may be written:

$$\mathbf{g} = \mathrm{H}\mathbf{f} + \mathbf{n}\ , \tag{57}$$

where $\mathbf{g}$ is an M-element column vector of measurements, $\mathbf{n}$ is an M-element column vector of noise values, and H is the MxN matrix with elements defined by Eq. (56). The matrix element H_{ij} gives the contribution of object component f_j to measurement g_i. In the voxel basis, H_{ij}, regarded as a function of i, is essentially the point spread function of the system; the jth column of H gives the "image" (set of detector responses) for a small, pointlike object at position j.

5.3 Singular-value decomposition

In solving Eq. (57), it is useful to have alternative representations of the imaging operator H. For example, in describing a linear shift-invariant system, the Fourier domain is valuable because the Fourier basis functions $\exp(2\pi i \boldsymbol{\rho}_j \cdot \mathbf{r})$ are eigenfunctions of H in that case (Barrett and Swindell, 1981). It is always useful to use as basis functions the eigenfunctions of a matrix of interest since then the matrix is diagonal.

For many practical problems, however, it is not possible to diagonalize H itself since it is a rectangular matrix. In such cases, we can create a square matrix by forming H^tH or HH^t, where the superscript t denotes a matrix transpose, such that $(H^t)_{ij} = H_{ji}$. Since H is an MxN matrix, H^tH is NxN and HH^t is MxM. Both matrices are symmetric by construction, and both are therefore Hermitian if we consider only real matrices.

The eigenvalue problem for H^tH may be stated as

$$H^tH\, \mathbf{u}_j = \lambda_j \mathbf{u}_j\ ,\ j = 1,2,....N, \tag{58}$$

where λ_j is the eigenvalue and $\mathbf{u}_j$ is the corresponding eigenvector in the form of an Nx1 column vector. (Note that the subscript j in this case specifies which vector we are considering, not which component of the vector. There should be no confusion regarding these two uses of subscripts since vectors are always written in boldface.)

Another way to write Eq. (58), which is really a set of N equations, is as a single matrix equation:

$$H^tH\,U = U\,\Lambda\,, \tag{59}$$

where the jth column of the matrix U is the eigenvector $\mathbf{u}_j$, and Λ is a diagonal matrix of eigenvalues, i.e. $\Lambda_{jk} = \lambda_j\delta_{jk}$. It is convenient to order the eigenvalues by decreasing magnitude, so that the elements of Λ decrease monotonically as one goes down the diagonal.

An important characteristic of any matrix is its rank R, which may be defined as the number of linearly independent rows or columns or, equivalently, as the number of nonzero eigenvalues. In the case of H^tH, which is the product of an NxM matrix and an MxN one, the rank R cannot exceed M or N, whichever is smaller. The first R diagonal elements of Λ are then nonzero, and the rest are zero.

Since H^tH is Hermitian, λ_j is real and non-negative. Also, again since H^tH is Hermitian, two vectors $\mathbf{u}_j$ and $\mathbf{u}_k$ for which $\lambda_j \neq \lambda_k$ are orthogonal. (If $\lambda_j = \lambda_k$, then we can always form linear combinations of eigenvectors belonging to the same eigenvalue in such a way that the new vectors remain eigenvectors but are orthogonal; this process is called Gram-Schmidt orthogonalization.) Furthermore, since $\mathbf{u}_j$ can be multiplied by an arbitrary constant without changing Eq. (58), we may normalize the eigenvectors to unit length to yield an orthonormal set. In equation form

$$\mathbf{u}_j{}^t\mathbf{u}_k = \delta_{jk}\,, \tag{60}$$

where the transpose of a column vector is a row vector, so that $\mathbf{u}_j{}^t\mathbf{u}_k$ denotes the product of a 1xN "matrix" and an Nx1 "matrix", or simply the scalar or inner product of the two vectors.

If we have solved the eigenvalue problem for H^tH, then we can get the solution to the corresponding problem for HH^t for free. Multiplying Eq. (58) from the left by H yields

$$HH^tH\,\mathbf{u}_j = \lambda_j H\mathbf{u}_j\,, \tag{61}$$

from which we see that the Mx1 column vector $H\mathbf{u}_j$ is an eigenvector of HH^t with the eigenvalue λ_j. This eigenvector is, however, not yet normalized. Therefore we define

$$\mathbf{v}_j = \lambda_j{}^{-1/2}\,H\mathbf{u}_j\,,\ j \le R, \tag{62}$$

where the condition on j assures that λ_j does not vanish. The reader may verify that the vectors $\{\mathbf{v}_j\}$ are orthonormal, i.e.

$$\mathbf{v}_j{}^t\mathbf{v}_k = \delta_{jk}\,. \tag{63}$$

The counterpart to Eq. (59) is

$$HH^t\,V = V\,\Lambda\,, \tag{64}$$

where $\mathbf{v}_j$ is the jth column of V, and Λ is the same matrix of eigenvalues as in Eq. (59).

The eigenvector sets $\{\mathbf{u}_j\}$ and $\{\mathbf{v}_j\}$ may be used to construct convenient representations of various matrices. It is straightforward to show that

$$H = \sum_{j=1}^{R} \lambda_j^{1/2}\, \mathbf{v}_j\, \mathbf{u}_j^t\,, \tag{65}$$

$$H^t = \sum_{j=1}^{R} \lambda_j^{1/2}\, \mathbf{u}_j\, \mathbf{v}_j^t\,, \tag{66}$$

$$H^tH = \sum_{j=1}^{R} \lambda_j\, \mathbf{u}_j\, \mathbf{u}_j^t\,, \tag{67}$$

$$HH^t = \sum_{j=1}^{R} \lambda_j\, \mathbf{v}_j\, \mathbf{v}_j^t\,. \tag{68}$$

Note here that $\mathbf{v}_j\, \mathbf{u}_j^t$ is an *outer* product of an Mx1 column vector and a 1xN row vector, yielding an MxN matrix. Outer-product matrices have only one independent row and hence have rank one, but a sum of R independent outer products has rank R. The decomposition of H or H^t into a sum of rank-one matrices as in Eqs. (65) and (66) is known as a *singular-value decomposition* or SVD, with $\lambda_j^{1/2}$ being the singular value for the jth term.

6. DISCRETE INVERSION METHODS

The basic matrix description of image formation, Eq. (57), is really a system of M equations in N unknowns f_j. It would seem that solution consists of finding the inverse matrix H^{-1}, but this is not always possible or desirable. Only square matrices have true inverses, so H^{-1} does not exist unless M = N. If M < N, the equations are underdetermined; there are fewer equations than unknowns and consequently there will be many vectors **f** that satisfy the equations equally well, and the solution is not unique. Conversely, if M > N, the equations are overdetermined and there will be no solution in general. Even if M = N, H^{-1} will not exist if the rank of H is less than its dimensionality N. Finally, even if none of the above objections apply and H^{-1} does exist, it may not be useful because of the noise term in Eq. (57). If we apply H^{-1} to both sides of Eq. (57), we obtain

$$H^{-1}\mathbf{g} = \mathbf{f} + H^{-1}\mathbf{n} . \qquad (69)$$

The last term may be quite large, even for relatively small amounts of noise in the data. For all of these reasons we say that the inversion problem defined by Eq. (57) is ill-posed, and we must seek solution methods that take these problems into account. Several such methods are reviewed in the sections below, but first we shall lay some groundwork by discussing the concept of null functions.

6.1 Null functions

One way to see a possible lack of uniqueness of the solution of Eq. (57) is to use the singular-value decomposition developed in Section 5.3. The eigenvectors of H^tH, found by solving Eq. (58), form a complete basis for expanding any Nx1 column vector. We may therefore write any object vector $\mathbf{f}$ as

$$\mathbf{f} = \sum_{i=1}^{N} \alpha_i \mathbf{u}_i . \qquad (70)$$

The action of our imaging system on this vector is given by

$$H\mathbf{f} = \sum_{i=1}^{N} \alpha_i H\mathbf{u}_i = \sum_{i=1}^{N} \alpha_i \lambda_i^{1/2} \mathbf{v}_i , \qquad (71)$$

where the last step follows from Eqs. (65) and (60).

Suppose now that the rank R of H is less than N. This means that the sum in Eq. (71) can be terminated at R, since the λ_i are zero for $i > R$. The sum in Eq. (70) must still run to N since it takes N basis vectors to expand a general Nx1 vector, but we can split it into two parts as

$$\mathbf{f} = \sum_{i=1}^{R} \alpha_i \mathbf{u}_i + \sum_{i=R+1}^{N} \alpha_i \mathbf{u}_i$$

$$= \mathbf{f}_{meas} + \mathbf{f}_{null} . \qquad (72)$$

The first term, $\mathbf{f}_{meas}$, specifies the components of $\mathbf{f}$ that can be measured by the system H. The second term, $\mathbf{f}_{null}$, is called a null vector since it makes no contribution to $H\mathbf{f}$, all of its components being multiplied by a zero eigenvalue in Eq. (71). Other names for null vectors are invisible objects or ghosts. Whenever such ghosts exist, the solution to Eq. (57) cannot be to **f** without affecting the data vector **g**, and there is no way to tell from the data which null vector is present.

6.2 Least-squares solutions and pseudoinverses

If no unique solution of Eq. (57) exists, we must have some rationale for choosing a "best" solution. A classic approach to this problem is to seek a least-squares solution $\hat{\mathbf{f}}_{LS}$, defined as the value of **f** that minimizes the norm of the error between the measured data vector **g** and our estimate of it, given by $\hat{\mathbf{g}} = \mathrm{H}\hat{\mathbf{f}}_{LS}$. In other words, $\hat{\mathbf{f}}$ is determined by

$$\| \mathbf{g} - \mathrm{H}\hat{\mathbf{f}}_{LS} \|^2 = \min , \tag{73}$$

where the norm squared of a vector is the scalar product of the vector with itself. By writing Eq. (73) out in component form and differentiating with respect to the components of **f**, we find that the error norm takes an extremum when

$$\mathrm{H}^t\mathbf{g} = \mathrm{H}^t\mathrm{H}\hat{\mathbf{f}}_{LS} \, . \tag{74}$$

Since the maximum error norm is unbounded, the extremum must be the desired minimum. Any least-squares solution of Eq. (57) must therefore also satisfy Eq. (74).

Equation (74) is not, however, sufficient to determine a unique solution, since once again we may add any null vector to $\hat{\mathbf{f}}_{LS}$. To get a unique answer we require that the null components of $\hat{\mathbf{f}}_{LS}$ be zero, which is equivalent to finding the solution of Eq. (74) with minimum norm. In other words, among all solutions of Eq. (74), we choose the one for which

$$\| \hat{\mathbf{f}}_{LS} \|^2 = \min \, . \tag{75}$$

The vector $\hat{\mathbf{f}}_{LS}$ satisfying this condition and Eq. (74) is the minimum-norm least-squares solution of our inverse problem. It is also known as the Moore-Penrose pseudoinverse solution.

The SVD of H may be used to find an explicit form for the Moore-Penrose pseudoinverse. Since $\{\mathbf{v}_j\}$, the eigenfunctions of HH^t, form a complete basis for any Mx1 vector, we may expand the data vector **g** as

$$\mathbf{g} = \sum_{i=1}^{M} \beta_i \, \mathbf{v}_i \, . \tag{76}$$

The expansion coefficients may be calculated from measured data values by exploiting the orthonormality of the $\{\mathbf{v}_j\}$, as given by Eq. (63), to yield

$$\beta_j = \mathbf{v}_j{}^t\mathbf{g} . \tag{77}$$

Thus the set $\{\beta_j\}$ is just another representation of the data.

It then follows from Eqs. (63), (71) and (77) that

$$\beta_i = \alpha_i \, \lambda_i{}^{1/2} . \tag{78}$$

If λ_i is not zero, we may readily solve this equation for the $\{\alpha_i\}$, which by Eq. (70) is tantamount to finding **f**. In real problems of interest, however, some of the $\{\lambda_j\}$, namely those corresponding to null components, will often be zero. In this case, we may solve Eq. (78) for the non-null components and simply set the null components to zero. Thus our estimate of f by this prescription is

$$\hat{\mathbf{f}}_{MP} = \sum_{j=1}^{R} \beta_j \, \lambda_j{}^{-1/2} \, \mathbf{u}_j \equiv H^+\mathbf{g} , \tag{79}$$

where H^+ is the Moore-Penrose pseudoinverse to H. The reader is invited to show that $\hat{\mathbf{f}}_{MP}$ is a least-squares solution that satisfies Eq. (74). Since the null components are zero by construction, it is also a minimum-norm least-squares solution.

It is a straightforward exercise in SVD to show that

$$H^+ = \lim_{\epsilon \to 0} \left[H^tH + \epsilon^2 I\right]^{-1} H^t , \tag{80}$$

where I is the NxN unit matrix. Equation (80) is a common alternative expression for H^+. Note that if H^{-1} exists, then $H^+ = H^{-1}$.

6.3 Regularization

Equation (79) avoids the embarrassment of dividing by zero since all of the $\{\lambda_j\}$ for $j \le R$ are nonzero, but nevertheless some of the nonzero values may be quite small. Since each component β_j of the data contains a noise contribution, the noise will be greatly amplified, with disastrous effect on the reconstructed image, if λ_j is small. A simple modification of Eq. (79) will, however, avoid this noise amplification. We define a *regularized* estimate of **f** by

$$\hat{\mathbf{f}}_{reg} = \sum_{j=1}^{N} \frac{\lambda_j{}^{1/2}}{\lambda_j + \epsilon^2} \beta_j \, \mathbf{u}_j , \tag{81}$$

where ϵ is some small but finite number known as a regularizing parameter. Note that the sum now runs to N, but terms with $\lambda_j = 0$ still vanish for any finite ϵ. In the limit of $\epsilon \to 0$, Eq. (81) approaches Eq. (79), but a finite ϵ prevents strong noise amplification. (The denominator cannot vanish since the $\{\lambda_j\}$ are never negative.)

An equivalent form of this regularized pseudoinverse is given by Eq. (80) without the limit. Still another way to express the same result is to say that the regularized least-squares solution is the value of $\hat{\mathbf{f}}$ that satisfies

$$\| \mathbf{g} - \mathrm{H}\hat{\mathbf{f}}_{reg} \|^2 + \epsilon^2 \| \hat{\mathbf{f}}_{reg} \|^2 = \min . \tag{82}$$

Once again, the Moore-Penrose solution is recovered by passing to the limit of zero ϵ.

The method of regularization we have described is sometimes known as Tikhonov regularization. Other regularization techniques exist, but all are equivalent to reducing noise by suppressing components of $\hat{\mathbf{f}}$ that correspond to small eigenvalues. Different regularization schemes amount to different choices for the weights to be given to small eigenvalues. A general regularized pseudoinverse has the form

$$\hat{\mathbf{f}}_{reg} = \sum_{j=1}^{N} W(\lambda_j)\, \beta_j\, \mathbf{u}_j , \tag{83}$$

where $W(\lambda_j)$ is the weight to be given to the jth singular vector. For large λ_j, $W(\lambda_j)$ should approximate the inverse filter $\lambda_j^{-1/2}$, but for smaller eigenvalues the maximum value of W must be controlled to avoid noise amplification. Since the small eigenvalues are almost always associated with fine details in the image, regularization amounts to sacrificing spatial resolution for lower noise. Another way to state the same point is that the Moore-Penrose pseudoinverse is pure inverse filtering in the SVD domain (except that we cannot divide by zero), while the regularized pseudoinverse is more akin to Wiener filtering in that it does not try to boost terms containing only noise and very little signal. (See, for example, Barrett and Swindell (1981))

The relationship between regularization and apodization should also be noted. For this purpose, we recall the discusssion of Section 4.7, where we showed that back-projection and summation over projection angles results in a shift-invariant blurring (see Eq. (47)). If the operator H corresponds to tomographic projection, then H^t corresponds to back-projection and summation, and Eq. (47) shows that the operation of $\mathrm{H}^t\mathrm{H}$ on an object $f(\mathbf{r})$ yields the convolution of the object with a $1/r$ blur function. In discrete form, SVD of a convolution operator is merely discrete Fourier analysis, since it is easy to show that the Fourier basis functions $\{\exp[-2\pi i \boldsymbol{\rho}_j \cdot \mathbf{r}]\}$ are eigenfunctions of a convolution operator. The eigenvalue λ_j associated with $\exp[-2\pi i \boldsymbol{\rho}_j \cdot \mathbf{r}]$ is $1/\rho_j$ (cf. Eq. (48)), and SVD calculation of the pseudoinverse form of $\hat{\mathbf{f}}$ is just a discrete version of Eq. (50). (It is a pseudoinverse rather than a true inverse only because the integral over ρ must be terminated when $1/\rho$ becomes very small and the eigenvalues essentially vanish.) The apodized inverse of Eq. (51) then corresponds to a regularized pseudoinverse.

6.4 Iterative methods

Though the SVD form of the regularized pseudoinverse, Eq. (83), is general enough to allow for any linear solution of any inverse problem, it is seldom used in practice for two reasons. First, if H is a large matrix, it may not be possible to find its SVD on computers of reasonable size. Second, and perhaps more importantly, *linear* solutions are often inferior to nonlinear ones that incorporate constraints on the estimate $\hat{\mathbf{f}}$. For example, it is often known *a priori* that the object cannot be negative, but a linear estimate of the form of Eq. (83) can certainly give negative values. It is therefore highly desirable to constrain the estimate to be positive. Other constraints that we may want to incorporate in particular problems include requiring the estimate to fit within some spatially bounded region (e.g. the patient's body) or to be less than some specified maximum value.

For these reasons, it is often convenient to find $\hat{\mathbf{f}}$ by successive approximations or *iterative algorithms*. There are many such iterative algorithms in the literature, but here we shall review only one generic algorithm, known variously as the Jacobi, van Cittert or Landweber method.

A convenient starting point for deriving this algorithm is the pseudoinverse form given by Eq. (80) which we rewrite here as

$$H^{\dagger} = (I - \Omega)^{-1} H^t , \tag{84}$$

where

$$\Omega = \lim_{\epsilon \to 0} [I - H^t H - \epsilon^2 I] = I - H^t H . \tag{85}$$

We note, however, that $(I - \Omega)^{-1}$ is the sum of an operator geometric series (provided, of course, that the series converges):

$$(I - \Omega)^{-1} = \sum_{n=0}^{\infty} \Omega^n , \tag{86}$$

where Ω^n is the same as the operator Ω applied n times in succession. To prove Eq. (86), we multiply the series by $I - \Omega$, obtaining

$$(I - \Omega) \sum_{n=0}^{\infty} \Omega^n = \sum_{n=0}^{\infty} \Omega^n - \sum_{n=0}^{\infty} \Omega^{n+1} = \sum_{n=0}^{\infty} \Omega^n - \sum_{k=1}^{\infty} \Omega^k$$

$$= \sum_{n=0}^{\infty} \Omega^n - \left[\sum_{k=0}^{\infty} \Omega^k - \Omega^0 \right] = \sum_{n=0}^{\infty} \Omega^n - \sum_{n=0}^{\infty} \Omega^n + I = I \,. \tag{87}$$

Hence the series is the operator inverse of $I - \Omega$, and Eq. (84) becomes

$$H^+ = \sum_{n=0}^{\infty} \Omega^n H^t \,. \tag{88}$$

This form of the pseudoinverse is convenient for devising an iterative algorithm. The first k terms in the series may be regarded as the kth approximation to H^+, and the $(k+1)^{th}$ term is a correction to the kth. If we take as a first approximation to $\hat{\mathbf{f}}$ just $H^t\mathbf{g}$, then subsequent approximations are generated by the rule

$$\hat{\mathbf{f}}_{k+1} = \sum_{n=0}^{k+1} \Omega^n H^t\mathbf{g} = \hat{\mathbf{f}}_k + \Omega^{k+1} H^t\mathbf{g}$$

$$= [I - H^tH]\, \hat{\mathbf{f}}_k + H^t\mathbf{g} = \hat{\mathbf{f}}_k + [H^t\mathbf{g} - H^tH\hat{\mathbf{f}}_k] \,, \tag{89}$$

where we have used the definition of Ω and algebra similar to that given in Eq. (87). The term in brackets in the last form is the correction added to the estimate at the $(k+1)^{th}$ step. Note that this iterative process reaches an equilibrium and no further corrections occur when $H^t\mathbf{g} = H^tH\hat{\mathbf{f}}_k$, which is just the condition for a least-squares solution, Eq. (74).

The algorithm expressed in Eq. (89) gives us a way of calculating the Moore-Penrose pseudoinverse solution of an inverse problem without performing an SVD or inverting any matrices, but its greater value lies in the fact that it can readily be regularized or modified to incorporate constraints. Regularization may be accomplished either by terminating the iteration or by introducing a *relaxation parameter* $\gamma \leq 1$, so that the new iteration rule is

$$\hat{\mathbf{f}}_{k+1} = \gamma\hat{\mathbf{f}}_k + [H^t\mathbf{g} - H^tH\hat{\mathbf{f}}_k] \,. \tag{90}$$

If $\gamma = 1$, then we have Eq. (89) again, but otherwise this recursion relation may be shown to be equivalent to the Tikhonov pseudoinverse with $\epsilon^2 = 1-\gamma$.

Constraints may be incorporated with a constraint-enforcement operator $\mathcal{C}$, such that Eq. (89) becomes

$$\hat{\mathbf{f}}_{k+1} = \mathcal{C}\{\hat{\mathbf{f}}_k + [\mathrm{H}^t\mathbf{g} - \mathrm{H}^t\mathrm{H}\hat{\mathbf{f}}_k]\} . \tag{91}$$

For example, in the case of a positivity constraint, $\mathcal{C}$ could be an operator that sets negative components of the operand to zero, thereby ensuring that the current estimate is positive after every correction. Since $\mathcal{C}$ is a nonlinear operator, it is difficult to predict analytically the characteristics of the final estimate with the algorithm of Eq. (91), but experience indicates that it is a substantial improvement over any of the linear algorithms.

6.5 Maximum likelihood

None of the inversion methods presented so far have been based on statistical models of the data, though some of the methods have implicitly acknowledged the presence of noise in the data and have attempted, through *ad hoc* methods such as apodization or regularization, to control the propagation of that noise through the reconstruction algorithm to the object estimate. In this section we take a somewhat more rigorous approach and consider specific probability laws for the data. For now, it is assumed that the object itself is not random, but this restriction will be removed in Section 6.6 where we consider Bayesian methods.

In diagnostic radiology and nuclear medicine, the basic source of noise in the data is the shot noise or photon noise stemming from the discrete nature of x-ray photons. If other noise sources are negligible, as they frequently are, then a measurement g_j (one component of the vector $\mathbf{g}$) is just the number of photons counted during some measurement time T. If the source of radiation and the medium between the source and the detector do not vary, it is almost always valid to assume that the number of detected photons is a random variable distributed according to the Poisson probability law (Barrett and Swindell, 1981). This means that the probability of detecting g_j photons during T is given by

$$P(\, g_j \mid \overline{g_j}\,) = \exp(-\overline{g_j})\, \frac{\left[\overline{g_j}\right]^{g_j}}{(g_j)!} \,, \tag{92}$$

where $\overline{g_j}$ is the mean number detected. Since Poisson random variables are inherently statistically independent, the overall probability law for the set of M measurements (or the vector $\mathbf{g}$) can be written as a product of expressions like Eq. (92):

$$P(\, \mathbf{g} \mid \overline{\mathbf{g}}\,) = \prod_{j=1}^{M} \exp(-\overline{g_j})\, \frac{\left[\overline{g_j}\right]^{g_j}}{(g_j)!} \,, \tag{93}$$

where $\bar{\mathbf{g}}$ is the vector of means. For the discrete model of Eq. (57), $\bar{\mathbf{g}}$ is simply Hf, so information about the object **f** is contained in $P(\mathbf{g} \mid \bar{\mathbf{g}})$, which we may also write as $P(\mathbf{g} \mid \mathbf{f})$.

The maximum-likelihood principle states that we should choose as our estimate of **f** the vector that maximizes the probability that the observed data would occur, given **f**. Mathematically,

$$P(\mathbf{g} \mid \mathbf{f}) = \max \text{ at } \mathbf{f} = \hat{\mathbf{f}}_{ML} . \tag{94}$$

This is a set of nonlinear equations for the unknown components of **f**, and iterative algorithms such as the EM algorithm must be used to actually find the maximum. Since this subject is covered in detail elsewhere in this volume, it will not be pursued here.

Considerable insight into the nature of the maximum likelihood solution can, however, be gained by using an approximate expression for $P(\mathbf{g} \mid \mathbf{f})$ that results from the central limit theorem. If the mean values $\bar{g}_j$ are fairly large, even 5 or so, it is reasonable to regard the random variables g_j as continuous and to approximate $P(g_j \mid \mathbf{f})$ by a normal probability *density* $p(g_j \mid \mathbf{f})$ of the form

$$p(g_j \mid \mathbf{f}) = (2\pi\sigma_j^2)^{-1/2} \exp[-(g_j - \bar{g}_j)^2/2\sigma_j^2] . \tag{95}$$

The variance σ_j^2 is equal to the mean $\bar{g}_j$ for the Poisson law. Once again, the different measurements g_j are independent, so we have

$$p(\mathbf{g} \mid \mathbf{f}) = \prod_{j=1}^{M} (2\pi\sigma_j^2)^{-1/2} \exp[-(g_j - \bar{g}_j)^2/2\sigma_j^2] . \tag{96}$$

To find $\hat{\mathbf{f}}_{ML}$ we must maximize this expression, or equivalently its natural logarithm:

$$\ln p(\mathbf{g} \mid \mathbf{f}) = \sum_{j=1}^{M} \left[-\tfrac{1}{2} \ln(2\pi\sigma_j^2) - (g_j - \bar{g}_j)^2/2\sigma_j^2 \right]$$

$$= \max \text{ at } \mathbf{f} = \hat{\mathbf{f}}_{ML} . \tag{97}$$

The first term, $-\frac{1}{2}\ln(2\pi\sigma_j^2)$, is slowly varying and may be safely ignored in the maximization. Thus the estimation principle becomes

$$\sum_{j=1}^{M} (g_j - \bar{g}_j)^2/2\sigma_j^2 = \min , \tag{98}$$

where we have dropped a minus sign and changed the maximum to a minimum. Equation (98) is reminiscent of Eq.(73), which was the basic minimization formula for least-squares estimation. The only difference is that here we have a *weighted* least-squares method, where each measurement is weighted by the reciprocal of its variance.

If the image data have low contrast, so that all variances are approximately the same, then maximum-likelihood estimation is the same as least squares (with the normal approximation to the Poisson). In this case, the sum in Eq. (98) is just a constant times the squared norm of $\mathbf{g} - \mathbf{H}\hat{\mathbf{f}}$, which is zero if $\hat{\mathbf{f}} = \mathbf{H}^+\mathbf{g}$. Thus maximum likelihood suffers from the same noise sensitivity as the pseudoinverse.

Even without the approximations above, the maximum-likelihood principle, like the least-squares principle, is insufficient by itself to yield a unique estimate since we may add any null vector to $\hat{\mathbf{f}}_{ML}$ without changing $\bar{\mathbf{g}}$ and hence without changing the likelihood.

Several modifications of pure maximum-likelihood estimation are commonly practiced. Iterative searches for the maximum are frequently terminated short of the true maximum in order to control noise in the reconstruction. Though this can be a useful expedient, it does introduce an element of arbitrariness and subjectivity in the outcome. Some workers have suggested stopping the algorithm when the error norm or χ^2 (the expression in Eq. (98)) reaches some predetermined value. This expedient minimizes the noise amplification but makes the final result strongly dependent on the specific algorithm and the initial estimate. Others have usefully incorporated smoothness constraints or "sieves" in maximum-likelihood algorithms to reduce the noise at the expense of spatial resolution in the manner of regularization. Finally, other constraints such as positivity are frequently used to good effect.

6.6 Bayesian methods

Pure maximum-likelihood and pseudoinverse methods strive to enforce consistency with the measured data. The results are almost always aesthetically unpleasing, with large noise spikes or abrupt discontinuities that we "know" are not present in the actual object. We attempt to cure these problems by introducing various regularization or smoothing tricks, but the root cause of the ailment is a philosophical inconsistency. In adopting a pseudoinverse or maximum-likelihood approach, we are saying, in effect, that we have no knowledge of the object except that which we obtain directly from the data; consistency with the data is the heart of the algorithm. Yet we do have certain prior knowledge, not just positivity but also the knowledge that the object is smooth in some sense and does not consist of sharp spikes This knowledge is inherently statistical in nature. We do not know the structure of any one object (unless it is a phantom or computer simulation), but we can say, for example, that neighboring voxels have a certain probability of being close in value. The goal of Bayesian reconstruction methods is to build this statistical knowledge into the algorithm in a rigorous and effective way.

To give a full statistical description of the object vector **f**, we need to regard it as a random variable and specify a prior probability density p(**f**). ("Prior" here means before any measurements have been made.) It is easiest to conceptualize p(**f**) if we imagine that there is an ensemble of objects, so that probabilities may be interpreted as frequencies of occurrence, but Bayesian theorists also admit *subjective priors* which in some ill-defined way specify one's belief about the nature of the object. Though this view may be useful, in medical applications we can always think of an ensemble of patients with some specified set of symptoms. Thus, though we may not always know p(**f**) with any precision, there would seem to be no conceptual problem in thinking of it in terms of frequencies of occurrence of various grey-level values in the object.

Bayesian methods begin by defining a *cost* associated with any decision or estimate. In an image reconstruction problem, the cost $C(\hat{\mathbf{f}},\mathbf{f})$ measures how well, in some sense, the estimate $\hat{\mathbf{f}}$ represents the actual object **f**. Large errors in the representation should "cost" more than small ones. Note that the cost is a random variable for two reasons: the object itself is random and, for any one object, the measured data and hence the object estimate $\hat{\mathbf{f}}$ will also be random. Thus, to construct a meaningful criterion for image reconstruction, we define a *risk* or average cost $\langle C(\hat{\mathbf{f}},\mathbf{f})\rangle$, averaged over the object ensemble and the noise processes. The goal of Bayesian reconstruction is to minimize this cost.

A familiar example of a cost function is the squared error between the object **f** and its estimate $\hat{\mathbf{f}}$:

$$C(\hat{\mathbf{f}},\mathbf{f}) = \| \mathbf{f} - \hat{\mathbf{f}} \|^2 . \tag{99}$$

This cost should not be confused with the error norm *in data space* that formed the basis for least-squares estimation (see Eq. (73)).

An advantage of the squared-error cost functional is that we do not need the full prior density p(**f**) in order to calculate the risk and devise an estimator. Rather, it suffices to know the average object $\bar{\mathbf{f}}$ and its covariance matrix K_f. The resulting estimator is the generalized Wiener filter (Pratt, 1978) given by

$$\hat{\mathbf{f}}_{MMSE} = \left[H^tH + K_f^{-1}K_n\right]^{-1}[\mathbf{g} - \bar{\mathbf{g}}] + \bar{\mathbf{f}}, \tag{100}$$

where MMSE denote minimum mean-squared error, and K_n is the noise covariance matrix.

The similarity between the generalized Wiener filter of Eq. (100) and the Tikhonov pseudoinverse, Eq. (80) without the limit, should not be overlooked. In both cases some positive-definite matrix is added to H^tH to guarantee that an inverse exists. In the Tikhonov case it is an *ad hoc* multiple of the unit matrix, while in the Wiener case there is more of a statistical basis for the choice, but the effect is much the same, namely to suppress noise at the expense of spatial resolution.

A drawback to the Wiener filter is that it is yet another linear method that does not guarantee positivity. Furthermore, incorrect choices for the covariance matrices can lead to serious artifacts (Smith, 1988).

Many different Bayesian cost functionals lead to the *maximum a posteriori* or MAP estimation principle, which says to choose as our estimator $\hat{\mathbf{f}}$ the vector that maximizes the probability density of $\mathbf{f}$ after the measurement:

$$p(\mathbf{f} \mid \mathbf{g}) = \max \text{ at } \mathbf{f} = \hat{\mathbf{f}}_{MAP} . \tag{101}$$

An equivalent form of this estimator may be found by using Bayes' rule:

$$p(\mathbf{f} \mid \mathbf{g})\, p(\mathbf{g}) = p(\mathbf{g} \mid \mathbf{f})\, p(\mathbf{f}) . \tag{102}$$

Since $p(\mathbf{g})$ is independent of $\mathbf{f}$, it does not affect the maximization, and the MAP rule becomes

$$p(\mathbf{g} \mid \mathbf{f})\, p(\mathbf{f}) = \max \text{ at } \mathbf{f} = \hat{\mathbf{f}}_{MAP} . \tag{103}$$

The first factor $p(\mathbf{g} \mid \mathbf{f})$ is just the likelihood, but here it is weighted by the prior object density $p(\mathbf{f})$. For this reason MAP estimation is sometimes referred to as weighted maximum likelihood.

Many different choices for the prior $p(\mathbf{f})$ may be used in the MAP formalism. If we choose to model the object as a multivariate normal distribution, then the Wiener filter is recovered. If we take for $p(\mathbf{f})$ the number of ways an object can be constructed from elementary building blocks or "grains", we obtain the maximum-entropy principle, but this approach takes no account of prior knowledge of correlations between neighboring voxels. Knowledge of correlations can be incorporated by Markov random field models, which lead to tractable forms for $p(\mathbf{f})$, or by other mathematical modelling approaches. Finally, one way of expressing total lack of knowledge about $\mathbf{f}$ is to take $p(\mathbf{f})$ = constant, in which case we are back to maximum likelihood.

REFERENCES

Barrett, H.H. (1988). Fundamentals of the Radon Transform. In: *Mathematics and Computer Science in Medical Imaging*, M. A. Viergever, A. Todd-Pokropek (eds), Springer-Verlag, Berlin, pp. 105-125.

Barrett, H.H. (1984). The Radon Transform and Its Applications, *Progress in Optics XXI*, E. Wolf (ed).

Barrett, H. H. and Swindell, W. (1981) *Radiological Imaging: Theory of Image Formation, Detection and Processing* (Vols. I and II) Academic Press, new York

Deans, S.R. (1983). *The Radon Transform and Some of Its Applications*, Wiley, New York.

Helgason, S. (1980). *The Radon Transform*, Birkhauser, Boston.

Radon, J. (1917). Ueber die Bestimmung von Funktionen durch ihre Integralwerte langs gewisser Mannigfaltigkeiten, *Ber. Saechs. Akad. Wiss.* (Leipzig) 69, pp. 262-278.

REGULARIZATION TECHNIQUES IN MEDICAL IMAGING

M. Defrise

Vrije Universiteit Brussel

Belgium

ABSTRACT

Many inverse problems in medicine, and in particular in tomography, are ill-posed, that is, the result of the inversion is unstable with respect to small perturbations of the measured data. This paper introduces the concept of ill-posed problem, and describes several classical regularization techniques, which allow one to recover stable solutions to ill-posed inverse problems. Applications to the problem of image reconstruction from projections are presented in the last section of the paper.

1. INTRODUCTION

Many diagnostics techniques in medical physics provide raw data which cannot be interpreted straightforwardly like an ECG or a planar radiography. The various tomographic modalities, for instance, have been discussed in the previous lecture; it has been shown how the measured projections (displayed e.g. in the form of the rather uninterpretable sinogram) undergo reconstruction procedures to yield an estimate of the clinically useful image. Other inverse problems in medicine deal with curves rather than images, and involve for instance the deconvolution of kidney time-activity curves measured from scintigraphic data to produce estimates of the kidney retention function. Such inverse problems often reveal a strong amplification of the measurement errors by the inversion algorithm. A closer investigation demonstrates that this behavior is due neither to some unfortunate propagation of the round-off errors during digital computations, nor to an inappropriate choice of the algorithm. The possibility that the inherently continuous physical problem has been discretized -sampled- too coarsely must also be discarded : indeed,a finer sampling of the reconstructed curve or image results in an even more dramatic noise amplification. It thus appears that the observed behavior is a property inherent to the inverse problem we are dealing with. This property is shared by many other problems in applied physics : the backward heat equation, the various inverse scattering problems (e.g. in radar detection), diffraction tomography (e.g. in seismology), the evaluation of the derivative of a curve are all referred to as *ill-posed problems.* Solving such problems using classical techniques (e.g. a Gauss-

Seidel inversion of the set of linear equations obtained by a straightforward discretization of the problem) usually leads to meaningless solutions.

The purpose of this paper is to introduce the mathematical definition of ill-posedness (section 2.1), to discuss some practical implications at the light of a few examples (sections 2.2 and 2.3) and to review *regularization techniques* (section 3) which can be used to cure the noise instability. This general presentation highlights concepts and techniques which are applicable to most ill-posed problems. The practical implementation of these mathematical methods may prove difficult of course; each inverse problem requires a solution specifically tailored to its peculiarities and the fine tuning of empirical "tricks" often prove as important as the underlying mathematical treatment. The last section briefly illustrates the application of a few regularization techniques to image reconstruction in emission and transmission tomography.

As will be seen in the next section, the concept of ill-posedness takes its full meaning only in a context of functional spaces, i.e. when the measured data and the solution we are looking for are described as functions of one or several variables. Such a description might seem inappropriate when using digital computers, which only process discrete sets of numbers (vectors) rather than functions. In such a discrete framework the instability symptoms appear through the emergence of ill-conditioned matrices. A description of the data and solution of the inverse problem as functions is often preferable, however : the solution of the inverse problem usually depends on variables (time or space coordinates for instance) which are inherently continuous, and forcing a priori a particular discretization scheme might bias the solution. The measurement process, on the other hand, is almost always discrete. Nevertheless the data are often described as a function in order to simplify the mathematical description, since many results are much easier to derive in a functional space framework than in matrix algebra. Furthermore this "continuous" description allows to separate the analysis of the problems related to the data sampling from the study of the underlying physical inverse problem.

2. ILL-POSED LINEAR INVERSE PROBLEMS

2.1 Definition

This paper is restricted to linear inverse problems of the type

$$g = Kf \tag{1}$$

where :

g, the measured data, belongs to a space of functions D equipped with a scalar product $(g,g')_D$ and with the induced norm $||g||_S = \sqrt{(g,g)_S}$.

f, the solution, belongs to a space of functions S equipped with a scalar product $(f,f')_S$ and with the norm $||f||_S = \sqrt{(f,f)_S}$.

K is a bounded linear operator mapping S onto D.

In many applications, for instance, D and S are sets of quadratic integrable functions on some interval [a,b], i.e. functions f such that

$$\int_a^b dx\, |f(x)|^2 < \infty \tag{2}$$

This set is denoted $L^2[a,b]$ and the scalar product between two functions f and f' is defined as :

$$(f,f')_{L^2[a,b]} = \int_a^b dx\, f(x)\, f'(x) \tag{3}$$

Other functional spaces can be defined by replacing $\int dx$ by a weighted integral $\int dx\; w(x)$ in eq. (2) and (3).
The choice of the scalar product is important because the distance between two solutions f and f' is represented by $||f - f'||_S$, similarly the measurement error is represented by $||g-g_0||_D$ where g and g_0 are respectively the measured and exact data. The scalar product should therefore reflect what we mean by "these two solutions (or these two data functions) are close to each other". In practice, however, the choice of the norm is also dictated by purely mathematical considerations. It is clear, furthermore, that no single number like a norm can faithfully reproduce our feeling of what the distance is between several curves or images.

We are now ready to define ill-posedness. The inverse problem (1) is ill-posed if it does not satisfy the three conditions given by Hadamard [1923] to define a well-posed problem:

1. The solution f exists
2. It is unique
3. It depends continuously on the data g

This definition requires a few comments. The first condition could seem superfluous in the case of a physical problem, since such a problem is *known* to have a solution. In practice, however, the data g include measurement errors, and it cannot be guaranteed that this noisy g belongs to the range of the operator K, i.e. it cannot be guaranteed that the equation (1) has a solution. In tomography for instance a given set of projections corresponds to a reconstructed image only if it satisfies certain consistency conditions (the Helgason-Ludwig conditions, cfr. e.g. Natterer 1986, pp. 36), which in general are not satisfied by noisy projections.
In order to overcome this non-existence problem, an approximate solution has to be defined using some optimum criterion. As the norm chosen in G is assumed to faithfully represent the distance between different data functions, it is natural to chose a solution f which best fits the measured data, i.e. which minimizes the quantity $||Kf - g||_D$.
Such a solution is called a *pseudo-solution* (or also a *least-square* s solution in L^2 spaces). Other criteria could be used as well. For instance, if the statistical properties of the measurement errors are known, one can define the maximum-likelihood solution which maximizes the conditional probability of measuring g given the solution f. This approach leads to non-linear criteria (unless the conditional probability distribution is gaussian) and will not be discussed here.

Let us now consider the second item of Hadamard's definition. When the null-space $N(K)=\{ f \in S \mid Kf=0 \}$ of the operator K is non-empty, the solution or the pseudo-solution of the problem is not unique because any element of N(K) might be added to some solution without perturbing the compatibility with the data. A criterion must therefore be defined to select one particular solution. As the data g provide no information about the projection of the solution onto the null-space it is natural to select the solution belonging to the subspace orthogonal to the null-space, this is equivalent to chosing the solution, or pseudo-solution, with the minimum norm $||f||_S$.

The pseudo-solution with minimum norm is called the pseudo-normal solution and is noted f^+. It is related linearly to the data by the generalized inverse operator K^+ :

$$f^+ = K^+ g \tag{4}$$

Note that the pseudo-normal solution depends on the norm chosen for the space S; this gives a certain freedom in selecting a criterion which adequately reflects the physics of the problem. Non-linear criteria on the other hand (e.g. the maximum "entropy" criterion) cannot be described in this framework and their discussion is beyond the scope of this paper.

Despite the non-existence and non-unicity problems, an approximate solution to the inverse problem has now been defined. In practice, however, this solution can be reliable only if it depends continuously on the input data, i.e. if small measurement errors produce small errors in the reconstructed solution (what "small" means here depends on the choice of the norm in the spaces D and S). Formally, this crucial continuity condition is expressed by

$$\lim_{||g-g_0||\to 0} ||f^+ - f_0^+||_S = 0 \tag{5}$$

where g_0 represents the noise-free data and $f_0^+ = K^+ g_0$. This is equivalent to requiring that the generalized inverse K^+ be a continuous (i.e. bounded) linear operator.

When the inverse problem is ill-posed the continuity condition is usually not satisfied and the pseudo-solution is meaningless; alternative inversion techniques (discussed in section 3) must then be used to reconstruct reliable approximations to the exact solution.

2.2 Example 1. The derivative of a function.

This first example is rather simple minded but highlights the relationship between ill-posedness and sampling, and gives a first idea of what regularization is.

Consider a function $g \in C^3[0,1]$ with three continuous derivatives. We measure this function and want to evaluate its derivative $f = dg/dx$ at a given point x. Using the supremum norm:

$$\|g\| = \sup_x |g(x)| \tag{6}$$

one checks that the problem is ill-posed by considering the following sequence of data functions:

$$g_n(x) = \frac{1}{\sqrt{n}} \sin nx \qquad \|g_n\| = \frac{1}{\sqrt{n}} \qquad n=1,2,3..... \tag{7}$$

and the corresponding solutions :

$$f_n(x) = \frac{d\, g_n(x)}{dx} = \sqrt{n} \cos nx \qquad \|f_n\| = \sqrt{n} \tag{8}$$

Note that $\lim_{n \to \infty} \|g_n\| = 0$ whereas $\lim_{n \to \infty} \|f_n\| = \infty$. Thus, it is possible to produce a set of noisy data $(g_0 + g_n)$ converging to the noise-free data g_0, corresponding to solutions f_0+f_n which do not converge to the exact solution f_0.This contradicts the continuity condition.

In practice the inverse problem can be solved e.g. by applying the definition of the derivative. Suppose we measure the function g_0 at $x+h$ and at $x-h$ and take as an approximation to the solution at point x the following quantity:

$$f_h(x) = \frac{1}{2h} \{ g(x+h) - g(x-h) \} \tag{9}$$

In order to evaluate the accuracy of this solution some knowledge is required over the measurement error and the nature of the exact solution. Let us assume an upper bound on the noise:

$$\|g - g_0\| = \sup_x |g(x) - g_0(x)| \leq \varepsilon \tag{10}$$

and a smoothness constraint on the solution :

$$\left\| \frac{d^2 f_0}{dx^2} \right\| = \sup_x \left| \frac{d^2 f_0(x)}{dx^2} \right| \leq M \tag{11}$$

Expanding g_0 in a 4 terms Taylor series then yields an error estimate which illustrates some properties common to most ill-posed problems :

$$|f_h(x) - f_0(x)| < \frac{\varepsilon}{h} + \frac{h^2 M}{6} \tag{12}$$

The first term in the RHS shows that the inversion amplifies the measurement error ε by a factor 1/h, which grows out of control if one samples the data on a finer and finer grid. Working with a fixed non-zero value of h stabilizes noise propagation but produces a systematic error represented by the second term in the RHS of eq. (12). This second contribution is the error which would arise if the inversion algorithm (with the finite value of h) were applied to noise-free data.This systematic error grows when the sampling distance h is increased to suppress noise propagation. An optimal value of the parameter h can be derived which minimizes the global error bound (12):

$$h_{opt} = \left(\frac{3\varepsilon}{M}\right)^{1/3} \tag{13}$$

The corresponding error bound is

$$|f_{h_{opt}}(x) - f_0(x)| \le \varepsilon^{2/3} M^{1/3} \frac{3^{2/3}}{2} \tag{14}$$

and converges to zero when the noise level ε tends to zero. The inverse problem is *regularized.* The rate of convergence of the regularized solution characterizes the degree of ill-posedness of the inverse problem. Problems with a power convergence ($\varepsilon^{2/3}$ e.g.) are referred to as mildly ill-posed. In the present case for instance, an improvement of the measurement error by 1/1000 leads to an improvement of the solution by a factor 1/100. A logarithmic convergence of the reconstruction error ($1/\log(1/\varepsilon)^{\beta}$ for some positive β) on the other hand characterizes more severely ill-posed problems.

2.3 Example 2. Convolution problems

Convolution problems constitute probably the most common class of ill-posed problems. Any problem which can be modelized by a linear and shift-invariant relationship (i.e. invariant for a time or space translation) between data and solution leads to a convolution equation of the type (in one dimension):

$$g(x) = \int_{-\infty}^{+\infty} dy\, k(x-y)\, f(y) \qquad -\infty < x < +\infty \tag{15}$$

where we assume that g, f and k belongs to $L^2[\mathbf{R}]$. The frequency representation of this equation is well-known :

$$G(\nu) = K(\nu)\, F(\nu) \tag{16}$$

where G, K and F are the Fourier transforms of g, k and f respectively :

$$\begin{bmatrix} G(\nu) \\ K(\nu) \\ F(\nu) \end{bmatrix} = \int_{-\infty}^{+\infty} dx \; e^{-2\pi i \nu x} \begin{bmatrix} g(x) \\ k(x) \\ f(x) \end{bmatrix} \tag{17}$$

The null-space of this convolution operator corresponds to the frequency regions where $K(\nu) = 0$, therefore the minimum norm solution is :

$$F^{+}(\nu) = \begin{cases} \dfrac{G(\nu)}{K(\nu)} & \text{if } K(\nu) \neq 0 \\ 0 & \text{if } K(\nu) = 0 \end{cases} \tag{18}$$

This solution is unstable when the Fourier transform $K(\nu)$ of the convolution kernel $k(z)$ takes values arbitrary close to zero; in such a case, the Fourier transform $G(\nu)$ of the data is multiplied by an unbounded factor $1/K(\nu)$ and f^{+} does not depend continuously on the data g.

The instability of deconvolution is well known in practical applications and can be cured by filtering off (apodizing) the high frequency components of the data; we will see in the next section how regularization techniques can be used to design apodizing filters in a systematic way.

3. REGULARIZATION TECHNIQUES

3.1 Definition

The purpose of regularization techniques is to produce reliable estimates of the solution of ill-posed problems in the presence of noisy data. To what extent this is possible depends largely on the degree of ill-posedness of the problem, but regularization attempts at least to extract a maximum of information from the available data. Satisfactory solution estimates can often be obtained using empirical methods which effectively restrict noise propagation. Regularization techniques, however, have a sounder basis :

- stabilization is based on prior knowledge over the solution, which supplements the insufficiency of the information contained in the noisy data.

- regularization techniques allow to guarantee the convergence of the solution estimate to the exact solution (or its projection onto $N(K)^{+}$) when the data converge to the exact data.

- error estimates can be derived to assess the quality of the inversion for finite values of the measurement error.

A regularized algorithm for equation (1) is defined by a family of bounded linear operators $U(\alpha)$ mapping D onto S, depending on a real positive *regularization parameter* α and approximating the generalized inverse operator in the sense that

$$\lim_{\alpha \to 0} U(\alpha)\, g = K^{+} g \tag{19}$$

for any g in the domain of the generalized inverse K^{+}. Regularization is then achieved by prescribing a choice $\alpha(\varepsilon)$ such that

$$\lim_{\varepsilon \to 0} \| U(\alpha(\varepsilon))g - K^{+} g_0 \| = 0 \tag{20}$$

where g and g_0 are the measured and exact data respectively, and satisfy $||g - g_0|| < \varepsilon$.

The definition can be illustrated by the example of the derivative studied in section 2.2 :

- the regularization parameter α is the sampling distance h (although this need not be the case in general),

- the operator $U(\alpha)$ is given by eq. (9),

- a possible choice of α ensuring convergence is the optimal choice eq. (13).

The example of the derivative also demonstrates how prior knowledge (in this case a smoothness constraint with parameter M) has been introduced to achieve regularization. The separation of the error estimate into two contributions, illustrated in the example of the derivative (eq. (12)) can also be derived in the general case, by writing :

$$\|U(\alpha(\varepsilon))g - K^{+} g_0\| \leq \|U(\alpha(\varepsilon))\,(g-g_0)\| + \|U(\alpha(\varepsilon))g_0 - K^{+} g_0\| \tag{21}$$

The first term in the RHS represents the amplification of the measurement error $(g-g_0)$ by the inverse operator U, the second term is the systematic error due to the difference between the regularized inverse U and the generalized inverse operator K^{+}. These two terms vary in opposite directions when the parameter α is modified, and an important and difficult issue when applying a regularized algorithm is to determine an optimal, or at least satisfactory value of the regularization parameter (this point will be further discussed in section 3.4).

The next sections introduce several techniques allowing the systematic construction of regularizing algorithms.

3.2 Regularization via constraints

3.2.1 Derivation

Most regularization techniques exploit linear constraints on the unknown solution [Tikhonov, Arsenine 1976]. The restriction to linear constraints (in practice, mainly smoothness constraints) is a major limitation since many constraints (like positivity, or an upper bound on f(x)) are not linear. We follow here the formulation due to Miller [1970].
Assume that the data are measured with a known accuracy ε:

$$\|g - g_0\| \le \varepsilon \tag{22}$$

When the problem is ill-posed, the set of candidate solutions f which are compatible with the measured data g within the error bound ε, i.e. which satisfy the inequality

$$\| Kf - g \| \le \varepsilon \tag{23}$$

is not bounded. Further information must therefore be used to restrict the set of admissible solutions. Let us assume that the available prior knowledge over the solution (or part of it) can be represented by an inequality [Miller 1970]

$$\| Bf \| \le E \tag{24}$$

where B is a linear operator mapping S onto S and having a continuous inverse B^{-1}, and E is a real positive number. The smoothness constraint in example 2.2 corresponds for instance to $B = d^2/dx^2$.

Miller now restricts the set *S* of admissible solutions to the intersection of the two sets defined by eq. (23) and (24):

$$S = \left\{ f \mid \| Kf - g \| \le \varepsilon \text{ and } \| Bf \| \le E \right\} \tag{25}$$

If the operator B, and the parameter ε and M faithfully modelize the physical situation, this set cannot be empty since it contains at least the exact solution f_0. Any element of the set *S* is regarded as an acceptable solution

to the inverse problem and the choice of one particular solution in S is rather arbitrary (unless further prior knowledge is available).
In order to construct a particular admissible solution, one considers the quadratic functional

$$\Phi(f) = \|Kf-g\|^2 + (\frac{\varepsilon}{E})^2 \|Bf\|^2 \tag{26}$$

It is easily checked that the sets
$S_1 = \{ f \in S \mid \Phi(f) \leq \varepsilon^2 \}$ and $S_2 = \{ f \in S \mid \Phi(f) \leq 2\varepsilon^2 \}$ are infinite dimensional ellipsoids which sandwich the set S of admissible solutions :

$S_2 \supseteq S \supseteq S_1$.

Thus, a particular admissible solution f_M is the center of the ellipsoid S_2, which is obtained by minimizing Φ:

$$f_M = C^{-1} K^* g \tag{27}$$

where K^* is the operator adjoint to K and the positive definite operator C is defined by

$$C = K^*K + (\frac{\varepsilon}{E})^2 B^*B \tag{28}$$

As B possess a continuous inverse, C does too and the solution f_M can always be defined (even when the operator K is not invertible). One should still check that f_M is an admissible solution, a sufficient condition for this is that

$$\Phi(f_M) = \| g \| - (g , Kf_M) \leq \varepsilon^2 \tag{29}$$

Note that the functional Φ can be rewritten in function of f_M as :

$$\Phi(f) = (C(f - f_M) , (f - f_M)) + \Phi(f_M) \tag{30}$$

So far, one has only constructed a solution satisfying certain constraints. In order to prove that regularization has been achieved, one should further check that the solution estimate f_M tends to the exact solution f_0 when the noise level ε tends to zero. To do this, an upper bound on the reconstruction error is derived by noting that the exact solution is lying somewhere in the set S, and a fortiori also in S_2. Using eq. (30), one gets :

$$\Phi(f_0) = (C(f_0 - f_M) , (f_0 - f_M)) + \Phi(f_M) \leq 2\,\varepsilon^2 \tag{31}$$

which leads to the following error bound :

$$\| f_0 - f_M \|^2 \leq \frac{2\varepsilon^2 - \Phi(f_M)}{\gamma(\varepsilon/E)} \leq \frac{2\varepsilon^2}{\gamma(\varepsilon/E)} \tag{32}$$

where $\gamma(\varepsilon/E)$ is the infimum of the spectrum of C. $\gamma(\varepsilon/E)$ is strictly positive since C has a bounded inverse, it should be evaluated for each particular problem to verify whether the problem has effectively been regularized, i.e. whether

$$\lim_{\varepsilon \to 0} \frac{2\varepsilon^2}{\gamma(\varepsilon/E)} = 0 \tag{33}$$

and also to evaluate the rate of convergence of the regularized solution f_M. It can be shown that convergence is guaranteed if the constraint operator B has a compact inverse [Bertero et al 1980].

3.2.2 Application to convolution problems.

Consider a convolution operator K defined on $L^2(\mathbf{R})$:

$$g(x) = \int_{-\infty}^{+\infty} dy\; k(x-y)\, f(y) \qquad -\infty < x < +\infty \tag{34}$$

Assume also that the linear constraint can be expressed in the form :

$$\|Bf\|^2 = \int_{-\infty}^{+\infty} d\nu\; |B(\nu)|^2\; |F(\nu)|^2 < E^2 \tag{35}$$

where $F(\nu)$ is the Fourier transform of f (eq.18). If the function $B(\nu)$ is a polynomial in ν, this corresponds to a smoothness constraint. We assume that B has a bounded inverse, which is the case e.g. if $B(\nu)$ is a polynomial having no zero in $(-\infty,+\infty)$. The Fourier transform of the regularized solution f_M is then easily written as (compare with eq.19):

$$F_M(\nu) = \frac{K^*(\nu)\; G(\nu)}{|K(\nu)|^2 + (\frac{\varepsilon}{E})^2 |B(\nu)|^2} \tag{36}$$

where $K(\nu)$ and $G(\nu)$ are the Fourier transform of the convolution kernel $k(x)$ and of the data $g(x)$ respectively. $K^*(\nu)$ is the complex conjugate of $K(\nu)$. Miller's technique thus defines an apodizing filter which corresponds to a well defined constraint on the solution.

3.3. Truncated eigenfunction expansion

A system of linear equations g=Af with a NxN symmetrical invertible matrix A can be solved by expanding the data vector g and the solution vector f onto the eigenvectors of the matrix A. This technique is particularly adequate when the matrix A is ill-conditioned and the data g are noisy because it allows a better control of the noise propagation in each reconstructed eigencomponent. This familiar concept can be extended to infinite dimensional inverse problems which can be modelized by a Fredholm integral equation of the first kind:

$$g(x) = (Kf)(x) = \int_a^b dy\, k(x,y)\, f(y) \qquad c \le x \le d \tag{37}$$

with a square integrable kernel k, i.e.

$$\int_c^d dx \int_a^b dy\, |k(x,y)|^2 < \infty \tag{38}$$

Assume for simplicity that K admits an inverse K^{-1}. It is then possible to construct a countably infinite set of positive numbers λ_n (called the singular values of K) and associated functions u_n and v_n (normalized on the intervals [a,b] and [c,d] respectively) such that, for every n, $Ku_n = \lambda_n v_n$ and $K^* v_n = \lambda_n u_n$. The singular functions u_n are thus eigenfunctions of the operator K^*K with eigenvalue $\lambda^2{}_n$ and they form an orthonormal basis of the solution space, so that any solution f can be written as :

$$f(x) = \sum_{n=1}^{\infty} (u_n,f)\, u_n(x) \qquad \text{with } (u_n,f) = \int_a^b dx\, u_n(x)\, f(x) \tag{39}$$

Similarly, the set of functions v_n is an orthonormal basis in the closure of the range of K, so that any *noise-free* data can be expanded on this basis. The singular values are usually ordered by decreasing magnitudes $\lambda_1 \ge \lambda_2 \ge \lambda_3$ ......; one shows that the λ_n tend to zero when n tends to infinity.

When the data g_0 are noise-free, the solution of the inverse problem (37) can be written as:

$$f_0(x) = \sum_{n=1}^{\infty} \frac{(v_n,g_0)}{\lambda_n}\, u_n(x) \qquad \text{with } (v_n,g_0) = \int_c^d dx\, v_n(x)\, g_0(x) \tag{40}$$

Thus each singular component is reconstructed by multiplying the scalar product (v_n,g_0) by a factor $1/\lambda_n$ which blows up when n tends to infinity. Therefore the inverse K^{-1} is not continuous and the problem is ill-posed. Regularization can be achieved by including in the reconstruction only the singular components corresponding to the N largest singular values:

$$f_N(x) = \sum_{n=1}^{N} \frac{(v_n,g)}{\lambda_n} u_n(x) \tag{41}$$

Note that the scalar products of g with the v_n eliminates all components of the noisy data g that do not belong to the range of the operator K since the v_n are a basis in this subspace. As in the previous sections, the difference between the solution estimate f_N and the exact solution f_0 can be separated into two terms : on the one hand the error due to the difference between the measured data g and the exact data g_0, on the other hand the systematic error due to the truncation of all higher order singular components n >N of the exact solution. Using the orthonormality of the singular functions and the fact that the singular values are ordered by decreasing magnitudes one obtains:

$$\|f_N - f_0\|^2 \leq \frac{\varepsilon^2}{\lambda_N^2} + \sum_{n=N+1}^{\infty} |(u_n,f_0)|^2 \tag{42}$$

assuming that $||g - g_0|| \leq \varepsilon$.

The number N plays here the role of the regularization parameter and should be chosen in function of the noise level to ensure the convergence of f_N to the exact solution f_0 when the noisy data g tend to g_0. A choice $N(\varepsilon)$ will guarantee regularization if:

- $\lim_{\varepsilon \to 0} N(\varepsilon) = \infty$ and

- $\lim_{\varepsilon \to 0} \varepsilon/\lambda_{N(\varepsilon)} = 0$.

Indeed, the second term in the RHS of (42) tends to zero whenever N tends to infinity since $||f_0||^2$ is finite. Several methods for chosing $N(\varepsilon)$ have been described in the literature. One can for instance include in the expansion all singular values larger than $\varepsilon^{1-\beta}$ for some small positive number β. Whatever the choice of $N(\varepsilon)$ it is clear from this analysis that the number of singular components which can be reliably retrieved from the data depends on the signal-to-noise ratio.

The truncated singular function expansion (41) is based on a sharp cut-off of the singular components n>N. This might introduce artifacts in the estimated solution, similar to the side lobes observed when sharp apodizing windows are used for Fourier filtering (e.g. the ramp filter in tomography). It is therefore preferable to multiply each singular component with a discrete apodizing filter h_n with a smooth transition from 1 (n=1) to 0 (at high values of n). The estimated solution is then :

$$f_h(x) = \sum_{n=1}^{\infty} h_n \frac{(v_n,g)}{\lambda_n} u_n(x) \tag{43}$$

In practice of course the series is interrupted when h_n/λ_n becomes negligible. Possible choices of the discrete filter $\{h_n\}$ which guarantee regularization are discussed by Bakushinski [1967]. Miller's technique for instance can be used to design a discrete filter corresponding to a well-defined constraint on the solution, provided the constraint operator B^*B has a compact inverse and commutes with K^*K, i.e. provided the singular functions u_n of K are also eigenfunctions of B^*B and, therefore, of the operator C (eq.28). Expanding the solution $f_M = C^{-1} K^*g$ on the singular system, one gets for h_n:

$$h_n = \frac{\lambda_n^2}{\lambda_n^2 + (\frac{\varepsilon}{E})^2 \beta_n^2} \tag{44}$$

where the β_n^2 are the eigenvalues of B^*B. In practice, realistic constraint operators are not likely to commute with K^*K and an empirical choice of the digital filter $\{h_n\}$ might be preferable.

3.4 Practical problems

The application of a regularization technique requires the choice of a particular value for the regularization parameter α. Depending on the method considered this parameter represents a number of iterations, a sampling length, a number of singular components, a frequency cut-off or any more abstract parameter. Whatever its nature, the parameter α should in principle be chosen in function of the noise magnitude and of the known properties of the solution. Miller's technique for instance does not involve any arbitrary coefficient : the value of the regularization parameter (ε/E) in eq. (27) is fixed by the ratio between the noise upper bound ε and the smoothness parameter E. Usually, however, the characteristics of the noise and of the solution are poorly known. An empirical evaluation of the regularization parameter might then be more efficient than a more "rigorous" choice based on an incorrect modelization of the problem (e.g. on a wrong value of E). A practical procedure consists in examining the solutions reconstructed using different values of α to determine the "best" one.

When the value of the noise magnitude ε is known with some confidence, but no prior knowledge about the solution is available, the regularization parameter can be determined using an empirical criterion known as the discrepancy principle [Morozov 1968, Vainikko 1983]: the approximate solution f to an inverse problem should not fit the data g with an accuracy greater than the accuracy of the measurement itself. In other words, it would be meaningless to have $||Kf-g|| \ll ||g-g_0|| \approx \varepsilon$. Assuming that the approximate solution is retrieved using a family of operators $U(\alpha)$, the regularization parameter α is then chosen in such a way that $||KU(\alpha)g - g|| = \varepsilon$. This prescription has been shown to guarantee convergence for some families of regularizing techniques [Morozov 1968, Defrise and De Mol 1987]. The same idea has been applied recently, although in a different framework, to prescribe the number of iterations for the maximum-likelihood algorithm in emission tomography [Veklerov and Llacer 1987].

4. APPLICATIONS IN TOMOGRAPHY

4.1 Inversion of the two dimensional Radon transform from a complete set of projections

The inversion of the 2D Radon transform has been analyzed in the previous lecture. We here demonstrate how Miller's technique can be applied to derive apodizing filters corresponding to well defined smoothness constraints. Furthermore, error estimates are obtained in this framework.

Consider the Radon transform operator R in $L^2(\mathbf{R}^2)$, and remember that the image obtained by backprojecting noise-free data is related to the original density $f_0(\mathbf{x})$ by a 2D convolution with $1/r = 1/\sqrt{x^2+y^2}$:

$$b(\mathbf{x}) = (R^* R\, f_0)(\mathbf{x}) = \iint_{\mathbf{R}^2} d^2\mathbf{y}\ \frac{f_0(\mathbf{y})}{|\mathbf{x}-\mathbf{y}|} \tag{45}$$

where b represents the backprojected image and R^* is the backprojection operator. Thus, the problem of retrieving the original image f_0 from the backprojected data is a convolution problem in this $L^2(\mathbf{R}^2)$ setting. Following the discussion in section 3.2.2, we chose a constraint operator B which is also diagonal in Fourier space. To be specific let us take :

$$F_2(Bf)(\nu) = (1 + |\nu|^2)^{\beta/2}\, F_2 f(\nu) \tag{46}$$

where F_2 is the 2D Fourier transform and β is positive. The constraint is then:

$$\|Bf\| = \{ \iint_{\mathbf{R}^2} d^2\nu \ (1+|\nu|^2)^\beta \, |F_2 f(\nu)|^2 \}^{1/2} \le E \tag{47}$$

for some positive number E. Note that some unit of length has been implicitly introduced in eq. (47). When β is a positive integer, this smoothness constraint involves bounds on the norm of the β first derivative of f. As previously, we assume that an upper bound on the measurement error is available :

$$\|g - g_0\| = \{ \int_0^\pi d\theta \int_{-\infty}^{+\infty} ds \ |g(\theta,s) - g_0(\theta,s)|^2 \}^{1/2} \le \varepsilon \tag{48}$$

where g_0 and g represent respectively the exact and the measured projections.

Miller's technique now leads to the regularized image estimate :

$$f_M = F_2^{-1} \{ \frac{1}{\frac{1}{|\nu|} + (\frac{\varepsilon}{E})^2 (1+|\nu|^2)^\beta} \} F_2 R^* g = C^{-1} R^* g \tag{49}$$

where $1/|\nu| = 1/\sqrt{\nu_x^2+\nu_y^2}$ is the 2D Fourier transform of the convolution kernel 1/r in eq.(45), and the two-dimensional inverse Fourier transform is represented by F_2^{-1}. This corresponds to the following apodizing filter :

$$W(\nu) = \frac{1}{1 + |\nu| \, (\frac{\varepsilon}{E})^2 (1+|\nu|^2)^\beta} \tag{50}$$

This filter is based on a rigorous approach, but does not in general lead to practical improvements with respect to the usual empirical filters (Hamming, Shepp-Logan etc) since the evaluation of the parameters β, E and ε should be empirical anyway. Miller's filter, however, has the merit of explicitly prescribing how the apodizing filter should depend on the signal-to-noise ratio. Furthermore, it allows the derivation of error estimates. This is done by evaluating the norm $\gamma^{-1}(\varepsilon/E)$ of the operator C^{-1}, which is straightforward using its Fourier space representation :

$$\|C^{-1}\| = \sup_\nu \{ \frac{1}{|\nu|} + (\frac{\varepsilon}{E})^2 (1+|\nu|^2)^\beta \}^{-1} \le \sup_\nu \{ \frac{1}{|\nu|} + (\frac{\varepsilon}{E})^2 |\nu|^{2\beta} \}^{-1}$$

$$= (2\beta)^{-\frac{1}{2\beta+1}} \, (1+\frac{1}{2\beta})^{-1} \, (\frac{E}{\varepsilon})^{\frac{2}{2\beta+1}} \tag{51}$$

According to equation (32) an upper bound on the reconstruction error is then given by :

$$\|f_M - f_0\| \leq (\beta + \frac{1}{2})^{-1/2} \; (2\beta)^{\frac{\beta}{2\beta+1}} \; \varepsilon^{\frac{2\beta}{2\beta+1}} \; E^{\frac{1}{2\beta+1}} \tag{52}$$

It should be emphasized that this error estimate does not only account for the propagation of the measurement errors but also for the systematic error due to the use of the smoothing filter (50). This result, which was first obtained in a similar form by Natterer, demonstrates that the image estimate f_M is regularized if β is strictly positive, indeed :

$$\lim_{\varepsilon \to 0} \|f_M - f_0\| = 0 \tag{53}$$

As expected, the convergence of the regularized image to the exact solution is faster when the parameter β increases, that is when the smoothness constraint becomes more restrictive. For instance, a bound on the norm of the Laplacian of the image corresponds to β=2; the solution estimate then converges as $\varepsilon^{4/5}$, which means in practice that the number of significant digits in the reconstructed image is 4/5 the number of significant digits in the data. On the other hand, if the class of admissible solutions is to contain images with sharp contrasts, the parameter β should be smaller than 1/2 [Natterer 1986, pp. 92] and the convergence in eq. (52) cannot be faster than $\varepsilon^{1/2}$. Even in this case noise propagation is not too dramatic, indicating that the inversion of the 2D Radon transform from a complete set of projections is a mildly ill-posed problem which can be easily regularized.

The previous analysis gives valuable insight into the nature of the instability problem in tomography, but its practical significance can be questioned :

a) a continuous sampling of the projections has been assumed,

b) equation (52) yields only a global L^2 error estimate,

Let us briefly discuss these two points.

a) <u>a continuous sampling of the projections has been assumed.</u>

In order to reliably describe the discrete nature of the measurement, one should consider a discrete Radon operator R mapping $L^2(\mathbf{R}^2)$ onto $\mathbf{R}^N$ where N is the total number of discrete data :

$$Rf_0 = \left\{ (Rf_0)_k = \iint_{\mathbf{R}^2} d^2\mathbf{x} \; f_0(\mathbf{x}) \, \Psi_k(\mathbf{x}) \quad | \; k=1,..,N \right\} \tag{54}$$

where Ψ_k is the characteristic function of the region (e.g. a strip) over which the detector number k integrates the unknown density f (Ψ_k may also incorporate the effect of a non-uniform sensitivity over this region). It is then possible to apply Miller's technique to this discrete Radon operator, noting that the adjoint of R is a discrete backprojection:

$$(R^* g)(\mathbf{x}) = \sum_{k=1}^{N} g_k \, \Psi_k(\mathbf{x}) \qquad \text{for } g \in \mathbf{R}^N \tag{55}$$

This approach reliably accounts for the discrete sampling of the projections but requires a numerical evaluation of Miller's image estimate (27) and of the stability estimate (32) since the operator C is no longer diagonal in Fourier space.

b) equation (52) gives only a global L^2 error estimate.

The global error estimate derived above does not convey much information over the reconstruction error at a given point, or in a small region around that point. In order to establish the relationship between spatial resolution and reconstruction accuracy, it is useful to consider the reconstruction of a blurred image f_w obtained from the regularized image f_M by a 2D convolution with some fixed function $w(\mathbf{x}) \in L^2(\mathbf{R}^2)$ representing the desired resolution (the integral of w over the plane is normalized to 1):

$$f_w(\mathbf{x}) = \iint_{\mathbf{R}^2} d^2\mathbf{x}' \; f_M(\mathbf{x} - \mathbf{x}') \; w(\mathbf{x}') \; = \; (f_M ** w)(\mathbf{x}) \tag{56}$$

For instance, one could choose for w the characteristic function of a square pixel in order to assess the accuracy with which the average of f_0 over one discrete pixel can be retrieved. It is then possible to obtain an upper bound for the local reconstruction error [Bertero et al. 1980]:

$$|f_w(\mathbf{0}) - f_{0w}(\mathbf{0})| \le \sqrt{2}\, \varepsilon \, (C^{-1}w,w) = \sqrt{2}\,\varepsilon \; \Big\{ \iint_{\mathbf{R}^2} d^2\nu \; \frac{|(F_2 w)(\nu)|^2}{\frac{1}{|\nu|} + (\frac{\varepsilon}{E})^2 (1+|\nu|^2)^{\beta}} \Big\}^{1/2} \tag{57}$$

where $f_{0w} = f_0 ** w$ is the blurred exact image. Splitting the integral into two integrals over the regions $|\nu| < 1/\sqrt{\varepsilon}$ and $|\nu| \ge 1/\sqrt{\varepsilon}$ one checks that $|f_w(0) - f_{0w}(0)|$ tends to zero when ε tends to zero, for any non-negative value of β. Thus, the blurred image estimate converges even when $\beta=0$ [Defrise and De Mol 1984] whereas strong convergence (in the sense of eq. (53)) is only valid for $\beta > 0$ and becomes exceedingly slow when β approaches zero.

In order to investigate the relationship between noise propagation and spatial resolution, let us now "shrink" the blurring function $w(\mathbf{x})$ by replacing it with $w_\sigma(\mathbf{x}) = 1/\sigma^2 \; w(\mathbf{x}/\sigma)$ for some $\sigma < 1$ (the factor $1/\sigma^2$ ensures that the integral of w is normalized to one). A value $\sigma=1/2$ for instance corresponds to a twofold improvement of the spatial resolution. For any fixed value of ε and E, one gets from equation (57) :

$$\sqrt{2}\,\varepsilon\,(C^{-1}w_\sigma, w_\sigma)^{1/2} \geq \sigma^{(-1+\beta)}\,\sqrt{2}\,\varepsilon\,(C^{-1}w, w)^{1/2} \tag{58}$$

Thus for any $\beta < 1$ (corresponding in particular to images with sharp contrasts) the error estimate diverges when the resolution σ tends to zero. This behavior is typical for ill-posed problems and has already been illustrated in section 2.2.

When $\beta=0$, i.e. if the regularizing constraint involves the norm of the solution, the error estimate (57) is proportional to $\sigma^{-3/2}$ in the limit of high signal-to-noise ratio. Thus a twofold improvement of the resolution increases the reconstruction error by a factor $2\sqrt{2}$, whereas the decreased signal-to-noise ratio per pixel only accounts for a factor 2. This agrees with the well-known formula for emission tomography [Budinger et al. 1979 pp.175, Barrett and Swindell 1981 pp. 602, Lim et al. 1982], giving the variance in the image of a uniform disk reconstructed with a pixel size σ.

4.2 The singular value analysis of the Radon transform in 2 dimensions

The knowledge of the singular values of an operator allows one to assess the degree of ill-posedness of the corresponding inverse problem. Indeed, we have seen in section 3.3 that the number of singular components which can be reliably recovered from noisy data depends on the rate with which the singular values λ_n tend to zero when n tends to infinity.

Louis [1980] and Davison [1983] have derived the singular value decomposition for the tomography problem with full- and limited-angle data. They consider the reconstruction of an image f defined, and square integrable, on the unit disk D : $f \in L^2(D)$. The projections $g(\theta,s)$ are defined on the interval $[-\Phi \leq \theta \leq \Phi]\otimes[-1 \leq s \leq 1] = I$ where the maximum projection angle Φ is smaller than $\pi/2$ for the limited angle problem. The scalar product in the data space is defined with a weight function $w(s) = 1/\sqrt{1-s^2}$:

$$(g_1,g_2) = \int_{-\Phi}^{+\Phi} d\theta \int_{-1}^{+1} ds \frac{1}{\sqrt{1-s^2}}\, g_1(\theta,s)\, g_2(\theta,s) \tag{59}$$

This rather exotic norm is introduced by Louis and Davison because it allows the factorization of the singular problem into a radial problem and an angular one. Indeed, using this norm, one checks that the eigenfunctions of the operator RR^* have the form:

$$v_{m,k}(\theta,s) = \left\{ \sum_{l=0}^{m} h_{k,l}\ e^{i(-m+2l)\theta} \right\}\ \sin\left\{ (m+1)\arccos(s) \right\} \qquad m=0,1,2... \quad k=0,..,m \tag{60}$$

where the vectors h_k are obtained for each value of m by diagonalizing the (m+1) by (m+1) matrix:

$$A_{k,l} = \frac{2}{m+1} \frac{\sin[2(k-l)\Phi]}{(k-l)} \qquad k,l=0,..,m \tag{61}$$

In the full-data case, $\Phi=\pi/2$ and the matrix reduces to $A_{k,l} = 2\pi\ \delta_{k,l}\ /(m+1)$. The (m+1) singular functions corresponding to each value of m are then degenerate and can be chosen e.g. as :

$$v_{m,k}(\theta,s) = e^{i(-m+2k)\theta} \sin\left\{ (m+1)\arccos(s) \right\} \qquad m=0,1,2.... \quad k=0,..,m \tag{62}$$

The corresponding singular values of the Radon operator only depend on the main index m:

$$\lambda_{m,k} = \sqrt{\frac{2\pi}{m+1}} \tag{63}$$

The gentle decrease of these singular values confirms the mild ill-posedness of the Radon transform with a complete data set, and can be related to the $1/\nu$ behavior of the Fourier transform of the convolution kernel in equation (45), where the Radon operator had been described in a $L^2(\mathbf{R}^2)$ setting.
In the limited-angle case, $\Phi < \pi/2$ and the diagonalization of the matrix A is no longer trivial. The singular values of the Radon transform operator can then be written as

$$\lambda^2_{m,k} = \frac{2\pi}{(m+1)}\ \sigma_k(m+1,\frac{\Phi}{\pi}) \qquad m=0,1,2... \qquad k=0,..,m \tag{64}$$

The numbers $\sigma_k(m+1,\Phi/\pi)$ have been studied by Slepian [1978]. For large values of m, a fraction of about $2\Phi/\pi$ of the numbers $\sigma_k(m+1,\Phi/\pi)$ is close to one, so that the corresponding singular values are close to their full-data value (63). Except for a few transition values, the rest of the singular values (i.e. a fraction roughly equal to $1-2\Phi/\pi$) are close to zero and decay exponentially with m; the corresponding singular components are therefore very difficult to recover, even if the signal-to-noise ratio is high, unless *very* restrictive prior knowledge over the solution can be introduced in the reconstruction (e.g. some symmetry property).

ACKNOWLEGEMENTS

It is a pleasure to thank Dr. C. De Mol for several suggestions concerning this paper.

The author is Research Associate with the National Fund for Scientific Research (Belgium).

REFERENCES

Bakushinskii, A.B.(1967) . A general method of constructing regularizing algorithms for a linear ill-posed equation in Hilbert space, U.S.S.R Comp. Maths. Math. Phys. 7, pp. 279-287.

Barrett H.H. and Swindell W. (1981). Radiological Imaging. Academic Press.

Bertero M., De Mol C., Viano G. (1980) . The Stability of Inverse Problems. In Inverse Scattering in Optics (H.P. Baltes ed.), Topics in Current Physics 20, Springer-Verlag, pp. 161-214.

Budinger T.F., Gullberg G.T. and Huesman R.H. (1979). Emission Computed Tomography, in Image Reconstruction from Projections (Herman G.T. ed). Topics in Applied Physics, Springer.

Davison, M.E. (1983). The ill-conditioned nature of the limited angle tomography problem. SIAM J. Appl. Math. 43, pp. 428-448.

Defrise M., De Mol C. (1984). Resolution limits in full- and limited angle tomography. Proceedings of the 8th conference on Information Processing in Medical Imaging, F. Deconinck ed., Martinus Nijhoff, pp.94-105.

Defrise M., De Mol C. (1987). A note on stopping rules for iterative regularization methods and filtered SVD , in "Inverse problems, an interdisciplinary study" (P.C. Sabatier ed.), Advances in Electronics and Electron Physics, supplement 19, Academic Press, pp. 261-268.

Hadamard J. (1923). Lectures on the Cauchy problem in linear partial differential equations. Yale University Press, New Haven.

Lim C.B., Han K.S., Hawman E.G. and Jaszczak R.J. (1982). Image noise, resolution, and lesion detectability in SPECT. IEEE Trans. Nucl. Sc. NS-29, pp.500-505.

Louis A.K. (1980). Picture reconstruction from projections in a restricted range. Math. Meth. in Appl. Sc. 2, pp. 209-220.

Miller K. (1970). Least squares methods for ill-posed problems with a prescribed bound. SIAM J. Math. Anal. 1, pp. 52-74.

Morozov V.A. (1968). The error principle in the solution of operational equations by the regularization method. USSR Comp. Maths. Math. Phys. 8, pp. 63.

Natterer F. (1986). The mathematics of computerized tomography. Wiley & Sons.

Slepian D. (1978). Prolate spheroidal wave functions, Fourier analysis and uncertainty. V. The discrete case. Bell Sys. Tech. J. 57, pp.1371-1430.

Tikhonov A. and Arsenine V. (1977). Solution of ill-posed problems. Winston &Sons.

Vainikko G.M. (1983). The critical level of discrepancy in regularization methods. USSR Comp. Maths. Math. Phys. 23, pp.1-9.

Veklerov E. and Llacer J. (1987). Stopping rule for the MLE algorithm based on a statistical hypothesis testing. IEEE Trans. Med. Imag. MI-6, pp.313-319.

NEW INSIGHTS INTO EMISSION TOMOGRAPHY VIA LINEAR PROGRAMMING

L.A. Shepp
R.J. Vanderbei

AT&T Bell Laboratories
Murray Hill, New Jersey 07974, U.S.A.

ABSTRACT

Suppose that each detector unit, d, of an emission scanner measures a count $n^*(d)$ which represents the number of emissions into d of an unknown emission density λ. The likelihood, $P(n^*|\lambda)$, is the (Poisson) probability of observing n^* under λ. The well-known EM algorithm starts with an estimate λ^0 of λ and produces a sequence λ^k, k=1,2,..., of estimators having increasing likelihood and which converge to an estimate λ^x with maximum likelihood as $k \to \infty$. It is well-known that for large k, λ^k becomes snowy or noisy in appearance, and various methods have been proposed to effectively smooth λ^k, but these all give up likelihood to get smoothness. We give a method using linear programming of smoothing λ^k which produces a smoother estimate $\tilde{\lambda}^k$ which nevertheless has the same likelihood as λ^k.

If λ^k is nearly unique among estimators with its likelihood then $\tilde{\lambda}^k$ cannot differ much from λ^k, while of there are many λ with the likelihood of λ^k, then $\tilde{\lambda}^k$ should be smooth. Experiments described here indicate that for large k, $\tilde{\lambda}^k$ is not much smoother than λ^k, so that λ^k seems to be nearly unique. This is surprising in view of the fact the problem considered is severely undetermined in the sense that there are typically 3 or more times as many unknowns as equations (measurements), but of course it is the inequality constraints that cause the uniqueness.

We survey the maximum likelihood approach to emission tomography using the iterative EM algorithm and discuss the known difficulty with the algorithm at high iteration numbers.

1. Introduction

In a typical but especially exciting new application of emission tomography (ET), Petersen et al., [P], assert that the reconstructed brain ET image varies reproducibly with the word or concepts being considered by the subject's brain during the experiment. Oxygen atoms in radioactive isotope form (^{15}O with half-life = 123sec.) are intravenously injected as H_2O into volunteers who words on a TV screen during an ET scan. This water is metabolically taken up in the active region of the brain where oxygen is needed at that moment. While it is there it decays and emits radioactivity which is detected in a bank of detectors. An image of the emission density is formed by an algorithm (the so-called EM algorithm) for ET reconstruction. A subtraction is made between two images, with and without the visual word stimulus, and, in order to increase signal/noise, several such images from different volunteers are added together after a registration shift and scaling. The resulting image shows [P] that different regions of the brain are active depending on the word flashed. It is difficult to perform this experiment on a single volunteer because of ET's inherently low signal/noise. It is indicated in [P] that it would be very useful to improve the understanding of ET algorithms so that the need for several volunteers could be avoided and such experiments could be made more efficient. Indeed if these results can be repeated by other groups the technique raises many possible directions for ET applications.

The purpose of this paper it to survey ET and the currently most important algorithm for ET, namely the EM (expectation-maximization) algorithm [SV], [VSK] which was used by [P] to obtain high likelihood estimates of the emission density. Here we focus on one of the shortcomings of EM in ET and the techniques suggested to avoid this problem. The problem, as shown in §3, is that estimators of very high likelihood tend to be too snowy to use. For newcomers to the field, we point out that ET is different from CAT scanning or transmission tomography [SK] which has much higher resolution (and counts and dose) than ET, but cannot be used for metabolic studies of this type. ET is used mainly to study function while CAT is used clinically.

In ET, there are typically (but not always, see below) a finite set D of detector units (we say units to cover the case of double photons or PET, cf. [SV]). For each $d \in D$ there is a received count $n^*(d) = 0, 1, 2, \ldots$ which is the number of counts of radioactivity recorded by unit d during the experiment. The geometric arrangement of the units d and their efficiencies determine the transition probability $p(d \mid x)$ that an event or emission at x which is detected will be detected in unit d. Thus $p(d \mid x)$ is known and given and

$$\sum_{d \in D} p(d \mid x) = 1 \qquad \text{for each } x. \tag{1.1}$$

If $\lambda(x)dx$ is the density of, say ^{15}O, emitter at x then the mean number of counts in unit d is

$$E n^*(d) = \int \lambda(x) p(d \mid x) dx \triangleq \mu(d) \tag{1.2}$$

where the integral is over all of 2 or 3 dimensional space where detected counts can be emitted. Of course $\lambda(x)$ is unknown but for each λ, $n^*(d)$, $d \in D$ are independent Poisson distributed variates whose means are given by $\mu(d)$ defined in (1.2). In [SV] we denoted $\mu(d)$ by $\lambda^*(d)$. The reader should prove the assertion that $n^*(d)$ are independent and Poisson provided that the emission process is a Poisson process with variable rate $\lambda(x)$.

Hint: If x-space is broken into small regions $b_1, b_2, \ldots$ the number, $n(b_i)$, of emissions from each b_i are not observable, but are independent and Poisson. With each emission a line is chosen independently and uniformly through x and so each emission independently reaches d from x with probability $p(d \mid x)$ and the emission from each b_i into each d are independent and Poisson. Finally, sums and limits of independent Poisson variables have the same property. For more details see [S] or [SV].

Thus for each λ there is a probability $\Lambda(\lambda)$ of observing the actual recorded counts $n^*(d)$ for $d \in D$ if λ is the true emission density. Indeed by independence and the Poisson formula, this probability of observing $n^*(d)$, $d \in D$ under λ is precisely

$$\Lambda(\lambda) = \prod_{d \in D} e^{-\mu(d)} \frac{\mu(d)^{n^*(d)}}{n^*(d)!} \tag{1.3}$$

where $\mu(d)$ is given by (1.2).

The maximum likelihood (ML) viewpoint is now very natural. Since (1.3) is an exact model for the physics (because the Poisson assumption that $n(b_i)$ are independent and Poisson for disjoint b_i is a perfect model of radioactive decay) it seems reasonable to estimate the unknown λ as any λ which maximizes the likelihood of the observed counts, i.e. the product (1.3). Such a λ is called a maximum likelihood estimate (MLE). Note that λ is an arbitrary density so that one would not expect $\Lambda(\lambda)$ to have a unique maximum. We will return to this point in §2, where we give evidence which indicates that *in typical cases the maximum is essentially unique* (of course it is not always unique).

Now we return to the question of the discreteness of D, the set of detector units. There are position-sensitive detectors which measure (up to an error) the point on the detector where the radioactivity strikes. For these detectors it becomes unnatural to think of detection as discrete. Also in time-of-flight (TOF) ET scanners [SV], [STT] a central location is measured for each detected count so that x itself is observed up to an error, usually taken to be a Gaussian displacement. Position-sensitive and TOF detection can be modelled by a finite D (in the case of TOF one often uses subtubes [SV] and assigns a count proportionally to more than one subtube although this seems to be a somewhat inelegant model. In the non-TOF case with error-free position detection, one can imagine that for each count arising at x we cannot observe x itself, but instead a random line uniformly distributed in angle is chosen through x and observed. It would then be desired to estimate λ given a finite set of observed or detected lines. It can be shown that an MLE must be concentrated on the set of intersections of the given lines. But such is not a density since this is a set of zero measure and so there is no maximum likelihood density. Of course exact detection of lines is not real, but we raise this limiting case to point out some difficulties with the ML criterion. Note Λ is bounded (by 1) since $\Lambda(\lambda)$ is a probability but the maximizing λ is achieved by a measure with no density.

Returning to the simpler discrete detector case, we may also discretize x into pixels or boxes, $b \in B$, and this was done in [SV]. Let $p(b, d) = p(d \mid b)$ be the probability that a detected emission in box b will be detected in unit d. >From (1.2) we have

$$(1.4) \qquad p(b, d) - \int_b \lambda(x) p(d \mid x) dx \Big/ \int_b \lambda(x) dx$$

so that $p(b, d)$ depends on λ, which is unknown. But if b has small diameter so that $p(d \mid x)$ is nearly constant for $x \in b$ then we may assume that $p(b, d)$ is known and say is approximately given by (1.4) where λ is say uniform over $x \in b$. For more on this see [SV].

The EM algorithm gives a simple update rule which starts with $\lambda^0(b)$, $b \in B$ usually taken to be flat, $\lambda^0(b) \equiv 1$, and assigns a new $\lambda^1(b)$, $\lambda^2(b)$, ... in order by the rule

$$\lambda^{k+1}(b) = \lambda^k(b) \sum_{d \in D} \frac{n^*(d)p(b,d)}{\mu^k(d)} \tag{1.5}$$

where

$$\mu^k(d) = \sum_{b \in B} \lambda^k(b)p(b,d). \tag{1.6}$$

As observed in [SV] the update rule (1.5) has elegant properties. First, the likelihood always increases,

$$\Lambda(\lambda^{k+1}) \geq \Lambda(\lambda^k) \tag{1.7}$$

with equality only if λ^k is a maximum of Λ. Moreover, it is clear from (1.5) that $\lambda^k(b) \geq 0$ and that for $k \geq 1$

$$\sum_b \lambda^k(b) = \sum_{d \in D} n^*(d) \tag{1.8}$$

so that λ^k has the same total number of recorded counts. Moreover $\lim_{k \to \infty} \lambda^k(b)$ always exists [CT, VSK] and this maximizes $\Lambda(\lambda)$.

The algorithm (1.5) is interesting in itself and can be used to approximately nonlinearly "deconvolve" or invert a matrix equation $n^* = \lambda P$ for vector λ given vector n^* and matrix P,

$$n^*(d) = \sum_{b \in B} \lambda(b)p(b,d), \quad d \in D \tag{1.9}$$

where $\lambda(b) \geq 0$ is sought from measurements of the left side where $p(b,d)$ is a known nonnegative matrix satisfying

$$\sum_{d \in D} p(b,d) = 1, \quad b \in B. \tag{1.10}$$

Our experience with EM in [SV], [VSK] shows that good numerical results can be expected from such a nonlinear inversion of positive matrices. It also suggests that methods from linear programming [K], [V] can be used to study ET and this will be done in theory (§2) and practice (§3).

Lemma. *If* (1.9) *has a solution* $\lambda(b) \geq 0$ *then it must maximize* Λ.

To prove this, call $\lambda^k = \lambda$, and note from (1.9) that $\mu^k(d) = n^*(d)$ and then from (1.5) and (1.10) that $\lambda^{k+1} = \lambda^k$. But (1.7) then holds with equality and so $\lambda = \lambda^k$ is a maximum of Λ. Q.E.D.

With low counts (small $n^*(d)$) there is lots of noise in (1.9) and there is typically no (nonnegative) solution λ in (1.9) as we will see from our experiments in §3. A theoretical analysis of noise as a function of total count is given in [JS].

Where did the EM rule (1.5) come from? Are there general principles underlying (1.5)? Are they useful elsewhere? The answers are yes and the history is as follows: L. Baum [B] studied a Markov chain model for breaking secrecy codes which led him to alternately: (a) use conditional expectation (E step) to obtain missing data and (b) use maximum likelihood (M step) to estimate the Markov chain parameters given this missing data, then go back to an E step to better estimate the missing data, and so on. This was generalized by Baum himself and further by [DLR], who coined the term EM for the two steps; the reader should consult [DLR] for more details. The EM algorithm is particularly elegant in this Poisson model because $\Lambda(\lambda)$ is logconcave and the above desirable properties of positivity, count preservation, and convergence can be obtained. This was first pointed out for ET in [SV]. To see how this E and M step works in the Poisson case see [VSK, p. 12]; but in a sense one does not really learn anything by "deriving" (1.5) in this way rather than by heuristics because the properties (1.7), (1.8) must then be proved anyway.

A major difficulty [SV, VSK] with the EM algorithm is that for large k, λ^k gets very noisy or snowy, i.e. has large oscillations. Although the likelihood continues to increase with k, at $k \approx 500$, one no longer likes the image. At $k \approx 5000$, one no longer even likes likelihood as a criterion, since as we show in §2, in a typical case, the image as given by λ^{5000} is ridiculous and unrecognizable. Why is this and what to do? If likelihood is a good thing why isn't more likelihood even better?

One explanation [VL, LV] is that likelihood as a criterion reflects noise in the data. Thus if one had direct observations, $n(b)$ of $\lambda(b)$, instead of indirect observations, $n^*(d)$, one would use $\lambda(b) = n(b)$ as the most likely value of $\lambda(b)$ since

$$e^{-\lambda} \frac{\lambda^n}{n!}$$

is maximized for a fixed n at $\lambda = n$. But this can be very noisy if $\lambda(b)$ is small, which is the case if the pixels b are too small.

As a way out of this problem, it has been suggested to smooth λ [SM], or to impose a penalty on λ for lack of smoothness, or equivalently to specify a prior distribution for λ and maximize posterior likelihood rather than absolute likelihood ([LiH], [GM], [LeH], [Rubin's comments to VSK]). Another way out was suggested by Veklerov and Llacer [VL], [LV] who propose to simply stop iterating EM using "a quantitative criterion with a simple probabilistic interpretation that allows the user to stop the algorithm just before this (bad) effect begins" [VL].

We prefer to use a Veklerov-Llacer stopping criterion rather than to arbitrarily smooth or to use an arbitrary questionable prior. Note that using (1.3) one easily computes the incremental likelihood in going from λ^k to λ^{k+1} and one could presumably design a quantitative stopping rule based on this alone (although the likelihood continues to increase with k). However most users of EM prefer other fixes for the problem.

Geman and McClure's method [GM], based on a carefully chosen Gibbs random field as prior, gives amazingly good reconstruction of the original λ in the simulation experiment they describe [GM]. We are however concerned that their choice of Gibbs prior is unfair since it places high probability on λ's which only take a few different values (piecewise constant) which was true for the original "phantom" used in [SV]. This seems likely to lead to errors in reconstruction when the true phantom is actually smoothly varying – which may indeed be the case for real emission densities. We have not checked it but have been assured by McClure that this is not a problem, and if this is so, their method of smoothing seems very useful.

Another way out of this likelihood paradox is to use a least squares [LeH] or other non-likelihood based criterion to reconstruct λ. However, starting from any λ and letting λ' be the new λ obtained by an iteration of (1.5), will increase the likelihood because of (1.7) and so long as this does not ruin smoothness it seems desirable to continue using EM iterations which leads us back to EM anyway.

In the discussion to [VSK], Gabor Herman claims that EM is similar to earlier algorithms called multiplicative ART. Although this is true and may be useful in suggesting variants of EM, we think it is nice that likelihood and EM is not nearly as arbitrary as the ART algorithms in that EM maximizes the meaningful quantity of likelihood. It does give reasonable reconstructions if one does not iterate too far.

It should be pointed out that EM has other troubles, mainly in slow execution times. However, this can and is being gotten around by using highly parallel computing [MBM]. Another problem, new to us, of "edge overshoot" is described in [PS]. We have never seen this phenomenon. Can it be due to a poor choice of $p(b, d)$'s which do not match the generated counts physically?

In §2 we study the question of uniqueness of λ maximizing $\Lambda(\lambda)$ and describe a linear programming approach to ET which exploits recent developments in linear programming.

In §3 we discuss the methodology used in doing computer simulation experiments to study the performance of algorithms for ET in general, and those experiments related to linear programming suggested by the observations in §2, in particular.

§2. Linear Programming and EM

It has long been believed [VSK, p. 13] that in typical cases of ET where the number of pixels greatly exceeds the number of detector units that $\lambda^\infty = \lim \lambda^k$, the maximum likelihood estimator obtained from λ^0, is far from unique and different $\lambda^0 > 0$ will produce different λ^∞'s. In the (absurd) special case when $p(b,d) \equiv 1/D$ for all $b \in B$, $d \in D$, there are indeed many maximizers, since if λ^∞ is any maximum of Λ then any other λ with

$$\sum_{b \in B} \lambda(b)/D = \mu^\infty(D) \tag{2.1}$$

has the same value of $\mu(d) = \sum \lambda(b) p(b,d) = \mu^\infty(d)$, $d \in D$ and since Λ in (1.3) only depends on λ through μ, $\Lambda(\lambda) = \Lambda(\lambda^\infty)$. Of course $p(b,d) = 1/D$ would make a very poor tomograph indeed since there is no information in the counts except for the total number! In realistic cases we will see that even if $|B| >> |D|$, λ^∞ should be expected to be unique, or at least essentially unique, in low count cases. The essential uniqueness despite the underdeterminedness of the problem in the sense that the number of unknowns $\lambda(b)$ greatly (factor of 3) exceeds the number of equations is surprising to us but is not totally unfamiliar [Bu].

The reason for this uniqueness is the following: Suppose instead that there are many $\lambda = \lambda(b)$, $b \in B$ say $\lambda \in L$ with

(2.2) $$\Lambda(\lambda) = \Lambda(\lambda^{\infty}) \quad \text{for all } \lambda \in L.$$

Then we may take a convex combination, or weighted mean

$$\bar{\lambda}(b) = \sum_{\lambda \in L} \alpha(\lambda)\, \lambda(b)$$

where $\alpha(\lambda) > 0$ for $\lambda \in L$ and $\sum \alpha(\lambda) = 1$, of $\lambda \in L$. Since $\Lambda(\lambda)$ is logconcave in λ, [SV], $\Lambda(\bar{\lambda}) = \Lambda(\lambda^{\infty})$ as well and so $\bar{\lambda}$ is an MLE but if L is large, $\bar{\lambda}$ will be smooth for some α weighting. That is, we can find a smooth estimate $\bar{\lambda}$ with exactly the same likelihood as λ^{∞}, an MLE. Thus either λ^{∞} must be unique (essentially unique) or there are smooth MLE's, which is not the case for low count cases. We remark that $\mu(d) = \lambda^*(d)$ is unique among all maximum likelihood estimators by a simple convexity argument.

We can carry the above observation a bit further in two ways. The first way is to note that by the lemma of §1, if we can find a solution of (1.9) then it is an MLE. But (1.9) is a linear programming (LP) problem since it is linear and $\lambda(b) \geq 0$ are the constraints. Recent developments in linear programming (see e.g. [K], [V]) have yielded efficient routines for deciding feasibility and for solving even such large LP's. We have tried the routine on a typical ET case of $p(b, d)$ and $n^*(d)$ in §3. Unfortunately, as was guessed in [SV], (1.9) fails to be feasible in typical cases (at least for the case of the phantom and $p(b, d)$ given in [SV]) as we will see in §3, unless the number of counts is enormously large.

There is another, potentially more fruitful way, to use LP in ET. Namely we will show that given any λ, not necessarily an MLE, say $\lambda = \lambda^k$ for some λ^0 and some k, we can use LP to find a new λ' with the same likelihood as λ and which is as smooth as possible in a Manhattan or L_1 metric norm. So suppose we have iterated EM, $k = 5000$ times, and obtained $\lambda = \lambda^k$. Now we form

$$\mu^k(d) = \sum \lambda^k(b) p(b, d), \quad d \in D \tag{2.3}$$

and regard $\mu^k(d)$ as $n^*(d)$ in (1.9). Now we know that (1.9) has a solution since $\lambda = \lambda^k$ is such. Thus the problem with constraints (1.9) and inequalities $\lambda(b) \geq 0$, $b \in B$, and objective, say

$$\sum_{b \in B} \sum_{b' \in N(b)} |\lambda(b) - \lambda(b')| \quad \text{is a minimum} \tag{2.4}$$

where $N(b)$ is a local neighborhood of b, i.e. the set of nearest neighbors of b, is a feasible LP. Note that objective of the form (2.4) can be viewed as a linear objective despite its nonlinear appearance because one can introduce variables $\epsilon(b, b')$ for each pair $b \in B$ and $b' \in N(b)$ which satisfy the constraints

$$\epsilon(b, b') \geq \lambda(b) - \lambda(b') \quad \text{and} \quad \epsilon(b, b') \geq \lambda(b') - \lambda(b) \tag{2.5}$$

and use as objective

$$\text{minimize} \sum_{b \in B} \sum_{b' \in N(b)} \epsilon(b, b') \tag{2.6}$$

subject to ϵ and λ satisfying (2.5) and (1.9) with $\lambda(b) \geq 0$. Unfortunately even the present LP package we have is not able to easily implement this LP because the introduction of the variables ϵ in (2.5) greatly increases the number of independent variables. Instead in §3 and Figure 3 we describe the solution to the LP with the objective

$$\text{minimize the maximum of } \lambda(b), \quad b \in B. \tag{2.7}$$

This is not exactly a smoothness criterion but should have the same effect since (as we shall see in §3) $\lambda^{5000}(b)$ has enormous values for some b.

In principle, and in practice too, LP can be used in conjunction with EM to give maximally smoothed (in the sense of (2.6) or (2.7)) estimators with the same likelihood as λ^k for any (even small) k. Whether this is enough of an advantage to consider implementing LP in an EM laboratory is another question that can only be decided by experimentation.

Another way to use LP to improve ET is somewhat ad hoc as follows. Suppose one wants to distort the true value $n^*(d)$ as little as possible so that the LP problem becomes feasible. Thus suppose that $\epsilon_d \geq 0$ are variables and we wish to find the least value of

$$\epsilon = \sum_d \epsilon_d \tag{2.8}$$

for which the problem

$$n^*(d) - \epsilon_d \sqrt{n^*(d)} \leq \sum_b \lambda(b) p(b, d) \leq n^*(d) + \epsilon_d \sqrt{n^*(d)} \tag{2.9}$$

is feasible, i.e. has a solution $\lambda(b) \geq 0$. Such a λ would then be maximum likelihood for some $n^*(d)$ problem (not necessarily integer counts) which differs as little as possible from $n^*(d)$. The factors $\sqrt{n^*(d)}$ were suggested by Y. Vardi. Unfortunately we shall see in §3 that this approach is also unsuccessful.

§3. Experiments

In each of our experiments we use the same approach as in [SK]. Namely we take an anthropomorphic phantom (perhaps one more realistic for ET can be chosen if a particular class of real densities can be envisioned), i.e. a particular $\lambda(x)$. As discussed in [SV] and in [VSK] the true counts $n(b)$ and $n^*(d)$ can be simulated perfectly (neglecting scatter and randoms [SV]) as described in [SV]. Thus Figure 1 represents $n(b)$ for 128×128 pixels b where 10^6 counts were generated. It is very important to understand that in generating $n^*(d)$ no use is made of the pixel boundaries or the shape of the boxes b. Instead each of the 10^6 points is chosen from $\lambda(x)$ by choosing a point x uniformly in the unit square and accepting it with probability proportional to $\lambda(x)$. If it is accepted, a random line l through x is chosen and the tube of l is incremented. It is easy to see (in principle) how to study effects of range, angulation, scatter, and random with this model (although we have not done so). Note that there are $\binom{128}{2} = 8128$ detector units in [SV], but that half of these have zero counts because they are not sufficiently opposing. Thus there are about $64^2 = 4096$ tubes and about $128^2 \cdot \pi/4 \doteq 12868$ pixels, so that there are more than 3 times as many unknowns $\lambda(b)$ as measurements, $\mu^*(d)$. Appendix 1 gives the program we used to generate the count data.

For $p(b, d)$ we use the value of $p(d \mid x)$ where x is the exact center of b. It is easy to see that the $p(b, d)$'s used in [SV] lead to infeasibility in (1.9). This is because the $p(b, d)$'s in [SV] are specially chosen for rapid computation and satisfy for example tubes in any (say vertical) direction,

$$\sum_{\text{vertical } d} p(b, d) \equiv c = \text{constant in } b\,. \tag{3.1}$$

But then for (1.9) to have a solution we must have

$$\sum_{\text{vertical } d} n^*(d) = c \sum_b \lambda(b) \tag{3.2}$$

which is an event of probability zero since $n^*(d)$ are independent. However, for $p(b, d) = p(d \mid \text{center of } b)$ as above there is no degeneracy as in (3.1) and (1.9) has a chance of being feasible. If there is less noise in the measurements (i.e. large $n^*(d)$) then (1.9) is more likely to be feasible. We determined roughly by experiment that if there are $nb \times nb$ pixels and $\binom{nd}{2}$ tubes then the range of the number nc of counts at which (1.9) has a feasible solution with reasonably positive probability is given in Table 1.

Table 1

nb	nd	nc
5	5	$[150, 300]$
10	5	$[150, 300]$
10	10	10^5
20	10	$[2 \times 10^4,\ 2.5 \times 10^4]$
30	10	$[2 \times 10^4,\ 2.5 \times 10^4]$
45	15	$[10^5,\ 3 \times 10^5]$
128	20	$\gg 10^7$

The evidence in Table 1 makes it very likely that for $nb = nd = 128$ as in [SV] the problem (1.9) is infeasible unless $nc \gg 10^{10}$, i.e. it is not going to be feasible under typical circumstances. This was guessed (without real evidence) in [SV], but now that we have a package to do large LP's we have empirically verified that (1.9) is typically not feasible at reasonable count rates.

Figure 2a-d shows the EM reconstruction after 20, 60, 500, 5000 iterations. The log likelihood values $V_k = C - \log \Lambda(\lambda^k)$ are approximately given in Table 2 where C is an arbitrary constant. The true value of the higher likelihood reconstructions is small. Smoothing the 5000th iteration with a crude average on 9 points restores a reasonable image (Figure 2e).

How can we use LP to get better estimates of $\lambda(b)$ without loss of likelihood? One way was described in (2.3) where $n^*(d)$ is replaced by $\mu^k(d)$. Figure 3a shows the result of using (2.7) to maximize smoothness under (2.3) with $k = 5000$. Unfortunately there is no improvement over Figure 2d (the large values if $\lambda(b)$ are in the same place) and we conclude that the method fails because the maximum likelihood estimator is essentially unique. We tried to start the EM algorithm off at a nonuniform λ^0 and again we got back to Figure 3a or 2d. Figures 3b, 3c show 3a smoothed once and twice.

The result of minimizing (2.8) subject to the feasibility of (2.9) yields the reconstruction in Figure 4a, again of the type of Figures 2d, 3a, with 10^6 counts. With 10^8 counts we obtain Figure 4b which is slightly smoother. Smoothing Figure 4b gives 4c. Single and double smoothing of 4a gives 4d and 4e.

Our conclusions from these experiments are that the likelihood criterion is flawed. More likelihood cannot be achieved without giving up smoothness. The true maximum likelihood estimator is essentially unique.

Table 2

Loglikelihood as a function of iteration number.

Iteration number k	$V_k = C - \log \Lambda(\lambda^k)$
10	5760
20	273
30	−512
40	−771
50	−901
60	−981
100	−1142
200	−1277
300	−1328
400	−1357
500	−1376
1000	−1430
2000	−1482
3000	−1512
4000	−1531
5000	−1545

```
c positron emission recon with e-m algorithm
c i=(i1,i2) indexes the ith box with center at -1+(2i-1)/(2b-1) i1=i1,i2
c b(i) = lambda(i);f(i)=sum(n(.,k)c(i,k)/ph(k))
c pnp(k)=no. counts in tube k; ph(k)=sum(b(j)c(j,k))
c nc=total n0. of counts,nd=no.detectors,.5nd(nd-1)=no.tubes
c t1k,t2k= distances from 0. to edges of tube k, -rd<t2k<t1k<rd
c w=width of the intersection of the kth tube and the ith sphere
c rd=rad. of detector circle div. by rad of patient circle(=1.)
        implicit real*8 (a-h,o-z)
        dimension x(128),st(-1:256),ct(-1:256),a(4)
c define constants
        open(9,file='nbnd')
        rewind(9)
        read(9,1234)nb,nd
1234    format(2i4)
        nd2 = 2*nd
        if (nb .gt. 128 .or. nd .gt. 128) then
                print *,'ERROR: problem too big'
                stop
        endif
        rho=1./nb
        rd=sqrt(2.)
        pi=3.14159265d+0
        pi2=2*pi
        cc=1./pi
        rnb2=.5*nb
        rndpi2=nd/pi2
        pi2nd=pi2/nd
        do 10 i=1,nb
                x(i)=-1.+(i-.5)*2./nb
10      continue
        do 11 k=-1,nd2
                th=.5*k*pi2nd
                st(k)=sin(th)
                ct(k)=cos(th)
11      continue
        icnt = 1
        write(14,997) icnt
997     format(i7)

        do 301 k1=0,nd-2
            do 302 k2=k1+1,nd-1
c ck=cos(thk),sk=sin(thk),thk=(alf+bet)/2 where x*ck+y*sk-t1k,t2k define
c the edge line of tube with centers at rdcosalf,rdsinalf & at
c (rdcosbet,rdsinbet), 0<alf<bet<2 pi.
                ck=ct(k1+k2)
                sk=st(k1+k2)
                t1k=rd*ct(k2-k1-1)
                t2k=rd*ct(k2-k1+1)
                if(abs(ck).ge.abs(sk)) then
c                   now run through boxes by rows
                    do 303 i2=1,nb
                        ia=((1.+(t2k-rho-x(i2)*sk)/ck)*nb+1.)*.5
                        ib=((1.+(t1k+rho-x(i2)*sk)/ck)*nb+1.)*.5
                        i11=max0(1,1+min0(ia,ib))
                        i12=min0(nb,max0(ia,ib))
```

```
                  do 304 i1=i11,i12
c                         tip=x(i1)*ck+x(i2)*sk+rho
c                         tim=tip-2.*rho
c                         w=amin1(amax1(tip,t2k),t1k)
c    *                            -amin1(amax1(tim,t2k),t1k)
                          a(1)=atan2(rd*st(2*k1-1)-x(i2),
     *                                rd*ct(2*k1-1)-x(i1))
                          a(2)=atan2(rd*st(2*k1+1)-x(i2),
     *                                rd*ct(2*k1+1)-x(i1))
                          a(3)=atan2(rd*st(2*k2-1)-x(i2),
     *                                rd*ct(2*k2-1)-x(i1))
                          a(4)=atan2(rd*st(2*k2+1)-x(i2),
     *                                rd*ct(2*k2+1)-x(i1))
                          ichk = 0
                          do 350 i=2,4
                              if ( a(i) .lt. a(i-1) ) then
                                  ichk = ichk+1
                                  do 351 j=i,4
                                      a(j) = a(j) + pi2
351                               continue
                              endif
350                       continue
                          if ( ichk .gt. 1 ) then
                                  print *,'ERROR: ichk .gt. 1',
     *  ' x k1',k1,'k2',k2,'i1',i1,'i2',i2,'a',(a(i),i=1,4)
                                  stop
                          endif
                          a(3) = a(3) - pi
                          a(4) = a(4) - pi
                          w = max(min(a(4),a(2))
     *                              -max(a(3),a(1)),0.0d0)
                          if ( w .ne. 0.0 ) then
                                  write(15,998) w*cc
                                  write(16,999) (i1-1)*nb+i2
                                  icnt = icnt + 1
998                               format(f6.4)
999                               format(i7)
                          endif
304               continue
303           continue
          else
c             now run through boxes by cols
              dk=sk*2./nb
              do 306 i1=1,nb
                  ia=((1.+(t2k-rho-x(i1)*ck)/sk)*nb+1.)*.5
                  ib=((1.+(t1k+rho-x(i1)*ck)/sk)*nb+1.)*.5
                  i21=max0(1,1+min0(ia,ib))
                  i22=min0(nb,max0(ia,ib))
                  do 307 i2=i21,i22
c                         tip=x(i1)*ck+x(i2)*sk+rho
c                         tim=tip-2.*rho
c                         w=amin1(amax1(tip,t2k),t1k)
c    *                            -amin1(amax1(tim,t2k),t1k)
                          a(1)=atan2(rd*st(2*k1-1)-x(i2),
     *                                rd*ct(2*k1-1)-x(i1))
                          a(2)=atan2(rd*st(2*k1+1)-x(i2),
```

```
     *                                  rd*ct(2*k1+1)-x(i1))
                              a(3)=atan2(rd*st(2*k2-1)-x(i2),
     *                                  rd*ct(2*k2-1)-x(i1))
                              a(4)=atan2(rd*st(2*k2+1)-x(i2),
     *                                  rd*ct(2*k2+1)-x(i1))
                              ichk = 0
                              do 352 i=2,4
                                  if ( a(i) .lt. a(i-1) ) then
                                      ichk = ichk+1
                                      do 353 j=i,4
                                          a(j) = a(j) + pi2
 353                                  continue
                                  endif
 352                          continue
                              if ( ichk .gt. 1 ) then
                                      print *,'ERROR: ichk .gt. 1',
     *   'k1',k1,'k2',k2,'i1',i1,'i2',i2,'a',(a(i),i=1,4)
                                      stop
                              endif
                              a(3) = a(3) - pi
                              a(4) = a(4) - pi
                              w = max(min(a(4),a(2))
     *                                 -max(a(3),a(1)),0.0d0)
                              if ( w .ne. 0.0 ) then
                                      write(15,998) w*cc
                                      write(16,999) (i1-1)*nb+i2
                                      icnt = icnt + 1
                              endif
 307                  continue
 306              continue
              endif
              write (14,997) icnt
 302      continue
 301  continue
      write(13,997) nb*nb
      write(13,997) (nd*(nd-1))/2
      write(13,997) icnt-1
      stop
      end

      double precision function artan2(x,y)
      double precision x,y
      double precision pi,pi2

      pi=3.14159265d+0
      pi2=2*pi

      if ( y .ge. 0.0 ) then
              if ( x .ge. 0.0 ) then
                      artan2 = atan(x/y)
              else
                      artan2 = pi2 + atan(x/y)
              endif
      else
              artan2 = pi + atan(x/y)
      endif

      return
      end
```

REFERENCES

[B]. Baum, L. E. An inequality and associated maximization technique in statistical estimation for probabilistic functions of Markov processes. In Inequalities, III: Proceedings of a Symposium, Oved Shisha ed., Academic Press, New York.

[Bu] Buchta, C. On nonnegative solutions of random systems of linear inequalities, Discrete Comput. Geom. 2, 1987, pp. 85-95.

[CT]. Csiszar, I. and Tusnàdy, G. Information geometry and alternating minimization procedures, Math. Inst. of Hungarian Acad. Sci.

[DLR]. Dempster, A. P., Laird, N. M. and Rubin, D. B. Maximum likelihood from incomplete data via the EM algorithm, J. Roy. Stat. Soc. B, 39, 1977, pp. 1-38.

[GM]. Geman, S. and McClure, D. E. Statistical methods for tomographic image reconstruction, Proc. of 46th Session of ISI, Bull. ISI, vol. 52, 1987.

[JS] Johnstone, I. and Silverman, B. W. Speed of estimation in PET, Tech. report 290, March '88, Statistics Dept., Stanford University.

[K]. Karmarkar, N. K. A new polynomial-time algorithm for linear programming, Combinatorica, 4,1981, 373-395

[LeH]. Levitan, E. and Herman, G. T. A maximum a posteriori probability expectation maximization algorithm for image reconstruction in emission tomography, IEEE Trans. Med. Imag., 1987.

[LiH]. Liang, Z. and Hart, H. Bayesian image processing of data from constrained source distributions I, II, Bull. Math. Biol. 49, 1987, 51-91.

[LV]. Llacer, J. and Veklerov, E. The maximum likelihood estimation method of image reconstruction: its fundamental characteristics and their origin, ms., May 1987.

[MBM]. McCarthy, A., Barrett, R. and Miller, M. I. Systolic implementation of the EM algorithm for emission tomograpy on mesh-connected processors, Proc. 22nd Conf. on Inf. Sci. and Systems, Princeton, University, 1988.

[P]. Petersen, S. E., Fox, P. T., Posner, M. I., Mintum, M. and Raichle, M. E. Position emission tomographic studies of the cortical anatomy of single-word processing. Nature 331, pp. 585-589, 1988, Feb.

[PS]. Politte D. G. and Synder, D. L. The use of constraints to eliminate artifacts in maximum-likelihood image estimation for emission tomography, monograph 489, October 1987, Washington University School of Medicine, St. Louis, MO.

[S]. Synder, D. L. Random Point Processes, Wiley, New York, 1975.

[SK]. Shepp, L. A. and Kruskal, J. B. Computerized tomography: The new medical X-ray technology, Amer. Math. Monthly, 85, 420-439.

[SM]. Synder, D. L. and Miller, M. I. The use of sieves to stabilize images produced with the EM algorithm for emission tomography, IEEE Trans. Nucl. Sci., NS-32, 1985, pp. 3864-3872.

[STT]. Synder, D. L., Thomas, L. J., Ter-Pogossian, M. M. A mathematical model for positron-emission tomography systems having time-of-flight measurements, IEEE Trans. Nucl. Sci., NS-28, 1981, pp. 3575-3583.

[SV]. Shepp, L. A. and Vardi, Y. Maximum-likelihood reconstruction for emission tomography, IEEE Trans. Med. Imag., vol. MI-1, pp. 113-122, 1982.

[V]. Vanderbei, R. J., Meketon, M., Freedman, B. A modification of Karmarkar's linear programming algorithm, Algorithmica,1,1986,pp.395-407.

[VL]. Veklerov, E. and Llacer, J. Stopping rule for the MLE algorithm based on statistical hypothesis testing, ms. February 1987.

[VSK]. Vardi, Y., Shepp, L. A., and Kaufman, L. A statistical model for positron emission tomography, J. Am. Stat. Soc. 80, 1985, pp. 8-37.

MATHEMATICAL MORPHOLOGY AND MEDICAL IMAGING

F. Prêteux

Ecole Nationale Supérieure des Télécommunications
Paris, France

ABSTRACT

This course deals with the foundations and principles of mathematical morphology. We present the consistent system of morphological operators in the framework of grey-tone images and detail their properties, characteristics and effects. All these theoretical and methodological aspects will be illustrated with various problems in medical imaging. Especially, from medical images obtained by different modalities, we will carry out organ segmentation and 2, 3 or 4-dimensional parameter quantification.

INTRODUCTION

To see better in order to better know, image analysis allows scene description at different abstraction levels :
For instance, in attributes extraction (from images), it consists in identifying regions verifying a given criterion (isodensity ...) and thus, homogeneous with repect to it.
In structural analysis (of images), the aim is to find relations between regions in order to constitute, if not entities of the image, at least parts of them.
In symbolic analysis (of images), the ultimate step, it tends to interprete objects as image parts verifying hypotheses put forward with reference to an a priori knowledge or to a model. In particular, these hypotheses have to take into account the relations between an object and its environment.
Image analysis, at a given step, is based on transformation, structuration, simplification and information reduction principles.

The more one ascends in the analysis hierarchy, the stronger and more restricted meaning this term gets : isolating significant objects. This leads to pattern comprehension systems where the main point is how to introduce an a priori knowledge. To some extent, it would be "necessary to take into account in an intelligent way all the knowledge related to the

considered field" (Haton, 1987). Performing image transformations becomes thus complex because of the combination of notions of very different nature. In this context, how to place mathematical morphology ?
Leaving aside the acquisition/pre-processing step for which the techniques are well-tried, we will consider the steps of attributes extraction and symbolic description for answering these questions.
First, we will remind the principles and foundations of mathematical morphology. Then, we will present, for grey-tone images, the consistent system of its primitives. Finally, all these theoretical and methodological aspects will be illustrated with several problems to be solved in medical imaging. From medical images obtained by different modalities (C.T., P.E.T.) and according to our mathematical morphological approach, we will carry out organ segmentation and 2, 3, or 4-dimensional parameter quantification.

MATHEMATICAL MORPHOLOGY

Mathematical morphology (Matheron, 1975), (Serra, 1982), (Coster and Chermant, 1985) is born more than twenty years ago, through the impetus given by G. MATHERON, at the Centre de Géostatistique et de Morphologie Mathématique (Ecole des Mines de Paris).
What is it about ? Strictly speaking, it means shapes and structures study, by means of mathematics. Beyond words, it is actually much more than a mere pattern recognition, it is a way of thinking image analysis. A school was formed, which has been enriched and diversified ever since, imposing, step by step, not only its terminology and notations, which are now generally acknowledged, but also its original approach, where "the concept precedes the measure".
What are the foundations, principles and concepts of mathematical morphology (M. M.) ?

1. THE METHOD : FUNDAMENTAL CONCEPTS

The concept of shape, which may be considered as being at the origin of M. M., assimilated to the notion of structure, was defined by the knowledge of a given family $\mathcal{R}$ of relations considered as possible and a given family $\mathcal{B}$ of geometrical sets, arbitrary but explicit, called structuring elements.
From a deterministic point of view, characterizing the shape of an object is to know whether or not, for each structuring element, the possible

relation is satisfied. Now, if we stick to this general principle we face infinite possibilities.
By emphasizing the relations of boolean logic and the context of lattices, M. M. is based on set theory, which makes it different from other methods of image processing, mostly based on relations of linear algebra. In fact, M. M. inherits all the properties resulting from this choice : increasing and anti-extensivity with respect to order relationship, convergence via idempotence, algebraic opening and closing...

The second concept which had to be defined concerned the topological nature of the objects to analyse. Indeed, let us consider the example of a porous medium with two components : the set A of the grains and the complementary set , the set of pores. Does a point situated on the boundary of A belong to the grains or to the pores ? If the structuring element B is a compact set or a relatively compact and open set, we cannot distinguish between two porous media if they have the same topological closure and interior.
This remark, suggested by experimental constraints, led G. MATHERON to define a topology, the "Hit or Miss Topology", on the subspace of all closed subsets (instead of the space of all subsets) of a given topological set, and to restrict the class $\mathcal{B}$ of stucturing elements to compact sets (Matheron, 1975). There again, the consequences of this choice are essential, since they induce semi-continuity or continuity properties of mappings in image analysis.

The next conceptual step consists to use the probability theory. MATHERON's Random Set Theory established that logical propositions such as "B is included in A" or "B is disjoined from A", were events of a certain σ-algebra and that we could assign them probabilities (Matheron, 1975). In other words, studied objects can be assimilated to random set realizations. The probabilities of the previous events are in a way the moments of a spatial distribution, with the fundamental remark, that they can in particular involve a neighbourhood, in the topological sense, and thus involve an uncountable infinity of points.

Thanks to these general mathematical foundations, the method, initially developed for binary images, could evolve when they gave place to grey-tone images, assimilated then to functions. The theory was generalized and adapted to this new functional context, richer and more complex.

2. THE TOOLS : MATHEMATICAL MORPHOLOGICAL PRIMITIVES

One of the original features of M. M. is to provide not only a collection of primitives, but a true construction, ordered according to the actions and effects of the transformations (Prêteux, 1987). So it answers the actual needs of practical image analysis.
To describe the original generation of the mathematical morphological primitives, let us imagine a pyramid of variable geometry built on its summit. This mathematical design is made of higher order primitives that become more and more complex, specific and efficient as we ascend in the pyramid.
Nevertheless, all of them can be written from a unique basic morphological primitive, the erosion.
In this course, we have chosen to present the morphological primitives in the framework of grey-tone images (more original and less known than the one of binary images). Because it is impossible to be exhaustive, only some primitives of higher order, chosen for their interest in practice, will be detailed here.

2.1 Erosion

2.1.1 Functional definition and geometrical interpretation

The choice of erosion as basic primitive is the result of successive orientations adopted in the principles and concepts of the M. M.. Indeed, in the framework of the lattice theory, we can consider the relation "$B \subset A$" and then define a new set C by :

$$C = \{x \,/\, B_x \subset A\}$$

where $B_x = \{ b + x \,/\, b \in B\}$ (B_x is the translate of B by the translation x).
C is called the eroded set of A with respect (w. r.) to the structuring element B. The previous interpretation of the erosion in terms of set theory being well-known, in this course, we will present it in a wider context : the general framework of semi-continuous functions.

2.1.1.1 Notations

We denote by :

- Φ the space of the semi-continuous functions from $\mathbb{R}^n$ into $\mathbb{R}$,
- Φ_s the space of the upper semi-continuous (u.s.c.) functions from $\mathbb{R}^n$ into $\mathbb{R}$,

- Φ_i the space of the lower semi-continuous (l.s.c.) functions from $\mathbb{R}^n$ into $\mathbb{R}$,
- Φ'_s, subspace of Φ_s, the set of functions of Φ_s such that :
  for any $t \in \mathbb{R}$, the section $\{x \in \mathbb{R}^n \,/\, g(x) = t\}$ is a compact set.

Let us remind that Φ_s and Φ_i are lattices, that is :
- Φ_s, Φ_i are ordered by the relation $\leq$ defined by :

$$f \leq f' \quad \Leftrightarrow \quad \forall\, x \in \mathbb{R}^n \quad f(x) \leq f'(x) \ ,$$

- Φ_s, Φ_i are closed under the "inf" and the "sup" for any finite family I,

that is :

$$\forall\, (f_i)_{i\in I} \subset \Phi_s \qquad \bigwedge_{i\in I} f_i \in \Phi_s \quad \text{and} \quad \bigvee_{i\in I} f_i \in \Phi_s \ ,$$

$$\forall\, (f_i)_{i\in I} \subset \Phi_i \qquad \bigwedge_{i\in I} f_i \in \Phi_i \quad \text{and} \quad \bigvee_{i\in I} f_i \in \Phi_i \ .$$

The necessity to consider the two subspaces Φ_s et Φ_i will be justified a posteriori, when we will define the duality principle.

2.1.1.2 Functional definition

Definition 1 :

The MINKOWSKI subtraction, denoted $\ominus$, is defined by the mapping from $\Phi \times \Phi$ into Φ, as follows :

$\forall\, (f,g) \in \Phi^2$, $\forall\, x \in \mathbb{R}^n$

$$(f\ominus g)(x) = \inf \{f(y) - g(x-y) \ / \ y \in \mathbb{R}^n\}$$

(with the convention $-\infty + \infty = -\infty$).

Proposition 1 :

For $g \in \Phi'_s$, Φ_s and Φ_i are closed under the MINKOWSKI subtraction :

$$\forall\, f \in \Phi_s \qquad f\ominus g \in \Phi_s \quad \text{and} \quad \forall\, f \in \Phi_i \qquad f\ominus g \in \Phi_i \ .$$

Definition 2 :

For any $g \in \Phi'_s$, the morphological erosion w. r. to the structuring element g is defined by the MINKOWSKI subtraction w. r. to $\check{g}$, where $\check{g}$ is given by :

$$\forall\, x \in \mathbb{R}^n \qquad \check{g}(x) = g(-x) \ .$$

The erosion is compatible under translation.

Corollary 1 :

The algebraic expression of the eroded function of a function f w. r. to $g \in \Phi'_s$ is given by :

$$(f\ominus\check{g})(x) = \inf \{f(y) - g(y-x) \ / \ y \in \mathbb{R}^n\} \ .$$

2.1.1.3 Geometrical interpretation

Let f be an arbitrary function from $\mathbb{R}^n$ into $\overline{\mathbb{R}}$. It is possible to associate it a subset of $\mathbb{R}^n \times \overline{\mathbb{R}}$ via the well-known concept of subgraph.

Definition 3 :

The subgraph of $f \in \Phi_s$, denoted SG(f), is defined by :

$SG(f) = \{(x, t) \in \mathbb{R}^n \times \overline{\mathbb{R}} \ / \ t \le f(x)\}$.

The subgraph of $f' \in \Phi_i$, denoted SG(f'), is defined by :

$SG(f') = \{(x, t) \in \mathbb{R}^n \times \overline{\mathbb{R}} \ / \ t < f'(x)\}$.

Proposition 2 :

If f is a u.s.c. (resp. l.s.c.) function, then SG(f) is a closed (resp. open) set.

In terms of subgraph, the eroded function of $f \in \Phi_s$ w. r. to $g \in \Phi'_s$ can be interpreted as the set of all positions of g's subgraph for which it is included in f's subgraph as shown in Fig. 1.

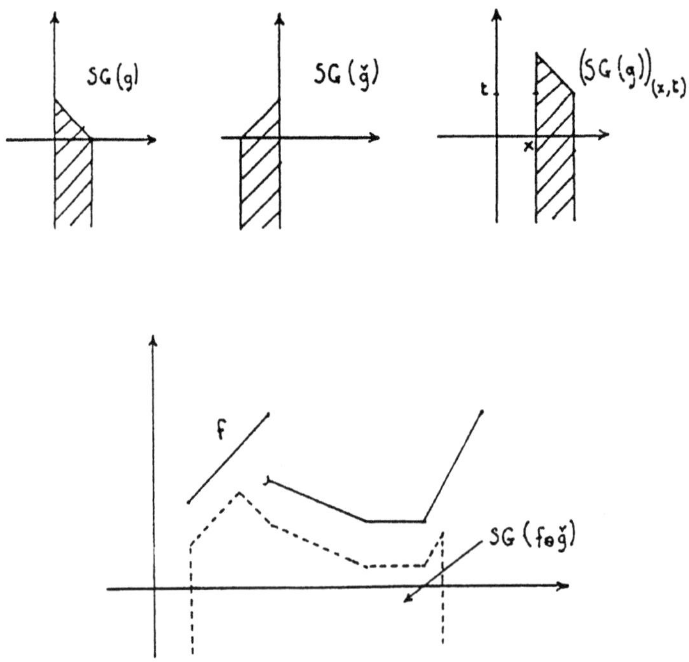

Fig. 1. Geometrical interpretation of the erosion of f w. r. to g.

This can be mathematically expressed by :

Proposition 3 :

$\forall f \in \Phi_s, \ \forall g \in \Phi'_s,$

$SG(f\ominus\check{g}) = \{(x,t) \in \mathbb{R}^n \times \bar{\mathbb{R}} \ / \ (SG(g))_{(x,t)} \subset SG(f)\}$.

Proof :

$(x,t) \in SG(f\ominus\check{g}) \quad \Leftrightarrow \quad t \le \inf\{f(y) - g(y-x) \ / \ y \in \mathbb{R}^n\}$

$\Leftrightarrow \quad \forall\, y \in \mathbb{R}^n \quad t + g(y-x) \le f(y)$

Moreover :

$(SG(g))_{(x,t)} = \{(y,u) \in \mathbb{R}^n \times \bar{\mathbb{R}} \ / \ (y-x,u-t) \in SG(g)\}$

$= \{(y,u) \in \mathbb{R}^n \times \bar{\mathbb{R}} \ / \ u-t \le g(y-x)\}$

$= \{(y,u) \in \mathbb{R}^n \times \bar{\mathbb{R}} \ / \ u \le g(y-x) + t\}$

Hence :

$(x,t) \in SG(f\ominus\check{g}) \quad \Rightarrow \quad (SG(g))_{(x,t)} \subset SG(f).$

That means that the translate of SG(g) by $(x,t) \in \mathbb{R}^n \times \bar{\mathbb{R}}$ is included in SG(f).

Conversely :

Let us consider $(SG(g))_{(x,t)} \subset SG(f)$.

For any $(y,u) \in (SG(g))_{(x,t)}$, we have :

$\forall\, y \in \mathbb{R}^n \quad u \le f(y) \quad$ and $\quad \sup\{u \ / \ (y,u) \in SG(g)_{(x,t)}\} \le f(y)$

thus : $\quad \forall\, y \in \mathbb{R}^n \quad t + g(y-x) \le f(y)$

or equivalently : $\quad (x,t) \in SG(f\ominus\check{g})$

The proven interpretation performed on functions is similar to the one for sets where the eroded set of a set A w. r. to a compact set K of $\mathbb{R}^n$ is the set of points x for which the structuring element K centered at point x is included in A :

$A\ominus\check{K} = \{x \in \mathbb{R}^n \ / \ K_x \subset A\}$

Let us now consider as a structuring element no longer an arbitrary function g but a function g_K associated to a compact set K of $\mathbb{R}^n$ and defined by :

$g_K(x) = 0 \qquad$ if $x \in K$

$= -\infty \qquad$ if not .

In functional terms, the eroded function of f w. r. to g_K is expressed by :

$\forall\, x \in \mathbb{R}^n \qquad (f\ominus\check{g})(x) = \inf\{f(y) - g(y-x) \ / \ y \in \mathbb{R}^n\}$

$= \inf\{f(y) \ / \ y \in K_x\}$.

This corresponds to the well-known notion of the erosion of a function w.

w. r. to a planar compact structuring element K, which consists in taking the "sup" in the neighbourhood defined by K as shown in Fig. 2.

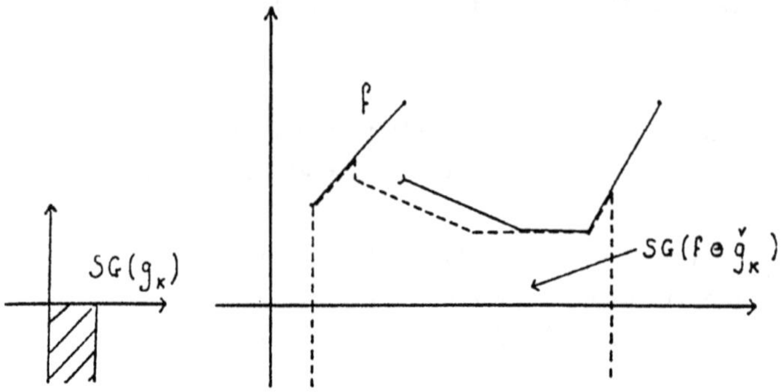

Fig. 2a

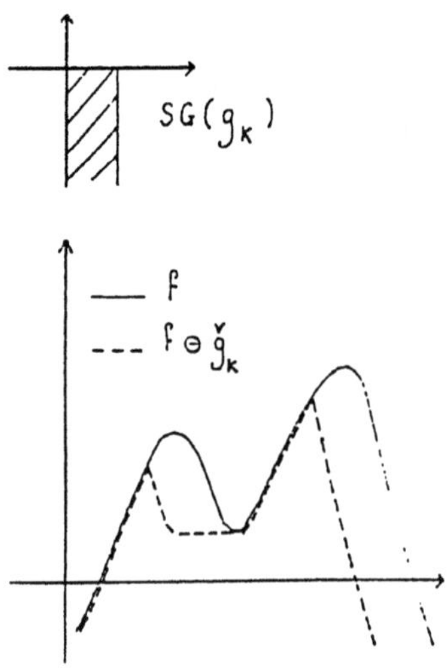

Fig. 2b

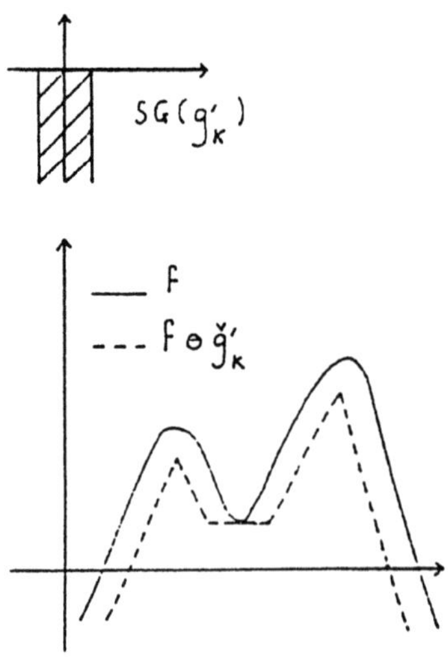

Fig. 2c

Fig. 2. Erosions w. r. to planar structuring elements.
2a. f is u.s.c. and g_K is not symmetric / $0 \in \mathbb{R}^n \times \overline{\mathbb{R}}$.
2b. f is continuous.
2c. f is continuous and g_K is symmetric / $0 \in \mathbb{R}^n \times \overline{\mathbb{R}}$.

2.1.2 Topological and algebraic properties

Proposition 1 :

The erosion w. r. to a function $g \in \Phi'_s$ is a u.s.c. mapping from Φ_s into Φ_s and a continuous mapping from Φ_1 into Φ_1.

As a general rule, the MINKOWSKI subtraction fulfils the following relationships :

Proposition 2 :

Let us take f, f', g, g' in Φ.

1 - If the origin O of $\mathbb{R}^n$ belongs to the support of g, $\ominus$ is anti-extensive, i.e. :

$f \ominus g \leq f$

2 - $\ominus$ is increasing w. r. to f and decreasing w. r. to g, i.e. :

$f \leq f' \quad \Rightarrow \quad f \ominus g \leq f' \ominus g$

$g \leq g' \quad \Rightarrow \quad f \ominus g' \leq f \ominus g$

3 - Moreover, $\ominus$ fulfils the relationships :

$f \ominus (g \vee g') = (f \ominus g) \wedge (f \ominus g')$

$f \ominus (g \wedge g') \geq (f \ominus g) \vee (f \ominus g')$

The morphological erosion fulfils these same properties for any g and g' belonging to Φ'_s.

Proof :

The first relation is an immediate consequence of the geometrical interpretation of erosion. However, the hypothesis : "the origin of $\mathbb{R}^n \times \overline{\mathbb{R}}$ belongs to the support of the structuring element g" is necessary as shown in the counter-example of Fig. 3.

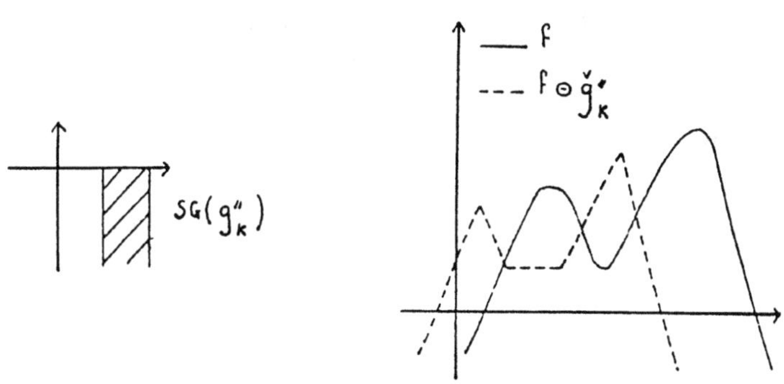

Fig. 3. Eroded function w. r. to a structuring element not including the origin.

The second and third properties are obvious. Fig. 4 shows that the inequality of the third property can be strict.

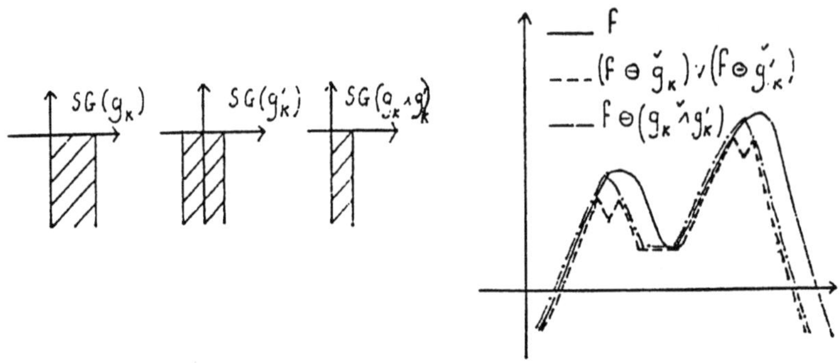

Fig. 4. Example where $f \ominus (g \wedge g') > (f \ominus \check{g}) \vee (f \ominus \check{g}')$.

2.1.3 Erosion and distance function

In this paragraph, we shall interprete the distance function in terms of erosion.

Let X be a closed set of $\mathbb{R}^n$. The distance function on X is defined by :

$$d_X(x) = 0 \qquad \text{if } x \notin X ,$$
$$= \inf\{d(x,y) \;/\; y \in X^c\} \qquad \text{if not} ,$$

where d is the euclidean distance in $\mathbb{R}^n$.

It is easy to prove that, for any $x \in X$, $d_X(x)$ is equal to the radius of the maximum hyper-sphere of $\mathbb{R}^n$ centered at x and included in X.

What is the interest of this function in image analysis ? Let us consider a binary image i.e. a closed set X of $\mathbb{R}^2$. We have :

$$X \ominus B_\rho = \{x \in \mathbb{R}^2 \;/\; d_X(x) \geq \rho\} ,$$

where B_ρ is a disc of radius ρ.

As the function d_X is obtained in two image scannings, the eroded function of any size ρ is computed in only three scannings (and not in ρ).

Let us put :

$$f_x(x) = 0 \qquad \text{if } x \notin X ,$$
$$= +\infty \qquad \text{if not} ,$$

Let us consider the structuring element g defined by :

$$\forall\, x \in \mathbb{R}^n \qquad g(x) = -\|x\| ,$$

where $\|\;\|$ is the euclidean norm in $\mathbb{R}^n$.

The distance function d_X can be interpreted as the eroded function of f_X w. r. to g, that is :

$$d_X = f_X \ominus g$$

Indeed, we have :

$$\begin{aligned} d_X(x) &= \inf\{d(x,y) \ / \ y \in X^c\} \\ &= \inf\{ f_X(y) + \|x-y\| \ / \ y \in \mathbb{R}^n\} \\ &= \inf\{ f_X(y) + g(x-y) \ / \ y \in \mathbb{R}^n\} \\ &= (f_X \ominus g)(x) \end{aligned}$$

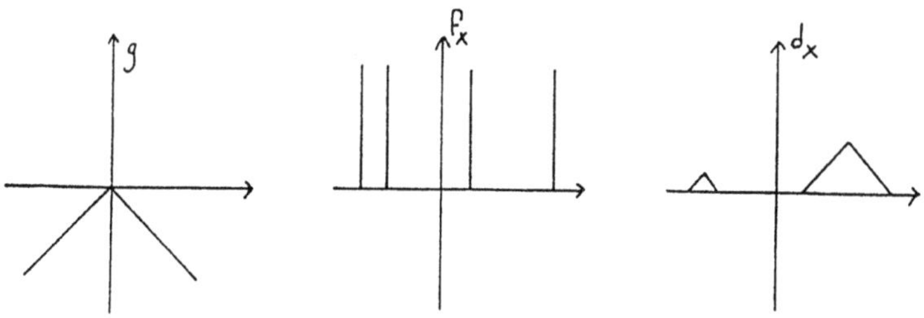

Fig. 5. Distance function interpreted from the erosion w. r. to the isotropic cone of summit O.

After this detailed presentation of the basic primitive of M. M., let us define a new morphological primitive, the dilation, which derives from the erosion via the duality principle.

2.2 Dilation

2.2.1 Duality principle

In the set framework, the duality w. r. to the complementation on the space of all subsets of a given set is expressed by :

ψ^* is the mapping dual to ψ iff ψ^* acts (in the same sense of ψ) on the complementary sets of the sets for which ψ is defined.

Mathematically, we have :

$$\psi^* = C o \psi o C,$$

where C denotes the complementation operator.

How to extend this concept to functions ?

First, the complementation notion is irrelevant on the space of the functions from $\mathbb{R}^n$ into $\overline{\mathbb{R}}$. That is why, we usually consider the following symmetry : $f \longrightarrow -f$ instead of the complementation C.

The transformation ψ^*, dual to the transformation ψ from Φ into itself, is thus given by :

$$\forall f \in \Phi \qquad \psi^*(f) = -\psi(-f)$$

Let us remark that if f is u.s.c. (resp. l.s.c.), -f is l.s.c. (resp. u.s.c.), which gives a sense to $\psi(-f)$.

Examples :

1 - Let us define the mapping dual to "$\wedge$", here denoted by $\wedge^*$.

$$\forall x \in \mathbb{R}^n \qquad (f\wedge^* g)(x-) = -((-f) \wedge (-g))(x)$$
$$= -\min\{-f(x), -g(x)\}$$
$$= \max\{f(x), (g(x)\}$$
$$= (f_\vee g)(x)$$

thus : $\qquad \wedge^* = \vee$

2- The mapping $f \longrightarrow -f$ is autodual, that is dual to itself.

Proposition 1 :

Let ψ be a mapping from Φ into Φ such that :

$\psi(\Phi_s) \subset \Phi_s$ and $\psi(\Phi_i) \subset \Phi_i$

If ψ is continuous then ψ^* is continuous.

If ψ is u.s.c. then ψ^* is l.s.c..

If ψ is l.s.c. then ψ^* is u.s.c..

2.2.2 Functional definition and geometrical interpretation

Definition-proposition 1 :

The mapping dual to $\ominus$ is the mapping denoted $\oplus$ from $\Phi \times \Phi$ into Φ, and defined by :

$\forall f \in \Phi$, $\forall g \in \Phi$, $\forall x \in \mathbb{R}^n$

$$(f\oplus g)(x) = \sup \{f(y) + g(x-y) \;/\; y \in \mathbb{R}^n\} .$$

$\oplus$ is called the MINKOWSKI addition.

Definition-proposition 2 :

The mapping dual to the morphological erosion w. r. to the structuring element $g \in \Phi'_s$, is called the morphological dilation w. r. to g, and is equal to the MINKOWSKI addition w. r. to $\check{g}$. It is expresed by :

$\forall f \in \Phi$, $\forall g \in \Phi$, $\forall x \in \mathbb{R}^n$

$$(f\oplus\check{g})(x) = \sup \{f(y) + g(y-x) \;/\; y \in \mathbb{R}^n\} .$$

Remark : Dilation is compatible with translation.

Let us consider the subgraph of the dilated function of $f \in \Phi$ w. r. to $g \in \Phi'_s$. We have :

$$(x,t) \in SG(f \oplus \check{g}) \quad \Leftrightarrow \quad t \leq \sup \{f(y) + g(y-x) \ / \ y \in \mathbb{R}^n\}$$

$$\Leftrightarrow \quad \exists\, y_0 \in \mathbb{R}^n \quad t - g(y_0 - x) \leq f(y_0) \ .$$

That means that the subgraph of f intersects the symmetric (w. r. to $0 \in \overline{\mathbb{R}}$) of the translated by $(x,t) \in \mathbb{R}^n \times \overline{\mathbb{R}}$ subgraph of f since this set is defined by :

$$[s_{0 \in \overline{\mathbb{R}}}(SG(g))]_{(x,t)} = \{(y,u) \in \mathbb{R}^n \times \overline{\mathbb{R}} \ / \ (y-x, -u+t) \in SG(g)\}$$

$$= \{(y,u) \in \mathbb{R}^n \times \overline{\mathbb{R}} \ / \ -u+t) \leq g(y-x)\}$$

$$= \{(y,u) \in \mathbb{R}^n \times \overline{\mathbb{R}} \ / \ t-g(y-x) \leq u\}$$

Moreover :

$$(x,t) \in SG(f \oplus \check{g}) \quad \Leftrightarrow \quad \exists\, y_0 \in \mathbb{R}^n \quad t-g(y_0-x) \leq f(y_0)$$

Hence :

$$SG(f \oplus \check{g}) = \{(x,t) \in \mathbb{R}^n \times \overline{\mathbb{R}} \ / \ [s_{0 \in \overline{\mathbb{R}}}(SG(g))]_{(x,t)} \cap SG(f) \neq \emptyset\}$$

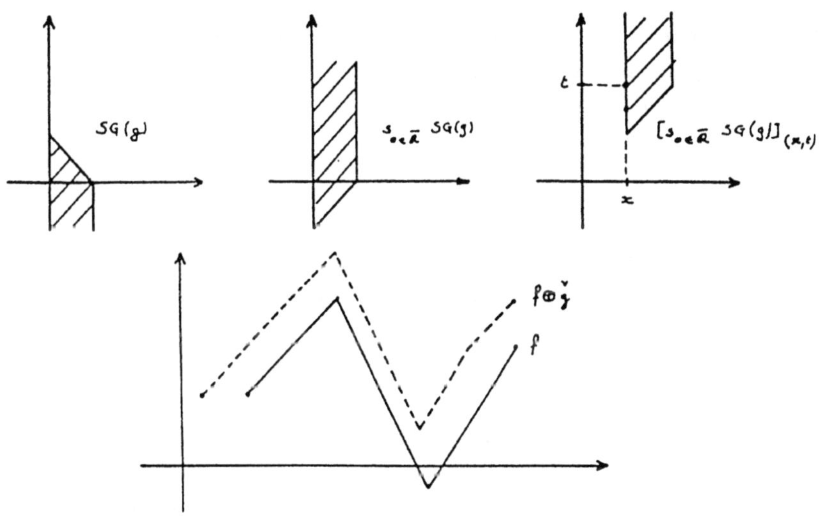

Fig. 6. Geometrical interpretation of the dilation.

Let us now consider the planar structuring element g_K, associated to the compact set K of $\mathbb{R}^n$.

In functional terms, the dilated function of f w. r. to g_K is expressed by :

$$\forall\, x \in \mathbb{R}^n \qquad (f \oplus \check{g}_K)(x) = \sup \{f(y) + g_K(y-x) \ / \ y \in \mathbb{R}^n\}$$

$$= \sup \{f(y) \ / \ y \in K_x\} \ .$$

Fig. 7 illustrates the dilation in this framework.

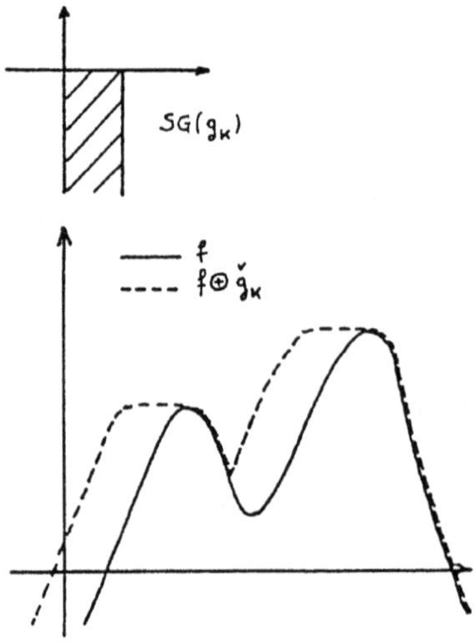

Fig. 7a

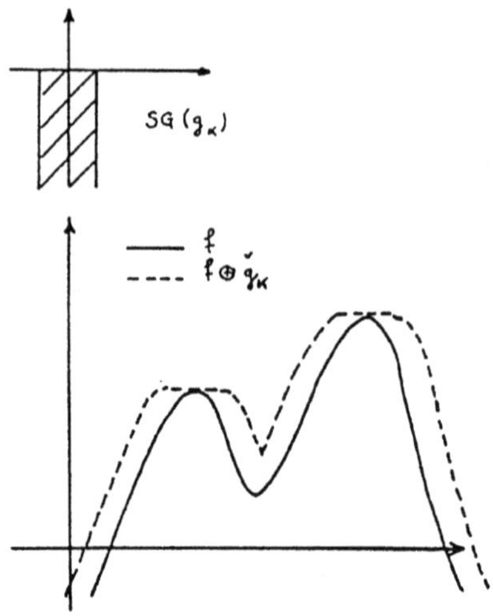

Fig. 7b

Fig. 7. Examples of dilation w. r. to planar structuring elements.

7a. $f \oplus \check{g}_K$ where K is not symmetric.

7b. $f \oplus \check{g}_K$ where K is symmetric.

2.2.3 Topological and algebraic properties

Thanks to the duality principle, all topological and algebraic properties of the MINKOWSKI subtraction and of the erosion can be thus transposed to the MINKOWSKI addition and to the dilation.

Corollary 1 :

1 - Φ_s et Φ_i are closed under dilation i. e. :

$\forall g \in \Phi'_s, \forall f \in \Phi_s, \forall f' \in \Phi_i \quad f \oplus \check{g} \in \Phi_s \quad \text{et} \quad f' \oplus \check{g} \in \Phi_i$

2 - The dilation is a continuous mapping from Φ_s into Φ_s and a l.s.c. mapping from Φ_i into Φ_i.

Corollary 2 :

Let us take $f, f', g, g' \in \Phi$. We have :

1 - If the origin of $\mathbb{R}^n$ belongs to the support of g, then $\oplus$ is extensive, i.e. :

$f \leq f \oplus g$

2 - $\oplus$ is increasing w. r. to f and to g i.e. :

$f \leq f' \quad \Rightarrow \quad f \oplus g \leq f' \oplus g$

$g \leq g' \quad \Rightarrow \quad f \oplus g \leq f \oplus g'$

3 - $\oplus$ is distributive w. r. to the "sup" i.e. :

$$f\oplus(g\vee g') = (f\oplus g) \vee (f\oplus g')$$

4 - $\oplus$ is associative i.e. :

$$(f\oplus g)\oplus g' = f\oplus(g\oplus g')$$

5 - Moreover, $\oplus$ fulfils the relation :

$$f\oplus(g\wedge g') \le (f\oplus g) \wedge (f\oplus g')$$

Let us interprete the property 3. The relation :

$$f\oplus(g\vee g') = (f\oplus g) \vee (f\oplus g')$$

means that it suffices to perform successively the dilations w. r. to the structuring elements from which the initial structuring element is the "sup". Thus, under certain conditions, the dilation can be factorized. Let us take an example. Let z be an element of $\mathbb{R}^n$ and g be the structuring element defined by :

$$g(x) = \begin{cases} 0 & \text{if } x = 0 \text{ or } z \\ -\infty & \text{if not} \end{cases}$$

g can be written as follows :

$$g = g_0 \vee g_1 \quad \text{with} \quad g_0(x) = \begin{cases} 0 & \text{if } x = 0 \\ -\infty & \text{if not} \end{cases}$$

$$\text{and} \quad g_1(x) = \begin{cases} 0 & \text{if } x = z \\ -\infty & \text{if not} \end{cases}$$

Thus :

$f\oplus\check{g}_0 = f$ and $f\oplus\check{g}_1 = f(z+.)$

(i.e. $SG(f\oplus\check{g}_1)$ is the translate of $SG(f)$ by z)

Hence :

$$f\oplus\check{g} = f \vee f(z+.)$$

which is the operating formula used to dilate a function w. r. to a bipoint (0,z).

The property 4 gives another factorization of the dilation which is very useful in practice. It means that the dilation of f w. r. to $g\oplus g'$ is performed in two steps :

1- dilation w. r. to g,

2- dilation of the result w. r. to g'.

Thus, to dilate a function f w. r. to the planar disc of radius $p \in \mathbb{N}^*$ is equivalent to perform p dilations w. r. to the disc of radius 1.

This property is very powerful, especially on the class of STEINER functions (written as finite MINKOWSKI sum of segments) since it allows to reduce the calculation complexity. To dilate w. r. to a STEINER function is equivalent to dilate w. r. to each segment which composes the STEINER function.

A contrario, the property 5 shows that it is impossible to factorize a structuring element w. r. to the "inf" as shown in Fig. 8.

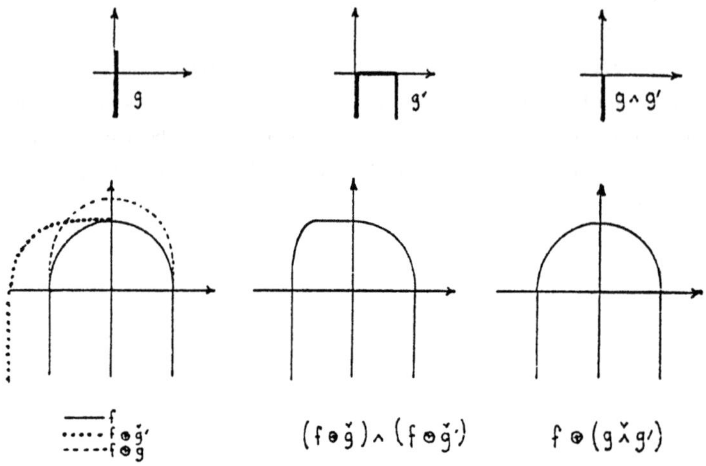

Fig. 8. Example where $f\oplus(g\check{\wedge}g') < (f\oplus\check{g}) \wedge (f\oplus\check{g}')$.

Finally, let us mention a new factorization formula of the erosion :

$$(f\ominus\check{g})\ominus\check{g}' = f\ominus(g\check{\oplus}g')$$

Let us now show how to build new morphological primitives from the erosion and the dilation, and how to combine information of "inf" and "sup" on the neighbourhood defined by the structuring element.

2.3 Some primitives derived from the erosion and the dilation

They are quite elementary because of both their low degree of complexity and the simplicity of the image characteristics they extract (intensity, contrast ...), but above all they are not robust.

The first primitive that we shall consider is the morphological gradient. It gives an approximation of the gradient vector module $\rho(x)$ at point x. Let f be an element of $\Phi_s \cap \Phi_i$:

$$\rho(x) = \lim_{\lambda \to 0} \frac{(f \oplus \lambda B) - (f \ominus \lambda B)}{2\lambda}$$

where B and λB are closed discs of respective radius 1 and λ.

In the restrictive framework of continuous functions, ρ has simple topological properties since it is a continuous mapping from $\Phi_s \cap \Phi_i$ into itself.

In the digital framework ($\mathbb{Z}^n$), the expression of the morphological gradient becomes :

$$\rho(x) = \frac{(f \oplus B) - (f \ominus B)}{2}$$

Thus, it is sufficient to dilate and to erode the grey-tone image since the gradient is the difference between a "sup" and an "inf" on the elementary disc.

This primitive is generally used to extract contours. Its robustness will be greater if artefacts elimination and noise reduction are done before.

The second primitive is an extension of the Kramer and Bruckner contrast enhancement. It gives information about the distance between the initial fonction and its dilated or eroded function.

Let us take $g, g' \in \Phi'_s$ ($0 \in supp(g) \cap supp(g')$) and $\alpha, \beta \in [0, 0.5]$.

Because of :

$$\forall f \in \Phi_s \cap \Phi_i \qquad f \ominus \check{g} \leq f \leq f \oplus \check{g}$$

let us define the mapping ψ from $\Phi_s \cap \Phi_i$ into Φ_s such that :

$$\psi(f) = \begin{cases} f \oplus \check{g}' & \text{if} \quad (f \oplus \check{g}')(x) - f(x) < \alpha\, M(x) \\ f & \\ f \ominus \check{g} & \text{if} \quad f(x) - (f \ominus \check{g})(x) < \beta\, M(x) \end{cases}$$

where $M(x) = (f \oplus \check{g}')(x) - (f \ominus \check{g})(x)$.

Intuitively, to sharpen the discontinuities in order to enhance contrast, the idea is to compare for each point x the values f(x), $(f \oplus \check{g}')(x)$, $(f \ominus \check{g})(x)$ and to affect $(f \ominus \check{g})(x)$ if f(x) is closer to $(f \ominus \check{g})(x)$ or $(f \oplus \check{g}')(x)$ in the contrary case (see Fig. 9a). Obviously, the effects of this contrast primitive are linked to the values of the derivative.

However, it happens that the result is not the expected one, as shown in Fig. 9b. The counter-example below is built from a continuous but not differentiable function.

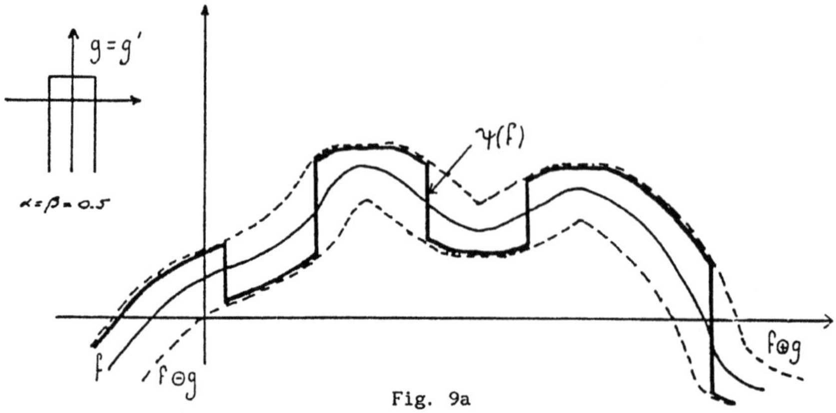

Fig. 9a

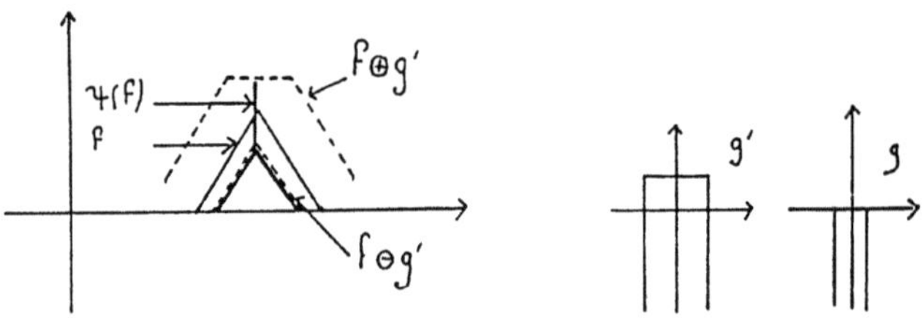

Fig. 9b

Fig. 9. Some effects of the contrast enhancement.

In practice, erosion and dilation are never used individually because of the lack of stability, robustness, and convergence properties : you have no control on iteration effects.
However, associated with the set intersection or with the "inf", dilation allows to define the primitive of particles or functions reconstruction. This higher order primitive conveys a geodesic information.
Let us present this primitive in the functional framework.
Let us take f, $g \in \Phi$ with $g \leq f$ (g is called "marker" of f). Let g_K be the planar structuring element associated with the compact set K of $\mathbb{R}^n$.
By definition, the reconstruction of f from g is the limit of the series $(g_n)_n$ where :

$$\begin{cases} g_0 = g \\ g_n = (g_{n-1} \oplus g_K) \wedge f \end{cases}$$

Let us emphasize the good property of convergence of this primitive, it is the first one with such a property. Indeed, it is idempotent : by iteration, it adds no effect.

Ascending through the pyramid of morphological primitives, we shall be able to build higher order primitives. These inherit all properties of basic primitives and are moreover idempotent.

2.4 Opening and closing

Morphological erosion and dilation are dual mappings, but not inverse one of the other. So, their functional combination allows to create new mappings : morphological opening and closing.

2.4.1 Functional definition and geometrical interpretation

Definition 1 :

Let us take $g \in \Phi'_s$.

We call morphological opening (resp. closing) w. r. to the structuring element g, the mapping from Φ into Φ, denoted ψ_g (resp. φ^g), and defined by :

$\forall f \in \Phi \qquad \psi_g(f) = f_g = (f \ominus \check{g}) \oplus g$

(resp. $\forall f \in \Phi \qquad \varphi^g(f) = f^g = (f \oplus \check{g}) \ominus g$)

Let us remark that ψ_g et φ^g are dual one of the other.

Remark : Morphological opening and closing are compatible with translation.

Geometrical interpretation :

We have already shown that for any function f and f' :

$$SG(f \ominus \check{g}) = \{(x,t) \in \mathbb{R}^n \times \overline{\mathbb{R}} \ / \ (SG(g))_{(x,t)} \subset SG(f)\}$$

$$SG(f' \oplus g) = \{(x,t) \in \mathbb{R}^n \times \overline{\mathbb{R}} \ / \ [s_{0 \in \overline{\mathbb{R}}} (SG(\check{g}))]_{(x,t)} \cap SG(f') \neq \emptyset\}$$

It follows :

$$SG(f_g) = \cup \{SG(g)_{(x,t)} \ / \ (x,t) \in \mathbb{R}^n \times \overline{\mathbb{R}} \text{ et } (SG(g))_{(x,t)} \subset SG(f)\}$$

Indeed :

$$SG(f' \oplus g) = \cup \{SG(g)_{(x,t)} \ / \ (x,t) \in SG(f')\}$$

Hence :

$$SG((f \ominus \check{g}) \oplus g) = \cup \{SG(g)_{(x,t)} \ / \ (x,t) \in SG(f \ominus \check{g})\}$$
$$= \cup \{SG(g)_{(x,t)} \ / \ (SG(g))_{(x,t)} \subset SG(f)\}$$

The subgraph of the opened f_g is the area swept by the subgraph of the structuring element g when it remains included in the subgraph of f.

2.4.2 Some properties

Proposition 1 :

Opening (resp. closing) w. r. to the structuring element g is a u.s.c. mapping from Φ_s into Φ_s, and a l.s.c. mapping from Φ_i into Φ_i.

Proposition 2 :

Morphological opening and closing are idempotent mappings :

$\forall\ g \in \Phi'_s \qquad \psi_g \circ \psi_g = \psi_g \qquad$ and $\qquad \varphi^g \circ \varphi^g = \varphi^g.$

Definition 2 :

Any mapping ψ which is increasing, anti-extensive (resp. extensive) and idempotent, is an algebraic opening (resp. closing), i.e. :

$f \leq f' \quad \Rightarrow \quad \psi(f) \leq \psi(f')$

$\psi(f) \leq f$ (resp. $\psi(f) \geq f$)

$\psi \circ \psi = \psi$

Corollary 2 :

A morphological opening(resp. closing) is an algebraic opening (resp. closing).

We have so far widely stressed on the property of idempotence. Let us now consider the increasing property. Although very simple, it is still powerful as it allows to characterize openings from their invariant sets and as it gives a procedure for creating new openings (Matheron, 1975).

2.5 Derived primitives from opening and closing

Let us firstly mention the so-called "top-hat" primitive (Meyer, 1979). At the begining, it was developed in the framework of binary images in order to extract contrast peaks with any height. Its interpretation is very easy in the functional framework : it gives the difference between the function f and its opened function f_{g_K} (g_K is the planar structuring element associated with the compact set K of $\mathbb{R}^n$).

$$\forall\ f \in \Phi_s \cap \Phi_i \qquad \psi(f) = f - f_{g_K}$$

The extraction of "valleys" is performed in a similar way : it is the difference between the closing w. r. to g_K and the initial function :

$$\forall\ f \in \Phi_s \cap \Phi_i \qquad \phi(f) = f^{g_K} - f$$

The "rolling ball" primitive (Sternberg, 1983) belongs to the same approach. It does not use a compact set of $\mathbb{R}^n$, but a structuring element whose subgraph holds a hyper-sphere B ($B \subset \mathbb{R}^{n+1}$) of radius ρ :

$$h_\rho(x) = \begin{cases} (1-\|x\|^2)^{1/2} & \text{if } \|x\| \le \rho \\ -\infty & \text{if not} \end{cases}$$

Then we have : $\psi 1(f) = f - f_{h_\rho}$ and $\phi 1(f) = f^{h_\rho} - f$

Fig. 10 illustrates these two transformations.

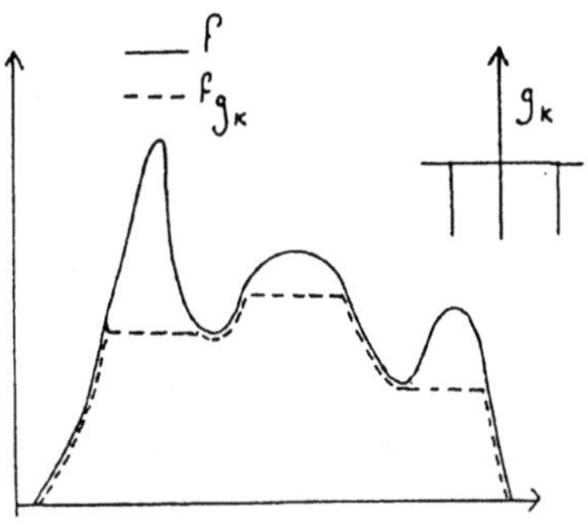

Fig. 10a

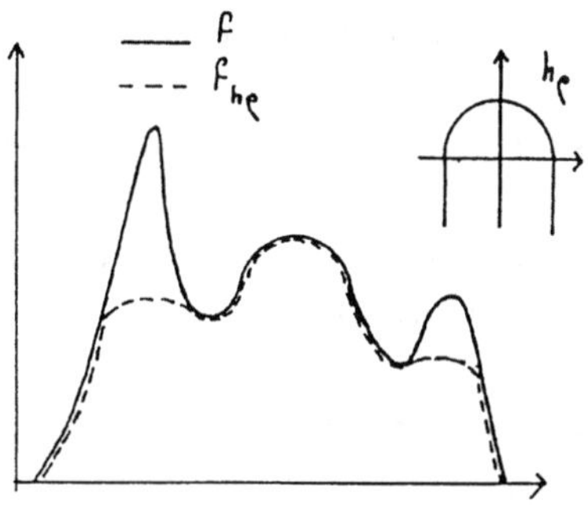

Fig. 10b

Fig. 10. Extraction of contrast peaks.
10a. by top-hat.
10b. by rolling ball.

With granulometry, second example of primitive derived from opening and closing, we tackle, for the first time in this course, the concept of quantification and measurement.
The mathematical status of granulometry was set by G. MATHERON (Matheron, 1967), (Matheron, 1975), in order to avoid the common confusion between concept and operating procedure (most of the time, granulometry is assimilated to the measurement of a sieve or of an image analyser).

Let us remind briefly the axiomatic of granulometries in a set framework.

Definition 1 :

Let E be a set and $\mathcal{A} \subset \mathcal{P} = \mathcal{P}(E)$.

A granulometry on $\mathcal{A}$ is a one-parameter family (ψ_λ) $(\lambda > 0)$ of mappings from $\mathcal{A}$ into itself such that : for any λ, $\mu > 0$ and for any X, Y of $\mathcal{A}$:

1 - $\psi_\lambda(X) \subset X$,

2 - $X \subset Y \Rightarrow \psi_\lambda(X) \subset \psi_\lambda(Y)$,

3 - $\mu \le \lambda \Rightarrow \psi_\lambda(X) \subset \psi_\mu(Y)$,

4 - $\psi_\lambda o \psi_\mu = \psi_\mu o \psi_\lambda = \psi_{\max(\lambda,\mu)}$.

We can complete the definition of a granulometry by putting :

$$\forall X \in \mathcal{A} \qquad \psi_0(X) = X$$

For example, the family (ψ_λ) of morphological openings whose structuring elements λK are homothetic to a given set K, defines a granulometry. It allows to quantify size and shape distributions from the granulometric repartition function of the studied set. This last function is given by :

$$\forall \lambda > 0 \qquad G(\lambda) = \frac{V(f) - V(\psi_{\lambda K}(f))}{V(f)}$$

where V( ) denotes the set volume.

The granulometry concept involves increase, anti-extensivity and idempotence.
A fine example of combined use of top-hat and granulometry can be found in (Prêteux et al, 1985 b).

Whereas granulometries are used commonly in binary imaging, others, like morphological filters, are completely efficient only in grey-tone imaging. These primitives of higher order, which we will study now, were developed especially for this purpose (Maisonneuve, 1982 a), (Matheron, 1983), (Serra, 1986).

Definition 2 :

A morphological filter is an increasing and idempotent mapping.

At once, such a filter can be distinguished from linear filters which are reversible.
A very easy way for creating morphological filters consists in considering a family (ψ_λ) of openings and a family (ϕ_λ) of closings and in defining :

$$\forall \lambda > 0 \qquad m_\lambda = \psi_\lambda o \phi_\lambda$$

Moreover, if we have :

$$\forall \lambda > 0, \forall \mu > 0 \quad \mu \leq \lambda \quad \Rightarrow \quad \psi_\lambda < \psi_\mu \quad \text{and} \quad \phi_\mu < \phi_\lambda$$

and the monotonous decreasing continuity of ϕ_λ and ψ_λ for any $\lambda > 0$, we can then build a family of filters whose limit defines the alterned sequential filter (Serra 1986). A fine example of use of such a filter is given in (Bloch and Prêteux, 1987).

2.6 An example of a higher order primitive : the r-h minima, r-h maxima (Schmitt and Prêteux, 1986)

The extraction of dark or clear "spots" is a common operation in image analysis, hence the importance of realising efficient algorithms to find out regional extrema.

Definition 1 :

We call regional maximum of f of value n, any connected component of X_n such that :

$C \cap (X_{n+1} \oplus H) = \emptyset$

where $X_p = \{x \in \mathbb{H} \,/\, p \leq f(x)\}$ is the threshold of f at level p.

Let us remind an efficient algorithm for regional maxima extraction. It is based on the notions of plateau, edge of plateau and marker of plateau (Maisonneuve, 1982 b) as shown in Fig. 11.

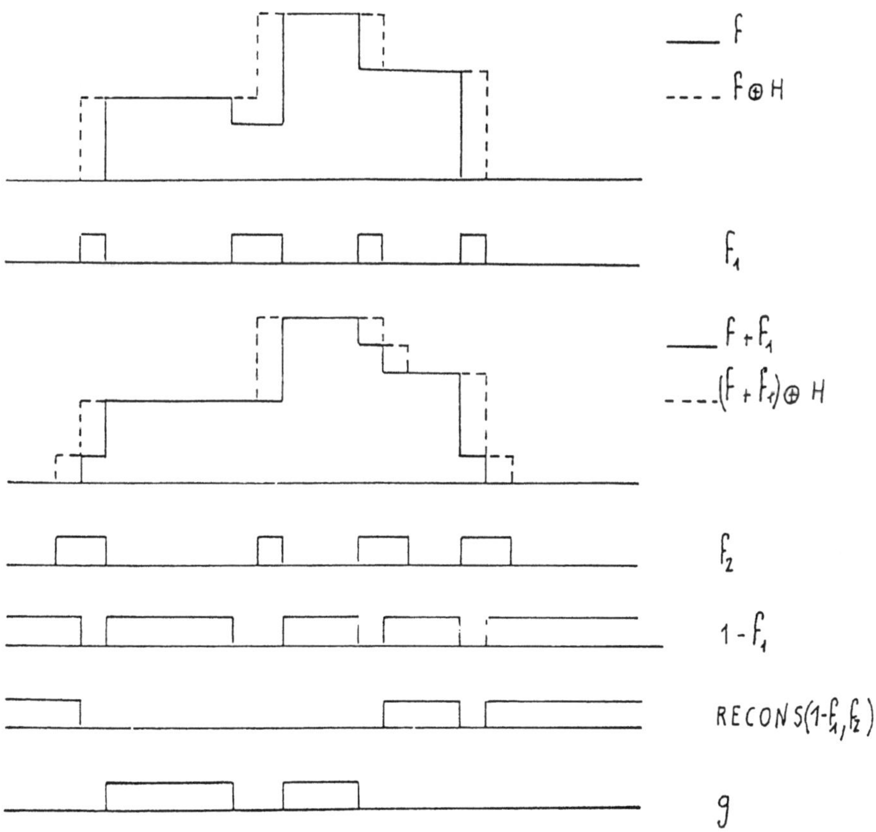

Fig. 11. Steps of regional maxima extraction.

Definition 2 :

We call plateau of level n any connected component of constant altitude n.

The binary mask associated with the edge of the plateau is defined by the function f_1 wich indicates where the initial function f is different from its dilated function, that is :

$$f_1 = (f \oplus H - f) \wedge 1$$

We call marker of plateau the set defined by :

$$f_2 = [(f + f_1) \oplus H - (f + f_1)] \wedge 1$$

Since a regional maximum is a plateau without any marker, the extraction algorithm of its binary mask is given by :

$$g = (1 - f_1) - [\text{RECONS}(1 - f_1 , f_2)]$$

where RECONS(h,k) denotes the reconstruction primitive of the binary set h from markers k.

The practice shows that the primitive of regional extrema is neither robust, nor discriminating for noised digital images, where it extracts a real cloud of unexploitable points.

So, we intend to define constraints in order to obtain a perfectly selecting primitive.

First, this primitive must be applied on a filtered image, the choices of the filter and of the structuring element (size, shape) being directly linked to the properties and characteristics of the maxima to be extracted.

Secondly, its principle is to keep only the maxima presenting, a given contrast in a given neighbourhood ; i.e. at a distance r around them, the grey levels of the image are inferior of at least a given quantity h. Hence the name of the primitive : r,h-maxima. Finally, the process must be parallel.

The algorithm is shown in Fig. 12 where f is the filtered digital function (memory M1) and g the digital function associated to the regional maxima (memory M2).

Proposition 1 :

The r,h-maxima primitive is anti-extensive, idempotent and increasing w. r. to r and deacreasing w. r. to h.

It is not a morphological filter since it is not increasing, but the family of r,h_0-maxima (where h_0 is fixed) defines a granulometry.

M1 ← f

M2 ← g

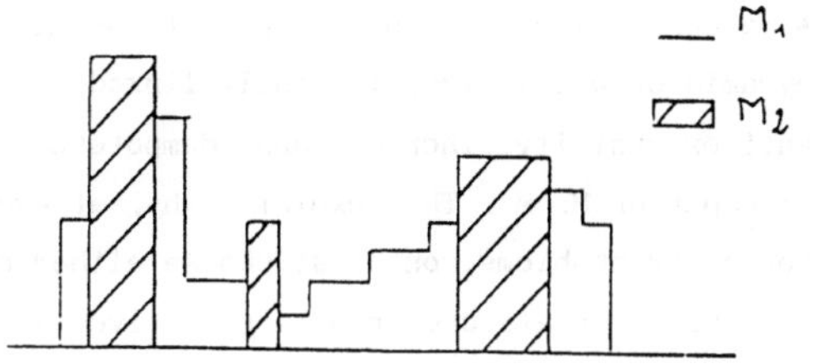

M2 ← [(M2 - h)⊕rH ; M1]

M3 ← 1 - min(1 ,M2)

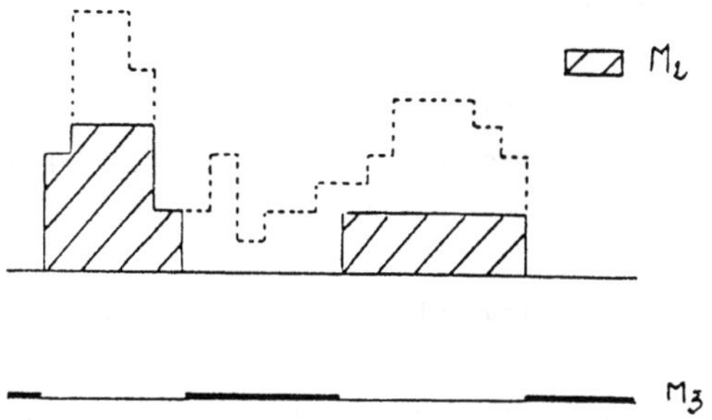

M4 ← M1 -M2

M4 ← min(1 , M4)

M5 ← RECONS(M4 , M3)

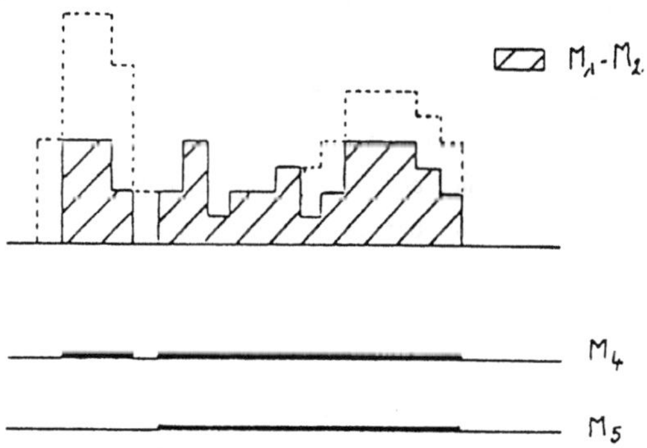

M6 ← (M4 -M5) * M2

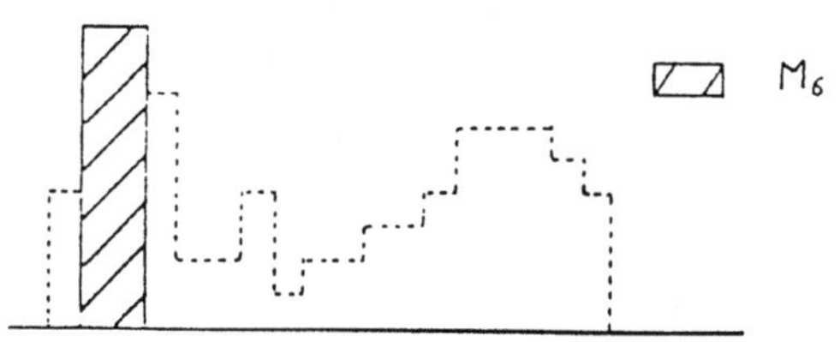

where [M⊕rH ; N] is the dilated function of size r of M conditionally to N. In M6 are the regional maxima verifying the criteria of height and contrast.

Fig. 12. Steps of r-h maxima extraction.

As a conclusion, we can notice that the hierarchy, on which is built the pyramid of primitives, is widely linked to the mathematical properties of anti-extensivity, increase and idempotence inherited from the fundamental concepts of M. M.. This explains why, at every steps of image analysis, for every problems, one must choose either a primitive that already exists, or create one or more primitives able to solve the problem. Let us see now some applications in medical imaging.

3. APPLICATIONS

The aim of image analysis is to give quantitative information and M. M. becomes a powerful tool for who only disposed, most of the time, of qualitative estimations.

3.1 Metacarpals study in Computed Tomography

The segmentation of the 2nd and 3rd metacarpals from C.T. slices (study realized at the Saint-Louis Hospital, Paris) (Prêteux et al, 1985 a) is only the preliminary step before the quantification of the cortical bone which requires the evaluation of :

- the area of the metacarpal diaphysis, the area of the medullar and their ratio,
- directional cortico-diaphysial indices,
- the ratio between outer and inner perimeters of the cortical bone,
- the mean density of the cortical bone.

All these measurements, fastidious and time consuming with traditional interactive methods, are here obtained in an automatic, rapid way. But what about measurements accuracy and reliability of our procedure in order to establish comparisons or clinical interpretations ?
It is a particularly ambiguous question, especially when the object is small, well contrasted, and generally segmented according to a thresholding technique.
Indeed, reconstruction algorithms in C.T. result from compromises between smoothing the discontinuities affecting now the spatial resolution and generating noise. Therefore, the response of a well contrasted object like a cortical bone, is a spline instead of a "step" function.
So, the problem is : where is the bone really located ? At the bottom of the slope, at the middle, or at the top ?

In order to answer objectively and to estimate experimental measurement accuracy, we used phantoms with known geometrical characteristics.
The chosen phantoms were calf bones, rubber pipes and glass tubes, with different sections and thicknesses. We have demonstrated that the measurement accuracy is of about 3 pixels. Under these conditions and considering the smallness of the lengths to be measured, only variations over 20% are statistically significant.
Therefore, the validation step has led to give up this study.

Moral : It is not enough to measure something. One also must know exactly what one measures and with which accuracy.

3.2 L_3 trabecular network analysis

An other example of quantification by M. M. comes from a study of the osteopenia in vertebral spongious bone (Experimental Radiological Laboratory of Paris VII).

After segmentation of the trabecular network, the aim was to quantify from 52 microradiographs of sagittal slices (1mm thick) of L_3 :
- the area of the calcified tissue,
- the size distribution of trabecular and inter-trabecular spaces.

The developed method (Prêteux, 1985 b) has allowed us :
- to find again well-known results about bone loss during aging,
- to quantify qualitatively established phenomena like the heterogeneity of the trabecular network,
- and moreover to bring new information :
  - the composition of the trabecular network by determining the contributions of weightbearing and transverse trabeculae, and their evolution during aging,
  - the maximal length of weightbearing and transverse trabeculae during aging.

With regard to the last point, the new information we have given (thinning of transverse and weightbearing trabeculae), were in contradiction with the qualitative descriptions in the scientific litterature. Only an objective validation could allow to conclude.

Thus, the analysis method was tested on 30 microradiographs of phantoms constituting aluminium networks. Wires with different thicknesses, regularly disposed or not, simulated the vertebral architecture. For each phantom, the surface of the wires and their size distribution were computed and then compared with the real values.

We could establish that the measurement accuracy is of 95%. The results, which were obtained by M. M., were thus validated. The prevailing ideas on the remodeling of the trabecular network during aging had to be corrected.

Moral : Image analysis has contributed to a better knowledge ofthe verbebral spongious tissue.

3.3 Lung volume in Computed Tomography

This example deals with the quantification of a tridimensional information not accessible to the physician. We present an automatic procedure for evaluating lung volume from C. T.

This algorithm consists in two steps :

- segmentation of the lungs on each slice and computation of their surface,
- estimation of the lung volume.

The image collection consists, for each patient, of about 30 slices (1mm thick, matrix 512^2, 10mm spaced). The main difficulty is that no criterion of shape or size could be used, considering the great variability from a patient to another one and from a slice to another one. We won't detail here the procedure which is described in (Prêteux, 1987) and (Merlet, 1987), but we will present the steps of the results validation.

In order to validate the algorithm, we have to evaluate the effect of each possible source of error on the final quantification.

A first source of error comes from the reconstruction algorithms used in tomography. It introduces a systematic bias, not easily quantified but not in contradiction with the notions of reproducibility and accuracy, necessary for comparisons and normalized scales.

The second one is linked to the segmentation procedure. How can we evaluate its effect ? By using the property of high gradient on the images, we establish that the maximal error does not exceed one pixel in a 512^2 matrix, inducing less than 3% error on the initially measured volume.

This is much less than variations considered as significant in diagnosis.

Finally, the last source of error is due to the method of linear interpolation used for computing the volume. This error, of an order of a centiliter, can be neglected.

However, another validation of our method can be established :
- absolutely, using phantoms, here, bottles of known volumes, (accuracy of 98%),
- relatively, on a statistically significant number of patients, by comparing our results with those obtained using plethysmography (reference clinical method). The measure of the total lung capacity of 15 patients (about 450 images) has led to a correlation factor of 0,92.

The importance of such a software is to use the information contained in the set of C.T. slices and to give access to volumes which could not be measured easily before such as left and right lung volumes ...

3.4 Radioactivity quantification of muscarinic receptors of acetylcholine in the myocardium in P.E.T.

We will finally give an example of image sequence analysis from the study of the radioactive distribution of muscarinic receptors in myocardium in P.E.T. (study undertaken at the SHFJ of Orsay).

After having extracted a mask of the "functional myocardium", (Boulanger et al, 1987), valid for the set of images of one patient, an absolute polary reference mark must be determined from the center of the cardiac cavity and from the long axis of the heart.

The pole is automatically determined as center of the convex hull of the myocardium without the myocardium.
The long axis is defined by the pole and the geometric center of the heart (also automatically obtained as the ultimate geodesic erosion of the cardiac cavity, conditionally to the convex hull of the myocardium).
In this reference mark, we define 30° sectors. For each one, the average activity in the myocardium is computed. Two kinds of information are then avalaible :
- a spatial one, which only depends upon the considered image. This information gives the cardiac activity for each sector at a given time and puts in a proeminent position the hypofixing areas,

- a temporal one related to the whole series of images which characterizes the activity for a given sector.

For this application, the validation was considered from three points of views :

- the one of the physician which is indispensable although not objective,
- the qualitative one, concerning the sensitivity of the measurements on the basis of Poisson noise of variable intensities. This has shown the robustness of the segmentation algorithm for an intensity up to 60%,
- the one which compares our method with a reference one based on the geometric modelisation of the muscarinic receptors (Travere et al, 1986).

If the same absolute reference (up to 3%) was obtained by the two methods, our is, up to now, the only one able to automatically detect hypofixing areas. Moreover, it allows the study of the transitional stage (corresponding to the uptake of radiopharmaceutical on the receptors sites), and finally the result is independent of the operator contrary to the reference method. So, physicians can have at their disposal new reliable and reproducible quantitative information.

CONCLUSION

The mathematical morphological method that we have validated on normal images, gives moreover access to quantifications in hypofixing areas, which is not possible with the reference method. This increases the interest of physicians, who will get then new quantitative information.

These several examples of quantification by M. M. show that the measurements which actually are obtained manually by physicians can be given automatically and rapidly by image analysis. With image analysis , measurements are more simple , robust and in particular reproducible. Moreover, in cases where the physician cannot realise measurements (volume, flows, tracking a radio-element ...), image analysis provides a solution which is sometimes sophisticated in its conception, but very easily put in practical realisation.

REFERENCES

Bloch, I. and Prêteux, F. (1987). Démarche morpho-mathématique de segmentation appliquée aux lésions cérébrales en imagerie par résonance magnétique, J. de Microsc. et Spectrosc. Electron., Vol 12, pp. 45-48.

Boulanger, F., Soussaline, F. and Prêteux, F. (1987). Un nouvel algorithme de segmentation en Morphologie Mathématique, in acts MARI 87, Vol. 2, pp. 342-349.

Coster, M. and Chermant, J.L (1985). Précis d'Analyse d'Images, Ed. CNRS, Paris.

Haton, J.P. (1987). Les systèmes à base de connaissances en reconnaissance et en interprétation de formes, in acts MARI 87, Vol. 2, pp. 73-80.

Maisonneuve, F. (1982 a). Ordinaux transfinis et sur (ou sous) potentes, N-780, C.G.M.M.

Maisonneuve, F. (1982 b). Extrema régionaux : un algorithme parallèle, N-781, C.G.M.M.

Matheron, G. (1967). Eléments pour une théorie des milieux poreux, Masson, Paris.

Matheron, G. (1975). Random Sets and Integral Geometry, Wiley, New York.

Matheron, G. (1983). Filters and Lattices, N-851, C.G.M.M.

Merlet, N. (1987). Détermination du volume pulmonaire dans le cadre du traitement de l'emphysème bulleux, S-227, C.M.M.

Meyer, F. (1979). Cytologie quantitative et Morphologie Mathématique, Thèse de Docteur-Ingénieur, E. N. S. M. P.

Prêteux, F., Laval-Jeantet, A.M., Roger, B. and Laval-Jeantet, M. (1985 a). New prospects in C.T. image processing via Mathematical Morphology, European Journal of Radiology, pp. 313-317.

Prêteux, F., Bergot, C. and Laval-Jeantet, A.M. (1985 b). Automatic quantification of vertebral cancellous bone remodeling during aging, Anat Clin., Vol. 7, pp. 203-208.

Prêteux, F. (1987). Description et interprétation des images par la morphologie mathématique ; Application à l'imagerie médicale, Thèse de doctorat d'Etat, Université Paris VI.

Schmitt, M. and Prêteux, F. (1986). Un nouvel algorithme en Morphologie Mathématique : les rh-maxima, rh-minima, in acts 2ème Semaine Int. de l'Image Electronique, Vol 2, pp. 469-475.

Serra, J. (1982). Image Analysis and Mathematical Morphology, Academic Press, London.

Serra, J. (1986). Eléments de théorie pour l'optique morphologique, Thèse d'Etat, Université Paris VI.

Sternberg, S.R. (1983). Biomedical image processing, Computer, Vol. 16, 1, pp. 22-31.

Travere, J.M., Charbonneau, P. and Syrota, A. (1986). Analyse automatique et modélisation graphique appliquées à l'étude des récepteurs muscariniques en tomographie par émission de positons, in acts 2ème Semaine Int. de l'Image Electronique, Vol. 2, pp. 497-503.

REFERENCES

[illegible] and [illegible], [illegible] (19[illegible]). Sur [illegible] de [illegible] approximation appliquée aux [illegible] de [illegible] magnétique [illegible] résonance magnétique [illegible] de [illegible] [illegible], 12, pp. [illegible]

[illegible], [illegible], Sommerfeld [illegible] and [illegible] (19[illegible]). [illegible] [illegible] et [illegible] mathématique [illegible], [illegible], [illegible]

[illegible] and [illegible] [illegible] (19[illegible]). [illegible] d'images [illegible], [illegible]

[illegible] (19[illegible]). [illegible] [illegible] [illegible] [illegible] [illegible] interprétation de [illegible] formes [illegible] [illegible], [illegible] pp. [illegible]

[illegible], [illegible] (1982). [illegible] [illegible] [illegible] [illegible] [illegible]

[illegible], [illegible] (1982). Extremes [illegible] [illegible], [illegible], [illegible]

[illegible], [illegible] (196[illegible]). [illegible] [illegible] pour une [illegible] [illegible] [illegible] [illegible]

[illegible]

[illegible] (19[illegible]). [illegible] [illegible]

[illegible] [illegible] [illegible] [illegible]

[illegible]

[illegible]

MULTISCALE METHODS
AND THE SEGMENTATION OF MEDICAL IMAGES

Stephen M. Pizer

Depts. of Computer Science, Radiology, and Radiation Oncology
University of North Carolina
Chapel Hill, NC 27599-3175, USA

ABSTRACT

Multiscale methods analyze an image via relationships between its properties at many different levels of spatial scale. Details and noise appear largely at small scale, and global properties of image objects appear at large scale. The segmentation of images into objects or coherent regions is therefore aided by viewing the image in multiscale terms. The methods and approaches that have been suggested for multiscale image analysis and for segmentation based on this analysis are summarized.

1. INTRODUCTION

Multiscale approaches for dividing an image into objects (segmentation) are based on the idea that the identifying features of an object or any coherent image region exist simultaneously at many levels of spatial scale. Global properties of the object are dominant at large scale, and at smaller scales details of size appropriate to that scale are principally represented. Thus, an oak tree consists of a trunk and a treetop at large scale, of limbs at a smaller scale, of branches and leaves at a yet smaller scale, and of leaf veins and indentations at a quite small scale. Defining the part of an image corresponding to an object therefore requires examining the image simultaneously at many levels of spatial scale. There is evidence that the human visual system operates in such a multiscale manner (Young, 1986).

Moreover, an image viewed at large scale is much simpler than one viewed at smaller scales. Therefore, for efficiency object recognition should occur in a sort of top-down manner, finding things tentatively at large scale where the image is simple, and then verifying and refining the recognition at smaller scales where details are better represented.

The computation involved in multiresolution segmentation thus requires a means of successive simplification of the image by increasing in scale, examining image features at various scales, and actually defining the objects in terms of the features found at the respective scales. The following chapters treat these matters in turn.

2. MEANS OF IMAGE SIMPLIFICATION

2.1. Pyramid methods

Historically the first multiresolution methods were pyramid approaches. These depend on forming new larger scale pixels each of which combines the information from image pixels into square groups of m X m, thus creating a summary of the image with fewer pixels (by $1/m^2$ -- as shown in Fig. 1). This larger scale image is in turn simplified by combining its pixels into groups of m X m, creating a yet larger scale image. This process is repeated to produce what can be viewed as a pyramid of image representations if the successive summaries are piled one on top of the other.

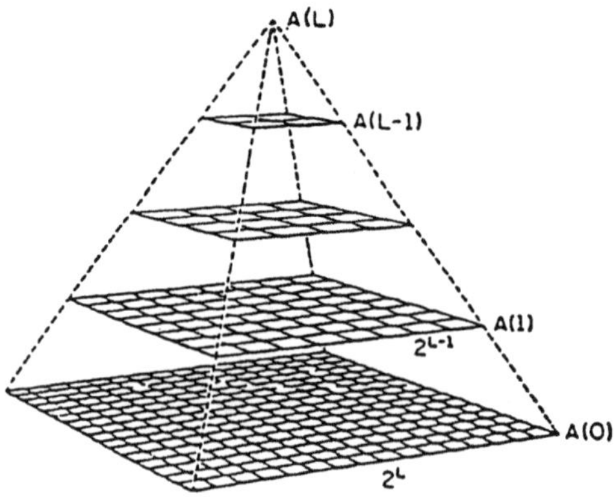

Figure 1: Pyramid. Compliments of Lawrence Lifshitz (1987)

Different pyramid methods vary by the means of combining image pixels into groups, i.e., by the information that is recorded in the parent pixels. Two general categories of pyramid methods can be defined: feature following and summary, and wavelet decomposition.

In feature following and summary (Burt, 1981; Rosenfeld, 1987) a parent pixel summarizes the information about some feature in the pixels that are its children at the next lower level of scale. For example, if the feature of interest was image intensity, the parent might record the average intensity of its children; or if the feature of interest was a line segment, segments described in some of the child pixels would be summarized into a global segment in the parent pixel.

In wavelet decomposition (Jaffard, 1986; Mallat, 1987) the information at smaller scale gives only the difference between the image analyzed at that scale and the image analyzed at the larger scale. This provides a space-efficient representation. By an elegant development, the information at each scale is represented by orthogonal functions that are band-limited in both the space and frequency domains.

2.2. Blurring methods

Another way to simplify the image is to blur it by convolution with a blurring kernel. A number of authors (Yuille, 1983; Witkin, 1983; Koenderink, 1984) have shown that the kernel that to the greatest degree limits the creation of new structure under blurring is the Gaussian; this kernel also has many advantages of compatibility with the measured receptive fields of the human visual system, being simultaneously isotropic and separable by coordinates, and holding its form under cascaded operation (Koenderink, 1989). As a result there has been considerable attention (Pizer, 1988b; Koenderink, 1984; Bergholm, 1987; Stiehl, 1988) to image segmentation by following features through Gaussian blurring.

It has been suggested that non-isotropic or non-stationary Gaussian blurring can allow the blurring to adjust to the orientations of and spacings between objects. This adds a considerable complication to the analysis and leads to iterative methods where tentative segmentation suggests the parameters of reblurring. The potential strengths of such iteration, as suggested by the human visual system, has led to initial consideration of such an approach (Lifshitz, 1988) and some trials indicating some success of this idea (Hsieh, 1988b). A related approach with some promise creates non-Gaussian summaries over areas grouped according to consistent geometric properties such as intensity gradient direction (Colchester, 1988).

Attention to edges suggests following the Laplacian of the image through Gaussian blurring. This idea can be combined with resampling with fewer pixels to produce the so-called Laplacian pyramid (Burt, 1984). Since D*(B*Image) = (D*B)*Image for any differential operator D and blurring kernel B, the Laplacian of a blurred image can be found by applying to the image the operator that is the Laplacian of a Gaussian. The Laplacian of a Gaussian is approximately equal to the difference of that Gaussian (DOG) with a Gaussian with standard deviation approximately 1.6 times as large. If one uses $\sqrt{2}$ in place of 1.6, one can efficiently produce a multiresolution image sequence from a series of DOG applications: G_i * Image - G_{i+1}* Image, where $G_{i+1} = G_i * G_i$ (Crowley, 1984). Crowley has used this idea to produce a series of images resulting from successive DOG filters, each with the same energy but covering a successively lower frequency passband.

2.3. Discretization and finiteness

In actual computation both the space and scale will be discrete, so the spatial discretization (pixel spacing) and scale discretization (spacing between levels of blurring: for Gaussian blurring, the value of s) must be decided. Sampling theory argues that the pixel spacing should be proportional to the scale, but a successive decrease in number of pixels covering an image forces the decision of how to map locations at higher scale to the new scale. Since the image at a larger scale will be simplifed, the representation of that image can frequently be made efficient of computer memory without increasing the pixel spacing.

Computational ease has led to many pyramid methods increasing the pixel spacing by 2 in each coordinate dimension at each scale step. This is the scale increase used in the wavelet decomposition. Crowley has shown how taking advantage of diagonal distances can allow efficient computing of the Laplacian pyramid with a scale increase of $\sqrt{2}$ at each scale step.

Koenderink (1984) and Pizer (Lifshitz, 1987) have suggested that the amount of blurring between successive discrete scales be determined by limiting the change in intensity of some basis function as it is blurred between successive discrete scales. They limit the intensity change to an amount proportional to the accuracy of the underlying floating-point intensities undergoing blurring.

Such analysis concludes that increasing the scale by a constant factor between steps is correct, but a factor of 2 or even $\sqrt{2}$ produces too much change in the basis function intensity between steps.

In contrast, some authors have recognized that the essential issue in scale discreteness is how it supports the following of image features that is involved in multiresolution methods. Lifshitz (1988) has suggested that if one is following features such as intensity extrema through scale space, with the intention that annihilations of one feature into another are to be discovered, large successive blurrings that are related to the spacing between the features should be used. Similarly, Bergholm (1987) has noted that if one is following a feature such as an edge from an optimal scale for recognition through successively smaller scales to its location in the in the original image, it is reasonable to limit the successive blurrings so that the feature movement is limited to a fixed distance. Bergholm assumes a fixed minimum corner angle of edges and then chooses his blurring to limit edge movement to 1 pixel.

The method for following features through scales can equivalently be thought of as the need to identify a multiscale form, e.g. (as shown in Fig. 2), an extremal path (a point extended through a part of the scale dimension) (Koenderink, 1984; Lifshitz, 1988) or a Laplacian zero crossing surface (a curve extended through part of the scale dimension). This form can be determined, either directly or by relaxation, after the corresponding single-scale form (e.g., extremum of intensity or zero of Laplacian, respectively) has been identified at each scale, but artifacts arising due to different discrete approximations at each scale can make difficult the determination of cross-scale coherences. Instead one can treat the coherence across scales as part of the identification process. In this approach, inspired by the "Snakes" approach of Terzopoulos, Witkin, and Kass (1987), a multidimensional form, such as a surface, with one of its dimensions being scale, is fit to the family of multiscale images in such a way as to maximize a combination of coherence of the form and fit to whatever image properties correspond to the feature being followed (e.g., the derivative properties corresponding to extrema of intensity or zero-crossing of the Laplacian, respectively) (Gauch, 1988b).

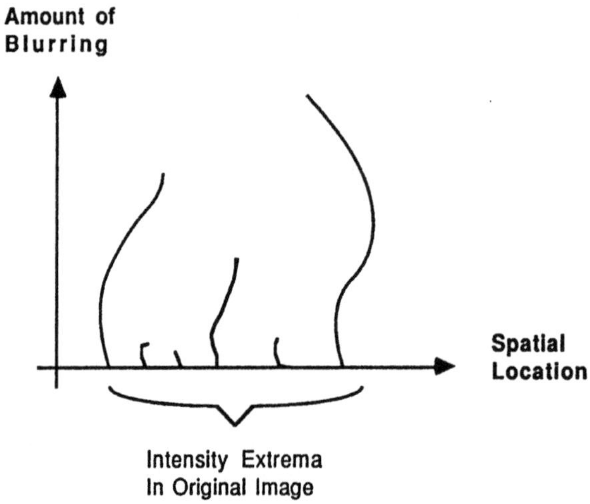

Figure 2: Extremal paths through scale. Compliments of Lawrence Lifshitz (1987)

In addition to discreteness, the finiteness of the image can cause problems when scale increase is accomplished by Gaussian blurring. The problem is to continue the finite image across infinite space so as to achieve desired behavior under scale reduction. Reflection across boundaries or the implicit wraparound continuation of frequency domain analysis does not allow the image to simplify fully as scale is increased without bound. Toet (1984) suggested that seeing Gaussian blurring as applying the diffusion equation leads to the solution. Based on this idea, Lifshitz (1987) modifies the image by subtracting a solution to the diffusion equation that at all levels of scale agrees with the original image at its boundary. Applying multiscale analysis to the result, extended everywhere outside the image by zeroes, can now go ahead.

3. USING VARIOUS SCALES

There are a number of strategies for using the images at multiple scales. Many authors (e.g., Coggins, 1986a; Neumann, 1988; Burt, 1981) try to find the optimal scale for a particular decision. For example, Laplacian zero-crossings can be used to find edges at an optimal resolution, even in 3D (Bomans, 1987). Similarly, in stereo matching, a scale that sets the stereo disparity at a given location to a fixed number of pixels is correct for that location (Hoff, 1988). In segmentation, the decision that a region is a sensible segment is best made at the scale of that segment. In these methods the decisions are frequently taken in decreasing order of scale, but it is also possible to go in reverse order, choosing the optimal scale when one comes to a scale where certain conditions first fail.

Another strategy for using images at multiple scales is to use each larger scale to summarize at each image location the information from the children at lower scales (Rosenfeld, 1987; Meer, 1986; Sher, 1987). In the simplest form of this method, intensity itself might be summarized by averaging over pixels that at the next lowest level of scale have been associated with the parent pixel being computed. The various methods (Burt, 1981) differ by the way in which these associations are determined. For example, one might associate a pixel with nearby potential children that are closer in intensity to its present intensity than any other potential parent. After these associations are determined, the parents' intensities are recomputed and the associations redetermined, and this process repeats until no associations change. After these associations have been set, a region can be determined as all the descendants of a pixel that is not close enough in intensity to any of its potential parents.

Instead of summarizing a feature as simple as intensity, one might follow a geometric feature such as an edge curve or an intensity trough, or a texture through scale space (Sher, 1987). Thus, for example, two approximately collinear line segments might be summarized into the larger line segment formed by their combination. In this case, also, there must be a means of choosing the pixels at the next lower scale that the present pixel is to summarize. Even given this, the approach frequently has the difficulty of finding a means of summarizing the information from the next smaller scale without the storage for a pixel increasing unmanageably with increasing scale.

Another class of methods takes advantage of the fact that as the image simplifies under increase in scale, component features can be expected to disappear into the more "important" features of which they are a part (if the features are correctly chosen to characterize image structure of interest).

The order of disappearance and the way in which image pixels collapse into structure-defining geometric forms can be used to define regions and impose a hierarchy on them (Koenderink, 1984; Pizer, 1988b). For example (as shown in Fig. 3), Lifshitz (1988) and Koenderink have followed intensity extrema to annihilation while following intensity values as they collapse into extrema, while Gauch (1988a,b) and Blom (1988a) are following to annihilation ridges and troughs defined by iso-intensity curve vertices.

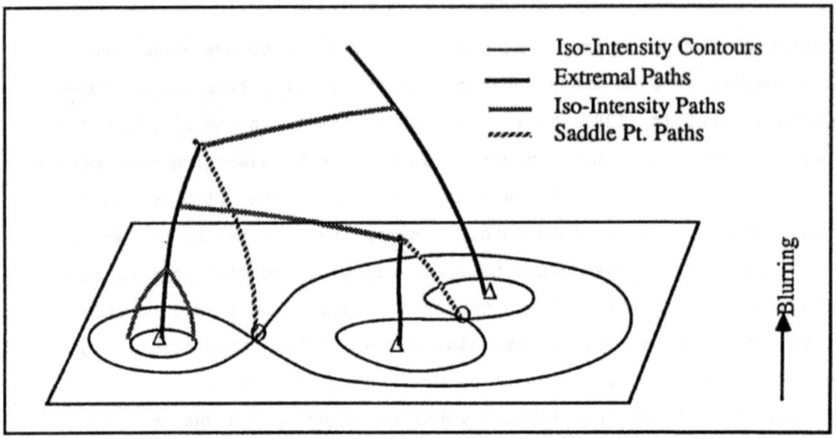

Figure 3: Extremal paths and iso-intensity paths through scale space. Note that maxima (Δ), and saddle points (O) move together and annihilate. The resulting non-extremal point (•) is then linked via an iso-intensity path to another extremal path.

The final member of our list of approaches to use multiple resolutions attempts to identify image features, such as edges or ridges, by their local behavior through the full family of resolutions. For example, a pixel on a linear edge can be identified by its sequence of values of intensity and derivatives as the image is blurred (Korn, 1985; Back, 1988; Blom, 1988a), or the occurrence of a particular geometric shape centered at a pixel can be identified by its response under a particular filtering as the scale parameter of that filter goes over some range (Coggins, 1986b). Furthermore, information about the extent of an edge, its curvature, and its distance from other edges can be obtained by the way in which this sequence of values varies from the sequence that would correspond to an infinite linear edge.

4. GEOMETRIC FEATURES IN OBJECTS AND INTENSITY AND TIME FAMILIES

The success of a multiresolution method depends not only on the strategy of using the multiple resolutions but also on the geometric forms that are examined at each resolution. The forms are chosen to capture image structure of interest while at the same time having appropriate behavior as resolution is lowered: e.g., capable of being summarized, or smoothly moving to annihilation. The choice of these forms requires a good understanding of the mathematical discipline of geometry (Koenderink, 1989). This geometry may well involve not only the two or three spatial dimensions of the image and possibly the dimension of time in a time-series of images, but also the dimensions of intensity and scale. The two-dimensional methods discussed below generalize to three spatial dimensions, though that generalization has in many cases not been worked out or tried out.

4.1. Intensity extrema

Bright and dark spots are features of importance in an image. Furthermore, Morse theory tells us that under Gaussian blurring these intensity extrema undergo a regular behavior: each extremum moves smoothly, with maxima decreasing in intensity and minima increasing in intensity as scale is increased. This behavior continues until the extremum annihilates with a saddle point. Unfortunately, it is also possible, though uncommon, for extremum-saddle point pairs to be created at some level of blurring (Lifshitz, 1987).

It seems attractive to use extrema as seeds for extremal regions, locally bright or dark image areas. Computation of extremal regions requires a means of associating pixels with nearby extrema. Furthermore, it seems desirable to use extremal annihilation to induce a hierarchy, by seeing one extremum as disappearing under scale increase into the hillside or pitside of another extremum. To this end a means of association of annihilating extrema as being part of the extremal region of another extremum must be found. Koenderink (1984) and Lifshitz (1988) have dealt with this problem by following constant intensity levels through scale space. They show that each such iso-intensity path must run into an extremum, thereby associating either the pixel or the just annihilated extremum at the source of the path with the extremal region that contains it. There are two difficulties with this means of association. First, it leads to iso-intensity curves as region boundaries. Image features such as edges and intensity ridges and troughs play no part. Second, all the pixels inside an iso-intensity curve that forms a partial boundary are not necessarily associated with the extremum in question (Lifshitz, 1988).

4.2. Edges

Object edges capture more geometry than extrema. Edge information is found in the derivatives of the image. Generally methods of edge calculation can be divided into those based on the gradient and those based on the Laplacian, though Koenderink and Blom have argued that it is useful to include information from derivatives of higher order than the second to determine edges (Koenderink, 1988b).

The gradient-based methods compute an edge strength from the magnitude of the gradient and a normal direction to the edge from the direction of the gradient. These values can be followed through Gaussian blurring (Koenderink, 1987). Hsieh (1988a) has combined the connectionist ideas of Grossberg (1985) for producing closed edges with a scheme that lets edges or edge continuations at one resolution support those at other resolutions, producing closed edge contours that frequently match what we see. Thus, in a multiscale fashion image locations cooperate and compete to select strong edges (and corners) with consistent edge directions, and continue edges across gaps (at multiple scales). Neumann (1988) adds line and channel features to edges in the Grossberg scheme, and attempts to identify these at optimal scale. Korn (1985) and Back (1988) attempt to find edges according to gradient-of-Gaussian edge strength by looking at the family of such edge strength measures across many scales. Bergholm 1987] is following the commonly used Canny edges, from a scale at which detection is straightforward but the edges are displaced, through decreasing scale to their actual location in the original image. Canny edges (1983) are locations with gradient magnitude above some threshold and of maximum gradient magnitude in the gradient direction. Pyramid methods have also been used with edge strengths.

Marr (1982), among others, has focused on the Laplacian as a direction-free indicator of edge strength. Near edges the Laplacian goes from a high positive value through zero to a high magnitude negative value. Marr has particularly focused on the closed contours given by Laplacian zero crossings, and it has been suggested to follow these contours through scale space to select important edges (Marr, 1980). While this frequently finds edges well, Blom (1987) has shown that all edges of interest do not correspond to a Laplacian zero crossing.

Edges, i.e. region boundaries, can be used in a multiscale way to describe objects, after these boundaries have been found. For example, Richards' (1985) codons can be followed through decrease in scale as one annihilates into the next (Gauch, 1988a). This type of analysis involves describing objects by deformation, the subject of section 4.4.

4.3 Central Axes

While edges provide important object information, that information is rather local. The axis down the center of an object, together with the behavior of the width of the object about its axis as a function of axis position, gives more global information about the object. When put in a form that applies to intensity-varying images, such axes can summarize information about edges, spatial shape, and the shape of variations in the intensity dimension. Two different categories of central axes have received attention. The first involves actual axes of symmetry, and the second involves intensity ridges or troughs, which tend to run down the center of objects.

Three symmetry axes of objects have been suggested (as shown in Fig. 4); all are related to the the family of circles tangent at two locations on the object boundary. The locus of the circle centers forms the *symmetric* or *medial* axis (Blum, 1978). The internal medial axis, made from the centers of tangent circles entirely within the object, has the advantage, for objects without holes, of being a tree which when divided into its branches subdivides the object into bulges. Each branch endpoint is the center of a circle which touches the object in a second degree way at one point, a point of maximum positive curvature of the object boundary. Similarly, the object's external medial axis, defined as the internal medial axis of the object's complement, is the center of a circle touching the object at a point of minimum negative curvature. A 3D object's axial surface, defined in terms of tangent spheres, has much the same properties listed above for the medial axis of a 2D object (Nackman, 1985; Bloomberg, 1988).

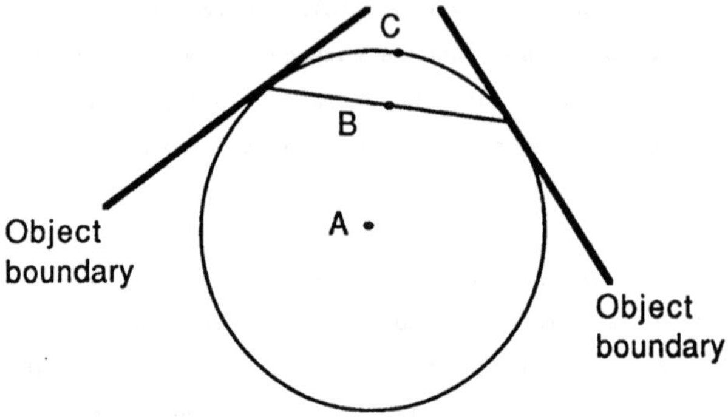

Figure 4: Points on axes of symmetry. Medial axis is locus of points of type A. Smooth local symmetry axis is locus of points of type B. Process-inferred symmetric axis is locus of points of type C.

The second axis, the locus of the centers of chords between the two tangent locations, forms the smooth local symmetry axis (Brady, 1984). It is not necessarily connected, even when it is restricted to internal circles. Brady has suggested combining this axis with the behavior of the object boundary curvature as it is followed through scale increase to produce object descriptions.

The third axis is the locus of center points of the smaller of the parts of the tangent circle between the two tangent points. Leyton (1988) calls it the *process-inferred symmetric axis* (PISA), because it touches the object on the side of the object where a deforming force can be thought to have been applied to make the object from a circle (or sphere, in 3D). We will return to the PISA in section 4.4, but here we will focus on the medial axis.

The problem is to find a form of the medial axis that applies to intensity-varying images, rather than to objects whose boundaries have already been determined (Gauch, 1988a,b; Pizer, 1988b). One can view the image as an intensity surface one dimension above the n-dimensional image. However, because intensity is incommensurate with the spatial dimensions, one cannot use a surface of symmetry in n+1 dimensions. Rather, think of the image as a single parameter family of intensity level curves, as in a terrain map. For each level curve, the medial axis can be computed and the result placed a its corresponding height. The resulting medial axis pile, that we will call the *intensity axis of symmetry* or *IAS*, can be shown to be made of branching sheets (as shown in Fig. 5), in some cases with loop branches. The IAS depends on the way the image was divided into an intensity family; in the above it was sliced along iso-intensity levels. An open area for study is the relative advantages of various means of slicing the intensity terrain.

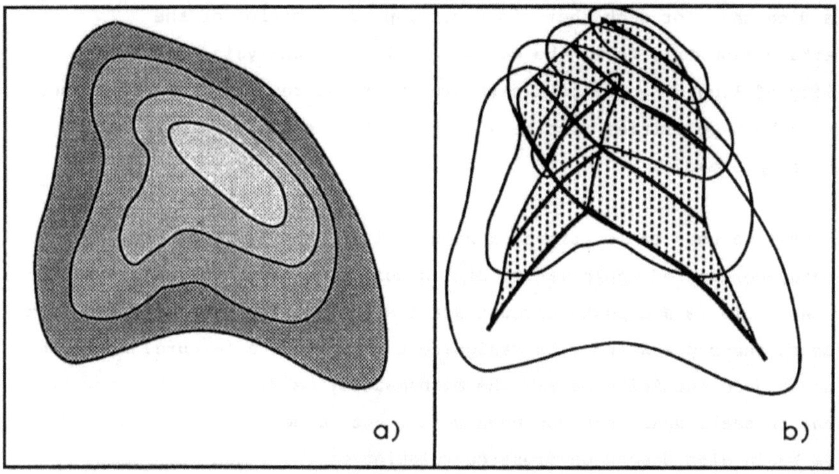

Figure 5: A simple intensity varying image and its IAS. a) 4 level curves of intensity; b) level curves and corresponding IAS.

Under Gaussian blurring, medial axis sheets annihilate into other sheets until a set of separate, simple (unbranching) sheets remains. Thus, the blurring process induces a set of hierarchies. Furthermore, associated with each subtree of sheets in this hierarchy is a subimage: at each pixel the intensity of the subimage is the maximum of the intensities of the IAS disks covering that pixel, where each IAS sheet point forms the center of a disk at the intensity of the point. That disk is the maximal disk corresponding to the medial axis point in question (for the particular level curve at that intensity). The way in which image regions should be formed by associating pixels absolutely or probabilistically with a sheet has yet to be determined. Presumably the association should take place via these subimages or geometric features such as surrounding intensity ridges or troughs It will not be surprising if the answer itself involves the multiscale following, back to the original scale, of a feature found near the annihilation scale of the IAS sheet determining the region.

Ridges and courses on the image terrain are features that not only tend to run down the middle of image objects but also have visual power. In the following, the many definitions of the general notion of ridge will be given, with the understanding that for each there is a corresponding notion of the complementary geometric form, the course. Full image analysis must lead to the possibility of light figures on a dark ground or the reverse, and simultaneous analysis in both polarities must be a part of any successful general segmentation process.

Crowley 1984] focuses on locations of high positive (or low negative) values of an energy-normalized Laplacian to find a sort of ridge in these values. He follows these ridges and peaks through scale space. Scale-space maxima in the magnitude of these peaks and pits designate the scale of a feature. While this method has a good intuitive basis, the mathematical behavior of the normalized Laplacians in scale space has not been worked out. Other methods with good intuitive basis also depend on cross-correlation of the image with some template function (Neumann, 1988).

Ridges can also be computed as watersheds, but the identification of watersheds is unfortunately nonlocal. That is, a change in the image at some distance away from a pixel can move a watershed from or to that pixel.

Other definitions of a ridge apply to any surface, independent of its orientation, and thus have the probably desirable property of being independent of the value of intensity or intensity slope. The best of these take advantage of the knowledge that the intensity surface forms a function, of space, i.e., is single-valued. One of particular interest is the locus of intensity level curve maxima of curvature, or *vertex curve,* because of its relation to the IAS: each vertex curve point is the touching point of the maximal circle corresponding to the endpoint of the medial axis of the intensity level curve at that intensity (as shown in Fig. 6). That is, there is a 1-1 relationship between vertex curves and IAS sheets. As a result, vertex curves can be followed through scale space to annihilation, a much simpler task computationally than following IAS sheets. The connectedness of the image is established by the branching of the IAS, which needs only to be computed for the original image, but the parent:child relationships are established by the order in which the vertex curves annihilate together with the relationship of the vertex curves to the IAS branches.

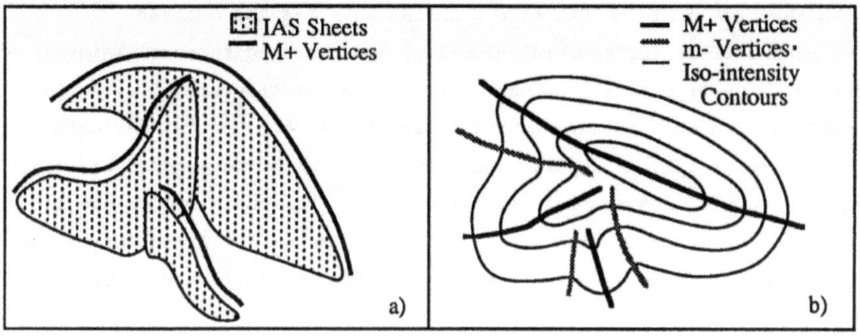

Figure 6: a) The relationship between the IAS and vertex curves; b) the same vertex curves shown superimposed on the intensity level curves. Vertices corresponding to positive curvature maxima (ridges) are indicated by M+, and vertices corresponding to negative curvature minima (courses) are indicated by m-.

4.4. Deformation

Time series of images make it necessary to follow the moving images, and their component objects, through scale space. In general, the motion involves non-rigid deformation, rotation, and translation of the image objects, and this compound motion can be used to help define these objects.

The same issues of multiple scales that are important in the spatial dimensions are also important in the time dimension. However, time is not simply another spatial dimension; it is asymmetric (the past affects the future); it is incommensurate with space; yet space and time are interrelated, by pixel velocities. The means for space-time multiscale analysis using these ideas has not been adequately developed. Certain approximations may nevertheless be useful.

For example, analyzing the time course as if it were independent of space can yield useful information. Koenderink (1988a] has treated the question of how time should be increased in scale to analyze a time series, e.g. at a single pixel, in a multiresolution fashion. He concludes that a present moment, t_0, should be chosen as a parameter and the transformation from the asymmetric variable $t \leq t_0$ to the symmetric variable $s = \log(t_0 - t)$ be carried out. Then increase in the time scale by Gaussian blurring in s has the desired properties of not violating temporal causality and treating temporal intervals in proportion to their time in the past. This transformation and blurring in s can be combined with the ordinary Gaussian blurring in the spatial variables to produce a form of spatio-temporal scale increase, with independent scale parameters in the time and spatial dimensions. With this scale change one is analyzing the time series of images as a single parameter family of images and looking at geometric structures such as bulges (ridges) and indentations (courses) in this "pile" of images.

With such a scheme, if regions are segmented to produce objects, the deformation of these objects can be described in terms of closest distances from an object boundary (or some other special object feature) at one point in time to that boundary (or feature) as time is increased. However, this does not lead to a description of the deformation of the whole object.

The description of this deformation can itself be described in multiscale terms if the object boundaries at successive times have already been found. The description is produced by minimizing a measure of energy to deform the object at between the successive points in the time series. If an appropriate energy measure is used (Bookstein, 1988; Terzopoulos, 1983), a multiscale deformation description is produced. Bookstein has shown how thin-plate spline deformation, generalized to be applied independently *within* each spatial co-ordinate, provides an energy function that is a quadratic form. The simplicity of the quadratic form makes the energy minimization mathematically tractable. Furthermore, the eigenvectors of the matrix involved in the quadratic form define principal warps, each concentrated in a particular region whose size is inversely related to the magnitude of the corresponding eigenvalues. Since the final deformation is made up of a linear combination of these basis principal warps, the deformation can be thought to be decomposed into multiscale components.

Oliver and Bookstein (1988) have suggested how this energy can be combined with a translational energy term to give a form whose minimization will specify, for all points in an object, the deformation between the object at two times, given the object boundaries at the two times. Oliver suggests that matching the medial axis endpoints which survive after increase in scale can provide a useful initial boundary match for the iterative minimization of energy that determines the deformation.

This approach, and the one to follow, depends on knowing the object boundaries. Its generalization to the situation in which one starts with intensity-varying images has yet to be investigated.

Deformation can also be used as a means of object description. The idea is that an object is described by the way it is obtained by deformation from a primal form, normally a circle or avoid (Koenderink, 1986; Leyton, 1988). Leyton has shown how the PISA gives locations of maximum positive or minimum curvature at which these deformational forces are to be applied, and gives the direction of each force, as well. Lee (1988) is investigating how a hierarchy can be imposed on this set of forces, by multiscale analysis with Gaussian blurring of the object boundary, i.e., convolving a Gaussian in arc-length s with the boundary function (x(s), y(s)). This approach might be applicable to grey-scale images by examining the PISA IAS, that is, the intensity family of PISA axes on intensity level curves.

Koenderink (1986) has investigated multiscale shape description by deformation with three-dimensional objects and blurring of the object's characteristic function rather than the boundary. This analysis focuses on the genesis, under decrease in scale (deblurring), of surface regions of specified curvature: convex elliptic, concave elliptic, hyperbolic (saddle-shaped), and parabolic (curves of flatness separating elliptic and hyperbolic regions). This idea of object analysis by treating the object layout in space as an intensity-varying image given by a characteristic function has much power. This approach can be used to describe models of images, but there is little experience in the use of such model descriptions for image segmentation, so characteristic function blurring will not be discussed further here.

5. OBJECT DEFINITION APPROACHES

The approaches described in chapters 1-3 produce image descriptions by multiscale analysis. In this section means are discussed of using these descriptions to determine objects or object probabilities for each pixel. This object definition must involve not only these image descriptions based only on image structure, but also semantics of the image, i.e., knowledge of what objects and object groupings appear in the real world. The two basic approaches to object definition, automatic and user-driven, differ in the way this semantic information is provided. In automatic object definition the information is provided in models or prototypes stored in the computer, whereas in user-driven object definition the user utilizes his own knowledge of semantics in creating an object definition from the computer-generated image description.

Automatic recognition of objects from multiscale image description can be divided into two subcategories: statistical and structural. These two methodologies need not be mutually exclusive.

5.1. Object definition by statistical pattern recognition

In an application of classical methods of statistical pattern recognition (Duda & Hart, 1973; Jain and Dubes, 1988), an m-vector of values for various local features is measured at each location, and objects of various types are associated with different regions of the m-space. This vector can include features computed at different scales, or a feature that is the optimal scale at that location, so this method can be implemented as a multiscale object definition strategy.

An example of methods of this type can be found in the work of Coggins 1985, 1986b]. Convolution with a sequence of m multiscale filters computes an m-vector for each pixel composed of the intensity levels of the m filtered images at corresponding locations in each filtered image. The pattern of responses to the filters at each pixel describes the pixel's relationship to its neighborhood. Scale extrema in the m-vector as well as spatial extrema in each filtered image provide useful information about the image's content. Some simple objects can be identified and measured directly from these patterns without an explicit image segmentation.

Coggins Coggins, 1986a; Packard, 1986] has shown not only the effectiveness of this idea, but also its generalization to maxima across other dimensions than scale, such as orientation. If the filter decomposition is nonspecific, more filters may be needed to obtain the same accuracy as with filters designed and tuned to the most critical aspects of the image. Also, the inference mechanism may need to be more complex since the object may be identified by the appearance of features at different scales and, in particular, relative spatial relationships to each other.

5.2. Object definition by structural pattern recognition

Structural methods represent objects by a graph whose nodes describe object substructures and whose arcs describe relationships between these substructures. The substructures may in turn be described by graphs, but at some stage they consist of feature measurements or coded representations of primitive structures. Structural methods recognize objects by matching a graph describing a prototype object with the graph obtained from the image. Thus, object recognition maps into a variant of the graph isomorphism problem (Read, 1977). Syntactic pattern recognition (Fu, 1982) is a variant in which the graph is reduced to a sequence and the prototype is represented as a grammar. This approach maps object definition into the problem of parsing a language.

Structural approaches often founder either by being excessively time-consuming (graph isomorphism is an NP-complete problem) or by being error-prone because of the required explicit representation of structural aspects of the object. Structural methods are likely to fail if some relevant structural feature is omitted from the model, or if the placement of structures does not quite conform to that expected in the model, or if the image introduces visual structure (e.g., shadows) that is not part of the model.

Multiscale descriptions for producing the graph or string can help with these problems. First, if matching proceeds top-down by scale, coarse matches can not only be efficient themselves but can guide matches at lower scales, limiting the computational complexity of the matching. Second, since important items in the graph can be expected to appear at large scale, minor details that may not match correctly can affect the matching at small scale, where their low importance can be weighted lightly in determining the match.

There is a problem with this scheme when the hierarchical descriptions have the property that a region corresponding to a parent node does not simply represent the union of the subregions that are its children, but also an additional part (as shown in Fig. 7). This is a property of the hierarchies produced by following geometric structures to annihilation. With such hierarchies a small change in an object in an image can exchange the role of two components, one being part of a larger scale component and the other being a subcomponent (as shown in Fig. 7). A solution to this problem, somehow allowing both components to be seen as subcomponents, at least with some probability, must be found by future research.

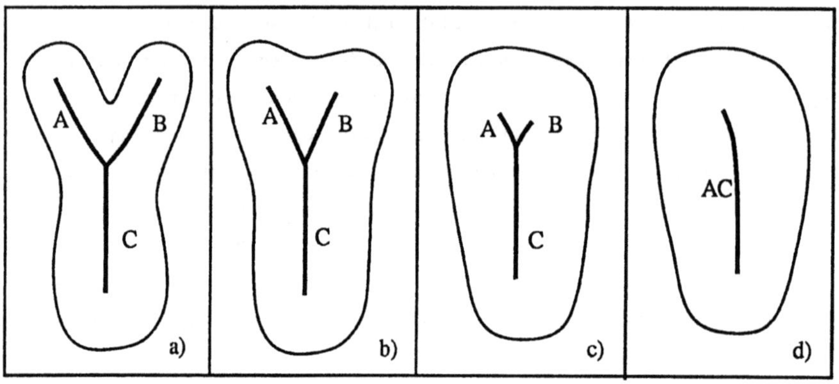

Figure 7: Effect of scale increase on figure 7a through stages b, c, and d illustrates the creation of a hierarchy with parent AC with child B. If region B is made slightly longer, scale increase will cause the branch A to annihilate sooner, so region A will appear in the description, and region B will not.

5.3. Object definition by a user via hierarchical descriptions

There seems to be more hope, at least in the short term, of letting the human user define objects than having the computer do it automatically. For user-based object definition to be quick, the image description should be to a reasonable degree in human terms, thus allowing the human easily to manipulate the components of the description into objects that match the user's understanding of semantics. The hope of multiresolution image descriptions is that they will generate such descriptions.

Therefore, a multiscale image description should generate both image regions that are visually coherent to the user and an organization of those regions into a hierarchy or simple graph. If such a description were fully successful, the user would be able to pick semantically meaningful regions simply by pointing to a pixel in them and then, if the region is a subregion, asking for the next larger containing region an appropriate number of times.

It is reasonable to hope for a multiresolution image description largely to meet these needs, but it is probably realistic to expect that it would at least have the following kinds of faults:

1. Regions might have small numbers of extra or missing pixels at their boundaries.

2. A semantically sensible region might not exist in the description but rather be joined, e.g. by a narrow isthmus, with another semantically sensible region (that may or may not be in the description) to form a region that does appear in the description.

3. A semantically sensible region does not appear as a region in the description but rather is a union of regions appearing at various positions of the hierarchy. The need to combine these regions probably reflects an error in the hierarchy.

Lifshitz (Lifshitz, 1988; Pizer, 1988a) has shown in prototype and Coggins et al 1988] are developing further an interactive display tool with the following functions that can allow the fast definition of objects, even from descriptions that have these faults.

1. Given a region or set of regions selected from the image description, it displays that region and its relation to the the original image data.

2. Given a pixel (or voxel) pointed to by the user, it displays the smallest region in the image description containing that pixel; given a displayed region, under command it displays the parent (containing) region in the description hierarchy.

3. Given two or more selected regions, it displays the union or difference of the regions, and under command modifies the hierarchy to reflect this logical operation.

4. It allows hand editing of regions, and the splitting of a region into two, given the painting out of joining pixels.

It remains to be demonstrated whether such a tool together with multiresolution image descriptions discussed in this paper can provide convenient 2D and 3D segmentation of medical images.

6. SUMMARY

A variety of types of multiscale methods for image segmentation have been reviewed. While these methods have shown great promise and deserve attention because of their attractive conceptual properties and basis in approaches of the human visual system, intensive research on these methods has a history of only a few years. The ultimate promise of these methods and the bases of choice among them remain to be brought out by further research.

ACKNOWLEDGEMENTS

I gratefully acknowledge the helpful comments of Dr. James Coggins, Mr. John Gauch, Mr. Cheng-Hong Hsieh, and Dr. William Oliver and the continuing collaboration with Prof. Jan Koenderink. I thank John Gauch, Carolyn Din, Sharon Walters, and Sharon Laney for help in manuscript preparation. This paper was written with the partial support of NIH grant number P01 CA47982.

REFERENCES

Back, S "Konturextraktion für Roentgen-Computertomogramme nach dem Korn'schen Multi-scale-Ansatz," Diplomarbeit, Informatics Dept., Technical Univ. of Berlin, 1988.

Ballard, DH and Brown, CM *Computer Vision*, Prentice-Hall, Englewood Cliffs, NJ, 1982.

Bergholm, F "Edge Focusing," *IEEE Trans-PAMI, 9(6)*:726-741, 1987.

Blom, H Personal communication. Department of Medical and Physiological Physics, State University of Utrecht, 1988a.

Blom, H "Geometrical Description With a Jet Space Upon a Multiresolution Base," Internal Report, Department of Medical and Physiological Physics, State University of Utrecht, 1988b.

Blom, H Personal communication. Department of Medical and Physiological Physics, State University of Utrecht, 1987.

Bloomberg, S "Three-Dimensional Shape Description Using The Multiresolution Symmetric Axis Transform." Ph.D. Dissertation, University of North Carolina, Department of Computer Science, 1989.

Blum, H, and Nagel, RN "Shape Description Using Weighted Symmetric Axis Features." *Pattern Recognition 10*:167-180, 1978.

Bomans, M, Riemer, M, Tiede, U, Höhne, KH, "3D-Segmentation von Kernspin-Tomogrammen", *Informatik-Fachberichte 149*: 231-235, 1987.

Bookstein, F "Principal Warps: Thin-Plate Splines and the Decomposition of Deformations." Accepted for publication in *IEEE Trans-PAMI*, 1988.

Brady, M and Asada, H "Smoothed Local Symmetries and Their Implementation." Tech Report, Artificial Intelligence Laboratory, Massachusetts Institute of Technology, 1984.

Burt, P, Hong, TH, Rosenfeld, A "Segmentation and Estimation of Image Region Properties Through Cooperative Hierarchical Computation." *IEEE Trans SMC 11*:802-809, 1981.

Burt, P "The Pyramid as a Structure for Efficient Computation." *Multiresolution Image Processing and Analysis*:6-35, Springer-Verlag, Berlin 1984.

Canny, JF "Finding Edges and Lines in Images." Artificial Intell. Lab., Massachusetts Inst. Technol., Cambridge, MA, Tech Report #720, 1983.

Coggins, J, Cullip, T, Pizer, S "A Data Structure for Image Region Hierarchies." Internal Report, Department of Computer Science, University of North Carolina, 1988.

Coggins, JM, Poole, JT "Printed Character Recognition Using an Artificial Visual System." *Proc of the 1986 IEEE International Conference on Systems, Man, and Cybernetics*, Atlanta, 1986a.

Coggins, JM, Fogarty, KE, Fay, FS "Development and Application of a Three-Dimensional Artificial Visual System." *Journal of Computer Methods and Programs in Biomedicine*, special issue of papers from the Ninth SCMAC, 22:69-77, 1986b.

Coggins, JM, Jain, AK "A Filtering Approach to Texture Analysis." *Pattern Recognition Letters 3*:195-203, 1985.

Colchester, A. et al: "Detection and Characterization of Blood Vessels on Angiograms Using Maximum-Gradient Profiles," this volume, 1988.

Crowley, J & Parker, A "A Representation for Shape Based on Peaks & Ridges in the Difference of Low-Pass Transform," *IEEE Trans. PAMI, 2(2)*:156-169, 1984.

Duda, R and Hart, P *Pattern Classification and Scene Analysis*, Wiley-Interscience Publication, John Wiley & Sons, 1973.

Fu, KS *Syntactic Pattern Recognition and Applications*, Englewood Cliffs, NJ: Prentice Hall, 1982.

Gauch, J, Oliver, W, Pizer, S, "Multiresolution Shape Descriptions and Their Applications in Medical Imaging." *Information Processing in Medical Imaging* (IPMI X, June 1987): 131-150, Plenum, New York,1988a.

Gauch, J and Pizer S, "Image Description via the Multiresolution Intensity Axis of Symmetry," *Proc. 2nd Int. Conf. on Comp. Vis.*, IEEE, 1988b.

Grossberg, S and Mingolla, E "Neural Dynamics of Perceptual Grouping: Textures, Boundaries, and Emergent Segmentations." *Perception and Psychophysics 38(2)*:141-171, 1985.

Hoff, W "Multi-Resolution Techniques in Stereo Vision." To appear in *IEEE Trans-PAMI*, 1988.

Hsieh, CH "A Connectionist Algorithm for Image Segmentation", Ph.D. Dissertation, University of North Carolina, Department of Computer Science, 1988a.

Hsieh, CH "A Connectionist Model for Corner Detection," Internal Report, University of North Carolina, Department of Computer Science, 1988b.

Jain, AK, Dubes, RC *Algorithms for Clustering Data*, Englewood Cliffs, NJ: Prentice Hall, 1988.
Jolion, J, Meer, P, Rosenfeld, A "Border Delineation in Image Pyramids By Concurrent Tree Growing." Tech Report #349, Center for Automation Research, University of Maryland, 1988.
Jaffard, S, Lemarie, PG, Mallat, S, & Meyer, Y "Multiscale Analysis," Internal Report, Centre Math., Ecole Polytechnique, Paris, 1986.
Koenderink, J.J., *Solid Shape*, MIT Press, Cambridge,1989.
Koenderink, JJ "Scale-Time," *Biological Cybernetics, 58*:159-62, January 1988a.
Koenderink, JJ "Operational Significance of Receptive Field Assemblies," *Biological Cybernetics, 58*:163-71, 1988b.
Koenderink, J & van Doorn, AJ "Representation of Local Geometry in the Visual System," *Biological Cybernetics, 55*:367-375, 1987.
Koenderink, J & van Doorn, AJ "Dynamic Shape," *Biological Cybernetics, 53*:383-396, 1986.
Koenderink, J "The Structure of Images," *Biological Cybernetics, 50* 363-370, 1984.
Korn, A "Combination of Different Space Frequency Filters for Modeling Edges and Surfaces in Gray-value Picctures." *SPIE 595*:22-30, *Computer Vision for Robots,* 1985.
Lee, SJ Personal communicaton. University of North Carolina, Department of Computer Science, 1988
Leyton, M "A Process-Grammar for Shape." *Artificial Intelligence 34* 213-247, 1988.
Lifshitz, L & Pizer, S "A Multiresolution Hierarchical Approach to Image Segmentation Based on Intensity Extrema," *Information Processing In Medical Imaging* (IPMI X, June 1987):107-30, Plenum, 1988.
Lifshitz, LM *Image Segmentation Using Global Knowledge and A Priori Information,* Ph.D. Dissertation, UNC Chapel Hill Technical Report 87-012, 1987.
Mallat, S "A Theory for Multiresolution Signal Decomposition: The Wavelet Representation," Tech Report MS-CIS-87-22, University of Pennsylvania, May 1987.
Marr, D *Vision,* Freeman, San Francisco, 1982.
Marr, D & Hildreth, EC "Theory of Edge Detection." *Proc R Soc Lond,* B 207:187-217, 1980.
Meer, P, Baugher, S, Rosenfeld, A "Hierarchical Processing of Multiscale Planar Curves." Tech Report #245, Center for Automation Research, University of Maryland, 1986.
Nackman, LR, Pizer, SM "Three-Dimensional Shape Description Using the Symmetric Transform I: Theory." *IEEE Trans-PAMI 7(2)*:187-202, 1985.
Neumann, H "Extraction of Image Domain Primitives with a Network of Competitive/Cooperative Processes." To appear in *Proc German Workshop on Artificial Intelligence,* W. Hoeppner, ed., Springer-Verlag, 1988.
Oliver, W & Bookstein, F Personal communication. University of North Carolina, Department of Computer Science, 1988.
Packard, C "Recognition of Neural Impulse Trains." MS Thesis, Computer Science Department, Worcester Polytechnic Institute, 1986.
Pizer, S, Gauch, J, Lifshitz, L "Interactive 2D and 3D Object Definition in Medical Images Based on Multiresolution Image Description," *SPIE 914*:438-444, 1988a.
Pizer, S, Gauch, J, Lifshitz, L, Oliver, W "Image Description via Annihilation of Essential Structures." UNC Chapel Hill Technical Report 88-001, 1988b.
Read, RC, Cornell, DG "The Graph Isomorphism Disease." *Journal of Graph Theory 1*:339-363, 1977.
Richards, W, Hoffman, DD "Codon Constraints on Closed 2D Shapes." *Computer Vision, Graphics, and Image Processing 31*:265-281, 1985.
Rosenfeld, A "Pyramid Algorithms for Efficient Vision." Tech Report #299, Center for Automation Research, University of Maryland, 1987.

Rosenfeld, A *Multiresolution Image Processing and Analysis*, Springer-Verlag, Berlin 1984.

Sher, A, Rosenfeld, A "Detecting and Extracting Compact Textured Objects Using Pyramids." Tech Report #269, Center for Automation Research, University of Maryland, 1987.

Stiehl, HS Personal communication. University of Hamburg, 1988.

Terzopoulos, D, Witkin, A, & Kass, M "Symmetry-Seeking Models for 3D Object Reconstruction," *Proc 1st Int. Conf. on Comp. Vis.* (London, England), IEEE Catalog #87CH2465-3, 1987.

Terzopoulos, D "Multilevel Computational Processes for Visual Surface Reconstruction." *Computer Vision, Graphics, and Image Processing* *24*:52-96, 1983.

Toet, A, Koenderink, JJ, Zuidema, P, Graaf, CN "Image Analysis - Topological Methods." *Information Processing in Medical Imaging* (Proc. IPMI VIII, 1983), Martinus Nijhoff, 1984.

Witkin, A "Scale-Space Filtering," *Proceedings of International Joint Conference on Artificial Intelligence*, 1983.

Young, RA "Simulation of Human Retinal Function With The Gaussian Derivative Model." *Proc. IEEE CVPR*, pp. 564-569, 1986.

Yuille, AL & Poggio, T "Scaling Theorems for Zero-Crossings," MIT A.I. Memo 722, 1983.

VOXEL-BASED VISUALIZATION OF MEDICAL IMAGES IN THREE DIMENSIONS

Höhne, K.H.; Bomans, M.; Pommert, A.; Riemer, M.; Schiers, C.; Tiede, U.; Wiebecke, G.

University Hospital Eppendorf, Hamburg
Federal Republic of Germany

ABSTRACT

Multi-slice images obtained from computer tomography and magnetic resonance imaging represent a three-dimensional image volume. For its visualization we use a ray-casting algorithm working on a gray scale voxel data model. This model is extended by additional attributes such as membership to an organ or a second imaging modality ("generalized voxel model"). It is shown that the combination of different surface rendering algorithms together with cutting and transparent display allow a realistic visualization of human anatomy.

1. INTRODUCTION

An ever-increasing number of medical diagnostic images are obtained from X-ray Computed Tomography (CT), Magnetic Resonance Imaging (MRI), and Positron Emission Tomography (PET), which produce sequences of two-dimensional (2D) cross-sectional slices. The current predominant method of analyzing these images is by sequential observation of individual 2D slices and the viewer's subsequent 'mental reconstruction' of three-dimensional (3D) relationships. Computerized reconstructions of CT scans have produced 3D perspective display of bony anatomy that have proved clinically useful in craniofacial surgery and orthopedics (Hemmy et al. 1983; Templeton et al. 1984; Vannier et al. 1984; Yasuda et al. 1984; Boecker et al. 1985; Chen et al. 1985; Herman et al. 1985; Witte et al. 1986). The procedures used in these applications are limited by the fact that only predefined surfaces (mostly bone) can be visualized and that all other information is not used or lost during processing. A few recent investigations have dealt with the software and hardware problems of displaying 3D tomographic volumes that preserve the entire original gray scale data and thus allow a detailed exploration of the volume (Goldwasser et al. 1985; Jackel 1985; Lenz et al. 1986; Kaufman 1986; Höhne et al. 1987). On the basis of data from computer tomography and MRI we have developed, investigated and compared a variety of methods for visualizing gray level volumes. This paper is a modified version of a previous paper (Höhne et al., 1989).

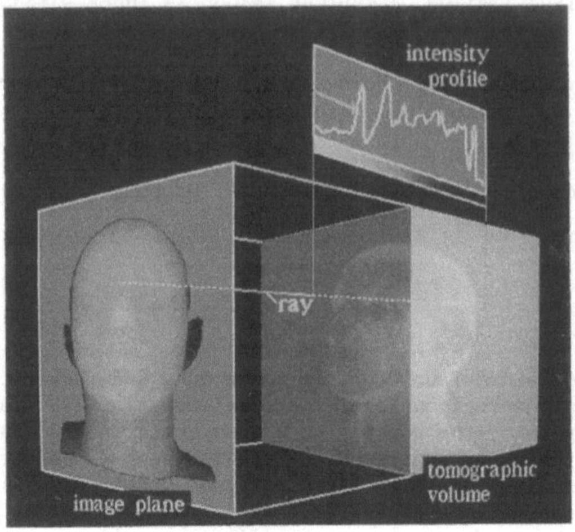

Figure 1. Principle of the ray casting algorithm. A parameter obtained from the intensity profile along the ray is projected to an image plane.

2. METHOD AND RESULTS

2.1 Data structure

The raw data is a spatial sequence of image matrices of 256^3 or 512^3 pixels. To save storage space the gray level data of the original volume is compressed to a dynamic range of 256 gray values. To achieve cubic volume elements a linear interpolation of the intensity values between the original slices is performed. The thus obtained 3D array ("voxel-model") is the basic data structure for most of the described algorithms. As an extension each voxel may not only contain a gray value but also further attributes such as membership to an organ or an intensity value delivered by an additional imaging modality ("generalized voxel-model"). In the present implementation each voxel can be described by up to 16 bits.

2.2 Projection strategy

When scanning the volume we basically have the choice between two classes of strategies: One class consists of the object space oriented methods that scan along lines or columns of the 3D-array and project a chosen aspect onto an image plane in the direction of view. These are known as

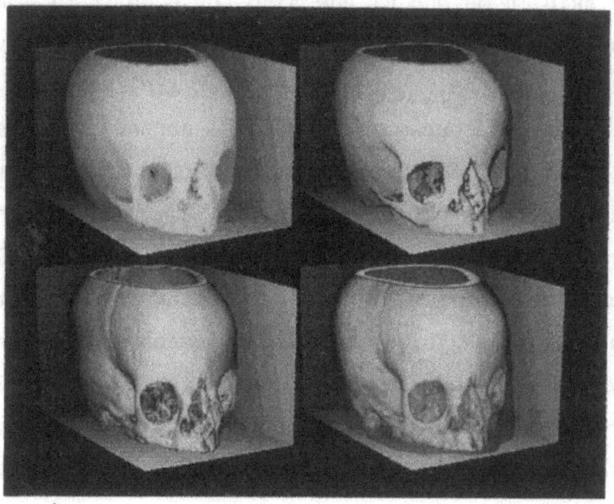

Figure 2. Skull of a child with a facial cleft reconstructed from 61 computer tomograms a) Z-buffer shading, b) Z-buffer gradient shading, (arrow points to the deformed nasal septum), c) gray level gradient shading (arrows point to a suture and the orbital cavity respectively), d) transparent gray level gradient shading.

back-to-front (BTF) or front-to-back (FTB) methods. They have proved to be reasonably fast when a pure surface display of a single object is required. When, however, volumetric properties such as translucency are to be visualized, image space oriented methods that scan the image volume along the viewing direction are unavoidable. The computing overhead of this algorithm referred to as ray casting (see fig. 1) is larger, especially when the whole data volume does not fit into the main memory of the hardware. Therefore we use in our project a front-to-back algorithm for the visualization of pure surfaces. For all other projections we rotate the whole volume such that scanning for a desired viewing direction can be done along the lines of the array. Thus the time consuming coordinate transformation has to be performed only once per viewing direction. The gray level assignment in the rotated volume is done by tri-linear interpolation. Optionally, the volume can be distorted such that a central projection is achieved. All pictures in this article have been produced this way.

2.3 Surface rendering

The classical approach of visualizing a volume is the display of surfaces contained in it. The easiest way of determining the voxels that represent a surface (segmentation) is intensity thresholding. This works quite well for the outer skin in MR and CT and also for bone in CT. As well known a raw image of the thus determined surface can be produced by imaging the negative distance to the observer (Z-buffer (Z-) shading, see fig. 2a). For realistic images, of course, the surface normal vectors have to be determined. A still not too time consuming method is the estimation of the surface normals from the Z-buffer (Chen et al. 1985) in our project further referred to as Z-buffer gradient (ZG-) shading. We have implemented a modified version of this algorithm (Tiede et al. 1987). This algorithm delivers fairly realistic images, but it suffers from the fact that the dynamic range of surface angles is low (fig. 2b). This is because they are computed on the basis of the position of surface voxels in a 3x3 neighbourhood, which does not allow a large number of choices.

Much better results can be obtained if the gray scale data are utilized for the determination of the surface normals. As a consequence of the tomographic data acquisition technique the gray values in the neighbourhood of a surface voxel reflect the relative average of the various (usually two) tissue types (air/skin, soft tissue/bone) in the voxels immediately adjacent to the voxel in question. These relative volumes are related to the surface inclination. Thus the gray level gradient can be considered as a measure for surface inclination. This idea of gray level gradient (GG-) shading has been described by (Barillot et al. 1985) and independently by (Höhne and Bernstein 1986; Tiede et al 1987 and 1988). The algorithm published by (Lorensen et al. 1987) is working on the same basis. Our procedure is as follows:

Given the gray level of a surface voxel at a location i,j,k the gray level gradient is computed as:

$$G_x = g_{(i+1,j,k)} - g_{(i-1,j,k)}$$
$$G_y = g_{(i,j+1,k)} - g_{(i,j-1,k)}$$
$$G_z = g_{(i,j,k+1)} - g_{(i,j,k-1)}$$

The components of the surface normals are normalized as

$$N_u = \frac{G_u}{\sqrt{(G_x)^2 + (G_y)^2 + (G_z)^2}}, u = x, y, z$$

The gradients are typically computed either from 6 central neighbours in the 3x3x3 neighbourhood or from all 26 neighbours. In the latter case the algorithm is identical to the Zucker-Hummel operator for edge detection (Zucker and Hummel 1979). The images of this paper have been computed with 26 neighbors. Using the normal vectors the object can be shaded according to any shading model such as Phong shading which has been used throughout this paper.

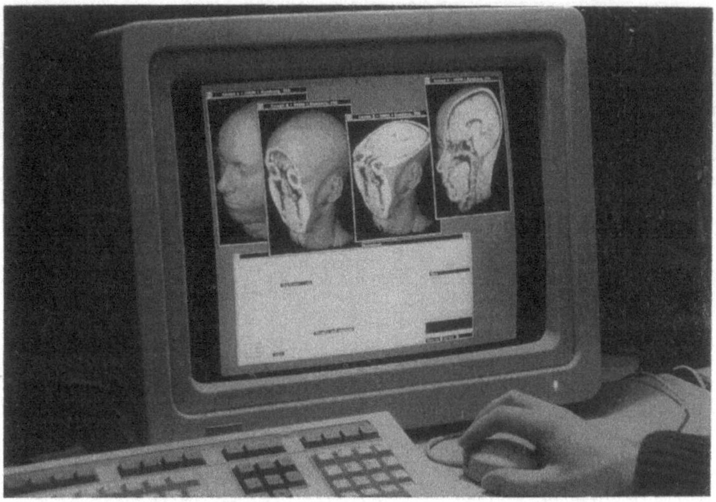

Figure 3. Menu surface of the program VOXEL-MAN with a demonstration of the possibilities of arbitrarily cutting MRI-objects.

Due to the high dynamic range of the gray levels a continuous shading is now possible that leads to a more realistic impression of the objects. Small details not visible with the Z buffer gradient shading such as the suture of a skull now become recognizable (fig. 2c). In addition, aliasing effects present in ZG-shading, especially during rotation, do not occur. In the case of small objects, however, such as a thin bone in the orbita or the nasal septum, this method leads to artefacts. This is due to the fact that in this case the gray level is no longer governed by the membership to an object class but by the thickness of the object. As a consequence neither the gray level threshold is representative for a surface nor the gradient for its inclination. It has been claimed (Levoy 1988) that in such cases where an exact segmentation is difficult a semi-transparent modeling of the object could improve the visualization. In his algorithm a transparency value is assigned to each voxel depending on its gray level and on its gray level gradient. Also here the shading is done using the gray level gradient method. Our preliminary experience with this transparent gray level (TGG-) shading method shows that small details on an otherwise compact surface (e.g. bone sutures) can show up even better (see fig. 2d). However, the really problematic regions such as the orbital cavity or the deformed nasal septum are not recognized better, although the image looks smoother. An additional problem of the TGG-method is the fact that two parameters can be choosen arbitrarily: the assignment of a transparency to the gray levels and the weight of the gradient. So a very broad spectrum of visual impressions can be obtained from a single object and it is hard to judge what is the truth.

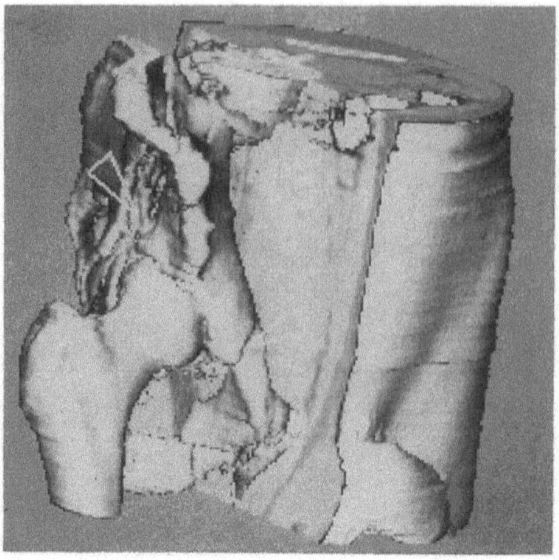

Figure 4. View of a male pelvis with multiple fractures (arrows) from 77 CT slices (bone and skin gray level gradient shading).

Since it is obviously difficult to judge surface rendering algorithms without knowing the exact reality we have started a study on quantifying image quality. Geometrically simulated objects undergo the same data acquisition and processing steps as the real ones. Global and local error measures for the surface location and inclination are computed. For further results we refer to (Pommert et al. 1989a and 1989b).

2.4 Cut planes

Once a surface view is available, a very simple and effective way of visualizing interior structures is cutting. In the ray casting environment it is easy to carry out. In our implementation arbitrary oblique cuts can be specified through three points on the visible surface or on an already existing cut (see fig. 3). A special case is selective cutting, where in the case of multiple objects surfaces may be excluded from cutting (fig. 4).

2.5 Transparent visualization

Our experience with 3D-visualization of tomographic volumes shows that if exact surfaces can be determined non-transparent rendering yields the best perception. This is especially true when no additional depth cue such as

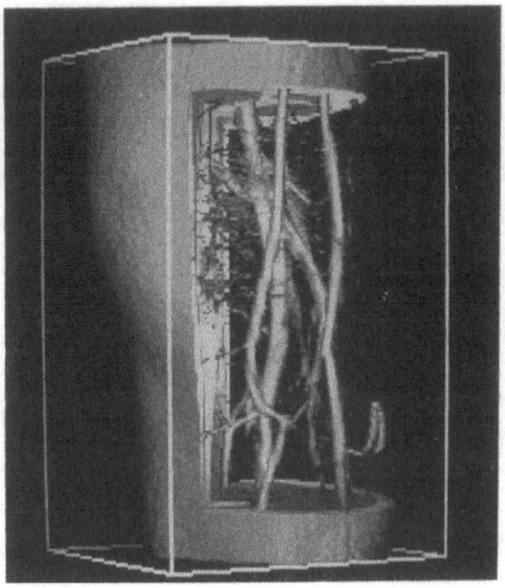

Figure 5. View of a human knee with its blood vessels reconstructed from 64 MRI slices. The vessels are rendered using transparent gray level gradient shading.

motion or binocular stereo can be used. If an exact surface definition by segmentation is not possible, because there is not enough contrast or the objects are too tiny (as in the case of blood vessels) the only way out is transparent visualization. In the case of the visualization of a beating heart from MRI data we used the TGG-algorithm because of segmentation problems. A manual segmentation would have been hopeless since the data set contained 28 slices for each of 12 heart phases. The visual impression (fig. 6) is fairly good, yet we do not know how precisely it reflects reality.

A similar problem arises in the case of small vessels. Even if their contrast is good, the neighbourhood is not large enough for the determination of a 3D-surface. Transparency, however, provides some kind of averaging leading to a satisfactory visualization as shown in fig. 5, where the vessels have been rendered using the TGG-algorithm. If the vessel diameters become of the order of one voxel width any shading must fail. In this case the visualization of the maximum intensity along the ray has proved to yield surprisingly good results.

2.6 Improvement through extended object descriptions

So far the definition of the objects displayed has been made by thresholding the intensity profile while traversing the volume. A more

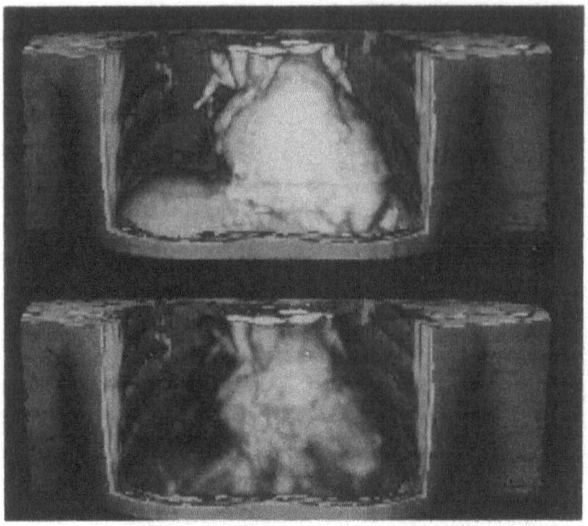

Figure 6. Views of a human heart reconstructed from 28 MRI slices. For the heart itself transparent gray level gradient shading has been used with low (upper image) and high (lower image) transparency.

detailed object definition cannot be achieved through the analysis of the intensity along the ray since we need a larger 3D-neighbourhood for the decision on whether a voxel belongs to a surface. Another possibility is to gain this information from an additional source. In the following subsections (2.6.1 and 2.6.2) we describe two kinds of extended descriptions, organ labels gained through segmentation and the addition of a second imaging modality.

2.6.1 Segmentation

For the determination of object surfaces other than bone and skin segmentation algorithms are necessary that take larger voxel neighbourhoods into account than is possible in the 1D-ray casting case. In order to find the intensity changes that represent surfaces we applied a 3D-extension of the Marr-Hildreth operator (Marr and Hildreth 1980), which is defined by

$$I'(x,y,z) = \nabla^2(I(x,y,z) * G(x,y,z,\sigma))$$

where ∇^2 is the Laplace operator, I the image volume, and G the Gaussian function. The zero crossings of I' are considered as surfaces. These are by definition closed which is important for 3D-display. We have applied this method to MRI data of the head (Bomans et al. 1987 and 1989). After binarizing the filtered volume I' the obtained regions are labeled

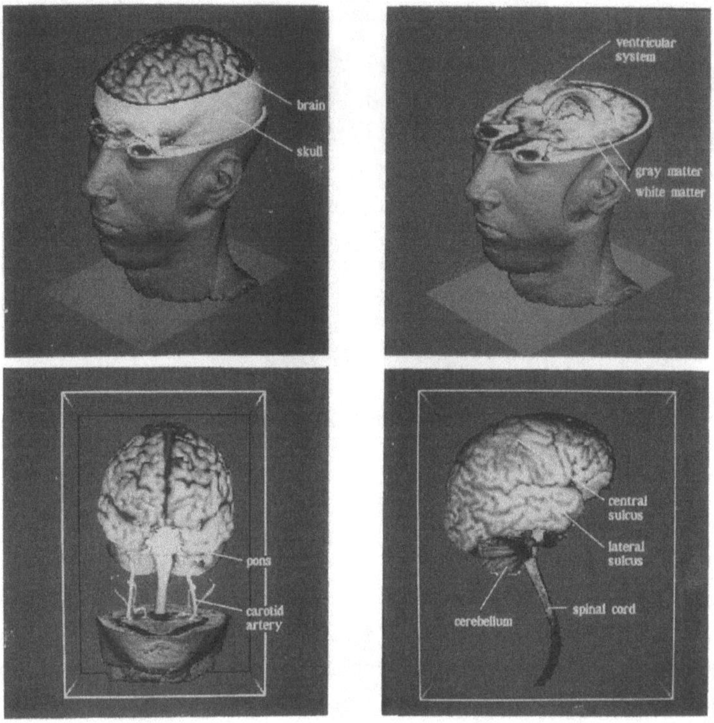

Figure 7. Views of a human head reconstructed from 128 MRI slices. The surfaces of bone, brain, and ventricular system have been determined by the Marr-Hildreth operator. Shading: skin, bone, and ventricle: gray level gradient shading; brain and carotid artery: "integral shading".

according to their correspondence to different constituents (skin, bone, brain etc.). In our current procedure this is done interactively for each slice. The regions found do not always correspond to anatomical structures. Errors had to be corrected by removing wrong connections or by inserting new surface elements. For the segmentation of the brain cortex, for example, we correct typically ten 2D-contours. For the 3D-contour of the cortex we apply in addition a 3D-dilation with a subsequent erosion. This procedure destroys small details of the contour. If we use, however, the thus obtained 3D-contour as a baseline for a semi-transparent display - here just an integration over 6 pixels ("integral shading") - we obtain images as shown in fig. 7.

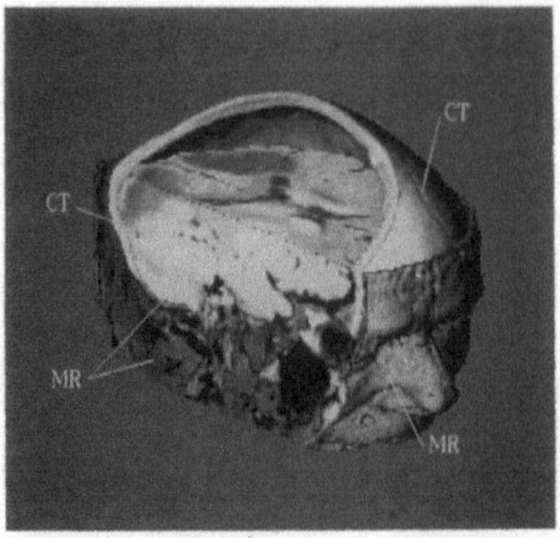

Figure 8. View of a head of a cadaver with bone taken from CT and tissue from MRI.

2.6.2 Multiple imaging modalities

A 3D-image if generated from a single imaging modality shows only special aspects of the reality. Improved information can be obtained by combining data of different modalities such as CT, MRI or PET. This way, e.g. bone structures may be described better by CT while soft tissue properties are better represented by MRI. Generally the different data sets do not match geometrically. It is therefore necessary to register the volumes to each other. As a basic tool we have developed a 3D-specifier: corresponding landmarks specified on surface images of MR or CT volumes serve for the computation of polynomials performing the required 3D-distortion (Schiers et al. 1989). Preliminary experience shows that this kind of specification needs a well trained user and sometimes more than one attempt to arrive at a satisfying match. Fig. 8 shows a thus obtained image with the combined properties of MRI and CT.

3. IMPLEMENTATION ASPECTS

The described algorithms are implemented within the program system VOXEL-MAN on a VAX 11/780 (24 mbytes of main memory) and VAX-station II/GPX (16 mbytes of main memory). The rotation of a volume of 256^3 voxels takes between 15 and 30 minutes. The described projections take between 10 and 60 seconds. Such times are certainly not sufficient for clinical work but can be tolerated in a research environment. In the meantime a subset of the algorithms has been implemented on a processor manufactured by the KONTRON Corporation. Here a view from any direction shaded with gray level gradient shading takes between 5 and 10 seconds. If we take into account the fast progress in hardware development, computing speed will not be a major issue in the future.

The main problem to be solved for a broad application is the design of an appropriate user interface. Presently VOXEL-MAN has two interfaces: One is a language interface that allows the specification of the desired action through a string of parameters. It has full flexibility, but it needs an expert to choose the right combination from a choice of more than 30 parameters. The other interface is a menu interface that is certainly suitable for the beginner. However, not all specifications can be expressed suitably in this form (see fig. 3).

4. APPLICATIONS

3D-imaging, especially of bone, has for several years proved to be useful in craniofacial surgery (fig. 2), traumatology (fig. 4), and orthopedics. The possibilities shown in this paper lend themselves to further applications.

In diagnostic radiology the assessment of arbitrary cross-sections is facilitated. Although of course radiologists are well accustomed to mentally projecting 3D-visualizations from multiple 2D-images, these techniques facilitate rapid and sure orientation particularly in instances where the angle of slice is unusual or unconventional. Once the outer surface and/or bone is visibly defined in three dimensions, the viewer can always be sure about the position of the cross-section he is looking at.

It is obvious that neurosurgical therapy will benefit. The relationships between regions to be treated (e.g. tumors) and those which have to be conserved (e.g. blood vessels) can be assessed prior to an operation (fig. 5). For broad application, however, the procedures still remain too complicated.

The high realism of the images gained from living persons make them suited for the computer aided instruction in anatomy. Here we can afford the high expenditure of work, because we have to process only a small number of specimens, from which we can make any number of copies. Since the times of Leonardo da Vinci, anatomical instruction material was painted. Computer graphics may replace those paintings to a large extent (see fig. 7).

5. CONCLUSION

We have demonstrated that by using a ray casting algorithm and the generalized voxel-model natural and detailed 3D-images of human anatomy can be obtained. We can draw the following detailed conclusions:

- Surface rendering is decisively improved by using gray level gradients for the determination of the surface normals.

- Semi-transparent surface rendering is recommended only in cases where the surfaces cannot be determined exactly (e.g. vessels).

- At least in the case of MRI more detailed object definitions can be achieved by segmentation or multi-modality imaging, thus enabling the display of multiple objects.

- The combination of multiple surfaces with different kinds of shading and cuts through the volume enables a detailed visualization of tomographic volumes, especially if it can be done interactively.

- As computing power will not be an issue in the future, the problem to be solved is the design of a user interface, that would allow one to easily and consistently specify the various possible manipulations we have described.

ACKNOWLEDGEMENTS

The authors would like to thank Prof. Dr. Dr. Wolf-Joachim Höltje (Dept. of Craniofacial Surgery), Prof. Dr. Wolfgang Schulze (Dept. of Anatomy), Dr. Jürgen Wening (Dept. of Traumatology), and Dr. Gerd Witte (Dept. of Radiology) for many discussions. We also thank Ellen Vaske and Rainer Schubert for their help. We are grateful to Siemens (Erlangen) for providing the original MRI-volume data from which the perspective views have been produced. The images of the mummy were produced in collaboration with the Institute of Anthropology and the Department of Neuroradiology, University of Tübingen. The tomograms of the cadaver have kindly been provided by the Dept. of Neuroanatomy, Medical University Hannover. The investigations were supported in part by the Deutsche Forschungsgemeinschaft and the Werner-Otto-Foundation Hamburg.

REFERENCES

Barillot C., Gibaud B., Luo L.M., Scarabin I.M. (1985). 3-D Representation of anatomic structures from ct examinations, Biostereometrics '85, Proc. SPIE 602:pp307-314.

Boecker F.R.P., Tiede U., and Höhne K.H. (1985). Combined use of different algorithms for interactive surgical planning, In: Lemke U. (ed) Computer assisted radiology. Springer, Berlin New York Tokyo, pp572-577.

Bomans, M., Riemer M., Tiede U., Höhne K.H. (1987). 3-D Segmentation von Kernspintomogrammen, 9. DAGM-Tagung Braunschweig, In: Informatik Fachberichte 149, Springer, Berlin, pp231-235.

Bomans M., Höhne K.H., Riemer M., Tiede U. (1989). 3D-segmentation of MR-images of the head for 3D-display, IEEE Trans. Med. Imaging, im press.

Chen L.S., Herman G.T., Reynolds R.A., Udupa J.K. (1985). Surface shading in the cuberille environment, Computer Graphics and Applications 5, pp33-43.

Goldwasser S.M., Reynolds R.A., Bapty T., Baraff D., Summers J., Talton D.A., and Walsh E. (1985). Physicians workstation with real time performance, Computer Graphics and Applications 5, pp44-57.

Hemmy D.C., David D.J., Herman G.T. (1983). Three-dimensional reconstruction of craniofacial deformity using computed tomography, Neurosurgery 13, pp534-541.

Herman G.T., Vose W.F., Gomori J.M., Gefter W.B. (1985). Stereoscopic computed threedimensional surface displays, Radio Graphics 5, pp825-852.

Höhne, K.H., Bomans, M., Pommert, A., Riemer, M., Tiede, U., Wiebecke, G. (1989). 3D-Visualization of tomographic volume data using the generalized voxel model, The Visual Computer, in press.

Höhne K.H., Bomans M., Tiede U., Riemer M. (1988). Display of multiple 3D-objects using the generalized voxel-model, Proceedings SPIE 914: pp850-854.

Höhne K.H., DeLaPaz R.L., Bernstein R., Taylor R.C. (1987). Combined surface display and reformatting for the 3D-analysis of tomographic data, Investigative Radiology 22:pp658-664.

Höhne K.H., Bernstein R. (1986). Shading 3D images from CT using gray level gradients, IEEE Transactions on Medical Imaging 5:pp45-47.

Jackel D. (1985). The graphics PARCUM system: A 3D memory based computer architecture for processing and display of solid objects, Computer Graphics Forum 4:pp21-32.

Kaufman A. (1986). Voxel based architectures for three-dimensional graphics, Proc. IFIP'86:pp361-366.

Lenz R., Danielsson P.E., Cronström S., Gudmundson B. (1986). Interactive display of 3D medical objects, In: Höhne KH (ed) Pictorial information systems in medicine, Springer, Berlin New York Tokyo, pp449-468.

Levoy M. (1988). Display of surface from volume data, IEEE Computer Graphics & Applications 8:pp29-37.

Lorensen W.E., Cline H.E. (1987). Marching Cubes: A high resolution 3D surface construction algorithm, Computer Graphics 21:pp163-169.

Marr D., Hildreth E.C. (1980). Theroy of edge detection, Proc. R. Soc. Lond. B 207:pp187-217.

Pommert A., Bomans M., Tiede U., Höhne K.H. (1989a). Investigations on image quality for 3D-display techniques, In: Lemke HU et al. (eds), Computer Assisted Radiology (Proc. CAR '89), Springer, Berlin, in press.

Pommert A., Bomans M., Tiede U., Höhne K.H. (1989b). Simulation studies for quality assurance of 3D-images from computed tomograms, In: Todd-Pokropek A, Viergever MA (eds): The Formation, Handling and Evaluation of Medical Images. NATO ASI Series F, Computer and Systems Sciences, Springer, Berlin, this volume.

Schiers C., Tiede U., Höhne K.H. (1989). Interactive 3D-registration of image volumes from different sources, In: Lemke HU et al. (eds), Computer Assisted Radiology (Proc. CAR '89), Springer, Berlin, in press.

Templeton A.W., Johnson J.A., Anderson W.H. (1985). Computer graphics for digitally formatted images, Radiology 152, pp527-528.

Tiede U., Höhne K.H., Riemer M. (1987). Comparison of surface rendering techniques for 3D tomographic objects, In: Lemke U (ed) Computer Assisted Radiology, Springer, Berlin New York Tokyo, pp599-610.

Tiede U., Riemer M., Bomans M., Höhne K.H. (1988). Display Techniques for 3D-Tomographic Volume Data, In Proc. NCGA '88, Vol. III, Anaheim, pp188-197.

Vannier M.W., Marsh J.L., Warren J. (1984). Threedimensional CT reconstruction images for craniofacial surgical plannning, Radiology 150, pp179-184.

Witte G., Höltje W., Tiede U., Riemer M. (1986). Die dreidimensionale Darstellung computertomographischer Untersuchungen kraniofacialer Anomalien, Fortschr. Röntgenstr. 144,4, pp24-29.

Yasuda T., Toriwaki J., Yokoi S., Katada K. (1984). Threedimensional display system of CT images for surgical planning, Int. Symposium on Medical Images and Icons. Silver Spring MD: IEEE Computer Society, pp322-327

Zucker S.W, Hummel R.A (1979). An optimal three-dimensional edge operator, McGill University, Report, pp79-10

PERCEPTION AND DETECTION OF SIGNALS IN MEDICAL IMAGES

A.E. Burgess

University of British Columbia, Vancouver
Canada

ABSTRACT

We are interested in how well the human visual system performs signal detection, discrimination, identification and recognition tasks given noise-limited medical images. This paper will discuss the use of statistical decision theory concepts to understand both fundamental and human limits of decision accuracy. In particular, the IDEAL OBSERVER concept will be emphasized.

The five major sections of the paper include an introduction to visual system design and operation; theoretical aspects of signal detection; methods of measuring observer performance; a review of the medical imaging literature on observer performance; and some example applications to system design problems.

INTRODUCTION

The study of human visual system function and performance covers a vast area of knowledge from neurophysiology to psychophysics. Most of the work concentrates on questions related to <u>how</u> the visual system functions at a detailed operational level. This subject is fascinating in its own right. However our main interest as medical image scientists is <u>how well</u> the system performs and <u>what</u> can we do with our images to improve performance. The performance we study is decision accuracy. We are particularly interested in noise-limited performance for two reasons. First, fluctuations due to noise represent the most fundamental limitation of decision accuracy. Limitations such as resolution and contrast can be overcome by technical improvements. Secondly, noise is the limiting factor in all medical imaging modalities - poisson noise due to quantized x-ray and gamma ray photons, speckle noise due to scattering in ultrasound and thermal noise in MRI systems. These noises can be reduced by longer imaging times but they cannot be eliminated.

This presentation will use the concepts of statistical decision theory in general and the narrower field of signal detection theory in particular.

Signal detection theory was developed for analysis of radar systems and was first applied to human sensory problems (vision and hearing) by a University of Michigan group in the 1950's. This work is summarized by two classical texts by Green and Swets (1966) and Swets (1964). This approach was first used in radiology by the University of Chicago group in the late 1960's and early 1970's (Lusted, Rossman Metz, Goodenough, Starr) with example references (Lusted, 1968; Starr et al, 1975; Metz, 1978).

Signal detection theory was used in analysis of a variety of psychophysics experiments during the 1960's. It fell out of favour in the vision research community and came to be regarded as a failed theory of human visual system operation. This was a misunderstanding of the purpose of the theory. It is in fact a correct theory of ideal observers for well defined tasks.

The approach emphasized in this presentation emphasizes the concept of the ideal observer. This hypothetical decision device makes the best possible use of prior knowledge together with new image data and reaches the highest possible decision accuracy given noise limitations. This approach was first suggested by Tanner and Birdsall (1958) which introduced two measures: the detectability index, d', and observer efficiency. The detectability index went on to wide-spread popularity and is very well known. The efficiency measure did not suffer a similar fate. It was used in a few audition papers (Simmons, 1973; Creelman, 1961) in the late 1950's where efficiencies were generally below 1%. The measure was revived by Barlow in the mid 1970's. Barlow was interested in using psychophysical techniques with random dot patterns to probe the visual cortex. He expected to encounter tuning to spatial patterns. This tuning had been seen in the retina using spatially continuous gray-scale signals at both a cellular level (electrophysiological measurement of cell firing) and at a global psychophysical level. Spatial tuning at the cellular level had been found by Hubel and Weisel (1977). Barlow used random dot patterns because the dots would by-pass the retinal spatial tuning. Barlow did not find evidence for cortical tuning at a global, psychophysical level (1978). However he did find very high efficiency (50%) for detection and discrimination of patterns. These experiments had been done using Poisson noise statistics. Barlow's work sparked my own interest in the efficiency measure. In 1979 we used random dot patterns with Gaussian statistics to separately measure inefficiency due to spatial tuning and inefficiency due to internal noise (Burgess and Barlow, 1983). We found that all human observer losses for the random dot tasks could be accounted for by internal noise. This internal noise would be the external manifestation of any form of observer inconsistency.

Section 1 will give an overview of the design problems and solutions of the front end of the visual system. Section 2 is a brief introduction to signal detection theory. Section 3 will briefly describe some experimental methods. Section 4 will provide a short survey of experimental results. Section 5 will describe some applications to system design.

1. OVERVIEW OF THE VISUAL SYSTEM

This overview will be given from the point of view of communications engineering. As a quick summary, one can state that in most cases the visual system deals very well with design problems. The presentation will begin with the optics of the eye, work backward through to the retina, then the optic nerve fiber to the visual cortex, and then off into cognitive processing. More than a decade of research has led to a general consensus that the human visual system contains a number of different classes of spatial mechanisms or channels (a concept due to Robson and Campbell) that process the retinal image in parallel. This hypothesis underlies much of the following discussion.

Four very good general references are suggested. A book by Uttal (1981) covers virtually everything one might contemplate about the visual system. The second book is the proceedings of a recent Royal Society of London symposium (Braddock and Sleigh, 1983) with contributions from physiology, psychophysics, and artifical intelligence. The third reference is a very readable review article by Regan (1982) on channels in the visual system. Finally there is an excellent book on cognitive aspects of vision by Gregory (1970).

1.1 Fundamentals

The visual system uses a narrow portion of the earth's atmospheric electromagnetic window in the 500 nanometer wavelength region. The low frequency end of the spectral sensitivity is limited to avoid excessive thermal noise triggering of the photoreceptors (Vos and VanNorren, 1984). The high frequency end is limited by corneal filtering to prevent ultraviolet damage to molecules (Ham et al, 1982). Hence the wavelength range of photoreceptor operation is determined. Efficient photoreceptor operation requires an optical waveguide design and this immediately restricts (Snyder, 1975) photoreceptor diameter to the 1 to 2 micron range (for all visual systems). Synder (Land, 1981) showed that the equations for diffraction limited lens operation, photoreceptor spacing distance, and the Nyquist sampling theorem can be combined to indicate that the eye's optics should have a focal length/aperture diameter ratio of 8. Most vertebrate eyes operate at f/8 under high light level conditions. Focal length is the parameter showing greatest diversity and eye size is approximately proportional to body length (Kirshfeld, 1976). This may be due to the fact that larger organisms must be able to see further for hunting or self-defense and therefore need better angular resolution. Scaling arguments can be used to show that compound lens design is best for eyes smaller than a few millimeters while a simple lens design is best for larger eyes (Kirshfeld, 1976). Hence one sees that the basic design of eyes is subject to severe physical constraints if optimum results are to be achieved. Nature seems to adhere to optimum design. One great uniformity of design is still unexplained. All visual systems use the same molecule (retinol) and the same chemical transition (11-cis to all-trans) as the basic photochemical transduction element.

Fig. 1.1 (from Barlow, 1983)

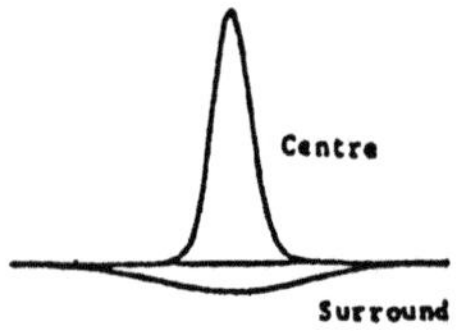

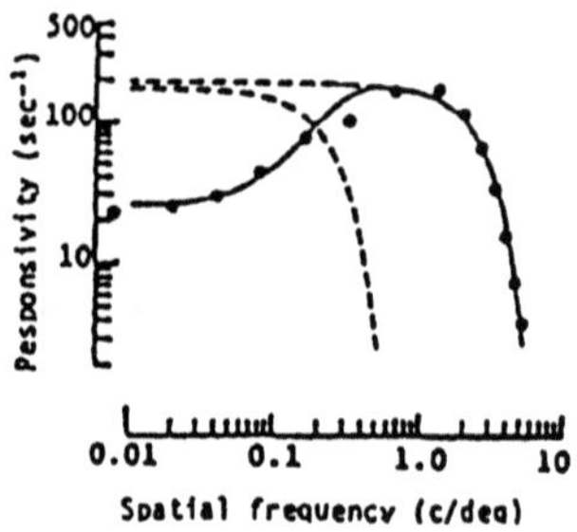

Fig.1.1A (left). Gaussian weighting functions of centre and surround of a retinal ganglion-cell receptive field.

Fig. 1.1B (right). The sine-wave responsivity function of a typical retinal X-cell fitted by the difference ot two Gaussian functions. The responsivity of the surround mechanism is shown by the lower dashed line, that of the centre by the upper dashed line and its continuation to the right as a full line. In this example the diameter of the surround is about ten times that of the centre and its responsivity at zero spatial frequence about 0.83 that of the centre.

Fig. 1.2 (from Pratt, 1978)

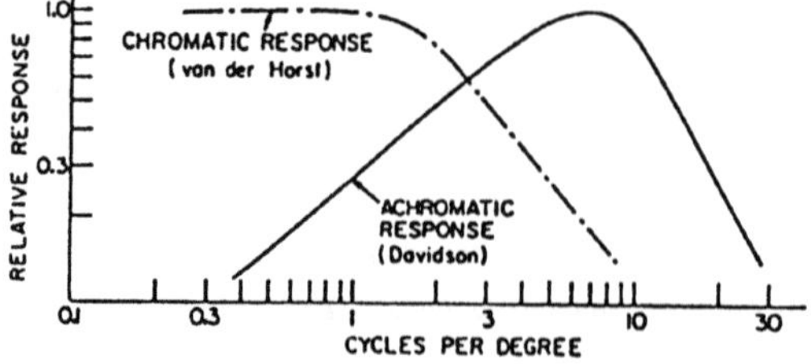

Spatial frequency response measurements of human visual system

1.2 Retinal Filtering and Sampling

The eye's optics has a focal length of about 17 mm with an f-stop range from f/2 with the iris wide open to f/8 under normally bright viewing conditions. With the iris open the optical quality of the eye is very poor. With the iris stopped down to f/8 the lens system is essentially diffraction limited. Under these relatively high light level conditions, the optical transfer function cut-off frequency is about 60 cycles/degree. The image at the retina is continuous and must be sampled. We know from communication theory that the optimum strategy is to sample at 2 samples/cycle at the highest spatial frequency present. Suppose for a moment that the retina were to attempt to sample the entire field of view at this sampling rate (120/degree). Using a field of view of 180 x 180 degrees, this would give a total of 4.6×10^8 samples. There are a total of 10^8 photoreceptors in the retina, so it could conceivably approach this ideal sampling. Of course, the retinal image is changing in time and the visual system is capable of responding up to a temporal frequency of 60 Hz which would require 120 samples/second at each photoreceptor. Therefore an optimally sampled retina would be sending a total of 5.5×10^{10} samples/second to the cortex. This rather large bandwidth is simply not available at the optic nerve. So data reduction strategies must be implemented. One is to use a high resolution fovea and low resolution periphery. The highest spatial acuity is available at the fovea where sampling is done at the Nyquist limit of 120 samples/degree. Outside the fovea, the spatial sampling distance increases rapidly and there is a potential for aliasing. The retina even has a strategy for anti-aliasing, by providing a random jitter to the location of the receptive fields. The high resolution region at the fovea consists of photoreceptors connected one per ganglion cell, with a total of about 25,000 ganglion cells within the central 1 degree. At the periphery of the retina roughly 100 photoreceptors are wired to one ganglion cell.

Each optic nerve fiber is driven by a separate ganglion cell and there are only about 10^6 ganglion cells in the retina. Therefore, the visual system must use some data compression strategies at the retina to reduce the amount of data being sent to the visual cortex. The strategies used include provision of several types of retinal cells and an inhomogeneous distribution of cells. Many retinal cells are spatially connected to ganglion cells through an on-center off-surround receptive field arrangement which has the effect of acting as spatial bandpass filter. This has the benefit of removing the DC and low spatial frequencies from transmissions by most of the ganglion cells. Some examples of retinal cell spatial frequency response are illustrated in figure 1.1 (from Watson, 1983). One final caveat--the contrast sensitivity curves one encounters (see figure 1.2 from Pratt, 1978) are not visual system MTF plots.

1.3 Dynamic Range Problems

The visual system faces a severe dynamic range problem in order to operate over a luminance range of a factor of 10^{13}. This is achieved by four separate processes. One is to use two different types of cells (rods and cones) varying in their sensitivity to light. Second, chemical gain control provides for most of the dynamic range but it is relatively slow in temporal response. Third, fast temporal response can be achieved by adjusting the iris opening to rapidly cover a range of brightnesses of about a factor of 64. In spite of these strategies there is still a dynamic range problem because, under the best conditions, the visual system can respond to spatial luminance variations of about 1 part in 500. This response can take place in a relatively short period of time, considerably less than 1 second. Ganglion cells transmit their information up the optic nerve fibers using a pulse code modulation method with a limit of about 300 pulses/second. This rate is clearly inadequate to transmit subtle variations in brightness that are changing quickly in time. Here, the benefits of spatial and temporal bandpass filtering (removing of low frequencies and DC) can be clearly seen because most ganglion cells need merely transmit changes in space or changes in time. This strategy is a form of predictive coding (Srinivasan et al). Of course, the visual system now faces a problem of having the visual scene disappear completely if it is static. So the rapid random motion of the eye, the saccades, are quite important. Spatial vision disappears rapidly in a few seconds if images are artifically stabilized on the retina and one only perceives a spatially unmodulated brightness under these conditions.

1.4 Visual Cortex

From experiments on cats, Hubel and Weisel (1977) found arrays of tuned cortical cells whose receptive fields are arranged in hierachies of complexity (figure 1.3 and figure 1.4 from Barlow, 1983). There is considerable redundancy of connections in the visual cortex. For example, the 25,000 photoreceptors in the fovea are connected to an equal number of ganglion cells and from there to an equal number of optic nerve fibers. However, when these fibers arrive at area 17 of the cortex there is an

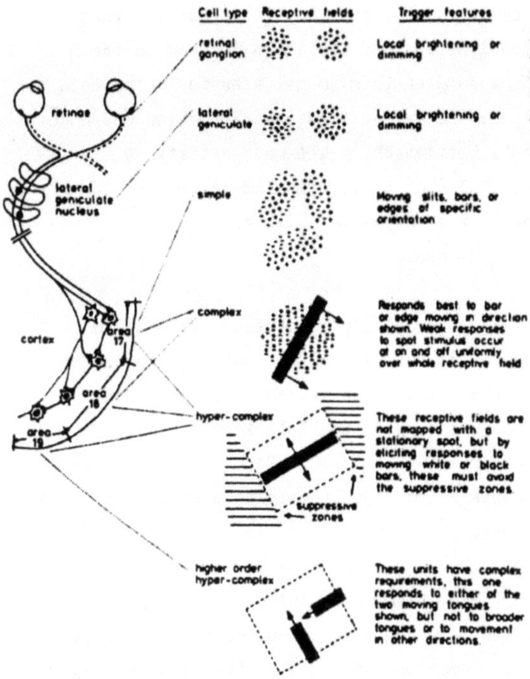

Maps of the receptive fields that can be obtained from neurones at different stages in the visual pathway.

Fig. 1.3 from Creelman (1961)

Another view of the cortical representation can be derived from considering somewhat more explicitly the receptive fields of the cortical cells

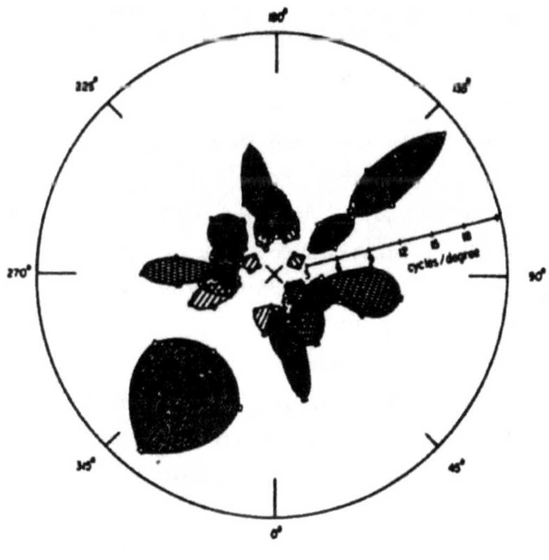

Half-maximum response contours of the most frequency-selective cells in the parafoveal projection of monkey striate cortex stimulated by drifting gratings of various spatial frequencies and orientations.

Fig. 1.4 from Creelman (1961)

expansion to about 2.5 million cells, i.e. about 100:1 redundancy. What are all these cells doing? Recall again that the image focused on the retina is continuous. The retina samples this image, transforms the data, and transmits it to the cortex. In the retina, the transformation was done because the optic nerve was a data bottleneck. The most satisfying suggestion from recent psychophysical and physiological measurements is that the data is being converted at the cortex to an internal representation for further use by the brain.

To make clear what is meant by this internal representation, one has to spend a few minutes talking about basis function sets. One example is the familiar concept of picture elements (pixels) to represent sampled data from images. These pixels are non-overlapping spatial elements that provide an orthogonal normalized basis function set. Each pixel is given a value (coefficient) which represents a gray level associated with that pixel.

The set of M x N pixels completely specifies the sampled image. However, this spatial element set is not the only basis function set that one might use. For example, one might take a Fourier transform of the image and use the coefficiencts of non-overlapping spatial frequency elements. In this case, the basis function set consists of sines and cosines of differing spatial frequencies. Physiological and psychophysical evidence suggests that neither of these alternatives is in fact used by the visual system.

What principle might the visual system use? A reasonable assumption is that the visual system gives equal weight to one's ignorance of where an object is in the visual space and how big it is. Therefore, one would expect basis functions that are localized both in space and in spatial frequency. Many of you will be familiar with a quantum mechanical analogy to this process. Arrays of coupled mass points act as harmonic oscillators. When one solves the differential equations for this set of oscillators, one gives equal emphasis to one's ignorance about the position of each mass and the momentum of each mass element (Heisenberg uncertainty), the solution is a set of Hermite polynomials. This is another orthonormal basis function set. The orthogonality condition has the effect of ensuring that the set has a minimum number of basis function elements so there is no redundancy in the values of the coefficients.

There have been a number of suggestions put forward as to how an internal representation might be achieved. Gabor in 1948 suggested a basis function set for hearing consisting of sine waves having gaussian envelopes. This set of basis functions would not be orthogonal; however, the overlap is only about 10% which is a tolerable redundancy. Marr (1982), taking a different approach, suggested that bandpass filters and zero-crossing detectors would be a suitable method of constructing a basis function set and determining coefficients. Sakitt and Barlow (1982) put forward a model for the first stage of cortical transformation of the visual image based on the assumption of a minimum number of channels to encode the retinal information. They proposed that the two spatial dimensions would be covered by a total of 256 basis functions with angular sensitivities ranging from 7.5 degrees at high spatial frequencies down to about 120 degrees at low spatial frequencies. From there, they estimated that a total of 1.36 million input samples from the retina would be required, which is in good agreement with modern estimates of the number of fibers in the optic nerve. Sakitt and Barlow go on to then ask the question "How does this process help us to see?". One cannot seriously contemplate a visual system that just repeatedly retransforms retinal data in a linear manner because this is not going to get the brain any closer to a final understanding of the visual scene. On the contrary, every step must be in the right direction and one should be able to pick up hints on what this direction is from each step. Why should the brain be organized in this way? Without knowing more about the next stages of processing, it is hard to say anything concrete about the virtues of this arrangement although it is similar in some respects to the scheme devised by Marr. Marr used the concept of a primal sketch which is used for subsequent analysis of pattern and form. Sakitt and Barlow suggested that perhaps by using general descriptive terms, the brain is performing an early step in recognizing the parts of an image that belong to each other.

Watson (1983) proposed a mathematically computable model which assumed that the observer makes optimal use of the available information. As feature sensors, he used two-dimensional spatial weighting functions that are sensitive to spatial frequency and orientation. Five feature dimensions were required for each spatial weighting function sensor. He took into account the fact that spatial sampling is not isoplanatic, and defined sensor gains for particular spatial frequencies. Finally, he assumed that an optimum Bayesian classifier was operating at the end of this system. The job of the classifier is to examine the feature vectors generated and to decide which of several possible images are in fact being presented. This optimal Bayesian classifier is of course able to use a priori information. Watson was able to achieve reasonably good agreement between theoretical predictions and a number of experimental results.

1.5 Cognitive Processing

The front end of the visual system delivers data to a very sophisticated and intelligent cognitive processing system. A great deal of auditory and visual experimental data is consistent with the view that the brain does Bayesian hypothesis testing to understand sensory information. It can be argued that infants spend a lot of time learning how to interpret the sensory world. Gregory (1970) is an eloquent advocate of this hypothesis testing point of view and applies it to the study of visual illusions. Research on noise limited visual signal detection and discrimination suggests that humans are very good at doing this "maximum likelihood" hypothesis testing.

2. THEORETICAL ASPECTS OF SIGNAL DETECTION

In order to understand the theoretical aspects of signal detectability one must first of all get an intuitive feel for what noise does to the visibility of signals. The first thing one notices in doing detection tasks with repeated displays of a signal in noise is that sometimes the signal is very easy to see and sometimes it is impossible to see. One also notices from time to time that there are patches of noise that look like the signal. This is, of course, due to chance correlations in the noise. Therefore for most experiments it is much easier to deal with noise that is intrinsically uncorrelated (i.e. "white"). It will be seen later that deterministic correlation in the noise is a very important consideration.

How does one go about representing this variation in signal visibility? The easiest approach is to use a physical measurement scale which, for the sake of a better word, I will call a "correlation value". This measure is a normalized correlation between the signal that was provided and the signal plus noise that occurs in the image (or in the case of chance correlation, the correlation between the signal and the region of noise only). This correlation value will have the same units as signal-to-noise ratio which makes it quite convenient. In fact, one should always remember that the <u>SNR is the average of this correlation value</u> over a very large number of display trials. The numerical value of the correlation for any particular trial represents in a physical way how easy it is to "see" the signal. Figure 2.1 shows a histogram of values for 400 trials in a typical experiment (SNR = 4.9). Because of the method of normalization, the standard deviation of the samples is 1.0. This paragraph has presented a method of representing signal "visibility" on a physical measurement scale. Of course this does not represent exactly what humans see.

Humans looking at the display also see a variation in visibility of the signal from trial to trial and presumably there is some relationship between physical measurements of correlation and the human interpretation of visibility. The assumption that one makes is that there is monotonic (and perhaps "noisy") mapping from a physical measurement scale to a sensory measurement scale. At the moment one does not know how to determine this monotonic mapping. The best that one can do is to infer characteristics of this mapping from measurements of human performance. A number of references on signal detection theory are listed in the references (Bunch et al, 1977; Hanson, 1979; Hanson, 1981; Raemer, 1969; Swets and Pickett, 1982; Vantrees, 1968; Wagner, 1977; Wagner et al, 1979; Wagner and Brown, 1985) in addition to those previously mentioned (Green and Swets, 1966; Swets, 1964).

2.1 Ideal Observers

The ideal observer uses a Bayesian decision strategy to calculate the a posteriori probability of a signal being present (VanTrees, 1968). This strategy allows the observer to combine known prior probabilities of possible signals with present image data. The procedure can also be considered to be a form of hypothesis testing. The general mathematical formuation is

$$p(H_i|\mathbf{x}) = \frac{p(H_i)p(\mathbf{x}|H_i)}{p(\mathbf{x})} \tag{1}$$

where $p(H_i|x)$ is the a posteriori probability of the ith hypothesis, H_i, being correct given the data, $\mathbf{x}$ (in vector notation). The other terms are the prior probability $p(H_i)$, the conditional probability of the data $p(x|H_i)$ in the case where the hypothesis is correct and the probability of the data, $p(x)$.

The observer then selects some decision rule in order to choose between alternative hypotheses. The problem is that if the image data is noisy the ideal observer will still sometimes make decision errors. If the observer knows the costs of various decision errors it can base its decision rule on some cost/benefit analysis (VanTrees, 1968). If the costs are not known then the optimum rule is to select the hypothesis with the maximum *a posteriori* probability (MAP). This MAP strategy requires knowledge of the prior probabilities. If the prior's are not known then the optimum strategy is based on the maximum likelihood ratio (ML) where the ratio is defined by

$$L_i(\mathbf{x}) = \frac{p(\mathbf{x}|H_i)}{p(\mathbf{x}|H_o)} \tag{2}$$

where H_o is the hypothesis that only noise is present.

On occasion the observer may be asked to make a binary YES/NO decision about whether a signal is present given one data vector ($\mathbf{x}$). The ideal observer then compares the *a posteriori* probability or the likelihood ratio to a "threshold" decision criterion. There are several ways of selecting a threshold (VanTrees, 1968). One example is the Neymann-Pearson method of maintaining the false-positive error at some pre-selected value.

The analysis of observer performance can be very complex (VanTrees, 1968). The discussion to follow will be restricted to the simple example of signals with copmletely specified parameters with no correlation between alternative signals. The noise will be assumed to be uncorrelated and has a gaussian probability density function. (As an aside, it should be carefully noted that these are two quite distinct properties of noise.) Noise can have a gaussian probability density function but be correlated. If one takes uncorrelated gaussian noise and passes it through a linear filter the noise PDF remains gaussian, but correlations are introduced due to the transfer function of the filter.

2.2 Signal known exactly (SKE) case

Suppose that one adds a signal with known properties to a known location in the uncorrelated noise. This has the effect of introducing local correlation in noise values. The ideal observer simply measures the cross-correlation between the expected signal and the actual data present at the signal location. In order to do this the ideal observer must make a number of independent samples of data in the region of the signal location then multiply the value of the sample found at each location by a weighting function which is equal to the expected value of the signal at that point. The observer then sums all of these weighted samples to obtain a number. This number I will refer to as the correlation value. This procedure is

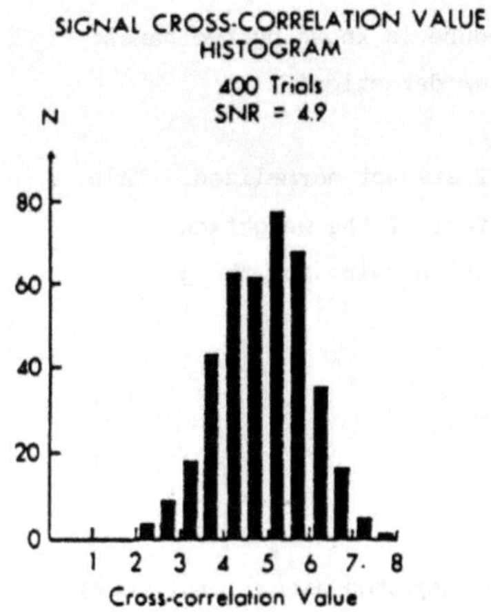

Fig. 2.1

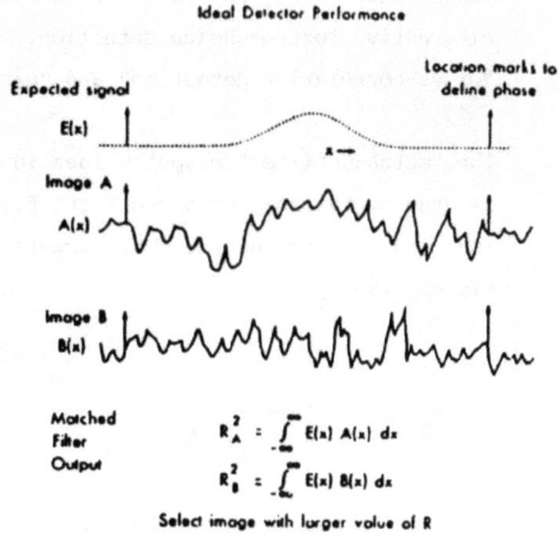

Fig. 2.2

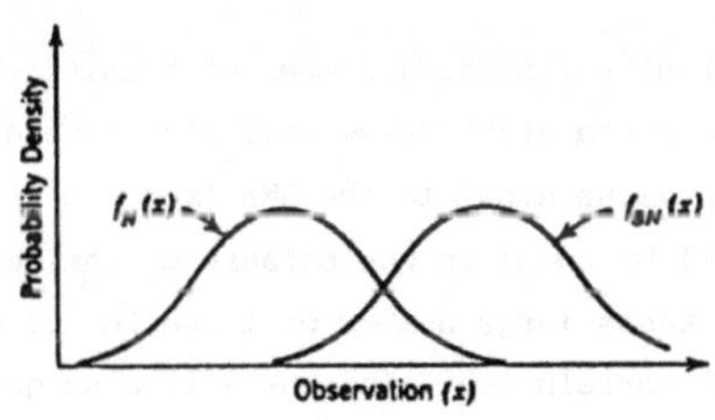

The probability density functions of noise and signal plus noise.

Fig. 2.3 from Swets (1964)

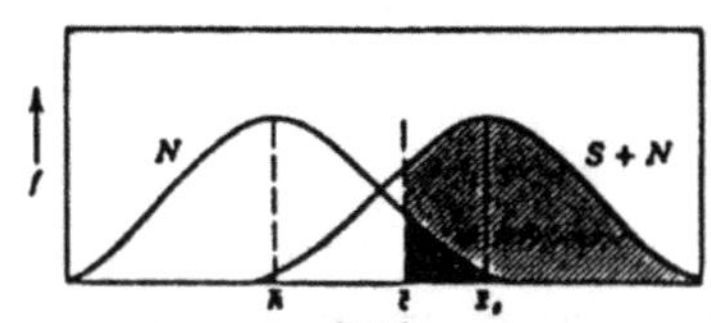

Probability density of the neural activity x, for noise alone and for signal plus noise; $\bar{c}$ is the criterion.

Fig. 2.4 from Swets (1964)

illustrated in figure 2.2 for a continuous one-dimensional example of two-alternative forced-choice detection. The procedure is known by two names: "cross-correlation detection" and "matched-filter detection".

The "matched-filter" output values in figure 2.2 are not normalized. This is done by taking into account the filtering effect of the weighting function upon the noise. The normalized correlation value for the jth signal display is

$$R_j = \frac{\Sigma_i e_i a_{ij}}{[N_o \Sigma_i e_i^2]^{1/2}} \quad (1)$$

where e_i is the weighting function (actually the expected signal amplitude) at the ith point, a_i is the image data value at that point ($a_i = e_i + n_i$, where n_i is a noise sample), N_o is the two-sided spectral density of the noise. Care is needed here since many authors [including Green and Swets (1966)] use a one-sided spectral density and factors of 2^k crop up -- k being the number of dimensions in the power spectrum.

Now consider the limit of an infinite number of trials. Because the noise has a gaussian PDF the correlation value will also have a gaussian distribution with mean value equal to the SNR (more on this below). The standard deviation will be equal to one because of the normalization in equation 1. If one takes a large number of image locations that have only noise and measures the correlation value one will also get a gaussian distribution. However since the noise has zero mean this gaussian distribution will be centered upon zero. Hence one has two gaussian distributions with an overlap that is determined by the SNR. These distributions are illustrated in figure 2.3 from Swets (1964). This double gaussian plot is the starting point for all simple discussions on signal detection theory. The subsequent analysis depends on details of the detection and discrimination tasks which the observer is asked to perform.

2.3 Signal-to-Noise Ratio

At this point it would be useful to define what is meant by SNR. The value of R_j in equation 1 is a random variable because of the noise component. The noise is defined to have zero mean (with reference to local mean luminance) so the ensemble average of normalized correlation over a large number of trials R_j is given by

$$SNR = \sqrt{\frac{E}{N_o}} = \langle R_j \rangle = \left[\frac{\Sigma e_i^2}{N_o} \right]^{1/2} \tag{2}$$

This definition of signal-to-noise ratio is a unique one in the sense that it represents precisely the detectability index, d', of the signal. Throughout all of these notes the term SNR means exactly what is stated in equation 2. For example, it is not the ratio of the highest possible signal amplitude with the standard deviation of the noise, a definition of SNR that appears from time to time in the literature but in fact should be referred to as "dynamic range". Another inappropriate SNR definition uses standard deviation per pixel as a noise measure. Early investigations of CT reconstruction algorithms used this noise measure and came to misleading conclusions about the efficacy of convolution kernals that introduced noise smoothing. Using this definition one also reaches silly conclusions--such as SNR "thresholds" less than one. Excellent discussions of SNR calculations are found in papers by Wagner et al (Wagner et al, 1979; Wagner and Brown, 1985).

2.4 2AFC Experiments

The simplest of all is a two-alternative forced-choice task. In this task the observer is given two noise fields. One of the noise fields contains a signal with a known size, shape and intensity placed in a known position. The observer is then asked to indicate which of the two noise fields contains the signal. With reference to figure 2.3, the observer on each trial is presented with two correlation values. (Since at this point we are dealing with ideal observers, they have no problem in measuring the values.) One value is from the signal plus noise distribution and one value is from the noise only distribution. The best strategy that the observer can use is to decide which alternative has the higher value on this physical measurement scale and answer that it is in fact the signal. However looking back again at figure 2.3, one can see that there is an overlap of the two distributions and so on occasion the value from the noise only distribution will be greater than the value from the signal plus noise distribution. On these occasions, the ideal observer will respond incorrectly. One can calculate the probability of incorrect response using a quite straight forward calculation. One goes to each point in the signal plus noise distribution and determines the probability at that point of having a noise correlation value that is greater. One then integrates over all possible values of the signal plus noise distribution and obtains the probability of correct response for that SNR. These relationships have been tabulated by Elliott in an appendix of Swets' book (1964). Only a few numbers will be given here to give you a feel for the variation.

SNR	Probability of Correct Response (2AFC)
0	50%
1	76%
2	92%
3	98.3%

2.5 Yes/No Experiments

In this type of experiment the observer is presented with one noise field rather than two. In this noise field there may or may not be a signal. The signal size, shape and intensity and possible location are also exactly known. Referring once again to the two probability density distributions in figure 2.3 the observer is presented now with one correlation value. This value has been taken either from the signal plus noise distribution or the noise only distribution. The observer does not know which one and his task is to decide. In order to make this decision the ideal observer has to select some criterion value on the physical measurement scale of visibility. Once this criterion value, which we shall call "c", has been selected the ideal observer's approach is straightforward. If the measured value is greater than c it says "yes, a signal is present". If the value is less than c it says "no, a signal is not present". There are two independent variables: the SNR which represents the average detectability of the signal and the criterion value, c, which is the dividing point between reponses of "yes" and responses of "no".

The decision situation is shown in figure 2.4. The value of c is shown and the shaded areas to the right of c represent the probability of a signal sample or noise sample being found with a correlation value greater than c. We see that the observer can make two kinds of mistakes. It can say "yes a signal is present" when in fact there is no signal, and it can say "no a signal is not present" when in fact a signal is present. One can see that as the value of c is changed, both the probability of correct response and the probability of false positive responses will change.

The covariation of these two response probabilities is shown in figure 2.5. This is the very well known receiver operating characteristic (ROC). There are solid curves in this diagram. One, the diagonal line from lower left to upper right, represents chance performance. This occurs when SNR = 0. The other curves represent the operating characteristic for different SNR's. Recalling that there are two independent variables; if only one (SNR) is specified then the ideal observer performance is only constrained to operate on one particular curve. The point of operation on this curve depends on the observer's selected decision criteria. There is very adequate discussion of ROC curves in the Radiological literature so I will not pursue this simple Yes/No experiment any further.

2.6 Multiple Alternative Forced Choice (MAFC) Experiments

The previous two examples (2AFC and Yes/No) were done under the condition referred to as signal known exactly (SKE). In those cases the signal was completely specified and the signal location was precisely known by the observer. We now go on to a slightly more complicated situation where there is some uncertainty about either signal parameters or signal location. Before doing this one very important constraint, orthogonality, must be considered. The purpose of this orthogonality constraint is to ensure that the alternative decisions are statistically independent. This is very important to ensure a reasonably simple calculation. Orthogonality is perhaps most easily understood by considering sharp-edged signals such as discs which will be orthogonal if there is no overlap between signals when they are placed in the alternative locations. Because the noise is uncorrelated and the signals do not overlap, there is no statistical dependence between alternative correlation calculations. A more precise definition of orthogonality is given by

$$\iint f_i(x,y) f_j(x,y)\,dxdy = 0 \qquad (3)$$

The calculation of observer performance is identical for either multiple alternative signals with one location or multiple alternative locations with one signal, or a combination of both. For sake of specificity let us consider the situation of one signal and multiple orthogonal locations and let the number of locations be equal to M.

In multiple alternative forced-choice (MAFC) experiments the observer is given a noise image with M possible signal locations. There is exactly one signal present in the image at one of the locations. The observer's task is to identify that location. Now refer back to the probability density functions in figure 2.3. The observer is presented with one sample from the signal plus noise distribution and M-1 samples from the noise

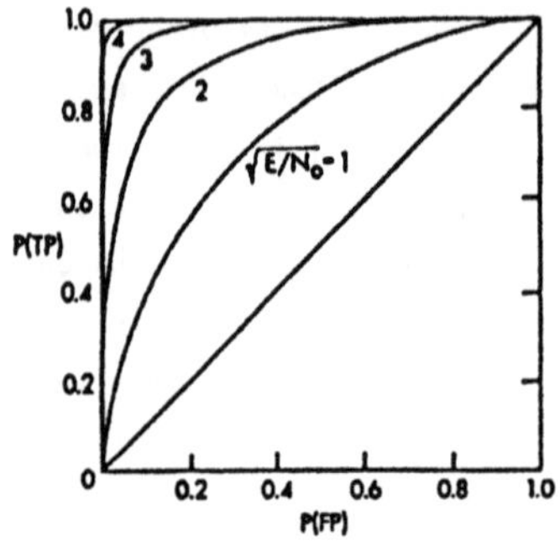

Fig. 2.5: A conventional ROC plot.

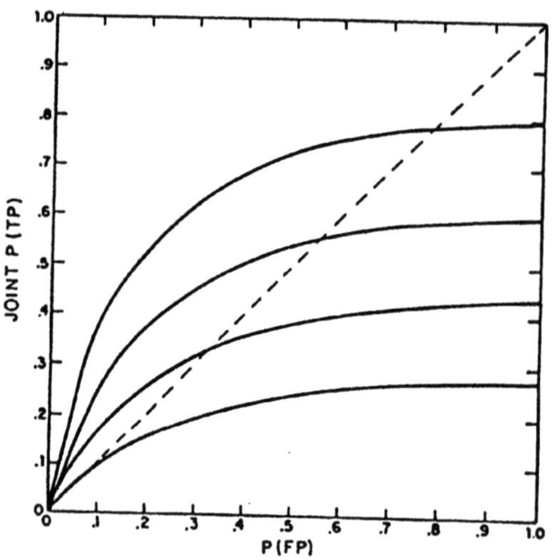

Fig. 2.6: An LROC plot of joint detection and localization probability on the ordinate and false positive probability on the abcissa. The curves are drawn for four different values of M. From Swets and Pickett (1982).

distribution. If the sample from the signal-plus-noise distribution has a larger correlation value than any of the samples from the noise-only distribution, the ideal observer will respond correctly and identify the correct signal location. If the correlation value associated with any one of the noise samples exceeds that of the signal plus noise sample then the ideal observer will respond incorrectly. Using the known distributions and elementary probability theory one can calculate the probability of correct response as a function of signal-to-noise ratio. Probability values for an number of different values of M have been tabulated by Elliott (Swets, 1964). A few of these values are presented in table 2 below for illustration.

TABLE 2

SNR Required For Specified Percent Correct

PC	M = 8	32	256	1024
0.50	1.34	2.09	2.90	3.20
0.75	2.13	2.85	3.64	3.90
0.90	2.84	3.53	4.30	4.53
0.99	4.06	4.70	5.44	5.61

We now briefly turn to multiple alternative Yes/No experiments. In this experiment the observer is once again presented with an image with M possible signal locations. There is a known signal which may or may not be present in the image and if it is present it is one of the M locations. The observer's task is to decide whether a signal is present and he might also be asked to indicate where it is. (It should be noted that analysis of results is different for these two cases.) Once again the observer must select a decision criteria in the same manner as the simple SKE Yes/No experiment. Once again the observer can make two kinds of errors (misses and false alarms). There are now three free parameters. There is the SNR, the decision criteria, and the number of alternative locations. Using the known behaviour for the ideal observer, one can generate multiple alternative ROC curves. These have been referred to in the radiological literature (Lusted, 1968) as LROC curves (Localization Receiver Operating Characteristic). An example of one of these is shown in figure 2.6 for 4 values of M.

2.7 Rose Model

The Rose model has been quite popular in radiology. It derives from his classic 1948 paper and is restated in his 1973 book. Unfortunately, it appears to have been misinterpreted by many users in radiology. Rose devised the model to discuss effects of statistical fluctuation of photon arrival at the retina and subsequently in photon statistics limited images. It is a combinatorial model that derives from work done before development of statistical decision theory and signal detection theory. Its main virtue is that it is easy to understand. However, it has some faults which render it not particularly valuable for assessment of human observer performance.

The Rose model uses an integration strategy. The signal component of the signal-to-noise ratio is determined by integrating the data over a defined area. The standard deviation of the noise component of the SNR is determined from the variance of the noise within the defined integration area. This yields equation 4 for the signal-to-noise ratio.

$$SNR = nA/(nA)^{1/2} \tag{4}$$

where n is the photon density per unit area and A is the signal area. This has been expanded by people in radiology using the form $SNR = \Delta N/N^{1/2}$ where N is the signal contrast. Rose then went on to assess the SNR requirements for certain false positive rates. He selected an arbitrary value of false positive rate and based on this suggested an SNR threshold which he referred to using the letter "k". He suggested that values of k lay in the range from 3 to 5.

A short critique of this model will now be given. It is clear from reading Rose's papers that he was originally presenting a model for an observer that is limited only by image noise. Recall from the previous discussion on signal detection theory that the ideal observer uses a cross-correlation strategy. Rose chose to use an integration strategy for the one special case where integration is equivalent to cross-correlation. The two mathematical approaches are only identical for the simple case of a signal with sharp boundaries, uniform intensity within the boundaries and uncorrelated noise. Therefore one cannot fault Rose for using the model. However, one can certainly fault people in radiology for using the model because it is unusual to find sharp-edged signals or uncorrelated noise. It is quite inappropriate to use the Rose model for computed tomography because of the very significant correlation in the noise. It is also inappropriate to use the Rose model for any signal that does not have sharp edges with uniform contrast within the edges. Finally, the concept of a threshold SNR is a difficult one to apply because, as one knows from statistical decision theory, the SNR requirements depend very much on the task and the degree of accuracy that one wants to achieve. Finally, since the Rose model is implicitly dealing with ideal observers, it is inappropriate to apply it to human observers because one is not certain that the human observer is collecting all the signal information and one knows for certain that the human observer is using a higher noise standard deviation than the ideal observer. However if you insist on using the Rose model it is suggested that you use values of "k" in the range from 5 to 10 for simple detection and 15 to 20 for signal identification tasks.

2.8 Filtered Noise

The discussion up until now has dealt with uncorrelated noise which is an uncommon situation in medical imaging. More commonly one finds either low pass filtered noise or ramp-like noise power spectra for images are reconstructed from projections. In both cases, the noise is correlated by filters and this introduces problems to the analysis. The determination of SNR for filtered noise is dealt with by Wagner et al (1979) and Hanson (1979) and will only be outlined briefly here. When presented with images that have filtered noise, the ideal detector is able to compensate for the correlations introduced by the filtering and it does so by first performing an inverse filter processing to remove the correlations. Once this is done the signal-to-noise ratio is defined by

$$SNR^2 = \frac{1}{N_o} \iint \frac{S^2(u,v)}{P(u,v)} du dv \qquad (5)$$

where S(u,v) is the Fourier transform of the signal and P(u,v) is the power spectrum of the noise normalized to a maximum value of one, u and v are spatial frequencies, and N_o is a spectral density constant.

Analysis of human performance given filtered noise power spectra is complicated by the fact that humans probably can not compensate for the correlations (i.e. humans cannot rewhiten the noise). In this case one can define a cross-correlation detector that cannot rewhiten the noise as a suboptimal pseudo-ideal detector (Wagner et al, 1979). For this cross-correlator that cannot rewhite the noise, SNR is determined by the following equation:

$$SNR^2 = \frac{[\iint S^2(u,v)dudv]^2}{N_o \iint P(u,v)S^2(u,v)dudv} \tag{6}$$

As an example of the effect of being unable to rewhiten noise, if one uses a typical problem of detecting sharped edge disc signals in CT noise one finds that signal-to-noise ratios with and without rewhitening differ by about 30%. The device that can rewhiten the noise achieves better performance.

2.9 Summary

This brings us to the end of the discussion of ideal observer performance for a small variety of tasks. The tasks can be classified in a number of ways. One classification distinguishs between forced choice and Yes/No experiments. In forced-choice experiments exactly one signal is always present in one of M possible locations and the observer's task is to decide in which location. For this type of experiment the observer does not need to hold a stable decision criterion. The optimum strategy in this task is to select the location that has the highest value of correlation with the signal. This decision criteria is known in several ways either as maximum *a posteriori* probability or maximum likelihood. In Yes/No experiments a signal may or may not be present. The observer is asked to decide which is the case and if he believes the signal to be present he may be asked to indicate where it is. For this type of task the observer must have a stable decision criteria. The decision criteria may be strict or it may be lenient. If the criteria is very strict the observer will make a small number of false alarms. If the criteria is very lenient the observer will make a lot of false alarms. The way the observer usually selects a decision criteria is decide first of all on a false alarm probability and then to calculate what decision criteria is needed to give this desired false alarm probability.

The experiments can also be divided into classifications of signal completely specified (SKE) or signal parameter uncertainty. The signal parameter uncertainty can include signal location, signal size, signal shape, etc. As might be expected observer performance is strongly dependent upon the amount of *a priori* information available about the signals.

3. MEASURING HUMAN SIGNAL DETECTION PERFORMANCE

Before considering the sorts of things that humans can do wrong, it is worthwhile considering once again what the ideal observer does correctly. The ideal observer is told what the signal looks like, its size, and shape and intensity, and possible signal locations. If it is necessary to use a decision criteria, the ideal observer selects this precisely and maintains it without any deviation. By contrast humans may not use exactly the correct expected function for correlation (the correct matching filter). The human might not be able to precisely use the location information and may have some residual signal location uncertainty (spatial "jitter"). The human may have some "internal noise" due to variability in conversion from a physical correlation scale to a sensory visibility scale. Finally, the human may not be able to adopt a completely stable decision criteria. There may be some variation in this decision criteria which will also show up as a form of internal noise.

As one would expect, human performance results depend very precisely on the task that the human is asked to do. Two-alternative forced-choice experiments are by far the simplest because the human need not maintain a stable decision criteria. Yes/No signal known exactly (SKE) experiments are more difficult because the human must remember the decision criteria and the experimenter must try to determine its value. ROC experiments are often done using a rating scale method with multiple decision criteria up to as many as six. This of course is even more difficult to do because of the memory requirements. Multiple-alternative forced-choice experiments (MAFC) are more difficult that two-alternative forced-choice because one must make decisions among many alternatives and select the highest value on some sensory scale. For multiple-alternative Yes/No experiments the observer must do multiple comparisons with the decision criteria which again must be held stable.

Anyone setting out to measure observer performance must think very carefully about <u>what</u> it is he wants to measure and <u>why</u>, before selecting a method. My experimental work to date has concentrated upon measurement of the ultimate limits of human sensitivity in signal detection. For this investigation the MAFC method is by far the most satisfactory because one does not have to worry about discovering what the human observer's decision criteria is. The most important aspect of this approach is that one can measure the <u>absolute</u> efficiency of human observer performance (Gould et al, 1980; Guignard and Salehi, 1983; Marr, 1982). I shall now discuss a number

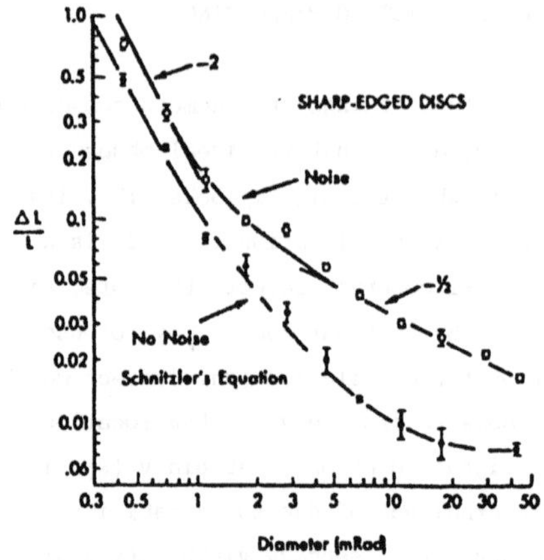

Fig. 3.1: Contrast threshold as a function of disc signal diameter. Done using method of adjustment (Burgess, Humphrey and Wagner, 1979)

Fig. 3.2A

"Ramp" Noise

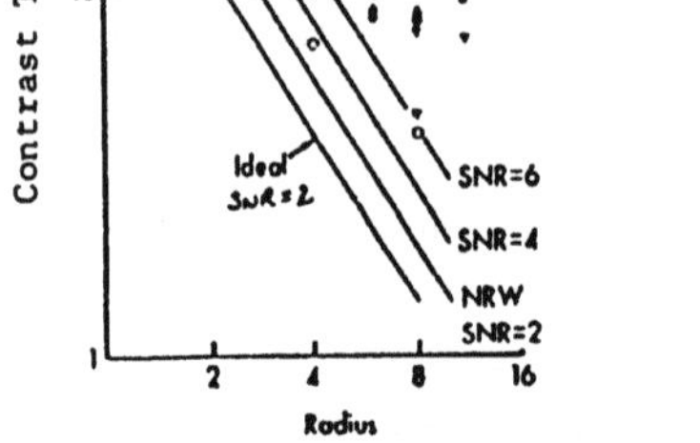

Fig. 3.2B

"White" Noise

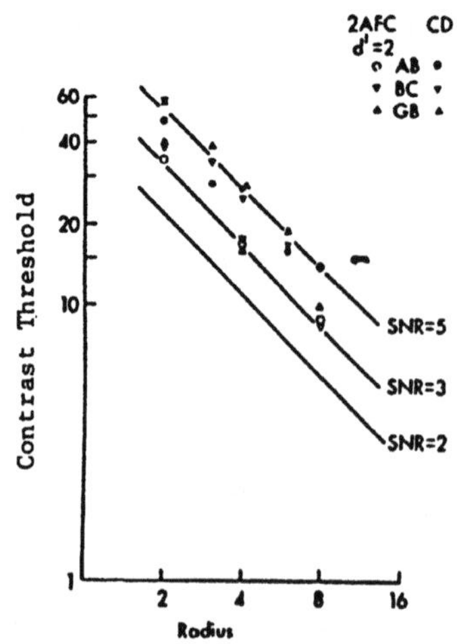

Fig. 3.2: Contrast threshold as a function of disc radius (in pixels). Done using "CD diagram" signal arrays (solid symbols) and by 2AFC (open symbols for $d' = 2$, i.e. 92% correct). The ramp noise SNR values are for a "re-whitening" cross correlator - with one exception identified as NRW SNR = 2.

of different experimental methods. These include the method of adjustment, the so-called contrast/detail diagrams, MAFC , Yes/No, and rating scale ROC methods.

3.1 Disc-Array Experiments (a.k.a. Contrast/Detail)

This type of experiment has become rather popular recently in radiology. The experimenter prepares some noise images containing an array of signals. The signals are usually discs with sharp edges and constant contrast inside the edges. In the display one finds rows of decreasing contrast and columns of decreasing diameter. The experimenter prepares a number of images, gives these to observers and asks them to indicate which discs they can see. The observer knows exactly where to look, exactly what to look for; and he knows that even if he may not be able to see discs they are in fact there. The experimenter uses observer responses to arrive at an estimate of a "contrast threshold". This type of experiment can be done very quickly and generates curves relating contrast threshold as a function of signal diameter, for example. Unfortunately the results are virtually impossible to analyze because of the large subjective component to the decision about whether signals are visible or not and also because the experimentor has no way of determining what decision criteria the observer is using.

The experimenter is relying on the basic honesty of the observer and is assuming that the observer is not lying about whether he can or cannot see something. Observers are basically honest. Naive observers tend to be very cautious and use a conservative decision criteria. The experimenter gets no information whatsoever about false positive rates. This experimental technique is useful only for the most crude of estimates of human observer performance. Example results obtained by this method are shown in figures 3.1 and 3.2.

3.2 Signal Known Exactly Experiments:

These experiments can be done either using two-alternative forced-choice (2AFC) or Yes/No methods. In either case one must be careful to ensure that the human observer is given a good opportunity to use signal parameter information. A reference copy of the signal should be placed in the image outside of the noise field so that the observer can refresh his memory as to signal size, shape and intensity as desired. The analyses of the experiments assume that the observer is aware of signal location so it is important to provide location marks to allow the observer to be precisely aware of the signal location.

There are two noise fields with a signal in one field. In standard methods of analysis it is assumed that signal and noise alternatives are equally likely and this convention should be observed. In order to keep sampling errors to a small value (5%) one must do a large number of trials, typically 200-400 at each signal amplitude. For 2AFC experiments one simply determines the percentage of correct responses over these blocks of trials. As a matter of experimental practise one should determine percentage of correct responses on occasions when signals are on the left and signals on the right separately to check for any observer bias. For Yes/No experiments one determines the percentage of correct responses and the percentage of false alarms. The determine of a detectability index is then straightforward. For 2AFC experiments one can determine the appropriate values using Elliott's tables (Swets, 1964). For Yes/No experiments one determines the normal deviate, z_t, for true positive responses and the normal deviate, z_f, for false alarms (see figure 3.7 and Green and Swets, 1966). These normal deviates are also referred to as "z-scores". The value of the detectability index, using $d' = z_p - z_f$. It has been demonstrated that these two approaches (2AFC and Yes/No) give identical results for detection of random dot signals in noise (Burgess and Barlow, 1983).

These procedures can then be repeated for signals of different amplitudes to generate a detectability plot (d' vs SNR). Example results of this sort of experiment are shown in figure 3.3 together with performance that would be obtained in the same experiment by an ideal observer. There are six experimental points each representing the average of 400 trials. Clearly this sort of experiment is rather tedious. However, it does have the advantage that experimental results can be analyzed quite precisely because they have been obtained on an absolute scale. The importance of this absolute scale should not be underestimated because if one is attempting to understand the fundamental limits of human signal detection this comparison is very, very important.

One can use a similar procedure to assess signal discrimination performance. The experimenter may be interested in the observer's ability to discriminate between signals of different amplitude, signals with different shapes, or signals with different edge gradients. For 2AFC discrimination experiments, the observer is presented with two noise fields and the two alternative signals simultaneously and is asked to select one particular signal. For Yes/No discrimination experiments the observer is presented with one of the two signals in a noise field and is asked to indicate which signal he believes is present. The percentage of correct responses and percentage of false positives (in Yes/No experiments) are determined and a value of d' is obtained. To generate a detectability plot one needs to know the SNR for the discrimination task. The appropriate SNR in this case is in fact the SNR of the difference signal. Given alternative signals, s_1 and s_2, the difference SNR is determined using:

$$SNR^2 = \frac{1}{N_o}\iint[s_1(x,y) - s_2(x,y)]^2 dxdy \tag{3-1}$$

Example results for this sort of experiment are illustrated in figure 3.4. The tasks include discrimination between a disc and a square (disc SNR = 15 at 90% correct), discrimination between sharp edged and gaussian discs (disc SNR = 7 at 90% correct), and disc amplitude discrimination.

3.3 Signals with Parameter Uncertainty

The extension to parameter uncertainty is straightforward. One simply introduces signal parameter uncertainty and makes the observer aware of the signal parameter limits. Examples of parameters that could be uncertain include signal amplitude, signal size, shape, and location.

The only occasion in which parameter uncertainty has no effect is signal amplitude uncertainty in forced-choice experiments. In this case the alternative signals are completely correlated and the observer is not required to use a decision criteria. We have obtained identical results for 2AFC detection with both known signal amplitude and randomly selected signal amplitudes. In all other cases signal uncertainty has a profound effect on observer performance. An example of MAFC experimental results using five different values of M (the signal location uncertainty parameter) are shown in figure 3.5. The solid lines through the data represent the performance for an observer working at 50% efficiency compared to the ideal observer.

The analysis of performance in parameter uncertainty experiments is not trivial and the experimenter is advised to consider carefully the method of analysis before even undertaking the experiment. Forced-choice techniques with one possible signal out of an orthogonal set are preferable because in this case the analysis is not particularly difficult. Conversion of

Fig. 3.3

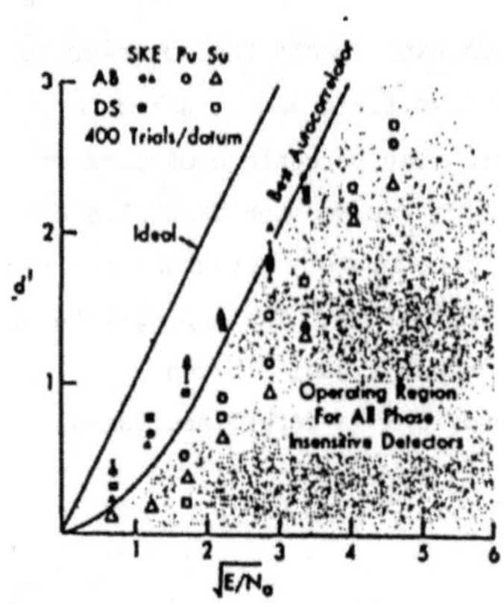

2AFC detection results for 2 cycle sine wave signals as a function of SNR. The closed symbols are for SKE experiments. The open symbols are for parameter uncertainty [phase (Pu) and starting location (Su)]. The data demonstrate that humans can be cross-correlation (i.e. phase sensitive) detection when given the required a priori information.

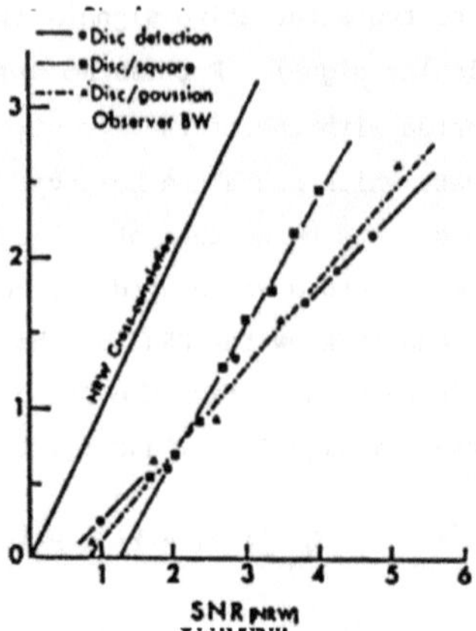

Fig. 3.4

2AFC results for disc detection, disc/square discrimination, and sharp-edged gaussian disc discrimination in CT-like "ramp" noise. The SNR values are calculated for a non-rewhitening (NRW) cross-correlation.

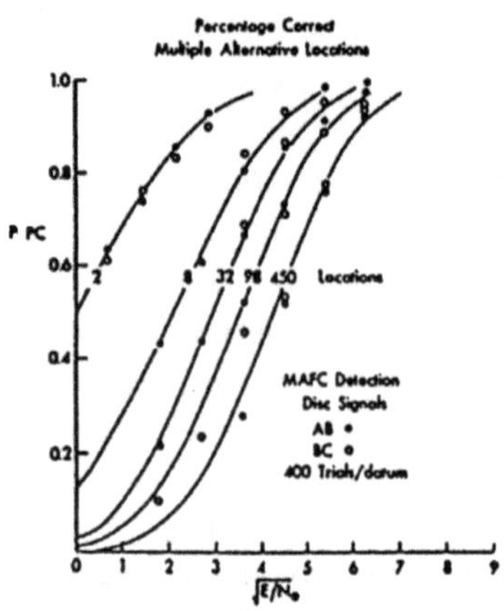

Fig. 3.5

MAFC detection results for disc signals in white noise as a function of SNR. The solid curves are theoretical values for an observer operating at 50% statistical efficiency.

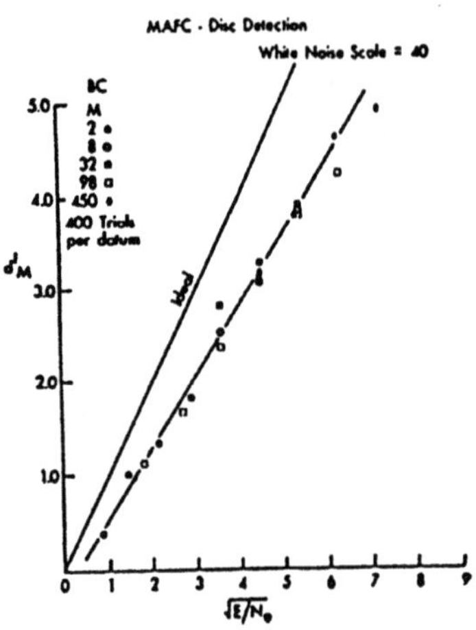

Fig. 3.6

The results of figure 3.5 with percentage correct converted to d' scores using Elliot's tables. The value of d'_M is the SNR required by the ideal observer to achieve the corresponding value of PC. The best fit line represents 50% statistical efficiency.

percentage correct responses into a detectability index, d'_M can be done using Elliott's tables (Swets, 1964). This procedure is illustrated in figure 3.6 where the results from figure 3.5 have been transformed into detectability index scores and are plotted as a function of SNR.

The MAFC experimental method has the advantage of having much better sampling statistics than 2AFC or Yes/No experiments and of allowing one to investigate performance at higher signal-to-noise ratios. Consider sampling error for 100 decision trials. The best 2AFC coefficient of variation ($s_{d'}/d'$) is 0.14 at 92% correct ($d' = 2$). For 128 AFC the coefficient of variation is 0.05 at 80% correct ($d'_M = 3$) and the minimum is very broad (40 to 95% correct). In many experiments it is no harder to use a large value of M (e.g. many possible locations) rather than a small one so the MAFC method can introduce considerable experimental economy (Burgess, submitted for publication).

3.4 ROC Methods

Receiver operating characteristic (ROC) techniques have been very popular in radiology. The method has been thoroughly discussed by a number of authors (Metz, 1978; Starr et al, 1975) and only a very brief description will be given here. The books by Green and Swets (1966) and Swets and Pickett (1982) are particularly worth reading carefully. The basic experimental approach is to produce noise images with precisely known signals in precisely known locations. Observers are carefully trained either to use multiple decision criteria or multiple rating scale criteria. For the multiple decision criteria approach the observer is asked to maintain one decision criteria for a large number of trials and the percentages of true and false positive responses are obtained to produce one point on the ROC curve. This procedure is then repeated for other decision criteria. For the rating scale method, the observer specifies a degree of certainty about signal presence on each decision. An example of a six-category scale might include the following possible responses: signal definitely present, signal probably present, signal possibly present, signal possibly not present, signal probably not present, signal definitely not present. For the purposes of analysis each of these rating values is assumed to correspond to a precise decision criteria. The probabilities of true and false positive responses for each decision criteria are then determined and points are plotted on the ROC curve.

Fig. 3.7 (from Swets and Pickett, 1982)

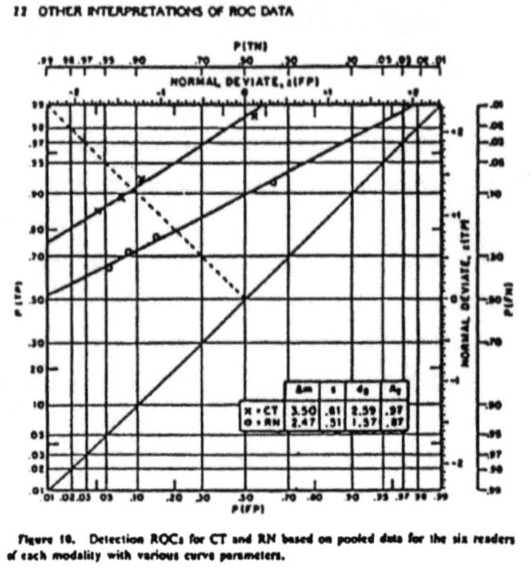

Fig. 3.8: Direct measurement of observer decision criterion (in this case a "strict" one). The observer made only 10 false positive responses from 50,080 noise-only locations. A noiseless observer using the same criterion would have demonstrated a step-function transition from zero to 100% correct.

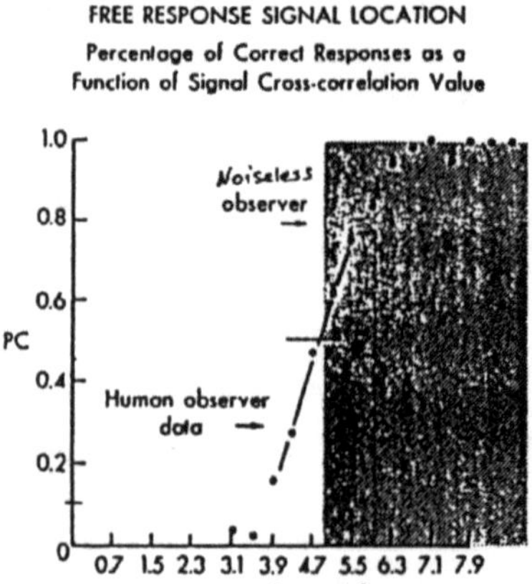

The results are most readily analyzed by plotting the data on double probability ROC paper. In this way the percentage of correct responses are converted into normal deviates (z - scores). This is illustrated in figure 3.7 [from Swets and Pickett (1982)] together with best fit straight lines to experimental results for CT and nuclear medicine images. The experimenter attempts to use the slope and intercept of the best fit line to make inferences.

3.5 Statistical Efficiency

Most of the methods described above provide experimental results in the form of a percentage of correct responses (and false positives if a forced choice method is not used). These results can be converted to the detectability index, d', measure. However, care must be taken here because the conversion is highly task dependent. In fact, d' is actually the SNR required by the ideal observer to achieve the recorded percentage scores.

Human performance can then be described in a task independent way using the statistical efficiency measure (Tanner and Birdsall, 1958) defined by

$$\eta = \left[\frac{d'_H}{d'_I} \right]^2$$

You may find it easier to think about efficiency in SNR terms. The ideal observer's index, d'_I, is actually the physical SNR of the signal and the human observer's index, d'_H, is actually the SNR required by the ideal observer to achieve the same accuracy as the human. Suppose for example the SNR is 3.0 and the human gets 92% correct on a 2AFC task. The ideal observer requires an SNR of 2.0 to get 92% correct. So the statistical efficiency is $\eta = (2.0/3.0)^2 = 0.44$.

3.6 Summary

The contrast detail method does not provide results on an absolute scale and is subject to decision criterion bias. However it is still very useful under carefully selected circumstances. One can collect "threshold" data very rapidly for a wide range of signal sizes. This is useful for comparing performance of imaging systems of fairly similar characteristics. However, one must be aware that differences in results may reflect bias changes rather than true performance differences.

The Yes/No and ROC methods allow one to detect and account for observer decision criterion. This is done at the penalty of collecting a prodigious amount of data in order to obtain adequate sampling statistics. Therefore it is not commonly used for studying human visual performance.

The forced choice methods by their nature exclude decision criterion effects and immediately yield a measure of observer sensitivity, d'. This result is often easily converted to the absolute efficiency scale. The MAFC method is especially economical from a data collection point of view.

4. EXPERIMENTAL RESULTS

In this section I shall attempt to give a rather brief review of experimental results under two subheadings. The first part will deal with the basic aspects of perception and equipment assessment, and the second section will deal with benefits of image processing. Due to limitations of space, I will not be able to cover every paper that has ever been published and I ask forgiveness in advance from anyone whose paper I miss. In some cases the references are not complete--a recent reference for each author is given to provide access to previous literature.

4.1 Basic Aspects

Any discussion of observer performance measurements in medical imaging must start with the classic work of Sturm and Morgan published in 1949. This paper did such a nice job of covering the basic issues that one would be hardpressed to improve upon it today. It is such a nice paper that it is really worthwhile making the effort to dig it out of the library to read. The work came about because of Dr. Morgan's interest in the then novel technology of image intensifiers for fluoroscopic imaging. Some radiologists had the rather naive hope that radiation doses would drop dramatically when the image intensifier was introduced because of the very high gain available. Sturm and Morgan were able to demonstrate that, in fact, only minor reductions in radiation dose would be available. They gave very clear demonstrations of the deleterious effects of statistical fluctuations in x-ray photon arrival (quantum noise). They borrowed from Rose's classical paper (1948) the concept of threshold SNR. This work on the limits of fluoroscopic systems continued with publications by Coltman and Anderson (1960) and Webster and Wipfelder (1964) and was the subject of a number of excellent publications in the 1960's including books by Moseley and Rust (1965) and by Gebauer, Lissner, and Schott (1967). An article by Webster in the Moseley and Rust book contains the first use of the term "contrast-detail diagram", in which one plots contrast threshold as a function of object diameter. Another fine book is titled "Perception of the Roentgen Image" (Tuddenham, 1969). It contains chapters dealing with many issues that still interest us today with chapter titles such as "Image Quality", "Computer Processing of the Roentgen Image", "Applications of Signal Detection Theory", and of course, "Mechanisms of Perception".

The early 1970's saw the introduction of statistical decision theory methods to radiographic perception problems. This work was done mainly at the University of Chicago by a very distinguished group including Lusted, Metz, Goodenough, and Starr. They introduced the concepts of receiver operating characteristics (ROC) to radiology and performed a large number of theoretical and experimental analyses (Lusted, 1968; Metz, 1978; Starr et al, 1975). Their findings can be summarized by the statement that experimental results agreed rather well with what one might expect from signal detection theory. With the advent of body CT, Swets and collaborators were commissioned to investigate its clinical efficacy. This led to excellent publications (Swets et al, 1979; Swets and Pickett, 1982) that clearly describe application of ROC methods to assessment of diagnostic systems.

Around the same time a number of significant studies were being done on the fundamental aspects of visual signal detection. Kundel (1972 and 1979) and Revesz (1974) began an examination of visual search patterns used by radiologists and made the important observation that normal anatomy can make it difficult to detect abnormalities. They coined the phase "structured noise" to describe these masking effects. Hay and Chester undertook a long series of experiments culminating in an integration model (1972, 1977) of signal detection. They proposed that all visual thresholds are ultimately determined by noise, that noise and signal are sampled by apertures at a hypothetical "decision site" in the visual cortex and that these apertures "match" the size of the signal as projected on the cortex. They suggested complete integration for circular signals less than 10 milliradians (0.5 degrees) in diameter and incomplete integration for larger signals. Similar experimental results were obtained by Rosell and Willson (1973) and a similar integration model was proposed by Schnitzler (1973).

The nuclear medicine area was also active during this time. Sharpe at the University of Aberdeen undertook a number of investigations related to radioisotope imaging (1982). He investigated the effects of a number of variables on signal detection including display intensity, noise level, and observing distance. Around the same time Cohen and collaborators (1982) undertook a large number of studies using contrast detail methods to investigate detectability of objects in CT scanner and digital radiography images. These experiments were undertaken to investigate the effects of a large number of parameters on signal contrasts required for detection. Cohen et al (1984) analysed the variation of results in CD diagrams. They concluded that it was more efficient to increase the numbers of replicated images and observers than to increase the number readings of each image by each observer. They found that sampling variations and between-observer variations were approximately equal. In a four-observer, four-image study they found a coefficient of variation of about 10% for threshold contrast.

By 1980 a large number of people had entered the area of observer performance measurement. Chew et al (1978) investigated the effect of CT noise on detectability of test objects. Edhome investigated the general principles for perception of boundaries in radiographic images (1981). Gould et al (1980) made objective performance measurements versus perceived image quality in intensified fluoroscopic images. Warren (1984) used CT phantom images and an ROC rating scale method to investigate the detectability of low-contrast disc signals. He found performance depended strongly on display window width, weakly on viewing distance and was independent of luminance. Guignard and Salehi (1983) found that the gaussian probability distribution function assumption was valid for some tasks but not others. They concluded that the gaussian assumption is not an intrinsic property of the eye-brain system but appears to be a consequence of the type of visual stimulus used.

Seeley, Roehrig and Hillman (1984) developed a computerized method for the measurement of conspicuity. They found this approach useful for comparing film and digital imaging techniques. Berbaum et al (1985) found that residents early in their training had increased lesion detection and diagnostic accuracy in chest radiography if a prior normal film was presented with each examination. Hillard et al (1985) presented preliminary results suggesting that radiologists may develop a visual presentation of a prototypical radiograph (a schema) in the course of clinical training. Observers with greater radiological experience exhibited poorer memory for normal films and better memory for abnormal films compared with less experienced observers. Kasel, Sharp and Dendy (1983) described experiments suggesting that observers could discriminate changes of 10% in image sharpness. Chester and Hay (1983) analyzed disc detection experiments using an SNR model based on a flat-topped, disc-shaped sampling aperture that matched the signal in spatial extent. They also concluded that threshold contrasts were proportional to the standard deviation of the background noise. They subsequently clarified this statement (1984) to point out that the threshold contrast is determined fundamentally not by the full power of the physical noise, but rather by the power obtained by sampling the noise through the sampling aperture used. Bollen (1986) presented results showing that when image noise is characterized by standard density fluctuation figures using a 400 micron densitometer aperture there is a good correlation with human subjective ranking.

The University of Chicago group did a number of observer performance studies as part of a larger research program on digital radiography. Loo, Doi and Metz (1984) compared eight physical image quality indices with observer performance in radiographic detection of nylon beads. They concluded that, for the conditions studied, human observer detection performance most closely resembles that of a sub-optimal statistical decision process. Giger and Doi (1985) investigated the effect of pixel size on SNR and threshold contrast using a statistical decision theory model including the observer's visual transfer function and internal noise. They calculated threshold contrasts from the SNR for various combinations of object size and shape, pixel size, resolution and noise. Predicted results agreed well with those determined from human observer experiments. Loo, Doi and Metz (1985) investigated the effect of unsharp masking on the detectability of simple patterns. The theoretical predictions agreed qualitatively with observer performance studies. Ohara et al (1986) investigated the detection of simulated low-contrast objects in digital subtraction angiographic images using an 18-AFC method. They found that the SNR calculated from the perceived statistical decision theory model could accurately predict experimental detectability results. A threshold SNR of 3.8 corresponded to 50% correct in the 18-AFC task. Giger, Ohara and Doi (1985) studied the effect of displayed grey levels on signal detection in an 18-AFC experiment. They found that noise can act as a carrier of the signal data so that a signal having contrast less than one displayed grey level could be detected and that signal contrasts as low as 0.1% of CRT dynamic range could be detected (albeit with only 15% accuracy). Kume et al (1986) investigated the effect of structure mottle of II-TV digital imaging systems. They found that the power spectrum of the structure mottle depended on the specific combination of components and that structure mottle could significantly reduce object detection in an image when II input exposure was greater than 0.1 mR. Chan et al (1987) did ROC studies of the effects of pixel size and unsharp-mask filtering on the detection of subtle microcalcifications in digital mammography. Observer studies indicated that 0.1 mm square pixels provide lower detectability than conventional film-screen mammograms.

A number of studies have been done on detection of signals in chest radiographs. Kelsey et al (1982) found slightly better nodule detection at 120 kVp rather than 70 kVp and attributed the improvement to wider latitude. Swensson, Hessel and Herman (1982, 1985) found that radiologists could distinguish much better between normal and abnormal features when searching chest radiographs for any abnormal findings rather than when explicitly evaluating each individual feature after having attention directed to it. They suggested that this improvement might be due to perceptual meachanisms that functioned only during the process of visual search. This result is not consistent with standard signal detection theory and their conclusion was subsequently challenged by Swets (1985) and by Kundel (1985) who suggested that their surprising results might be due to a methodological artifact. Parker et al (1982) compared nodule detection in chest radiographs under free search and directed search (by which chest quadrant to examine). They found significantly better performance in directed search. Revesz, Kundel and Bonitatibus (1983) compared observer diagnostic accuracy of three different radiographic techniques against several methods of defining the correct response. They concluded that definitions based on the diagnostic tests under evaluation are inadequate and that the truth must be determined by clinical or pathological proof and follow-up data. Kundel et al (1987) studied scanning performance of radiologists using chest films and an eye-tracker. They suggested that radiologists used a strategy designed to cover the lung images in a minimum time using the smallest possible visual field (3.5 degrees visual angle). Berbaum et al (1988) found that prior availability of clinical history did not improve detectability of lung nodules but did improve performance for more diverse lesions. Sorenson and colleagues (1987) used ROC methods to study detection of lung nodules. They concluded that detection in poorly penetrated areas was limited by display contrast while detection in well-protected lung fields was limited primarily by display contrast. They reported that unsharp masking can significantly improve performance in the contrast-limited areas. Kundel et al (1985) used an ROC method to demonstrate that nodules are much more detectable on uniform backgrounds than on chest images.

4.2 Efficiency

There has been a rapid increase since 1980 in the use of efficiency as a method of evaluating the human visual system. There have been three special issues of the Journal of the Optical Society of America in which the approach played a major role (Sept. 85, Dec. 87 and April 88).

Barlow and colleagues (Barlow, 1978; Burgess and Barlow, 1983) started using this measure for detection and discrimination of random dot patterns and found 50% efficiency for a variety of tasks. Burgess et al (1981) extended this to amplitude discrimination of simple test patterns in grey-scale images.

My own work on observer performance began in 1975 on a project related to optimization of magnification angiography. My first work (Burgess and Humphrey, 1977; Burgess et al, 1979) was done using method of adjustment with bar and disc signals optically superimposed on radiographic noise images. The purpose of these experiments was to investigate the effects of signal size, shape and edge gradient as well as noise amplitude on signal detectability. In 1979 I began experiments on absolute performance measures, first with Barlow (Burgess and Barlow, 1983) then Wagner and Jennings (Burgess et al, 1981; Burgess et al, 1982) using digital image display systems. The approach is based upon the fact that given complete knowledge of the signals and noise, one can calculate upper limits of observer performance using the concept of the ideal observer, and then one can measure human performance (statistical efficiency) on an absolute scale for a variety of tasks under a variety of experimental conditions. Our findings can be summarized very briefly by the observation that for a wide variety of tasks and a wide variety of experimental situations, human observers perform at a statistical efficiency of about 50%. This can be converted into SNR terms and one finds that humans typically need $\sqrt{2}$ more SNR than the ideal observer to perform at the same level of accuracy. This decrement in human performance is clearly rather small as one appreciates by comparing two CT scans that differ by a factor of $\sqrt{2}$ in mAs ($\sqrt{2}$ in SNR). One notices immediately that these two images are not particularly different. The message for image processing is pessimistic in that the perfect image "enhancement" algorithm will only be as beneficial as a factor of 2 increase in mAs.

The high human efficiency results suggested perhaps that humans tried to perform many of the procedures used by the ideal observer. My colleagues and I then did a series of experiments to test this hypothesis. We first checked on the ability to use phase information in sine-wave detection since this ability is necessary in order to do cross-correlation detection (Burgess and Ghandeharian, 1984). We found that performance when phase information was provided was better than the best possible phase-insensitive detector. There was a marked reduction of performance when phase information was not available. We then measured the effect of signal-location uncertainty on signal detectability using an M-alternative forced choice technique (Burgess and Ghandeharian, 1984). We found the same statistical efficiency (50%) for all values of M so that 2AFC results can be used to predict all other results. This indicates that signal location uncertainty does not reduce efficiency and suggests that humans can use the maximum a posteriori decision strategy. Next we used signals from the 2-dimensional Hadamard basis set (Burgess, 1985). Experiments were done on 2AFC detection with and without prior knowledge about which signal might be present. As in the sine wave case above, there was a marked reduction in performance when prior information was not given. The decrement in performance was the same as that predicted for the ideal observer given the same uncertainty and suggests that no additional inefficiency is introduced by signal profile uncertainty. It was also found that signal identification efficiency was the same as 2AFC detection efficiency. These results suggested that humans can use a Bayesian maximum a posteriori probability decision strategy available using prior information and cross-correlation (or matched filtering) of expected signal profiles with those present in the image display. Finally we did a detailed investigation of internal noise (Burgess, 1986; Burgess and Colborne, 1988) but chose to describe this phenomenon as by its observable manifestation, namely "observer inconsistency". Previous authors (Legge et al, 1987; Nagaraja, 1964; Pelli, 1981) had postulated a constant internal noise. We did the experiments several ways and found the novel result that internal noise has a component proportional to external image noise. This so-called induced internal noise component dominates when external noise is easily visible. The ratio of induced internal and external noise standard deviations was found to be 0.75 setting an upper limit of 64% for human observer efficiency. We interpreted the results using sub-optimal observer model developed by Wilcox (1968) that included sensory encoding fluctuations, intrinsic signal parameter uncertainty, and decision variable fluctuations. We concluded that human observers probably have all three sources of variability.

We also considered a number of side issues. One issue was the effect of long-range anti-correlation in CT noise. An ideal observer would re-whiten this noise before doing signal detection. Wagner, Brown and Pastel (1979) and Hanson (1979) suggested that perhaps human observers cannot do this re-whitening procedure. We presented experimental results (Burgess et al, 1982) suggesting that this was probably the case. Myers et al (Myers, 1985; Myers et al, 1985) did a more extensive investigation and demonstrated that simple disc detection performance was best described by the non-rewhitening assumption. They subsequently (Myers and Barrett, 1987) proposed a physiological channels-based model to explain human inability to rewhiten. Judy and Swensson (1981) measured the detectability of disc signals superimposed on CT water phantom images. Their results were inconsistent with a model based on an ideal ramp power spectrum for the CT noise. They had previously found departures from an ideal ramp in CT power spectra and suggested that these deviations affect the detectability of large discs.

Judy, Swensson and colleagues have done a number of fundamental studies on observer performance. They compared human performance with fundamental signal detection limits (1981) and found that there was a good correlation between SNR and human d' scores. They measured detectability of signals superimposed on CT images and found that changes in detectability are closely related to the cross-correlation SNR predictions (Judy and Swensson, 1985) based on object size, CT contrast and noise levels and noise correlations. They also did disc size discrimination experiments and found that observer performance could be predicted by the task SNR (Judy and Swensson, 1986). They investigated the effect of display thresholding of images (Judy and Swensson, 1987). They found very good agreement between a cross-correlation model and human results. Judy et al (1987) found that observer efficiency for disc detection did not depend on whether the disc contrast was positive or negative. They also studied size discrimination in CT images (Seltzer et al, 1988) and found that discrimination performance was nearly independent of display window width in marked contrast with window effect on feature detection results.

Hanson has done theoretical calculations of ideal observer performance for some higher order tasks. He demonstrated that there is increased emphasis on high spatial frequency components (Hanson, 1983; Hanson, 1984) when the observer is asked to do object edge localization, width determination and measurement of separation between objects. Burgess (1985) presented human observer results showing efficiencies in the 20-40% range for single bar position identification task.

Wagner, Insana, and Brown (1985) considered the possibility that there may be higher-order statistical information in some medical images that cannot be extracted by the human eye-brain system. They suggested that this may be the case for ultrasound speckle patterns and that machine analysis performance might be superior to that of humans.

Myers et al (1985) found that images with equal pixel SNR but different correlation properties give quite different observer performance measures for a simple detection experiment. They found constant observer efficiencies for humans relative to a quasi-ideal observer that was not able to prewhiten the noise but was otherwise ideal.

Caelli and Nawrot (1987) measured the ability of humans to locate signals in white noise, low pass noise and digital breast images. They found that performance was consistent with that predicted by a cross-correlator that normalizes for background variability and assumes some uncertainty as to the precise location of the signal.

Van Meeteren and Valeton (1988) measured human detectability of sine-wave gratings in white noise and low-pass filtered noise. They found threshold contrasts elevated by factors of 3 to 10 over that required for ideal detection. This implies efficiencies between 1 and 10%. However most of the reduction in efficiency may be an artifact of their experimental method. They used a 0.2 second display time and compared human performance without a phase reference against ideal performance with a phase reference.

Jaques et al (1985) compared human and computer estimates of stenoses in digital subtraction angiography phantoms. With few exceptions, human estimation accuracy was about 5% and was superior to the computer estimation.

Kersten (1984) measured the statistical efficiency of the detection of a variety of sizes of gaussian-modulated sine-waves in one-dimensional dynamic white noise. He found that the spatial summation resembled the performance of a cross-correlator whose template is matched to a signal with one cycle between i/e points. He found peak efficiencies of 20 to 40% near 1 cycle per degree. Kersten (1987) also measured statistical efficiency for the detection of one-dimensional static noise patterns in one-dimensional dynamic noise. He found the same 4th-root of bandwidth dependence for the ideal observer and for human data. Efficiencies were in the 30 to 60% range. He concluded that the results could be interpreted either in terms of a channel with an adjustable bandwidth or as very efficient combination of information between spatial frequency channels.

Most of the theory and experiments described in this work relate to well-defined observer tasks. Clinical radiology diagnosis, on the other hand, extends to a much more complex task range. Barrett, Myers and Wagner (1986) considered the theoretical aspects and suggested several approaches to finding a surrogate d' for statistically defined tasks. Wagner and Barrett (1987) continued the theoretical investigation of higher order tasks. They considered two examples of tasks that require a decision function that is quadratic in the data. One example was the case where a null hypothesis is stated but no signal alternative is specified. The other example is detection of a specified object whose possible positions are randomly distributed over a specified image area. Fiete et al (1987) used simulated liver images and an ROC method to compare Hotelling trace criterion (HTC) classification performance with that of human observers. They found a correlation of 0.988 suggesting that HTC is a matched-filtering method adapted from the pattern-recognition field for discriminating between statistically defined classes of objects. The HTC can include effects of image blur, noise and object variability. They also noted that while, in principle humans could use nonlinear features and hence outperform the linear HTC, this did not seem to be the case.

4.3 Image Processing

In the 1960's Robert Selzer (1968) at the Jet Propulsion Lab at Cal Tech began investigation of computer processing of x-ray images using techniques developed for enhancement of satellite photographs. This work continued over a number of years. However, he eventually reached the conclusion that there was really nothing much that one could do with x-ray images that was beneficial. This conclusion of course did not deter other people from trying to find adequate enhancement techniques. This area became rather popular with the advent of computed tomography because of the relative ease of digital processing of the images. There were a number of proposals related to supposed benefits of smoothing on detectability. Joseph was an early advocate of this possibility for CT (Joseph, 1977). The literature on smoothing is quite entertaining and begins with a paper by Kuhl et al in 1972. They concluded that observer performance was not improved by smoothing. The controversy started immediately with a rebuttal by Metz and Goodenough in a subsequent issue of the same journal (Metz and Goodenough, 1973). The two groups of authors seem to reach opposite conclusions on the basis of the same experimental data. Herath and Sharpe (1976) used the ROC method to investigate "matched filter" smoothing of nuclear medicine images and concluded that smoothing is beneficial. Investigation of processing continued with papers by Hemmingsson et al (1981) who studied noise and noise texture in CT images before and after image processing. They concluded that the RMS noise level could be reduced by a factor of 8 using various filters (more on this point below). Pullin et al (1980) investigated the effect of smoothing brain scans using ROC techniques with radiological observers and concluded that there was no significant change in lesion detectability after smoothing. Kuzima and Uchida (1982) investigated the effects of noise smoothing on image improvement.

One might wonder why such an apparently simple topic should lead to controversy. On the surface the problem appears to be simple. If one looks at an image with very little smoothing it appears to be very noisy and as the image is smoothed it appears to be much less noisy. Naively one might expect that the smoothing should definitely improve signal detectability. One certainly sees that the RMS variation in noise decreases dramatically as the image is smoothed. This sort of reasoning led to suggestions that smoothing would be beneficial. Unfortunately this line of reasoning is incorrect. The reasoning is based on the fact that the standard deviation (also known as the RMS deviation) of the noise is decreasing. Unfortunately this is not the relevant parameter for describing the "amount of noise" in signal detection calculations. Starting back at the matched filter approach to signal detection one can see that it is the standard deviation of the noise as seen through the signal as a filter that is the appropriate parameter and one finds that the underlying important parameter is the noise spectral density (i.e.,noise power per unit frequency bandwidth) in the frequency band of the signal. When one goes through this analysis one reaches the opposite conclusion, namely that smoothing should have no effect whatsoever (at best) or a detrimental effect (at worst) if humans can do precise "matched filter" detection. In fact one finds human performance lies in between these two extremes. My own unpublished experiments on smoothing effects demonstrate a rather small and subtle improvement in performance for signal detection in both white noise and CT noise. Increases in detectability index, d', are only about 10-15 percent and this requires a rather large number of display trials to reliably determine. In my experiments, one's subjective impression is that signal detection is very much easier with smoothed images. However, there is also a much greater difficulty in avoiding false positives because patches of filtered noise also end up looking like signals. On balance, the net improvement is small.

4.4 Image Enhancement

While the controversy on benefits of smoothing were taking place, there was a parallel attempt to apply edge enhancement techniques (e.g. unsharp masking) in medical images. The rationale for this is based on the observation that the retina appears to do bandpass filtering and therefore may be doing edge enhancement. Techniques such as unsharp masking use additional edge enhancement and the image's apparent sharpness is increased. It is presumed by some authors that this process is beneficial. There is certainly evidence that edge enhancement is beneficial for images such as radiographs where contrast and noise levels are inherently low but there is not evidence to demonstrate that edge enhancement helps in situations such as computed tomography where observers are allowed to adjust display windows and levels freely. This discrepancy can be explained by the observation that dynamic range requirements for radiographs is very large and unenhanced radiographs have low contrast gain. After bandpass filtering, the dynamic range requirements are reduced and one can then increase the contrast gain (i.e. film gamma). Given a higher contrast gain, the human intrinsic limits on contrast sensitivity are less important and the human is more likely to be limited by the SNR of the data. If this explanation for benefits of edge enhancement is correct, it demonstrates that the benefits do not come from the edge enhancement _per se_ but from the higher contrast amplification that is possible once the low frequency components of the data have been suppressed.

There are a number of other image processing techniques which may be beneficial under some circumstances for the same reason that edge enhancement is occasionally helpful. These techniques include histogram equalization, homomorphic filtering and a whole array of non-linear techniques. In most cases one is redistributing data values in a way that allows higher contrast gain to be applied. Ishida et al (1984) measured the detectability of simulated low-contrast square signals in radiographs. They found that signal detectability was significantly improved by unsharp mask filtering and by overall contrast enhancement. Chan, Metz and Doi (1985) developed an optimal filter that maximizes the perceived SNR based on a statistical decision theory model for a human observer. This filter is designed to serve as a prewhitening filter that compensates for both the image noise Wiener spectrum and the observer's visual system response, thus allowing the observer to perform matched-filtering in a white noise background during signal detection. Penney et al (1987) investigated fidelity measures to high and low pass filters. They found that the convolution mean-squared error can predict with reasonable accuracy which images an average observer will prefer.

4.5 Summary

Experimental results to date allow one to make a number of substantial statements about human observer performance. It seems clear that the main limitation with medical images is the image noise. If observers are given complete freedom to control display conditions, they definitely rank images in order of increasing SNR and measured human performance for detection and discrimination tasks seems to be rather close to the fundamental limits determined by statistical decision theory. Humans do have fundamental contrast sensitivity limits which can be represented theoretically as a source of observer internal noise. This inherent contrast sensitivity causes problems for images displayed with low contrast gain. However, as the display gain is increased the observer contrast sensitivity limit becomes less and less important and observer performance approaches a limit that appears to be a factor of $\sqrt{2}$ higher than that of the ideal detector. It appears that for relatively small signal areas humans can do maximum likelihood signal detection. However there is an upper limit to signal size for this sort of processing. However, this is usually not a problem because observers automatically adjust their viewing distance to compensate. Another human limitation seems to be the inability to compensate for inherent correlations in image noise. This inability causes a significant decrement in performance for images that are produced by reconstruction from projections as is done in CT and in some modes of NMR imaging. For this reason it has been observed (Wagner et al, 1983) that NMR two-dimensional Fourier transform techniques will possibly give superior images to those obtained by back projection reconstruction.

We seem to be reaching the end of the road for improving signal perceptibility in x-ray imaging. CT and digital subtraction methods have virtually eliminated the masking effect of superimposed patient structure. Digital imaging systems have decoupled data acquisition from display and allow adaptive adjustment of display conditions. There is certainly no convincing evidence that image processing techniques will be beneficial for enhancing display of digital medical images. The one possible exception to this is contrast transformation techniques.

5. APPLICATION TO SYSTEM DESIGN

In this section two common design questions will be discussed. How should data quantization be done (how many grey levels) and how good should resolution be?

5.1 Number of Grey Levels

This topic has not been an important display issue until recently. CT imaging systems all have window and level controls so the user is free to select the most appropriate video output transformation. The present standard for video digital to analogue convertors (DAC's) is 8 bits giving 256 grey levels. This is usually adequate for noise-limited images. The quantization may be an issue for digital projection images if the aim is to record and display the full dynamic range of the data in one image. I will discuss human capabilities some common viewpoints, and describe some experimental results.

How many shades of grey can the human visual system detect? A number of answers have been given and the answer depends on details of how the measurement is made. To take an extreme example suppose that 2 signals are placed side by side on a uniform noiseless background (luminance L_B) and the observer is asked to identify the brighter. Many experimenters have found that this can be done with 75% accuracy using contrasts (L/L_B) of 0.005 to 0.002 depending on signal shape. That means discrimination of differences of 1 part in several hundred. Pizer (1981 and 1982) used a detection measurement technique to conclude that about 100 grey levels (on a logarithmic scale) are necessary. This sort of conclusion is related to the standard Weber-Fechner law. Pizer's estimate is, in a sense, a lower limit because he used a detection measure while it is known that under optimum conditions amplitude discrimination is about a factor of 2 better than detection.

The other extreme view is encountered in about one paper per conference, namely "humans can distinguish 8 shades of grey". Sometimes 16 or 32 replaces 8. The speaker has usually needed a number and "read it somewhere". Where do these numbers come from? The value of 8 is probably based on absolute brightness judgements, that is naming of grey levels. The observer is shown grey level cards and is asked to specify the level. Humans are clearly very poor at this task and we don't even have much of a vocabulary for performing it. The human eye is not a very good photometer. By contrast, we are much better at naming colours. The television industry is an alternative source for the number 8. TV cameras are frequently described as being capable of eight (or 10) shades of grey. This refers (Biberman, 1973) to the available number of steps of $\sqrt{2}$ in luminance in a "resolution element". This is really a statement of the camera dynamic range using a particular definition of SNR (namely peak signal/noise standard deviation) and has nothing whatsoever to do with the human visual system. In this method of analysis 8 shades of grey requires a dynamic range of ($2)^8$ or 16 to 1. By this reckoning a very high quality digital fluoroscopy TV camera (1000:1 SNR) would give only 20 shades of grey.

So far, we have values from 8 to several hundred. The high values were for featureless backgrounds which is not a common situation. Values of 32 to 64 can be obtained by the following experiment. A complex, continuous tone image is quantized with several numbers of bits and observers are asked to judge quality based on artifical contouring. Pratt (1978) reports that a number of experiments (e.g. Goodall, 1951) found 64 levels for good quality and 32 levels for moderate quality. Roberts has demonstrated that this contouring can be removed by adding random noise.

This brings us to the rôle of noise on "grey level perception". Castleman, in an excellent book (1979), gives a rule of thumb that the "effective grey level" resolution is equal to the RMS noise level. Since the RMS noise equals the noise standard deviation per spatial resolution element, I will adopt the notation s_p (per pixel) to represent it. Hence Castleman suggests a grey level step $g = s_p$. The practitioners of digital radiography (Tesic et al, 1984 for example) propose $g = 2s_p$. Rimkus and Baily (1984) use a pixel detection model to propose $g = 3s_p$. My experimental results (Burgess, 1985) demonstrated that observer efficiency is significantly reduced with $g = 0.8s_p$ or more (see Figure 5.1).

You may have noticed that the noise was described as uncorrelated. This is a very important point. As mentioned several times in this chapter, the important noise measure is the standard deviation measured over the signal area. Alternatively one can use the noise spectral density. RMS noise and s_p are really derived from integrals over the entire noise spectrum and can be highly misleading. In smoothing experiments one can reduce s_p by a factor of 10 with no significant change in signal detectability. If the noise is correlated for neighbouring pixels then the value of s_p always gives an under-estimate of the s_p for uncorrelated noise. This may be the source of the rather optimistic estimates of g equal to or greater than s_p.

To put all this in perspective, consider CT scanning image data. A typical scanner gives s_p on the order of 5 Hounsfield units with 1024 steps from air to water. The above proposals for g of 1, 2, and $3s_p$ suggest 200, 100, and 65 grey levels respectively. By contrast, the origonal EMI scale had 500 levels and this was considered to be inadequate. Hence the increase to 1024 levels. The grey level discussion can be summarized as follows. Given 8 bit video DAC's, the number of display grey levels to use is rarely an issue. Clearly proposals of 8 or 16 grey levels are just silly. Ultrasound systems, for example, use 32 levels for extremely noisy images.

Fig. 5.1 Experimental results for two human observers showing the reduction in signal-detectability index, d', as the quantization-step size, Q, is increased. There is a significant reduction in performance for step size greater than $0.8\sigma_p$, where σ_p is the data-noise standard deviation per sample (σ_p = 40 and 12.5 for these experiments). The error bars represent the coefficient of variation for sampling statistics error using 256 trials per datum. The solid line is the theoretical result for the ideal observer.

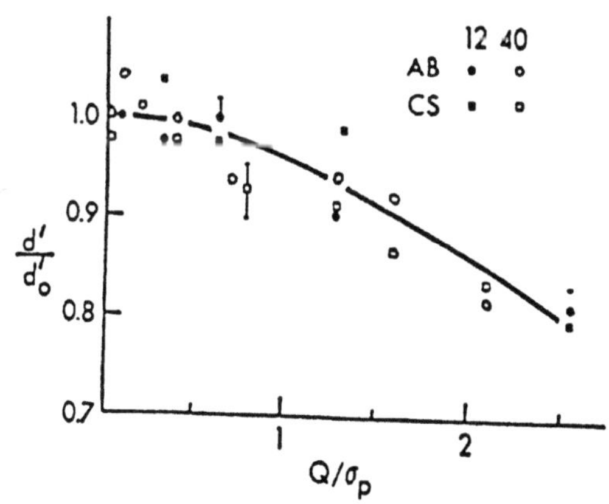

The display of medical images involves transforming recorded data intensities such as CT numbers into perceivable intensities such as combinations of color and luminance. For the viewer to extract the most information about patterns of decreasing and increasing recorded intensity, the display designer must pay attention to three issues: (1) choice of display scale, including its discretization; (2) correction for variations in contrast sensitivity across the display scale due to the observer and the display device; and (3) contrast enhancement based on the information in the recorded image and its importance, determined by viewing objectives. Pizer (Pizer et al, 1981 and 1982) has performed a number of psychophysical experiments and has used these results to propose several contrast transformation tables. Pizer's approach is designed to allow one to simultaneously display the entire dynamic range of the data by using spatially adaptive contrast transformation tables that attempt to maximize the transmitted information to the human. This approach is likely to be very useful for display of digital radiography images. Pizer et al (1986) have developed an automatic histogram equalization (AHE) method that gives simultaneous contrast enhancement in all regions of an image. They have designed a VLSI device that allows calculation of AHE in less than one second per megapixel. They also reported results of an ROC study demonstrating that for CT chest images, AHE introduces no measurable loss of diagnostic performance compared to standard windowing (which is much more time-consuming).

McAdams et al (1986) developed a histogram-based method of contrast enhancement of chest images. They found dual-slope and V-shaped grey-scale transformations to be useful. Sezan, Yip and Daly (1987) did a theoretical investigation of digital radiography amplitude quantization requirements using human psychophysical data. Then they compared simulated digital images produced from sampled film radiographs. They found at least 512 quantization levels (9 bits) are needed to produce images free of false-contouring artifacts. Giger et al (1985) had found that there was a small probability (15%) that signals with an amplitude of 1 part in 1024 (10 bits) could be detected. Baxter et al (1982) proposed a model based on physiological mechanisms to predict observer contrast thresholds for disc signals in both uniform and non-uniform backgrounds. Alter et al (1982) found that using low ambient room lighting and restricting light from surrounding viewboxes significantly improved observer signal detection performance. Rogers, Johnston and Pizer (1987) concluded that the ambient-light levels that typically occur in CRT-based radiographic reading rooms do not appear to alter brightness-discrimination thresholds as measured by their experimental paradigm.

5.2 Resolution

In this secton data acquisition and display system resolution requirements will be discussed. The data acquisition aspect is very important because once patient content information is spatially pooled (blurred or locally averaged) signal phase information is irretrievably lost. In addition if noise is injected after this pooling, any image processing attempts to do a "high frequency boost" to restore image "sharpness" will have detrimental effects. To make this clear consider the Dolby audio technique. It is known that tape recording systems inject high frequency noise. To suppress this noise, the Dolby strategy is to boost signal high frequencies <u>before</u> recording and then de-emphasize them on play back. Hence the SNR in the high frequency range is improved. It would be nice to be able to do this with x-rays. In fact, legend has it that Dolby started out investigating noise suppression in x-ray fluoroscopy and then went on to a more feasible (and lucrative) application.

Resolution requirements are always determined by what needs to be seen. So patient content and diagnostic problems must be considered. Noise plays a rôle in the sense that for a given photon fluence and detection SNR, smaller objects require higher contrast in order to be visible (the well known CD diagram). Other imaging system parameters need to be considered because usually increased resolution can only be obtained by trading off other benefits. Finally, in a digital system costs usually go up as the inverse square of the resolution distance.

Consider the contrast-size diagram in figure 5-2. It is assumed that the detected photon fluence is 10^4 or 10^5 per mm^2, that image blur due to the x-ray focus is 0.25 mm, that the desired SNR is 10, and that the objects to be detected are spherical. Similar analysis for linear objects is straight forward. The goal of an SNR of 10 is justified on the following grounds. If object location is unknown and a search is necessary then an SNR of 4 gives only about 50% detection probability (see figure 3.5), an SNR of 7 gives nearly 100% detection probability, and an SNR of 10 gets one into the range where crude object recognition becomes possible. Now a few definitions. A simple photon counting SNR is defined, the signal being $(N_o - N_1)$ where N_o is the total number of photons that would have been detected in the signal area if the signal were absent and N_1 is the number detected with the (radiopaque) object in position. Define contrast, C, as (1 - exp(-dt)) where d is the attenuation coefficient increment due to the object displacing background material and t is the object thickness. The object will be assumed to be a cube to simplify arithmetic. Let the background photon fluence per mm^2 be n_o. Then SNR = $Cn_oA/ \sqrt{n_oA}$ = $C \sqrt{n_oA}$. Note that this is the <u>object</u> detection SNR and that the highest possible contrast is unity (when the object transmits no photons).

With reference to figure 5-2, lines have been drawn for 2 photon fluences (10^4 and 10^5 per mm^2). The lines have two slopes. The change in slope occurs when the projected object size is equal to the blur (PSF) due to the focal spot. The precise point of inflexion will of course depend on object geometric magnification. The lower limit of detectable object size is at the intersection of the line of increased slope and the line for contrast equal to 1. These small objects will, of course, appear to be larger because of focus blur (0.25 mm for this example).

What can one now say about imaging system resolution requirements? The most obvious point is that, everything else being equal, one would like to have resolution as high as possible (see Hanson, 1983, for example). Increased resolution, <u>per se</u>, will never have a detrimental effect on signal detectability. It would be interesting to know if a radiologist has <u>ever</u> complained about too much resolution. Unfortunately, there is usually a cost of some sort paid in order to obtain increased resolution. This trade-off may be the source of the common view that <u>resolution</u> and <u>noise</u> are related variables.

How are resolution and noise related? This question can be answered several ways. First consider the above CD diagram. This size of the smallest <u>detectable</u> object depends on the spectral density of the noise (or photons per unit area). Note, however, that resolution and small object detectabiliy are <u>not</u> the same thing. System resolution is usually defined without reference to noise and is measured at very low noise levels. Another answer is that the noise standard deviation (or RMS value) per resolution element increases as the size of a resolution element decreases. This statement is obviously true. Note, however, that this represents an incorrect measure of noise effects. Object detectability is determined by noise standard deviation measured over the signal <u>area</u> (not over a resolution element area) and is independent of element size as long as the object is large compared to a resolution element. These two views of the relationship between resolution and noise are not very useful because they are really just issues of semantics. They do not illustrate any real trade-off.

There are two simple, and obvious, relationships. One is that to obtain higher resolution one often must use a thinner phosphor layer at the x-ray to light conversion stage. Reduced thickness gives a lower absorption efficiency and noise goes up (if photon fluence is held constant). A second relationship is due to electronic noise in the imaging system. Consider a two-dimensional array of crystal-photodiode-amplifier detectors and assume that there are M detectors per mm^2. Let s_o^2 be the detected photon variance per mm^2 and let s_e^2 be the electronic noise variance per detector (measured in equivalent photons). The total noise variance per mm^2 is $s_T^2 = s_o^2 + Ms_e^2$, hence the noise penalty for more detectors per unit area. As an example, a CsI-photodiode-amplifier channel (with a noise floor of 1 pA) gives electronic noise equal to photon noise at a detection rate of 3000 photons per channel per second (Haque, 1981).

So we see that there is an engineering relationship between resolution and noise. This relationship is not fundamental in the sense that, given enough money, one could always design around the problem by using more sophisticated technology. However, there is an engineering problem so the question reduces to "How much resolution is enough?".

The engineering answer to resolution requirement is usually given by appeal to the Nyquist theorem (i.e. two samples per cycle at the highest frequency present). This represents a lower limit and is only valid for sinc function interpolation of sample values in the spatial domain. Since no one ever does this correct interpolation, oversampling is usually desirable. As a rule of thumb for aperiodic signals, one should take at least 2 samples across the smallest detectable object. However, this lower limit is not a desirable one. If possible, 4 samples per smallest object would be a better choice. With reference to the above CD diagram these proposals mean sampling distances of 0.05 and 0.025 mm respectively if focus blur is neglected. Consideration of focus blur would relax these requirements by at least a factor of 2. Another reduction in requirement can be gained by observing that small objects in patients are rarely radio-opaque so attenuation depends on object size. Assume for example a dense, spherical calcification as illustrated in figure 5-2. If its diameter is less than 0.8 mm, the reduced contrast puts it into the detection threshold range for 10^4 photons/mm^2. Hence a sampling distance of 0.2 mm may be justified. By this sort of reasoning one can obtain a more realistic estimate of minimum object detectable size and then devise an adequate sampling strategy.

Fig.5-2: A theoretical CD diagram for detection of a spherical object (SNR = 10) and 10^4 and 10^5 photons per mm^2. The dotted line is the calculated contrast for a spherical calcification with $\Delta\mu = \mu(ca) - \mu(water) = 1.75\ cm^{-1}$.

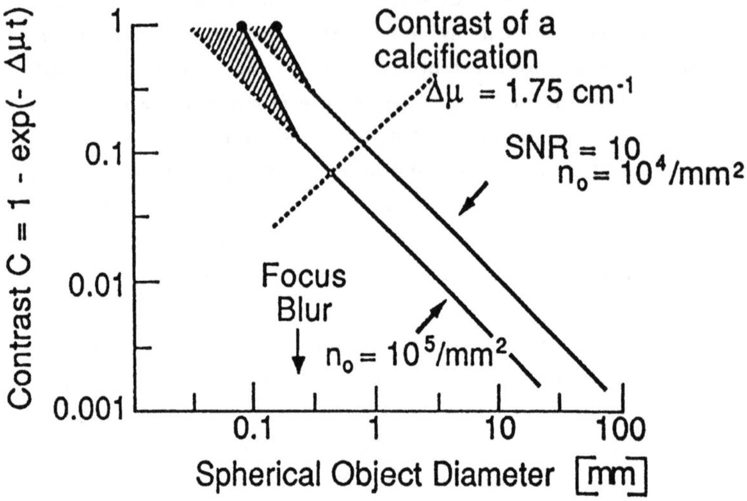

In summary, there are usually real engineering trade-offs between resolution and noise. The best choice of sampling distance depends on the details of the imaging situation and analysis must be done for each individual case. There is no general rule setting a precise upper limit to resolution requirement. Lorre and Gillespie (1980) give an excellent discussion of artifacts due to inadequate resolution.

The resolution issue has recently become a research issue with the advent of digital radiography techniques. Carterette, Fiske and Huang (1986) used an ROC method to compare nodule (9.5 mm diameter) detection accuracy in low (512 x 512) and high (1024 x 1024) digital chest images with film images. There was a significant difference in accuracy between low resolution digital images and the other two (which were not significantly different). Giger and Doi (1987) compared results of an SNR model based on statistical decision theory and human psychophysics with 18AFC observer results. They concluded that a 0.2 mm pixel size may be adequate for visual detection of necessary detail in most radiographic examinations. However they also reported that 0.1 mm pixel size was inadequate for mammography (Chan et al, 1987). Fiske et al (1987) found that the diagnostic image quality of digital images (1024^2) displayed on a video monitor was equivalent to images presented on laser-printed film. Seeley et al (1987) did an ROC study and found that a 1024 by 1536 pixel electronic display console delivers equivalent diagnostic information to that found on film. They also found that contrast manipulation plus roam and zoom capabilities were essential for diagnosis.

REFERENCES

Alter, A.J., Kargas, G.A., Kargas, S.A., Cameron, J.R., and McDermott, J.C. (1982). The influence of ambient and viewbox light upon visual detection of low-contrast targets in a radiograph, Invest. Radiol. 17, pp. 402-406.

Barlow, H.B. (1978). The efficiency of detecting changes in density in random dot patterns, Vision Research 18, pp. 637-650.

Barlow, H.B. (1983). Understanding natural vision. In: Physical and biological processing of images, Braddock, O.J. and Sleigh, A.C. (eds.), Springer-Verlag, New York, pp. 2-14, and J.G. Robson, Frequencty domain visual processing, pp. 73-87).

Barrett, H.H., Myers, K.J. and Wagner, R.F. (1986). Beyond signal detection theory, S.P.I.E. 626, pp. 231-239.

Baxter, B., Ravindra, H., and Norman, R.A. (1982). Changes in lesion detectability caused by light adaptation in retinal photoreceptors, Invest. Rad. 17, pp.394-401.

Berbaum, K.S., Franken, E.A., Dorfman, D.D., and Barloon, T.J. (1988). Influence of clinical history upon detection of nodules and other lesions, Invest. Rad. 23, pp. 48-61.

Berbaum, K., Franken, E.A., and Smith, W.L. (1985). The effect of comparison films upon resident interpretation of pediatric chest radiographs, Invest. Radiol. 20, pp. 124-128.

Biberman, L.M. (1973). Perception of displayed information, Plenum Press, NY, p. 73.

Bollen, R. (1986). Correlation between measured noise and its visual perception, S.P.I.E. 626, pp. 251-258.

Braddock, O.J. and Sleigh, A.C. (1983). Physical and biological processing of images, Springer-Verlag, New York.

Bunch, P.C., Hamilton, J.F., Sanderson, G.K., and Simmons, A.H. (1977). A free response approach to the measurement and characterization of radiographic observer performance, S.P.I.E. Proceedings 127, pp. 124-135.

Burgess, A.E. (1985). Detection and identification efficiency: an update, S.P.I.E. 535, pp. 50-56.

Burgess, A.E. (1985). Effect of quantization noise on visual signal detection in noisy images, JOSA A2, pp. 1424-1428.

Burgess, A.E. (1985). Visual signal detection. III. On bayesian use of prior knowledge and cross correlation, J. Opt. Soc. Am. A2, pp. 1498-1507.

Burgess, A. (1986). On observer internal noise, S.P.I.E. 626, pp. 208-213.

Burgess, A.E. (submitted). Sampling error in MAFC experiments. Submitted for publication in Perception and Psychophysics.
Burgess, A.E. and Barlow, H.B. (1983). The efficiency of numerosity discrimination in random dot images, Vision Research 23, 811-829.
Burgess, A.E. and Colborne, B. (1988). Visual signal detection IV. Observer inconsistency, J. Opt. Soc. Am. A5, pp. 617-627.
Burgess, A.E. and Ghandeharian, H. (1984). Visual signal detection. I. Ability to use phase information, J. Opt. Soc. Am. A1, pp. 900-905.
Burgess, A.E. and Ghandeharian, H. (1984). Visual signal detection. II. Signal-location identification, J. Opt. Soc. Am. A1, pp. 906-910.
Burgess, A.E. and Humphrey, K. (1977). Visual perception limits in angiography, S.P.I.E. Proc. 127, pp. 51-59.
Burgess, A.E., Humphrey, K., Wagner, R.F. (1979). Detection of bars and discs in quantum noise, S.P.I.E. Proc. 173, pp. 34-40.
Burgess, A.E., Jennings, R.J., and Wagner, R.F. (1982). Statistical Efficiency - a measure of human visual signal detection performance, Journal of Applied Photographic Engineering 8, pp. 766-78.
Burgess, A.E., Wagner, R.F., and Jennings, R.J. (1982). Human signal detection performance for noisy medical images, Proc. International Workshop on Physics and Engineering in Medical Images. IEEE Computer Soc. Cat. No.82CH1751-7.
Burgess, A.E., Wagner, R.F., Jennings, R.J., and Barlow, H.B. (1981). Efficiency of human visual signal discrimination, Science 214, pp. 93-94.
Caelli, T., and Nawrot, M. (1987). Localization of signals in images, J. Opt. Soc. Am. A4, pp. 2274-2280.
Carterette, C.E., Fiske, R.A., and Huang, H.K. (1986). ROC evaluation of a digital viewing station for radiologists, S.P.I.E. 626, p. 441-446.
Castleman, K.R. (1979). Digital image processing, Prentice-Hall, Englewood Cliffs, N.J., p.14.
Chan, H.P., Metz, C.E., and Doi, K. (1985). Digital imaging processing - optimal spatial filter for maximization of the perceived snr based on a statistical decision theory model for the human observer, S.P.I.E., 535, pp. 1-11.
Chan, H.P., Vyborny, C.J., MacMahon, H., Metz, C.E., Doi, K., and Sickles, E.A. (1987). Digital mammography: ROC studies of pixel size and unsharp mask filtering on the detection of subtle microcalcifications, Invest. Rad. 22, pp. 581-589.
Chester, M.S. and Hay, G.A. (1983). Quantitative relation between detectability and noise power, Phys. Med. Biol. 28, pp. 1113-1125.
Chester, M.S. and Hay, G.A. (1984). Concerning the quantitative relation between detectability and noise power, Phys. Med. Biol. 29, pp. 602-604.
Chew, E., Weiss, G.H., Brooks, R.A., and Di Chiro, G. (1978). Effect of CT noise on detectability of test objects, A.J.R. 131, pp. 681-685.
Cohen, G., McDaniel, D.L., and Wagner, L.K. (1984). Analysis of variations in contrast-detail experiments, Med. Phys. 11, pp. 469-473.
Coltman, J.W., and Anderson, A.E. (1960). Noise limitations of resolving power in electronic imaging, Proc. IRE 48, pp. 858-865.
Creelman, C.D. (1961). Detection of complex signals as a function of signal bandwidth and duration. J. Acoust. Soc. Am. 33, pp.89-94.
D.J. Goodenough, D.J. (1981). Psychophysical perception in CT images. In: Technical aspects of computed tomography. Newton and Potts (eds.), C.V. Mosby Company, St.Louis and also Ph.D. Thesis, Univ. of Chicago (1972).
Dooley, R.P. (1977) Advances in the psychophysical and visual aspects of image evaluation. Proc. SPSE Technical Section Conference.
Edhome, P. (1981). Boundaries in the radiographic image, Acta Radiologica 22, pp. 457-473.

Fiete, R.D., Barrett, H.H., Smith, W.E., and Myers, K.J. (1987). Hotelling trace criterion and its correlation with human-observer performance, J. Opt. Soc. Am. A4, pp. 945-953.
Fiske, R.A., Valentino, D.J., Huang, H.K., and Blume, H. (1987). The effect of digital image display format on perceived image quality, S.P.I.E. 767, pp. 631-638.
Gabor, D. (1946). Theory of communication, J.IEE (London), 93 (III), pp. 429-457.
Gebauer, A., Lissner, J., and Schott, O. (1967). Roentgen television, Grune & Stratton, New York.
Giger, M.L. and Doi, K. (1985). Investigation of basic imaging properties of digital radiography. 3. Effect of pixel size on SNR and threshold contrast, Med. Phys. 12, pp. 201-208.
Giger, M.L. and Doi, K. (1987). Effect of pixel size on detectability of low-contrast signals in digital radiography, J. Opt. Soc. Am. A4, pp. 966-975.
Gigar, M.L., Ohara, K., and Doi, K. (1985). Investigation of basic imaging properties of digital radiography. 9. Effect of displayed gray levels on signal detection, Med. Phys. 13, pp. 312-318.
Goodall, W.M. (1951). Television by pulse code modulation, Bell Syst. Tech. J.
Gould, R.G., Balongier, B., Goldberg, H.I., and Moss, A. (1980). Objective performance measurement versus perceived image quality in intensifier fluoroscopic or photospot images, Radiology 137, pp. 783-788.
Green, D.M., and Swets, J.A. (1966). Signal detection theory and psychophysics, Wiley, New York.
Gregory, R.L. (1970). The intelligent eye, Weidenfeld and Nicholson, London.
Guignard, P.A., and Salehi, N. (1983). Validity of the gaussian assumption in the analysis of ROC data obtained from scintigraphic-like images, Phys. Med. Biol. 28, pp. 1409-1417.
Ham, W.T. Jr., Mueller, H.A., Ruffolo, J.J., Guerry, D. III, and Guerry, R.K. (1982). Action spectrum for retinal injury from near-ultraviolet radiation in the aphakic monkey, Am. J. Ophthalmol 93, pp. 299-306.
Hanson, K.M. (1979). Detectability in CT images, Med. Phys. 6, pp. 441-451.
Hanson, K.M. (1981). Noise and contrast discrimination in CT. In: Technical aspects of computed tomography. Newton and Potts (eds.), C.V. Mosby Company, St. Louis.
Hanson, K.M. (1981). Noise and contrast discrimination in computed tomography. In: Technical aspects of computed tomography. Newton and Potts (eds.), C.V. Mosby Company, St.Louis, chapter 113.
Hanson, K.M. (1983). Variations in task and the ideal observer, S.P.I.E. 419, pp. 60-67.
Hanson, K.M. (1984). Optimal object and edge localization in the presence of correlated noise, S.P.I.E. 454, pp.-17.
Haque, P. (1981). In Technical Aspects of CT, Vol.5 of Radiology of the Skull and Brain. Newton and Potts, (eds.), C.V. Mosby, St. Louis, p. 4131.
Hay, G.A. and Chester, M.S. (1972). A model of visual threshold detection, J. Theor. Biol. 67, 221-240 and also J. Opt. Soc. Am. 62, 990 (1972).
Hemmingsson, A., Jung, B., Naslund L., and Ytterbergh, C. (1981). Perceptability of experimental and clinical lesions in the CT image with and without image processing, Acta Radiologica 22, pp. 67-76.
Herath, K.B., and Sharp, P.J. (1976). The effects of matched filter smoothing as measured by ROC curve, Phys. Med. Biol. 21, pp. 442-446.
Hillard, A., Myles-Worsley, M., Johnston, W., and Baxter, B. (1985). The development of radiological schemata through training and experience, Invest. Radiol. 20, pp, 422-425.

Hubel, D.H., and Wiesel, T.N. (1977). Ferrier Lecture, Proc. Roy. Soc. B. 198, p. 1.
Ishida, M., Doi, K., Loo, L.N., Metz, C.E., Lehr, J.L. (1984). Digital image processing: effect on detectability of simulated low-contrast radiographic patterns, Radiology 150, pp. 569-575.
Lythgoe, J.N. (1979). Ecology of vision, Clarendon Press, Oxford.
Jaques, P., DiBianca, F., Pizer, S., Kohout, F., Lifshitz L., and Delany, D. (1985). Quantitative digital fluorography - computer vs. human estimation of vascular stenoses, Invest. Rad. 20, pp. 45-52.
Joseph, P.M. (1977). Tradeoff between resolution and density discrimination in CT scanners.
Judy, P.F., and Swensson, R.G. (1985). Detection of small focal lesions in CT images: effects of reconstruction filters and visual display windows, Br. J. Radiol. 58, pp. 137-145.
Judy, P.F., Swensson, R.G. (1985). Detectability of lesions of various sizes on CT images, S.P.I.E. 535, pp. 38-42.
Judy, P.F. and Swensson, R.G. (1986). Size discrimination of features on CT images, S.P.I.E. 626, pp. 225-230.
Judy, P.F. and Swensson, R.G. (1987). Display thresholding of images and observer detection performance, J. Opt. Soc. Am. A4, pp. 954-965.
Judy, P.F., Swensson, R.G. and Kijewski, M.F. (1987). Observer efficiency and feature polarity, S.P.I.E. 767, pp.310-316.
Judy, P.F., Swensson, R.G., and Szulc, M. (1981). Lesion detection and signal-to-noise ratio in CT images, Med. Phys. 8, pp. 12-23.
Kasal, B., Sharp, P.F., and Dendy, P.P. (1983). Relationship between objective and subjective assessment of gamma ray image sharpness. Phys. Med. Biol. 28, pp. 1127-1134.
Kelsey, C.A., Mosely, R.D., Mettler, F.A., Garcia, J.F., Parker, T.W., and Briscoe, D.E. (1982). Comparison of nodule detection with 70 kvp and 120 kvp chest radiographs, Radiology 143, pp. 609-611.
Kersten, D. (1984). Spatial summation in visual noise, Vision Res. 24, pp. 1977-1990.
Kersten, D. (1987). Statistical efficiency for the detection of visual noise, Vision Res. 27, pp.1029-1040.
Kirshfeld, K. (1976). The resolution of lens and compound eyes. In: Neural Principles in Vision. Weiler, R. (ed.) Springer, Berlin.
Kozima, K., and Uchida, S. (1982). Noise smoothing in image improvement, Optica Acta 29, pp. 371-376.
Kuhl, D.E., Sanders, T.D., Edwards, R.Q., and Mecker, P.T. (1972). Failure to improve observer performance with scan smoothing, J. Nucl. Med. 13, pp. 752-757.
Kume, Y., Doi, K., Ohara, K., and Gigar, M.L. (1986). Investigation of basic imaging properties of digital radiography. 10. Structure mottle in II-TV digital imaging systems, Med. Phys. 13, pp. 843-849.
Kundel, H.L. (1979) Images, image quality, and observer performance. Radiology 132, pp. 265-271 and also Radiology 103, 523-528 (1972).
Kundel, H.L., Nodine, C.F., Thickman, D., Carmody, D., and Toto, L. (1985). Nodule detection with and without a chest image, Invest. Radiol 20, pp. 94-99.
Kundel, H.L., Nodine, C.F., Thickman, D., and Toto, L. (1987). Searching for lung nodules - a comparison of human performance with random and systematic scanning models, Invest. Rad. 22, pp. 417-422.
Land, M.F. (1981). Optics and vision in invertebrates. In: Handbook of Sensory Physiology, Vol.VII/6B, Autrum, H. (ed.), Springer-Verlag, Berlin.
Legge, G.E., Kersten, D., and Burgess, A.E. (1987). Contrast discrimination in noise, J. Opt. Soc. Am. A4, pp. 381-404.
Livingstone, M., and Hubel, D. (1988). Segregation of form, color, movement, and depth-anatomy, physiology, and perception, Science 240, pp. 740-749.

Loo, L.D., Doi, K. and Metz, C.E. (1985). Investigation of basic imaging properties of digital radiography. 4. Effect of unsharp masking on the detectability of simple patterns. Med. Phys. 12, pp. 209-214.
Loo, L.D., Doi, K. and Metz, C.E. (1984). A comparison of physical image quality indices and observer performance in the radiographic detection of nylon beads, Phys. Med. Biol. 29, pp. 837-856.
Lorre, J.L. and Gillespie, A.R. (1980). Artifacts in digital images, S.P.I.E. Proc. 264, pp. 123-135.
Lusted, L.B. (1968). Introduction to medical decision making, Thomas, Springfield, IL.
Marr, D. (1982). Vision, W.H. Freeman, San Francisco.
McAdams, H.P., Johnson, G.A., Suddarth, S.A., and Ravin, C.E. (1986). Histogram-directed processing of digital chest images, Invest. Rad. 21, pp. 253-259.
Metz, C.E. (1978). Basic principles of ROC analysis, Semin. Nucl. Med. 8, p. 283.
Metz, C.E. and Goodenough, D.J. (1973). On failure to improve observer performance with scan smoothing: a rebuttal, J. Nucl. Med. 14, pp.873-876.
Moseley, R.D., and Rust, J.H. (1965). Diagnostic radiological instrumentation, Charles C. Thomas. Springfield, Illinois.
Myers, K.J. (1985). Visual perception in correlated noise, Ph.D. Thesis, Univ. of Arizona.
Myers, K.J., Barrett, H.H. (1987). A channels-based model. J. Opt. Soc. Am. A4, pp.2447-2457.
Myers, K.J., Barrett, H.H., Borgstrom, M.C., Patton, D.D., and Seeley, G.W. (1985). Is ideal-observer signal-to-noise ratio a good predictor of human performance? S.P.I.E. 535, pp. 12-15.
Myers, K.J., Barrett, H.H., Borgstrom, M.C., Patton, D.D., and Seeley, G.W. (1985). Effect of noise signals in medical imaging. J. Opt. Soc. Am. A2, pp. 1752-1759.
Nagaraja, N.S. (1964). Effect of luminance noise on contrast thresholds, J. Opt.Soc.Am. 54, pp. 950-955.
Ohara, K., Chan, H.-P., Doi, K., Gigar. M.L., and Fujita, H. (1986). Investigation of basic imaging properties of digital radiography. 8. Detection of simulated low-contrast objects in digital subtraction angiography, Med. Phys. 13, pp. 304-311.
Parker, T.W., Kelsey, C.A., Moseley, R.D., Mettler, F.A., Garcia, J.F., and Briscoe, D.E. (1982). Directed versus free search for nodules in chest radiographs, Invest. Radiol. 17, pp. 152-155.
Pearson, D.E. (1975). Transmission and display of pictorial information, Pentech Press, London.
Pelli, D. (1981). Effects of visual noise, doctoral dissertation (University of Cambridge, Cambridge).
Penney, B.C., King, M.A., Schwinger, R.B., Stritzke, P., Doherty, P.W., and Baker, S.P. (1987). Convolution squared error versus observer preference, S.P.I.E. 767, pp. 317-323.
Pizer, S.M., Austin, J.D., Perry, J.R., Safrit, H.D., and Zimmerman, J.B. (1986). Adaptive histogram equalization for automatic contrast enhancement of medical images, S.P.I.E. 626, pp. 242-250.
Pizer, S.M., Zimmerman, J.B., and Johnston, F.E. (1981). Contrast transmission in medical imaging display. Proceedings ISMIII'82, IEEE Computer Society. Also see S.P.I.E. Proceedings 271, 21-27 (1981).
Pratt, W.K. (1978). Digital image processing, Wiley & Sons, New York, p.154.
Pullin, B.R., Tichings, C.T., Isherwood, I., and Adams, J.E. (1980). Effect of smoothing brain scans. J.C.A.T. 4, pp. 91-93.
Raemer, H.R. (1969). Statistical communication theory and applications, Prentice-Hall, New York.

Ravesz, G., Kundel, H.L., and Graber, M.A. (1974). The influence of structured noise on the detection of radiological abnormalities, Invest. Rad. 9, pp. 479-486.
Regan, D. (1982). Visual information channeling in normal and disordered vision, Psych. Review 89, pp. 407-444.
Revesz, G., Kundel, H.L., and Bonitatibus, M. (1983). The effect of verification on the assessment of imaging techniques, Invest. Radiol. 18, pp. 194-198.
Rimkus D., and Baily, N.A. (1984). Patient exposure requirements for high contrast resolution in digital radiographic systems. A.J.R. 142, pp. 603-608.
Rogers, D.C., Johnston, R.E., and Pizer, S.M. (1987). Effect of ambient light on elecltronically displayed medical images as measured by luminance discrimination thresholds, J. Opt. Soc. Am. A4, pp. 976-983.
Rose, A. (1948). Sensitivity of the eye on an absolute scale, J. Opt. Soc. Am. 38, pp. 196-208.
Rose, A. (1973). Vision -- human and electronic, Plenum Press, New York.
Rosell, F.A., and Wilson, R.H. (1973). Recent psychophysical experiments and the display SNR concept. In: Perception of displayed information. (Chapter 5). L.M. Biberman (ed.), Plenum Press, New York. Also see A.D. Schnitzler in the same book (Chapter 4).
Sakitt, B., and Barlow, H.B. (1982). A model for the economical encoding of the visual imaging in cerebral cortex, Biological Cybernetics 43, pp. 97-110.
Second International Conference on Visual Psychophysics and Medical Imaging (1981). IEEE Engineering in Medicine and Biology Society, IEE Cat. No. 81Ch1676-6.
Seeley, G.W., Robles-Sotelo, E., Cannon, G., Bjelland, J.C., Ovitt, T.W., Standon, J., Capp, M.P., Fisher, H.D., and Dallas, W.J. (1987). The use of psychophysics as a system design aid - comparison of film-screen with an electronic review console, S.P.I.E. 767, pp. 639-643.
Seeley, G.W., Roehrig, H. and Hillman, B.J. (1984). A computerized method for the measurement of conspicuity, Invest. Rad. 19, pp. 583-586.
Seltzer, S.E. Swensson, R.G., Judy, P.F., and Nawfel, R.D. (1988, in press). Size discrimination in CT images: effect of feature contrast and display window, Invest. Radiol.
Selzer, R. (1968). The use of computers to improve biomedical image quality, Proc. Face Joint Computer Conference 33, pp. 817-834.
Sezan, M.I., Yip, K.L., and Daly, S.J. (1987) An investigation of the effects of uniform perceptual quantization in the context of digital radiography, S.P.I.E. 767, pp. 622-630.
Sharp, P.F., Chesser, R.B., and Mallard, J.R. (1982). The influence of picture element size on the quality of clinical radionuclide image, Phys. Med. Biol. 27, pp. 913-926.
Simmons, J.A. (1973). The resolution of target range by echo locating bats, J. Acoust. Soc. Am. 54, p. 157-173.
Snyder, A.W. (1975). Photoreceptor optics - theoretical principles. In: Photoreceptor Optics, Springer-Verlag, Berlin.
Sorenson, J.A., Mitchell, C.R., Armstrong, J.D., Mann, H., Bragg, D.G., Mann, F.A., Tocino, I.B., and Wojtowcyz, M.M. (1987). Effects of improved contrast on lung-nodule detection. A clinical ROC study, Invest. Rad. 22, pp. 772-780.
Srinivasan, M.S., Laughlin, S.B., and Dubs, A. Predictive coding: a fresh view of inhibition in the retina. Proc. Roy. Soc. (London).
Starr, S.J., Metz, C.E., Lusted, L.B., and Goodenough, D.J. (1975). Visual detection and localization of radiographic images, Radiology 116, pp. 533-538.
Sturm, R.E., and Morgan, R.H. (1949). Screen intensification systems and their limitations, A.J.R. 52, pp. 617-634.

Swensson, R.G., Hessel, S.J., and Herman, P.G. (1982). Radiographic interpretation with and without search - visual search aids in the recognition of chest pathology, Invest. Radiol. 17, pp.145-151.
Swensson, R.G., Hessel S.J., and Herman, P.G. (1985). Author's Response, Invest. Radiol. 20, pp.111-113.
Swets, J.A. (1964). Signal detection and recognition by human observers, Wiley, New York.
Swets, J.A. (1985). On the supposed value of searching films, Invest. Radiol. 20, pp. 108-109.
Swets, J.A., and Pickett, R.M. (1982). Evaluation of diagnostic systems --methods from signal detection theory, Academic Press, New York.
Swets, J.A., Pickett, R.M., Whitehead, S.F., Getty, D.J., Schnur, J.A., Swets. J.B., and Freeman, B.A. (1979). Assessment of diagnostic technologies, Science 205, pp. 735-759.
Tanner, W.P., and Birdsall, T.G. (1958). Definitions of d' and η as psychophysical measures, J. Acoust. Soc. Am. 30, pp. 922-928.
Tesic, M.M., Sones, R.A., and Morgan, D.R. (1984). Single-slit digital radiography, A.J.R. 142, pp. 697-702.
Tuddenhamm W.J. (1969). Perception of the roentgen image, Radiologic Clinics of North America 7, W.B. Saunders Co., Philadelphia.
Uttal, W.R. (1981). A taxonomy of visual processes. L. Erlbaum Associates, Hillsdale, NJ.
Van Meeteren, A., and Valeton, J.M. (1988). Effects of pictorial noise interfering with visual noise, J. Opt. Soc. Am. A5, pp. 438-444.
VanTrees, H.L. (1968). Detection, estimation and modulation theory. Volumes I, II and III, John Wiley, New York.
Vos, J.J., and VanNorren, D. (1984). Limits in the visual spectrum. In: Limits in perception, Van Doorn, A.J., Van de Grind, W.A., and Koenderink, J.J., (eds.), VNU Science Press, Utrecht, pp. 66-84.
Wagner, L.K. and Cohen, G. (1982). Energy dependence of contrast detail dose and objective detectability dose curves for CT scanners, J. Comput. Assist. Tomogr. 6, pp. 378-382.
Wagner, R.F. (1977). Toward a unified view of radiological imaging systems. Part II: Noisy Images, Med. Phys. 4, pp. 279-296.
Wagner, R.F. and Barrett, H.H. (1987). Quadratic tasks and the ideal observer. In Medical imaging, R.H. Schneider and S.J. Dwyer, (eds.), Proc. Soc. Photo-Opt. Instrum. Eng. 767, p. 306-309.
Wagner, R.F. and Brown, D.G. (1985). Unified SNR analysis of medical imaging systems, Phys. Med. Biol. 30, pp. 489-518.
Wagner, R.F., Brown, D.G., and Pastel, M.S. (1979). Application of information theory to the assessment of CT, Med. Phys. 6, pp. 83-94.
Wagner, R.F., Brown, D.G., Burgess, A.E., and Hanson, K. (1983). The observer SNR penalty for image reconstruction from projections, Magnetic Resonance in Medicine 1, pp. 76-77.
Wagner, R.F., Insana, M.F., and Brown, D.G. (1985). Progress in signal and texture discrimination in medical imaging, S.P.I.E. 535, pp. 57-63.
Warren, R.C. (1984). Detectability of low-contrast features in computed tomography, Phys. Med. Biol. 29, pp. 1-13.
Watson, A.B. (1983). Detection and recognition of simple spatial forms. In: Physical and biological processing of images, Braddock, O.J., and Sleigh, A.C. (eds.), Springer-Verlag, New York, pp. 100-114.
Webster, E.W., and Wipfelder, R. (1964). Contrast and detail perception in television and ciné systems for medical fluoroscopy, J. Soc. Motion Picture Television End. 73, p. 617.
Wilcox, G.W. (1968). Inter-observer agreement and models of monaural auditory processing in detection task, doctoral dissertation (University of Michigan, Ann Arbor, Mich.).

ARTIFICIAL INTELLIGENCE IN THE INTERPRETATION OF MEDICAL IMAGES

E. Backer

Delft University of Technology
Department of Electrical Engineering
Delft, The Netherlands

ABSTRACT

This article presents an overview of the field of knowledge-based approaches in the processing and interpretation of medical images. It describes the major developments in artificial intelligence and image processing that have led up to the current potential to further the increasing interest in automated processing (knowledge-based systems) and interpretation (expert systems) of medical images. After a brief discussion of the principal scientific and engineering issues in the field of knowledge-based image processing, the process of building knowledge-based systems, and the frontiers of research and development, a number of reported systems for processing and interpretation of medical images are briefly reviewed. The development of an expert system for the analysis of scintigraphic images is then presented, emphasizing knowledge acquisition, implementation and evaluation. Finally, a perspective view on reasoning under uncertainty is presented.

1. INTRODUCTION

This paper is devoted to the problem of automatic processing and interpretation of medical images using techniques from artificial intelligence. In short, relevant knowledge about the task domain is represented in a system in an efficient way aiming at using that knowledge either to facilitate processing of the input image (knowledge-based system approach) or to derive a diagnostic interpretation from the pixels of an input image (expert system approach). Medical imaging has become an important source of diagnostic information, in particular since new and powerful imaging devices like X-ray and MR tomograms, scintigraphic imaging, and ultrasound tomograms have become available. The diagnostic interpretation of these images requires experts having the relevant competence in the field. For various reasons there is - and will be - a strong need and a challenge to built support systems which have capabilities for and performance comparable to human experts in a reasonably complex area like medical image interpretation. Likewise, processing of those images may require domain dependent knowledge to achieve interpretable results. Those systems rely on explicitly represented knowledge about the task domain, and can handle imprecise and uncertain situations. Both, knowledge manipulation and reasoning under uncertainty naturally leads to artificial intelligence techniques and particularly to knowledge-based systems and expert systems.

Before turning to the field of knowledge-based systems and expert systems in processing and interpretation of medical images, we should consider briefly some background material on artificial intelligence and the design and engineering issues of knowledge-based systems and expert systems, as well as related issues of image understanding in general.

1.1 Artificial intelligence

The practice of designing systems that possess and acquire knowledge and reason with that knowledge is perhaps the ultimmate intellectual challenge. The prospect of machine intelligence may have produced a widespread scepticism[1] and the field of study - Artificial Intelligence (AI) - may need several decades to materialize its promise and show practical feasibility. Nontheless, the AI-research community is rapidly growing and the development of AI-tools has proven to have commercial and practical value[2].

Broadly defined, Artificial Intelligence is a field of study that encompasses computational techniques for performing tasks that apparently require intelligence when performed by humans. Such problems include diagnosing in medicine, understanding scenes by machine vision systems at a human level, assembling and inspecting products in factories, to name just a few. Each of them involves knowledge representation, search, perception and inference as the fundamental issues. Two major branches of research can be distinguished: a branch related to cognitive science aiming at human modelling, and a branch related to the design of intelligent machines aiming at construction of computer programs that perform at high levels of

---

1 Hubert and Stuart Dreyfus (1986): "After 25 years artificial intelligence has failed to live up to its promise and there is no evidence that it ever will be".

2 Dana S. Nau (1983): Artificial intelligence is no longer science theory. A variety of "thinking" systems are out of the laboratory and successfully solving problems using AI knowledge representation techniques.

competence in cognitive tasks without any attempt to mimic the internal structure of human behaviour.

Artificial intelligence was born in 1956 with heuristic search as founding principle. In the sixtees, AI-research shifted to the knowledge based paradigm: how to acquire and how to represent knowledge. To date, the majority of AI-research is concerned with the design and understanding of knowledge representation schemes. Knowledge can be made available in many forms: collection of logical assertions, heuristic rules, procedures, statistical relations, and so on. The representation should be such that it can be easily used in reasoning, can be easily examined and updated, and can be easily judged as relevant or irrelevant to particular problems.

Search is a key issue because it should help us to avoid the 'combinatorial explosion' that swamps all brute-force attempts. Finally, inference is the process of creating explicit representations of knowledge from implicit ones: so to speak, the creation of knowledge itself. Deductive inference proceeds from a set of assumptions to new statements that are logically implied by those assumptions. Inductive inference typically starts with a set of facts (observations) and produces generalizations which account for the given information and which may have the power to predict new facts or observations.

Excellent and easy reading has been provided by Tanimoto (1987), Rich (1983), Winston (1977) and Nilsson (1981), to name just a few books from the large amount of AI-literature.

1.2 Knowledge representation

We think of knowledge as an often more general kind of information than that found in a conventional data base consisting of raw figures and measurements. But it may be incomplete or fuzzy as well. We may also think of knowledge as a collection of related facts, procedures, models and heuristics that can be used in solving a problem. Finally, knowledge may be regarded as information in context, as information organized so that it can be readily applied to specific (sub-) problems.

Thus, knowledge varies widely in both its content and appearance. It may be specific, general, exact, fuzzy, procedural, declarative and so on.

Production rules are a general method to represent knowledge that is particularly appropriate when knowledge is action-oriented. In many expert systems much of the knowledge is represented within production rules.

The left hand side of a production rule (the condition part) expresses the characteristics of a situation in which it is appropriate to perform the action (the right hand side) of the rule. The production rules do not embody all the knowledge in a system. The structure of the knowledge base itself enbodies part of the available (contextual) knowledge.

Frames provide such an organizational scheme for knowledge bases. By providing the knowledge in modules (frames), the designer makes life easier for the algorithms that will access the knowledge. As such, a frame is a collection of knowledge relevant to a particular object, situation, or concept, providing a representation in terms of a set of attribute names and values for the attributes (declarative knowledge).

In semantic networks each object or concept is represented only once and all interrelations between objects are established by pointers for cross reference. As such, it is a very efficient knowledge representation scheme. If each object or concept is treated as a node, all associations between nodes are represented as arcs in a graph. As a consequence, it is possible to use efficient graph-search methods to locate desired information.

1.3 Search

Exhaustive techniques can only be appropriate for (very) small problems. For larger problems we need a 'clever' search method to circomvente the exhaustive search. A good search method typically uses some particular information about the problem or some general knowledge to focus the search for a solution on areas of the state space that have a reasonable chance of containing the solution. Based on problem knowledge we either guide the search by some evaluation function or we prune[3] the search-tree beforehand.

---

3 In expert systems, pruning refers to the process whereby one or more branches of a decision tree are "cut off" or ignored. In effect, when an expert system consultation is underway, heuristic rules reduce the search space by determining that certain branches (or subsets of rules) can be ignored.

As such, the choice of a search method is often a critical choice in a system.

Common strategies like 'depth-first' and 'breadth-first' are 'blind' in the sense that they do not have any sense of where the goal-node lies until they find it. They generally spend a lot of time searching totally fruitless directions. They only differ in the way they go through the search-tree. Heuristic search methods are guided by some evaluation function that maps each node to a real number, and which serves to estimate the relative benefit ofcontinuing the search from that node. The 'best-first' strategy is guided by the evaluation function such that successive nodes are ordered on the basis of the evaluation function and the most favourable is treated as the next node; and so on.

The A*-algorithm performs usually better if some additional heuristic information is available such that the total cost upto the goal can be estimated. Successive nodes are the ordered on the basis of their actual costs added to the estimated goal costs. The node with the smallest 'overall' cost is then selected.

1.4 Reasoning under uncertainty

As mentioned before, the knowledge emploid in a machine is always limited. It is incomplete, and not always exact. Yet, we want them to use for inference problems. The phenomenon of uncertainty has been widely studied on the theory of probability and theories of evidence and degrees of belief. To an arbitrary statement, any one who knows what he believes can lend a judgement: 'sure','impossible', 'may be', 'unlikely but possible'. If we require that the person choose a number in some range, say, 0 to 10, to indicate his degree of belief in the truth of the statement, we could interprete 0 as his certainty that the statement is false, 5 his belief that it may just as well be true as false, and 10 his certainty in its truth. The value he chooses represents his (subjective) belief in the statement.

Probabilities are numerical values between 0 and 1 that represents 'ideal' certainties of statements, given various assumptions about the circumstances in which the statements are relevant. The concept of probability (Bayes theory) has been studied through the controlled circumstances of mathematically simple situations. There are other systems for assigning values to degrees of belief that are arguable more appropriate than probability or systems equivalently to probability. One of these, commonly known as 'Dempster-Shafer theory' receives much attention recently. If we always had accurate general knowledge for such inference problems, we could make simple and clean machines to compute probabilities for various things considering all the evidence. Unfortunately, we usually do not have accurate knowledge of the conditional probabilities of sets of evidence given the hypothesis, so that the ideal, all-bayesian system cannot be successfully built. Heuristic modelling tools can be used to represent known relationships between evidence and conclusion. A variety of approximate bayesian systems has been developed. Among them, the 'Shortliffe-Buchanan' model is most widely used. Also, the fuzzy set theory is more and more become an appropriate candidate formalism to model the propagation of uncertainties through the inference procedure.

In Section 4, we will come back to problem of reasoning under uncertainty in much greater detail.

1.5 Knowledge acquisition

The process of locating, collecting, and refining knowledge is called knowledge acquisition. Or phrased differently, knowledge acquisition consists of the collection and processing of data on the functioning of expertise in some domain, in order to design, build, extend, adapt or modify a knowledge-based system (Breuker and Wielinga, 1984) The question of where knowledge comes from is an appropriate one to begin with before examining any processes of knowledge acquisition (from problem identification to testing and refining the system). One of many possible classifications of knowledge sources is the following:

1. experts (note that from a philisophical point of view only a vague model exists of what an expert makes an 'expert');
2. books (although written by experts, book knowledge appear differently);
3. direct experience.

The acquisition process highly depends on the source of knowledge and on the representation to be used. For each representation method considered, problems of forming a knowledge base may differ:

1. production system: to turn new knowledge into rules, and to incorporate the new rules into the production system;
2. semantic nets: adding knowledge to a semantic net typically requires creating new nodes and linkes to represent the new knowledge, and then tying the new structure into the existing network by either adding links, merging old, or both;
3. frames: to expand the knowledge of a frame-based system, it may a simple matter of filling in more slots of existing frames or creating new frames.

In most cases, a knowledge acquisition system is designed to transfer knowledge and to update the data structure automatically.
Knowledge engineering preceeds the ultimate implementation of the knowledge base. In that, the knowledge engineer plays an important role. His task can be seen as a sequence of five stages:

1. identification: discussion with the expert aiming at a global view upon the subject matter, straight forward recipies and simple testing;
2. conceptualization: familiarizing with the subject jargon, case studies, and formalizing common structures and reasoning;
3. formalization: formalizing a - possible still naive - structure of a reasoning process; this stage may iterate towards a final structure;
4. implementation: focussing on the process of drawing conclusions (heuristics); what is being concluded, under what conditions and how certain are the conclusions; as a result we obtain the production rules;
5. testing and validating: testing and validating the knowledge base in practice; experts should be asked to tune the knowledge base; if more experts are involved, additional sessions may be needed to resolve conflicting issues.

The above process is iterating until test consultations yield stable and satisfying results.
However, knowledge elicitation is still the bottleneck to AI-progress because it is difficult and very time consuming. Medical image interpretation is in that no exception:

- knowledge of an image reader is complex and messy and often ill-formulated;
- image readers find it difficult to articulate what knowledge they have and how they use that knowledge to interprete their images;
- the more expert an image reader becomes, the more "unconscious" his or her knowledge becomes.

The data one gets from using knowledge elicitation techniques are in the form of an expert's verbal comments or his actions. These need careful, even skilled, interpretation as to what underlying knowledge they imply.

In Section 3, some of the above problems are exemplified in greater detail.

1.6 Knowledge based systems

As said in the beginning, knowledge based systems generally incorporate knowledge both in conventional procedures and in knowledge bases, and are aiming at using knowledge in a highly structured manner. Such systems are said to have the following desirable properties (Hayes-Roth, 1984; Buchanan and Duda, 1983):

- powerful : intelligent behaviour in a complex task;
- robust : capable of handling incomplete and inprecise data;
- opportunistic: capable of handling knowledge at the right time;
- transparant : capable of reasoning and supporting conclusions.

The backbone of the anatomy of a knowledge based system is formed by the control mechanism, the data structure and the system architecture.

Very often the system organization is such that procedures have access to both the data memory (short term memory) and a knowledge memory (long term memory). The data memory contains input data, parameter values, modified data, results and so on. The long term memory contains all production rules for processing, control and strategy, as well as object models and interrelations.

The control strategy is typically focussed upon how and in which order production rules should be asked, and how dynamically the focus of attention can be modified.

In a rule-based system, the inference engine typically is that portion of the system that contains the inference and control strategies (forward and backward chaining)[4].

In the blackboard architecture (Erman et al., 1980), the problem complexity is typically reduced by deviding the solution space (the common memory, the blackboard) into an hierarchy of solution (abstraction) levels and knowledge into independent knowledge sources. Inferences take place in the blackboard. Very often, the control is agenda-based[5].

Within the wide scope of applications, knowledge-based systems in the interpretation of medical images has proven to be most challenging (and difficult). There are several reasons why interpretation of medical images is such a challenging kind of artificial intelligence to achieve:

- the human body is three-dimensional, and the images from which a description must be formed are only two-dimensional;
- each pixel of an image represents the interaction of many processes, and it is difficult to separate these influences (image formation, reflections, absorptions, distortions and so on);
- the volume of data is very large, and the consequently need for special architecture is felt strongly;
- in order to interprete a medical image intelligently, much knowledge is needed about the phenomena that may appear in the scene.

Intelligent interpretation is known to be an act of decision that is goal-oriented, arrived at by an understandable chain of symbolic analysis and reasoning steps, and is one in which knowledge of the human body and medical phenomena informs and guides the processing and reasoning. The use of expert systems and AI-methods has just started to be applied to medical

4 Inference is the process by which new facts are derived from known facts; forward chaining is one of several control strategies that regulate the order in which inferences are drawn, initiated by those rules whose IF clauses are true; backward chaining is one of several control strategies that regulates the order in which inferences are drawn, initiated by a goal-rule.

5 An agenda is an ordered list of actions; an agenda-based control system continually examines all of the possible pending actions and chooses the one to try next.

image data. In what follows, a distinction is made between between an expert system and a knowledge-based system in such a way that "expert system" is reserved for systems that truly rival human experts in interpretation of medical images and "knowledge system" is used when speaking of small processing systems developed by means of AI-techniques. This distinction is clarified in Figure 1; Section 2 is primarily devoted to category A and Section 3 is exemplifying category B.

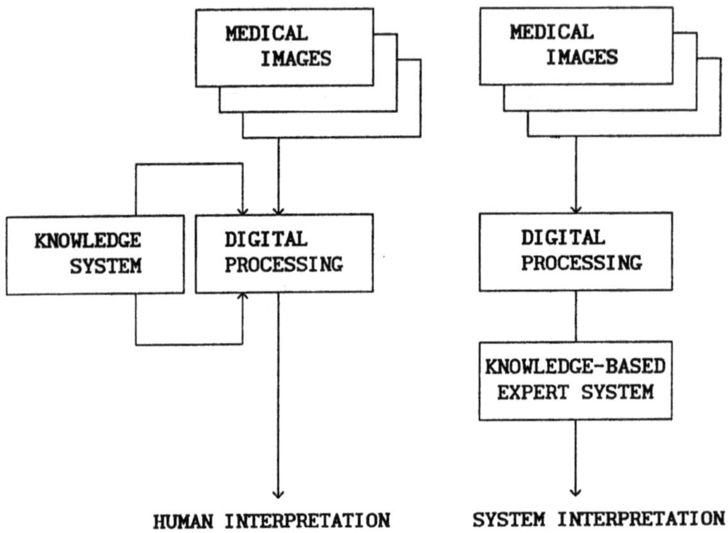

Figure 1: Knowledge system versus expert system;
Category A: knowledge guided processing;
Category B: expert system image interpretation.

1.7 Knowledge guided image processing

Image processing includes a wide variety of actions upon images for different purposes. Figure 2 shows a general diagram in which those actions are indicated, going from the raw data (image sensor) upto the image interpretation at the very end (Rosenfeld and Kak, 1982; Castleman, 1979).

Research in image understanding is focussed upon automatic segmentation of images in meaningful components and interpreting these components in their relationship. Two fundamental steps in the analysis are to be distinguished:

- segmenting the image in regions which - though not exact - should relate to objects, surfaces of parts of objects in the original scene represented by the image: SEGMENTATION;

- interpreting the segmented results and assigning to them some meaningful labels including their relationships: INTERPRETATION.

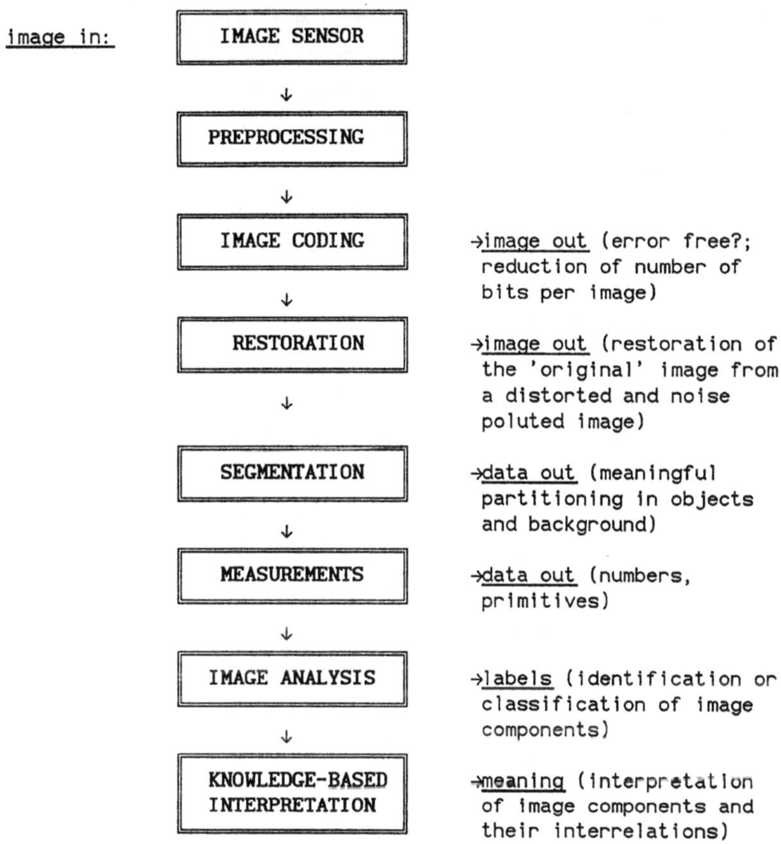

Figure 2: Schematic diagram of actions in digital image processing.

Traditionally[6] , the difference between these two levels in image analysis is due to the fact that different kind of available knowledge is utilized. Segmentation applies general knowledge about the nature of images, the criteria how pixels are to be grouped, the properties of local neighbourhoods and alike. Interpretation, however, applies problem dependent

---

6 Nazif and Levine (1984): A low-level expert system for image segmentation; one of the early attempts to incorporate segmentation heuristics coded into rules.

knowledge or domain knowledge about the scene and the nature of origin, i.e. the medical phenomena. Today[7], there is a tendency to use domain knowledge also at low-level processing.

Integrating knowledge is as old as image processing itself. A large diversity of procedureal algorithms[8] has been developed, e.g. contour detection, region growing, split and merge, and so on for a variety of applications, each of which applying numbers of adhoc and heuristic rules. As a consequence, algorithms which came to existence are more often very domain specific and rarely comparable. There does not exist a well developed method for deciding which algorithm should be most appropriate for a given input image. Also, at each processing level, knowledge about processing itself and knowledge about the problem domain has not been incorporated in a structured manner.

Knowledge guided operation is conceptually multi-layered in the sense that at each layer (going from raw data to full understanding and interpretation) the appropriate knowledge can be applied, and at each layer the appropriate algorithm can be selected. Due to feedback of detected anomalities, reliability, certainty or inexactness, bottom-up and top-down reasoning may take place at each pair of layers. In figure 3, knowledge guided operation is schematized in the form of conceptual abstraction pyramids.

A very general model for implementation is schematized in Figure 4. In this model, knowledge-free procedures are communicating with a data memory (Short Term Memory) and a knowledge memory (Long Term Memory). The data memory contains input image data, parameter values, segmentation data, output results and so on. Whereas in the long term memory all production rules for processing, control and strategy are stored together with object models and relations. All processing modules do have access to both memory blocks.

---

7 Suetens et al. (1987): Knowledge-based segmentation of subtraction angiograms. In a bottom up / top-down fashion they use geometrical knowledge (problem-independent) at the lowest level, general knowledge (bloodvessels) at intermediate level, and domain-specific knowledge (angiograms and related diseases) at the highest level. See Section 2.

8 An algorithm is a systematic procedure that, if followed, guarantees a correct outcome. In developing a conventional program, the programmer must specify the algorithm that the program will follow.

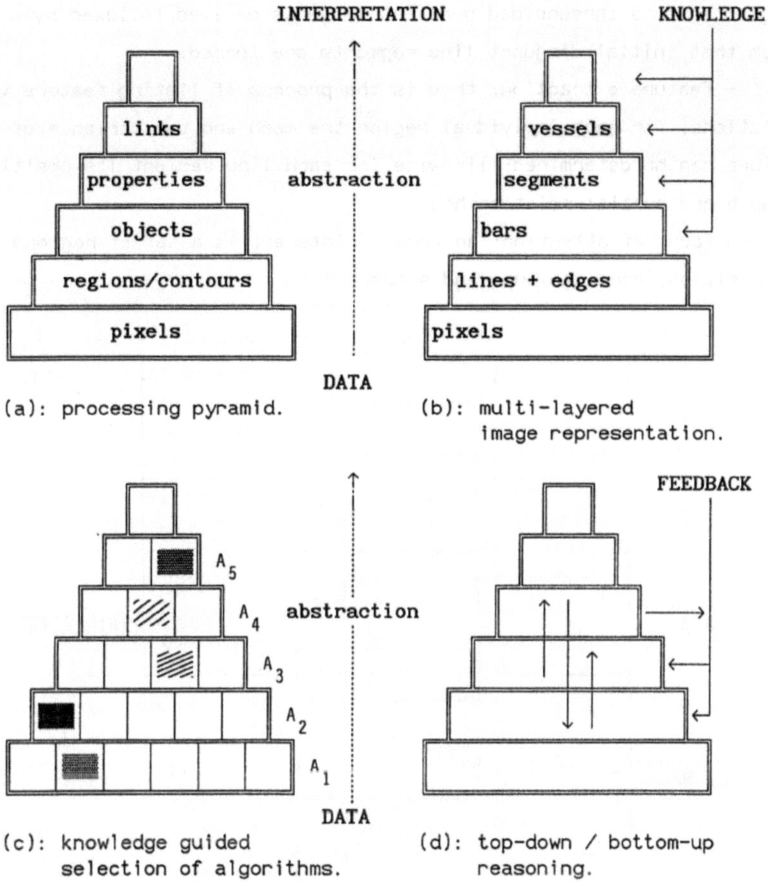

Figure 3: Conceptual pyramids for knowledge guided operation.

The expert system approach to low-level image segmentation of Nazif and Levine (1984) is going along the above system block diagram. Due to the fact that segmentation results are not to be found uniquely, evaluation will be necessary in order to keep the number of interventions as small as possible. In this sense, the segmentation problem can be defined as an optimization problem in which a set of evaluation parameters has to be extremized. The control strategy is focussed upon the order in which different heuristics are applied, and the path along which the heuristics are tested in the image (as such, both the rules and the processing path are dynamically selected).

The processing steps are the following:

- initialization: this is the process of generating a random region partitioning (non-overlapping connective regions); from the initial segmentation a thresholded gradient image is derived followed by a grouping such that initial disjunct line segments are formed.

- feature extraction: this is the process of listing feature values and relations; for each individual region the mean and the variance of the gray values can be determined; likewise for each line segment its position, length and spatial relationship.

- focus of attention: an area of interest is a set of regions described by position, area, texture and shape.

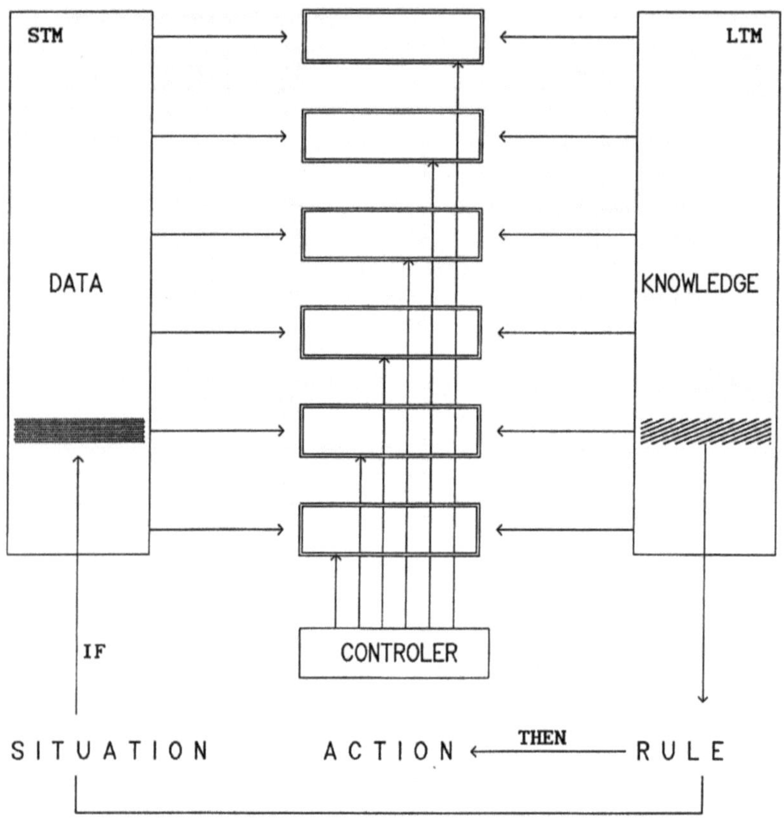

Figure 4: Knowledge guided operation in a rule-based system.

Now, the segmentation procedure will be driven by situation-action rules (based upon the properties of regions, line segments and areas of interest, actions are foreseen like splitting, merging, joining, deleting, extending and modification of the area of interest), control rules (specifying the order in which rules are used, specifying the next action and evaluation), and strategy rules (selection of control rules and dynamic decision making).

A situation is characterized by one of the following quantifications: *very low*, *low*, *medium*, *high*, and *very high*. As an example, a rule reads as follows:

IF
- the line length is *very low*
- the line average gradient is *low*

THEN
- delete line.

2. KNOWLEDGE-BASED SYSTEMS IN PROCESSING AND INTERPRETATION OF MEDICAL IMAGES

As we have reviewed the major engineering issues in artificial intelligence and knowledge-based approaches in image processing, we review in this section four major contributions[9] on the basis of which we are able to identify all scientific and engineering issues which we have encountered in Section 1.

One of the early papers on AI methods and systems for medical consultation is due to Kulikowski (1980). Although the paper does not consider medical image processing or interpretation, it highlights a number of underlying problems and compares the pattern recognition approach with the artificial intelligence approach.

The second contribution is from Stansfield (1984): ANGY: a rule- based expert system for automatic segmentation of coronary vessels from digital subtracted angiograms. In this paper three levels of processing are considered: preprocessing, low-level processing and high-level processing in which segmentation knowledge and knowledge about cardiac anatomy and physiology is used.

---

9 We note that the number of valuable contributions in literature is still increasing substantially; they are listed in the REFERENCES. The majority of them are devoted to general methods and/or cardiology.

In the third contribution, Niemann et al. (1985) also considers low-level, high-level processing and rule inference. However, the knowledge about the image sequences taken in nuclear medicine from the human heart is represented in a semantic net (declarative knowledge, where nodes are the objects or events, and edges are the relationships between them).

Finally, the contribution of Cleyenbreugel et al. (1987) describes a real bottum-up and top-down reasoning scheme applying geometrical knowledge at the lowest level, general anatomic knowledge at the medium level, and domain-specific knowledge of angiograms and related diseases at the highest level of processing. All approaches have in common that they are very much goal-driven, but essentially knowledge-based as far as the processing is concerned (category A). Full interpretation in the sense that medical reasoning is involved (category B) towards medical causality, medical diagnosis and so on is not present in these cases. That will be the subject of Section 3.

2.1 Decision-making methods in medical consultation

A significant component of human consultative reasoning is often characterized as being judgemental. In designing computer-based systems an immediate question arises as to how best to simulate such judgements, if indeed they are to be simulated at all.

A variety of formal methods of decision-making have been applied either by using objective methods, usually of a statistical decision-theory type, or by attempting to capture the expertise of human consultants in the form of reasoning rules that directly incorporate judgemental elements. Much about the above can be found in the paper by Kulikowski (1980).

Two different, though related, ways of encoding medical reasoning in a computer program - to be discussed here within the limited scope of this paper - are pattern recognition approaches and knowledge-based artificial intelligence approaches.

a) pattern recognition methods

In the late 1960's and early 1970's various pattern recognition methods began to be applied to medical decision making. Figure 5 shows a schematic diagram of a typical statistical pattern recognition system for medical consultation. Such a pattern recognition system is intrinsically suffering from the curse of statistically small sample sizes, and is from the viewpoint of medical reasoning suffering from being a "black box" approach. Usually, the use of well-chosen heuristics is indispensable to be able to do any judgemental recommendation.

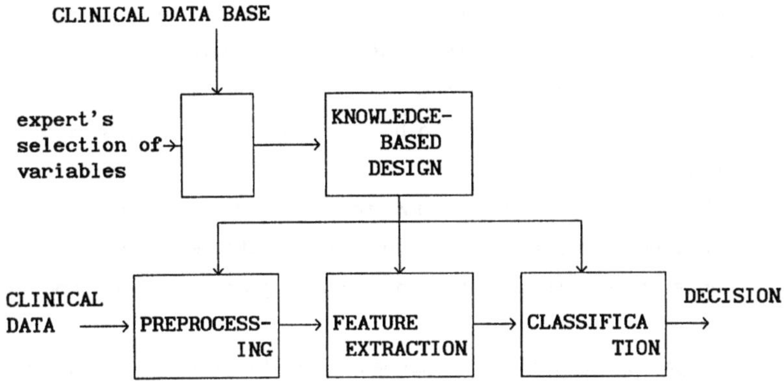

Figure 5: Schematic diagram of a typical pattern recognition system for medical consultation.

The upper part of the block diagram in Figure 5, typically belongs to the learning or design phase of the system. Given the scope of the problem, the medical expert defines the relevant features to be used, and the data base of well-established diagnostic cases is used to learn and to generalize the appropriate decision rules.
The sequence of operations in the lower part of the block diagram is operational. The preprocessing is to make medical data assessible for numerical interpretation and manipulation, i.e. the feature extraction, on the basis of which the classifier statistically infer a predefined interpretation.

The application of artificially intelligence methods sought to remedy the "black box" situation by introducing a structure of knowledge familiar to the physician into the decision-making scheme.

b) artificial intelligence methods

All the AI approaches use heuristic measures for scoring the weight of confidence or credibility that they assign to a hypothesis as an explanation of the patient's condition. These measures are typically computed from uncertainty weights attached by the human experts to the various reasoning rules in the consultation model. Much about this in Section 4.

The reasoning strategies of all the systems, however, rely as much on the structure of connectivities among concepts and facts, as on the scoring mechanisms themselves. This provides the systems with a natural way of supporting explanations, and often allows for the use of alternative and sometimes redundant lines of reasoning to be pursued, giving a measure of flexibility to their behaviour.

In Kulikowski (1980), the major AI-based systems are reviewed [10]. They all share certain characteristic properties. Figure 6 illustrates some of the principal components of the systems, and the resulting consultation process.

In contrast to the pattern recognition approach, knowledge manipulation is incorporated in the opertational process, and deliberately separating domain-specific knowledge, general mechanisms of evaluation, and the control strategies.

The control strategies specify how the different reasoning mechanisms are to be invoked, either automatically, or in response to interactive commands given by the user.

After the clinical data has been presented to the program, the control strategies can lead it to generate local interpretations (directly data-related hypotheses), request more data, proceed to a global interpretation (evaluating and comparing the partial interpretations, and selecting the most likely and coherently structured groupings of hypotheses), and generate conclusions (integrating various hypotheses). An explanation facility (not drawn in the block diagram) finally may produce explanations for any of the preceding steps. Modern approaches allow for going back and forth through the inference process from local to global interpretation. In building an AI consultation system, we rely more heavily

---

10 CASNET: Claucoma Consultation System (1978);
MYCIN: Infectious Disease Treatment Consultant (1976);
INTERNIST: Diagnostic Consultant in Internal Medicine (1975);
PIP: Present Illness Program (1978).

on the knowledge of medical experts than in pattern recognition systems.

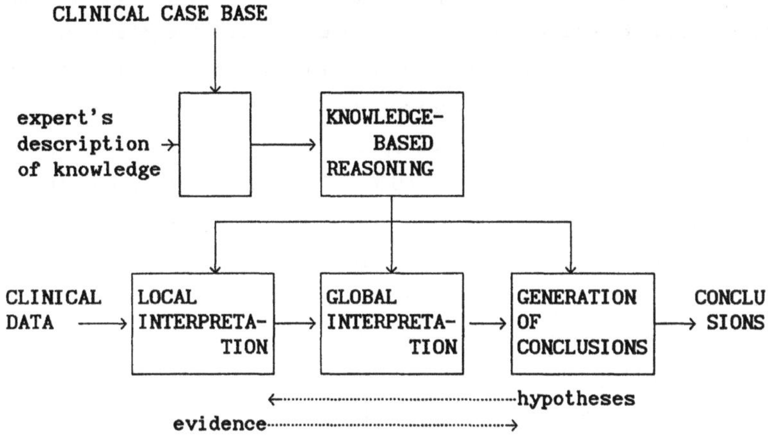

Figure 6: Artificial Intelligence system for medical consultation.

2.2 Knowledge-guided processing of medical images

In view of this section, we discuss three systems that appeared in literature, on the basis of which we trace the evolution of the knowledge-based schemes that have been developed to the present.

1) ANGY: A rule-based expert system for automatic segmentation of coronary vessels from digital subtracted angiograms (Stansfield, 1984). This system combines the traditional methods of image processing with techniques from the field of artificial intelligence to solve the problem of identifying and isolating the coronary vessels in digital angiograms. The expert system itself is separated into two independent stages. In the low-level stage, a domain independent knowledge of image segmentation, grouping, and geometric relations is applied to the segmented image created by the preprocessing stage. Rules are applied to join line segments, to merge regions, and to establish such relations as adjacent and parallel. The result is a refined segmentation and a set of relations between the objects in the segmented image[11].

---

11 Note that the low-level specialist of ANGY can be compared with the low-level expert system approach of Nazif and Levine (1984).

The high-level stage then applies a domain-dependent knowledge of cardiac anatomy and physiology to interpret this segmented image. The system recognizes only two classes of objects: vessel and noise. Rules are applied to label unknown objects as belonging to one of these two classes. The labels are then used to either retain or eliminate the labelled object and to guide the system in further processing. Anatomical knowledge is embodied within the system in the form of the spatial relations between objects and the expected characteristics of the objects themselves. As such, the high-level stage produces a labeled structure in the segmented image[12].
The reported implementation suffered from two major failures:

i) the labeling of some noise structures as vessel if they are of the right shape and are too close to a vessel or the aorta;

ii) the labeling of some portions of vessel as noise if they are not elongated and the segmentation process has resulted in their being isolated from any vessel.

2) A knowledge-based system for analysis of gated blood pool studies (Niemann et al., 1985). The system consists of four main modules as indicated in Figure 7.

Problem specific knowledge about the image sequences under consideration and their possible interpretations is contained in a model which consists of declarative and procedural knowledge. The declarative part contains a description of structural properties of the heart and the left ventricle, of a heart cycle and its motional phases, and of medical evidence which can be inferred from them. A semantic net structure was chosen to represent this knowledge: about 170 concepts arranged in a "necessary-part-of" hierarchy (eight levels deep), and on each level a "specialization" hierarchy (between one and three levels deep). The lowest level in the necessary-part-of hierarchy is the interface to the results of low-level processing methods; the highest level is the description of medical evidence.
Inferences about medical evidence are made by a set of production rules. The low-level processing incorporates standard algorithms for image smoothing and contour detection. Finally, the control module creates instances and

12 We note that separation of the low-level and high-level knowledge is in principle a good idea, however, lacking feedback from the high-level into the low-level stage will give rise to a large amount of "false" data.

We also note that the system can be regarded as a knowledge-guided pattern recognition system rather than an AI-based interpretation system.

initializes the appropriate procedural knowledge in the model. Competing instances are evaluated by a tree search.

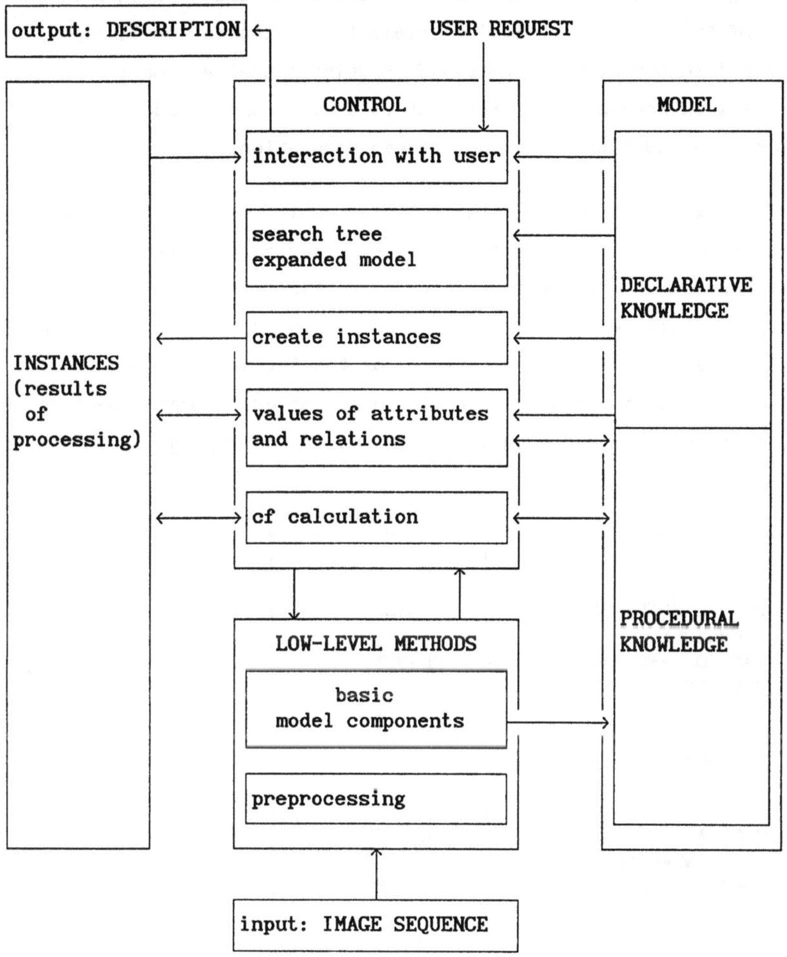

Figuur 7: System structure for knowledge-based analysis of image sequences in gated blood pool studies, Niemann et al (1985)[13].

The authors report that experiments show encouraging performance. A high agreement has been observed between a physician's result and those of the system, with respect to the location of contours and diagnostic interpretations as well.

---

13 Compare with Figure 4: knowledge-guided operation in a rule-based system.

3) Knowledge-based segmentation of subtraction angiograms, (Van Cleynenbreugel et al., 1987). The major characterization of their approach is its multi-layered image representation. Using that paradigm, the image segmentation gradually moves from the raw image data towards a structured description of the relevant anatomical elements. At the bottom level a primal sketch-like structure is obtained by using classical image processing. A knowledge-based system - structured in three modular stages -, is used to refine and extend this primal representation.

The system's knowledge base consists of three modular sections. The first phase (from the pixels to the lines and edges) uses only geometrical and topological knowledge in order to improve the primal image representation. The second phase exploits general blood vessel knowledge (e.g. the width of a normal blood vessel gradually decreases, or the intensity strongly increases where vessels cross). The third phase includes domain-specific knowledge, obtained from radiologists: the type of the angiogram (carotid or

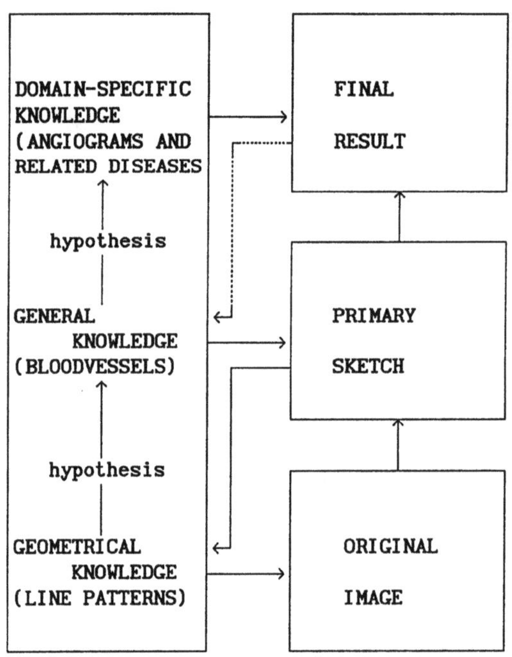

Figure 8: Schematic diagram of the three phases of processing in the multi-layered segmentation of Van Creynenbreugel et al., 1987.

coronary), important regions in the image, the medical history of the patient, the probability of a certain disease.

The system architecture is illustrated in Figure 8. Intentionally, the architecture allows bottom-up / top-down reasoning. Conceptually, the suggested structure is almost equivalent to the AI-based decision-making proposed by kulikowski (1980).

3. DEVELOPMENT OF AN EXPERT SYSTEM FOR THE INTERPRETATION OF SCINTIGRAPHIC IMAGES

In this section, we pay attention to the kind of problems which we have referred to as category B problems. In other words, after the input images have been processed digitally, the result must be interpreted automatically by a computer system that could perform near the level of a human expert. The ultimate goal of building such a system is not to replace the cardiologist by any computer system, rather than to provide him with software tools for an objective interpretation to obtain a more consistent and reproducible diagnosis.

3.1 Identification of the problem

Sequential thallium-201 (Tl-201) scintigraphy has achieved widespread use for the assessment of exercise-induced myocardial perfusion abnormalities[14]. In this nuclear imaging technique, images of the myocardium are acquired by means of a scintillation camera after intravenous administration of Tl-201. Thallium-201 is a radiopharmaceutical that is distributed over the body as a function of the distribution of cardiac output. Consequently, the Tl-201 accumulated in the myocardium reflects regional myocardial blood flow. An area with reduced thallium-uptake early after exercise may be due to either transient uptake abnormalities (ischemia) or to myocardial infarction (scar). These possibilities can be distinguished by comparing the immediate postexercise images with the late postexercise or redistribution images taken between three and four hours later. An ischemic area is characterized

14 Detailed references can be found in Reiber etal. (1987): An expert system for the objective interpretation of Thallium-201 scintigrams.

by the thallium-uptake being reduced in the early postexercise image, but being normal in the late postexercise image. In the case of a scar, the uptake is low in both images. To minimize variations in the interpretation of the scintigrams and to quantify the perfusion abnormalities, the early and late postexercise images must be compared in detail. This is only feasible by computer processing. Over the past years, an extensive software package for the quantitative analysis of thallium-201 scintigrams has been developed[15].

In clinical practice, it has been observed that the cardiologists still rely heavily on the visual interpretation of the original images, in addition to the interpretation of the quantitative results. Apparently, experience plays an important role in the analysis of Tl-201 scintigrams. The existing software package may be described as a tool to obtain quantitative measurements from the scintigrams. The development of an expert system as an additional software tool aiming at the objective interpretation of scintigrams forms the central issue of this section.

3.2 Quantitative analysis

image acquisition

The thallium scintigrams are acquired following standard techniques. The patient performs a maximal or symptom limited exercise test on a bicycle ergometer. One minute prior to maximal exercise, 1.5 mCi Tl-201 is administered intravenously. Static thallium scintigrams are collected successively in three orientations: left-anterior-oblique under 45 degrees and 65 degrees (LAO45 and LAO65), and anterior (ANT). After four hours late postexercise imaging is performed in the same sequence. The images are stored in the DEC Gamma-11 computer system in 64x64 matrices. After collection of all three views from the early and late postexercise studies, the analysis is performed independently for each view.

The early and late images of a specific orientation are automatically brought into registration by means of some geometric calibration points. Next, the centre of the left ventricular activity structure in the early image is determined automatically. Subsequently, a closed myocardial contour

15 See 14.

is searched for by applying a sequential edge-detection technique. The obtained segmentation into the myocardial region and background creates the possibility for background subtraction, absolutely necessary due to the fact that background activity within the myocardial region substantially differs in early and late studies. As a result, we obtain images which are in a sense normalized and well suited for comparisons.

circumferential profiles

To quantify the radionuclide activity distribution in the myocardium, circumferential profiles are calculated for the early and late postexercise images of all views. These profiles are defined by the maximal activity levels along 60 radii through the centerpoint, after appropriate smoothing. Each profile thus constitutes the maximal activity level as a function of the radial segment number. From the early and late postexercise profiles, a washout circumferential profile is derived, indicating the degree of redistribution of the thallium activity from the time of exercise until the second image acquisition. For the objective quantitative interpretation of a patient's thallium uptake, the computed circumferential profiles are to be compared with standard ranges obtained from groups of normal individuals. The circumferential profiles of a patient's study are aligned with the standard limits from the data base on the basis of the manually indicated position of the apex. A profile is said to be abnormal if a segment of at least 18 degrees (3 contiguous radii) falls below the normal limits. Figure 9 gives an example.

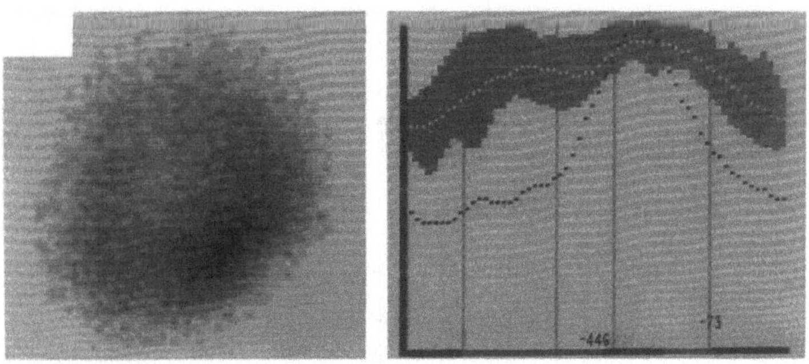

Figure 9: Comparison of the circumferential profile with a band of 'standard limits'.

interpretation
The left ventricle is divided into a number of anatomically defined regions relative to the user-indicated apex location. For the anterior view these regions are antero-lateral, apex, inferior and postero-septal, as indicated in Figure 10.

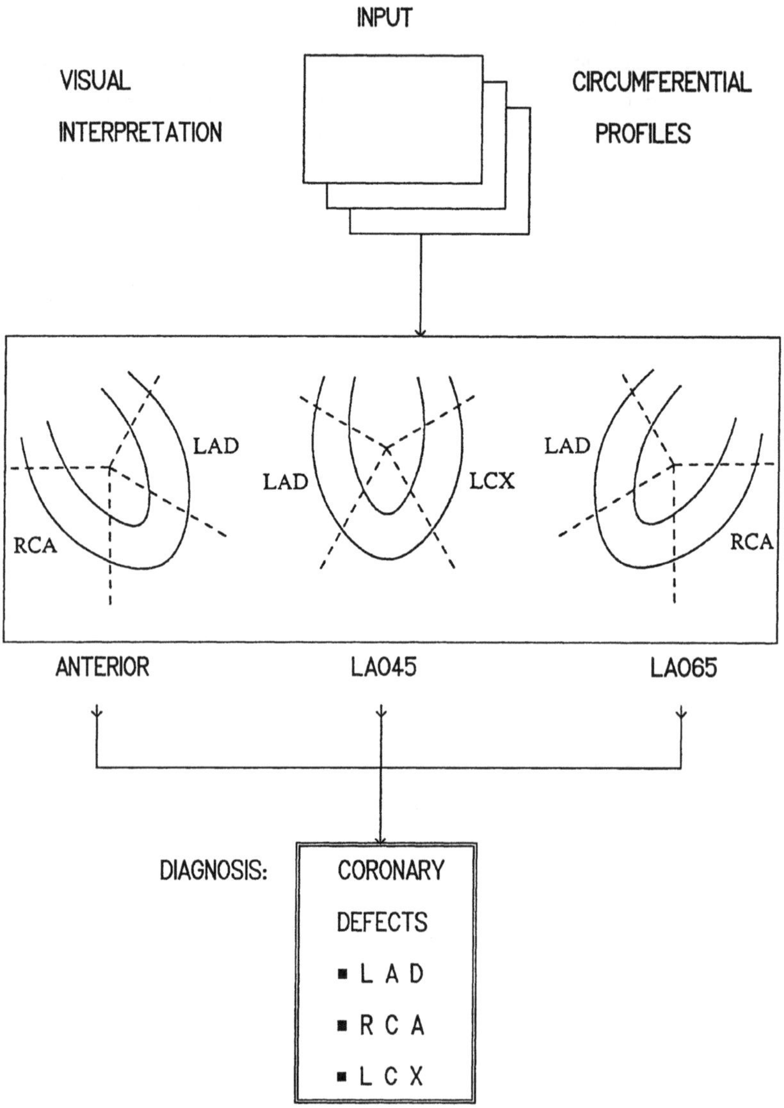

Figure 10: Relational scheme for inference of coronary insufficiencies.

For each region, the areas of the abnormalities in the early and late postexercise and washout circumferential profiles are computed, under the constraint of a minimum of three contiguous radii mentioned before. Finally, the detected abnormalities are compared with the entries of a decision table (Table 1), to define the kind of abnormality present.

diagnosis

It has been pointed out that the basic quantitative evidence is provided by computing the differences between the computed circumferential profiles and the standard profiles. As a result we obtain labelled segments with respect to the type of defect. Segments with the same labels are merged into prediagnostic areas. Small areas are then merged with larger neighbouring areas in accordance with some pre-given label-priority scheme. The priorities are based on existing clinical experience.

This relabelling process continues until no small areas are left. By doing so, we achieve the final diagnostic areas. Next, three relevant parameters are computed which characterize the resulting diagnostic areas. Those parameters are the depth (maximum difference), extent (the width of the diagnostic area) and the severity (the total difference). See Table 2.

Table 1: Decision table on the type of abnormality based on circumferential profiles.

ECP	LCP	WCP	Type of abnormality
N	N	N	normal
N	A	N	rapid washout
N	N	A	slow washout
N	A	A	undefined
A	N	N	ischemia
A	A	N	infarction
A	N	A	ischemia
A	A	A	partial redistribution

Abbreviations: ECP = Early post-exercise Circumferential Profile, LCP = Late post-exercise Circumferential Profile, WCP = Washout Circumferential Profile.
(N = normal, A = abnormal)

Finally, based on the relationships between the anatomic regions on the one hand, and the main coronary arteries which account for the supply of blood of each of the anatomic regions on the other hand, the insufficiency of the coronary arteries is inferred by reasoning from the detected and labelled diagnostic areas, such as:

due to a detected diagnostic 'infarct' area,
insufficiency of the LAD is concluded with CF = 935
insufficiency of the RCA is concluded with CF = 134
insufficiency of the LCX is concluded with CF = 134

Example of a system diagnosis; CF = Certainty Factor.

Table 2: Summary of quantitative findings in LA045 view as an example.

begin	end	diagnosis	ext	dep	sev	anatomic regions
31	56	normal	20	0	0	apical, outflow, postero-lateral
57	14	ischemia	18	16	141	outflow, antero-septal
15	30	infarct	16	23	278	apical, antero-septal

3.3 Identification of the system

As pointed out, the ultimate goal of building a system (the system under development is called ESATS: an Expert System for the Analysis of Thallium Scintigrams) is to provide the cardiologist with a tool enabling him to quantify reproducibly the exact location, the extent, and the type of any existing perfusion defect, and to relate these observable defects to specific coronary insufficiencies. This is illustrated in Figure 11.
As such, the system must to be capable of handling:

- a large solution space;
- imprecise data;
- multi-input data;
- different types of uncertainty;
- external procedures.

The above requirements are fulfilled within the system concept of HYDRA[16], Groen et al. (1985). The essential part is the control mechanism which

16 HYDRA is a parallel expert system shell developed at Delft University of Technology.

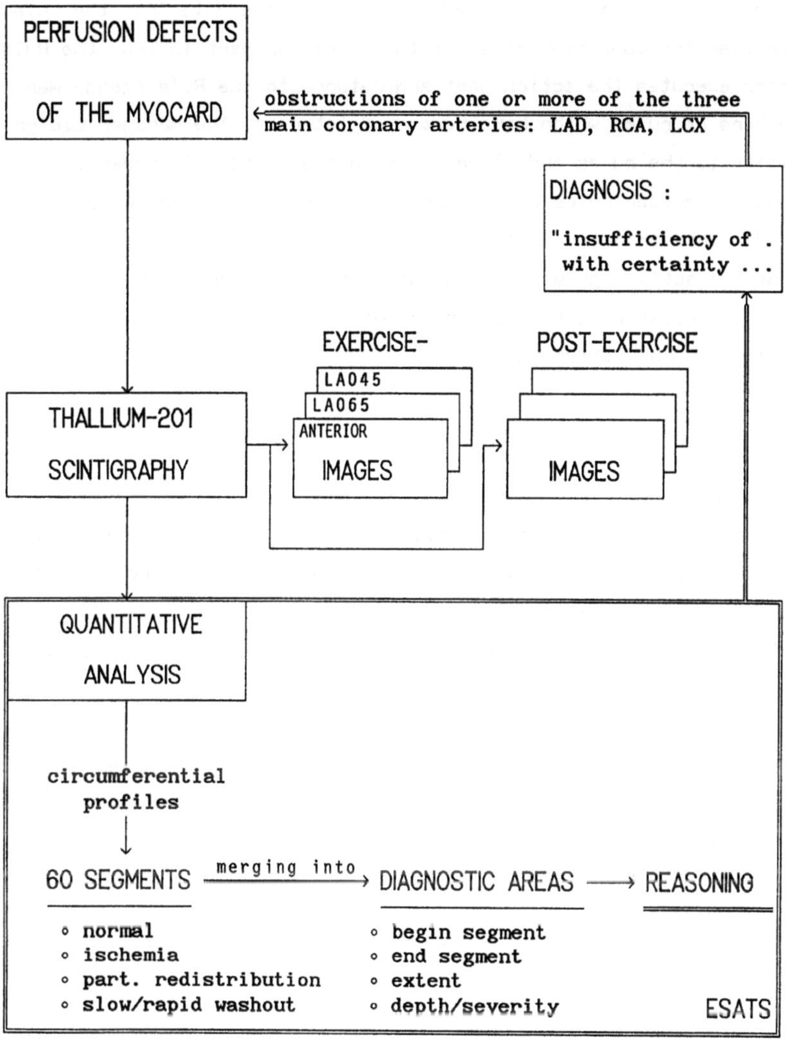

Figure 11: The impact of ESATS on the diagnosis about coronary insufficiencies in relation to the observable thallium perfusion defects.

should be able to perform a bottom-up as well as a top-down reasoning process. The basic idea is that there is a Rule Agenda containing the numbers of the rules to be processed. Each rule is assigned a priority. The priority assignment mechanism can be designed in such a way that both top-down and bottom-up analysis is supported. The agenda itself is controlled by the Rule Agenda Manager, and the rules are processed by the Rule Processor. The Rule Processor receives from the Rule Agenda Manager the number of the rule to be processed. It collects the rule from the rule base and verifies the condition part. If the condition part is *true* the Rule Processor executes the action part and returns to the Rule Agenda Manager the numbers of rules which have become of interest. The manager assigns priorities to the rules and places them on the agenda. In order to distinguish between rules of the current context and other rules, the agenda is divided into an active rule agenda and a nonactive rule agenda. The rules of the active agenda are processed first. If the active agenda is empty, the context with the highest priority from the nonactive agenda is then placed on the active agenda.

3.4 Prototype evaluation

It can be argued that the process of knowledge acquisition and system evaluation somehow goes hand in hand. At present, the Thallium-201 knowledge base is based on interviews with three scintigram-experts and on some case studies, undoubtedly incomplete and possibly still naive. A small preliminary evaluation then has been performed to validate the above prototype. For these purposes, 35 patients with known single vessel coronary artery disease were selected[17]. The Tl-201 scintigrams of these patients were analysed by the system; for each patient the one vessel that was said to be obstructed with the highest certainty factor was compared with the vessel designated by quantitative coronary arteriography. The results are summarized in Table 3.

Table 3: Confusion matrix of ESATS versus catheterization results.

		results of ESATS:		
		LAD	RCA	LCX
catheterization results:	LAD	15	7	1
	RCA	2	7	0
	LCX	1	1	1

17 A full scale validation will be performed on 500 cases of coronary angioplastic treatment of which 248 cases incorporate both catheterization results and scintigraphic studies. The interpretation of the Thallium Scintigraphy studies with respect to the specific coronary insufficiencies can be evaluated on the basis of the "golden standard" due to cardiovascular angiographic measurements.

In 66% of the cases the coronary artery with the highest certainty factor corresponded with the coronary artery with a percent narrowing larger than 50%. This relatively low sensitivity demonstrates the fact that manipulation with certainty factors might be a very subtle issue. This is precisely the subject of the next section.

4. MODELLING OF UNCERTAINTY IN EXPERT SYSTEMS

This section is entirely devoted to the problem of representing uncertainty and how to manipulate these uncertainties.

It is needless to argue that processing medical images generally provide uncertain or imprecise information. Moreover, the expert knowledge itself about the medical phenomena tends to be rather imprecise or uncertain. It is a characteristic of human reasoning that a reader of medical images can cope wich such imprecision or uncertainty. As knowledge based systems are meant to imitate or represent subjective human reasoning at least in terms of input-output performance, study of uncertainty models and propagation of uncertainty is inevitable.

The meaning of uncertainty, however, is far from clear. It can be related to inaccuracy, incredibility, inexactitude, unreliability, indistinctness, fuzziness etc. Generally speaking, uncertainty itself is restricted to propositions of which the truth cannot be determined definitely. As an example, a proposition is said to be fuzzy if its meaning cannot be determined with sufficient precision on a given scale. On the other hand, the term credibility is used in order to express the degree of which a proposition is assumed to be to be true. Note that those concepts do not coincide necessarily: a precise proposition can be uncertain, whereas a certain proposition can be fuzzy etc.
Moreover it is possible that within a given context more than one type of uncertainty is apperent at the same time. In those cases we have to cope with so-called multitype uncertainty.

4.1 The uncertainty model

With respect to reasoning with uncertainty, we distinguish an uncertainty calculus and an inference calculus. A calculus is meant to be a set of mathematical rules on the basis of which calculations can be performed. An uncertainty calculus deals with the mathematical representation of uncertainty whereas an inference calculus is dealing with combination and propagation of uncertainties through the inference process. Uncertainty calculus and inference calculus form together the uncertainty model. As mentioned before, the application or context prescribes which uncertainty models are most appropriate for that given application or context. Note that control over uncertainty models is part of the knowledge involved seen from a meta-view. In Figure 12, the uncertainty model and the uncertainty model control is schematized.

Note that the characteristics of uncertainty in a given application together with knowledge about uncertainty model assumptions (domain, range, precision, parameter independency or numerical completeness) is input for the uncertainty model control in order to select the optimal model. Hereby, optimallity is considered as the closest correspondence between the systems' performance and that of an human expert[18].

ever-false proposition	t_0
ever-true proposition	t_1
propositions	u, v
evidence	e
hypothesis	H
probability function	$P(.)$
belief function	$B(.)$
credibility function	$Cr(.)$
plausibility function	$Pl(.)$
necessity function	$N(.)$
possibility function	$F(.)$
increased belief	$MB(.)$
increased disbelief	$MD(.)$
certainty factor	$Cf(.)$
membership function	$\mu(.)$

Table 4: Notations.

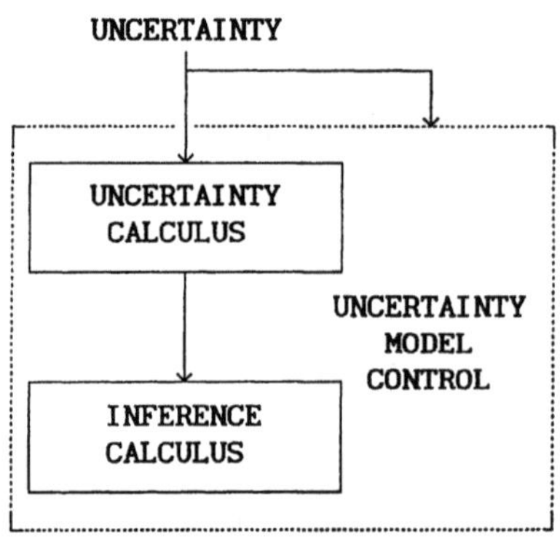

Figure 12: The uncertainty model and the control for model selection.

18 In fact, we want to model the medical expert in diagnosis making such that if that model is used in the knowledge system, the input-output performance matches. However, as such, uncertainty modelling appears to be quite limited.

4.2 Uncertainty calculi

Depending on the type of uncertainty a variety of uncertainty calculi can be distinguished each of which is dealing with the representation of the degree of uncertainty of a proposition or event. Some of them are considered; for greater detail, see Prade (1985).

In the following we will use the notations as given in Table 4.

probability

By definition, $P(u)$ and $P(v)$ are the probabilities that the propositions u and v are true. The probability function satisfies the following requirements: $P(t_0) = 0$, $P(t_1) = 1$, and

if $u \wedge v = t_0$ then $P(u \vee v) = P(u) + P(v)$.

These axioms imply: $P(u) + P(\neg u) = 1$, and

if u implies v ($u \Rightarrow v = t_1$) then $P(v) \geq P(u)$.

The above axioms correspond to the Bayes model inwhich the probability of an hypothesis H given evidence e is expressed as $P(H|e)$.

Apart from computational complexity and numerical completeness, the major drawback of the Bayes model is that the assignment of $P(u)$ implies the assignment of $P(\neg u)$ which seldomly corresponds to the expert's judgement.

credibility and plausibility

By definition, $B(u)$ expresses the belief in the truth of a proposition u. Hereby, the following axioms hold:

$B(t_0) = 0$, $B(t_1) = 1$, and

if u implies v then $B(v) \geq B(u)$.

A particular belief function, satisfying the last axiom, is the credibility function $Cr(u)$ for which it holds:

$$Cr(u) + Cr(\neg u) \leq 1.$$

We note that $Cr(u) = 0$ does not imply $Cr(\neg u) = 1$, whereas $Cr(\neg u) = 1$ just implies $Cr(u) = 0$. The plausibility function $Pl(u)$ can de defined directly on the basis of $Cr(\neg u)$, as follows:

$$Pl(u) = 1 - Cr(\neg u).$$

This simply expresses the fact that the plausibility that u is true corresponds to the incredibility that u is false.

The Shortliffe-Buchanan model (Buchanan and Shortliffe, 1984) is closely related to the above approach. In this model, the measure of increased belief in an hypothesis H based on evidence e is given by:

$$MB(H|e) = \begin{cases} 1 & \text{if } P(H) = 1 \\ \dfrac{\max\{P(H|e), P(H)\} - P(H)}{1 - P(H)} & \text{elsewhere} \end{cases}$$

In fact $MB(H|e)$ can be considered as a credibility function. It can easily be shown that: $MB(H|e) + MB(\neg H|e) \leq 1$.

Apart from $MB(H|e)$, also a measure of disbelief $MD(H|e)$ is introduced. However, this is not an analogon of the above mentioned plausibility function. In fact $MD(H|e)$ is also a credibility function since it can be shown that:

$$MD(H|e) = MB(\neg H|e).$$

A definition of a plausibility function would be

$$Pl(H|e) = 1 - MB(\neg H|e) = 1 - MD(H|e).$$

The Shortliffe-Buchanan model offers an overall certainty factor:

$$Cf(H|e) = MB(H|e) - MB(\neg H|e).$$

Rewriting yields:

$$Cf(H|e) = MB(H|e) + Pl(H|e) - 1$$

from which it can be concluded that the certainty factor is a combination of credibility and plausibility. Furthermore it holds that:

$Cf(H|e) + Cf(\neg H|e) = 0$.

The uncertainty calculus of Dempster-Shafer (Shafer, 1975) is also based on credibility and plausibility functions. Consider Q being the exhaustive set of all mutually exclusive hypotheses. The influence of each piece of evidence on the subsets of Q is expressed by the so-called basic probability assignment in Dempster-Shafer's theory of evidence. The basic probability assignment $m(H)$ represents the degree of belief in H. As a result, in the case of three hypothesis H_1, H_2 and H_3, we have $m(H_1)$, $m(H_2)$, $m(H_3)$, $m(H_1,H_2)$, $m(H_1,H_3)$, $m(H_2,H_3)$, $m(H_1,H_2,H_3)$, for which:

$$\sum_{A \in Q} m(A) = 1$$

Then the credibility function (measure of total belief in A)

$$Cr(A) = \sum_{B \subseteq A} m(B)$$

satisfies $Cr(A) + Cr(\neg A) \leq 1$.

From $Cr(A)$, the plausibility function $Pl(A)$ follows easily:

$$Pl(A) = 1 - Cr(\neg A) = \sum_{A \cap B \neq \emptyset} m(B).$$

Clearly, both models relax the Bayes model in a different way. In Shortliffe-Buchanan we have the credibility heuristically defined whereas in Dempster-Shafer the credibility is defined in basic probabilities.

possibility and necessity

Necessity functions $N(.)$ are credibility functions satisfying the following additional requirement: $\mu(u \wedge v) = \min(N(u), N(v))$.
Likewise, a possibility function $F(.)$ is a plausibility function satisfying additionally: $F(u \vee v) = \max(F(u), F(v))$.
The above implies $F(u) < 1 \Rightarrow N(u) = 0$ meaning that if the possibility that u is true is less than 1, than v is not necessarily true. Likewise, it holds that:

$$N(u) > 0 \Rightarrow F(u) = 1, \text{ and } N(u) = 1 - F(\neg u).$$

It is remarked here that Shortliffe-Buchanan's $MB(.)$ is also a necessity function since:

$$MB(H_1 \wedge H_2 | e) = \mathrm{Min}(MB(H_1 | e), MB(H_2 | e)).$$

Likewise, $Pl(.)$ appears to be a possibility measure.
According to Dempster-Shafer's theory $Cr(.)$ and $Pl(.)$ are necessity and possibility functions, respectively, if the subsets of Ω are nested. It is important to note that the uncertainty of a proposition u is represented by a pair $\{N(u), F(u)\}$ rather than by one number.

fuzzy set membership functions

Whereas the above mentioned calculi are related to the uncertainty of the truth of a proposition, it is also possible to consider the imprecision or fuzziness of its contents [Dubois and Prade (1980)]. The membership function $\mu_L(u)$ can be interpreted as the compatibility of u with the concept labelled L defining the degree of possibility of the proposition "u has the property L". The major outcome is that fuzzy membership functions follow the necessity and possibility functions:

$$\mu(u \wedge v) = \min(\mu(u), \mu(v)) \text{ and } \mu(u \vee v) = \max(\mu(u), \mu(v)),$$

but still satisfy the probabilistic property: $\mu(u) + \mu(\neg u) = 1$.
In Figure 13, the above perspective is schematized.

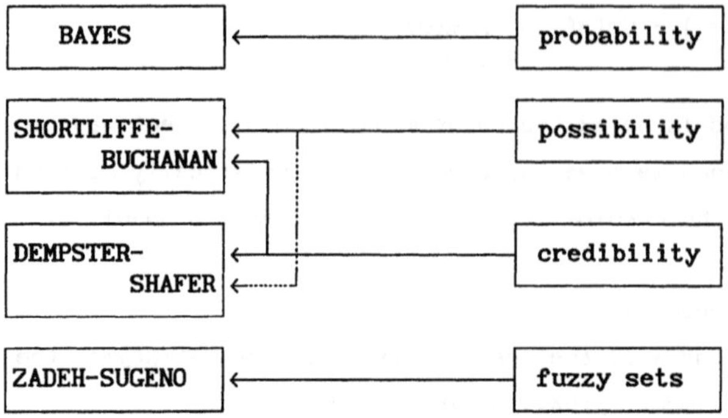

Figure 13: The representation of uncertainty and inexactness and their corresponding calculi.

4.3 Inference calculi

Whereas the uncertainty calculi are mainly related to the mathematical representation of uncertainty or inexactness of propositions, inference calculi provide us with the framework where reasoning with uncertainty can be performed: combination and propagation of evidences. Wellknown inference calculi are the ones introduced by Bayes, Shortliffe-Buchanan, Dempster-Shafer and Zadeh-Sugeno .

As in most applications, it is very unlikely that one has to deal with just one type of uncertainty. For that reason an inference calculus on the basis of an uncertainty vector is proposed as illustrated in Figure 14. It has the advantage of learning and tuning but it suffers from increasing dimensionality.

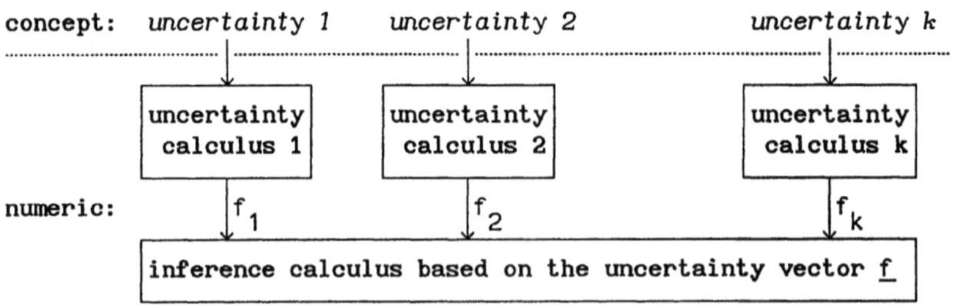

Figure 14: Uncertainty model using multitype uncertainty.

In those cases where reasoning using *rule inference* like

$$\left.\begin{array}{l} f_1(\text{premisse}),\ f_1(\text{rule}) \Rightarrow f_1(\text{conclusion}) \\ f_2(\text{premisse}),\ f_2(\text{rule}) \Rightarrow f_2(\text{conclusion}) \\ \cdot \\ f_k(\text{premisse}),\ f_k(\text{rule}) \Rightarrow f_k(\text{conclusion}) \end{array}\right] \longrightarrow \underline{f}(\text{conclusion})$$

can hardly be justified (and this includes many applications in medical diagnosis making), *classification inference* might be much more appropriate, i.e.:

given:

$\underline{f}_1(\text{premisse}),\ \underline{f}_1(\text{rule}) \Rightarrow \underline{f}_1(\text{conclusion})$

$\underline{f}_2(\text{premisse}),\ \underline{f}_2(\text{rule}) \Rightarrow \underline{f}_2(\text{conclusion})$

$\cdot$

$\underline{f}_N(\text{premisse}),\ \underline{f}_N(\text{rule}) \Rightarrow \underline{f}_N(\text{conclusion})$

where N is the number of training diagnostic cases,

determine:

$\underline{f}(\text{conclusion})$?, based on $\{\underline{f}(\text{premisse}),\ \underline{f}(\text{rule})\}$.

Such an approach has been suggested by Backer (1988). In that paper, the underlying mathematics of classification inference by a fuzzy nearest neighbour decision scheme is presented.
Much research is still be needed to solve the entire uncertainty modelling problem when such modelling implemented in an expert system should meet the human performance.

5. CONCLUDING REMARK

An overview of the field of knowledge-based approaches in the processing and interpretation of medical images has been presented. Recent developments in artificial intelligence and image processing have shown their potential value in automated processing and interpretation of medical images. However, it should be realized that applying artificial intelligence methodsdoes not guarantee solving the entire problem perse. In other words, it is just part of the solution or it may lead to a better understanding of the problem itself. In spite of the fact that knowledge-based approaches are a challenging kind of artificial intelligence to achieve, much of research is still be needed in almost all of its underlying principles and mechanisms. A wonderful pioneering area has just become to existence.

ACKNOWLEDGEMENT

The author would like to thank J.J.Gerbrands and J. van der Lubbe (both from Delft University of Technology), H.Reiber (from Erasmus University, Thoraxcenter, Rotterdam) and J. van den Herik (University of Limburg, Maastricht), as well as their co-workers. Without their continuous cooperation over many years, this paper could never have been written.

REFERENCES

Backer,E., Gerbrands,J.J., Reiber,J.H.C., Reijs,A.E.M., Krijgsman,W.,and Herik,J. van den, (1988). Modelling uncertainty in ESATS by classification inference. Pattern Recognition Letters, 8,2, 103-112.

Breuker,J.A., Wielinga,B.J. (1984). Techniques for Knowledge Elicitation and Analysis. Report 1.5 Esprit Project 12, University of Amsterdam.

Buchanan,B.G., and Duda,R.O. (1983). Principles of rule-based Expert Systems. Advances in Computers, Vol. 22, 163-216.

Buchanan,B.G., and Shortliffe,E.H. (1984). Rule-based expert systems. Massachusets.

Castleman,K.R. (1979). Digital Image Processing. Prentice-Hall Englewood Cliffs.

Cleynenbreugel,J. van, Fierens,F., Smets,C., Suetens,P.,and Oosterlinck,A. (1988). Knowledge-based segmentation of subtraction angiograms. Proc. IPMI 1987, 307-314.

Dreyfus,H., Dreyfus,S. (1986). Why computers may never think like people. Technology Review, MIT, 42-61.

Erman,L.D., Hayes-Roth,F., Lesser,V.R., and Reddy,D.R. (1980). The Hearsay-II Speech-understanding system: Integrating knowledge to resolve uncertainty. Computing surveys, 12, 2, 213-252.

Groen,A., Herik,H.J. van den, Hofland,A.G., Kerckhoffs,E.J.H., Koppelaar,H., Stoop,J.C., and Varkevisser,P.R. (1985). Performance model of a parallel knowledge-based system. In: Systematica, Delft University Press, Delft.

Hayes-Roth,F. (1984). The knowledge-based Expert System: A tutorial. IEEE Computer, 11-28.

Hayes-Roth,F., Waterman,D.A., Lenat,D.B. (1983). Building Expert Systems, Addison-Wesley, Reading, Massachusetts.

Kulikowski,C.A. (1980). Artificial Intelligence Methods and Systems for Medical Consultation. IEEE Trans. PAMI, 2,5, 464-474.

Nau,D.S. (1983). Expert computer systems. Computer, 63-83.

Nazif,A.M., Levine,M.D. (1984). Low-level image segmentation: An Expert System. IEEE Trans. PAMI, 6, 5, 555-577.

Niemann,H., Bunke,H., Hoffman,I., Sagerer,G., Wolf,F., and Feistel, H. (1985). A knowledge-based system for analysis of gated blood pool studies. IEEE Trans. PAMI, 7, 3, 246-259.

Nilsson,H. (1977). Principles of Artificial Intelligence. Springer, Berlin.

Prade,H. (1985). A computational approach to approximate and plausible reasoning with applications to expert systems. IEEE Trans. PAMI, 7, 3, 260-283.

Reiber,J.H.C., Bloom,G., Gerbrands,J.J., Backer,E., Herik,J. van den, Reijs,A.E.M., Feltz,F. van der, and Fioretti,P. (1988). An Expert System approach for the objective interpretation of Thallium-201 scintigrams.

Rich,E. (1983). Artificial Intelligence, McGraw-Hill, Singapore.

Rosenfeld,A., Kak,A.C. (1982). Digital Picture Processing. Academic Press, New York.

Shafer,G. (1975). A mathematical theory of evidence. Princeton University Press.

Stansfield,S.A. (1986). Angy: A rule-based expert system for automatic segmentation of coronary vessels from digitial subtracted angiograms. IEEE Trans. PAMI, 8, 2, 188-199.

Tanimoto,S. (1987). The elements of Artificial Intelligence.

READING MATERIAL

artificial intelligence and knowledge-based systems in general:

Backer,E. (1986). Knowledge-based image processing. Expert Systems, 77-88.

Backer,E., Lubbe,J.C.A. van der, Krijgsman,W. (1988). On modelling of uncertainty and inexactness in expert systems. Proc. 7th Symp. Information Theory, 105-111.

Baldock,R., Towers,S. (1988). First steps towards a blackboard controlled system for matching image and model in the presence of noise and distortion. Proc. 4th Int. Conf. on Pattern Recognition (Kittler, J. ed.), 429-438.

Ballard,D., Brown,C. (1982). Computer vision, Prentice-Hall, Englewood Cliffs, NJ.

Cleary,J.G. (1987). Acquisition of uncertain rules in a probabilistic logic. Int. Journal Man-Machine Studies 27, 145-154.

Dubois,D., Prade,H. (1988). Possibility Theory and fuzzy logic in inference systems: A perspective view. Busefal'88, 151-157.

Duda,R.O., Hart,P.E. (1973). Pattern Classification and Scene Analysis, Wiley, New York.

Feigenbaum,E.A. (1983). Artificial intelligence. IEEE Spectrum, 77-78.

Gordon,J., and Shortliffe,E.H. (1985). A method for managing evidential reasoning in a hierarchical hypothesis space. Artificial Intelligence 26, 323-357.

Hanson,A.R., Riseman,E. (1978). Computer Vision System. Academic Press, New York.

Harmon,P., King,D.(1986). Expert Systems - Artificial Intelligence in Business, Wiley, New York.

Kidd,A. (1985). Human factors in expert systems. Data Processing, Vol. 27, 4, 15-17.

O'Keefe,R.M., Balci,O. and Smith,E.P. (1987). Validating Expert System Performance. IEEE Expert, Winter 1987, 81-90.Pearl,J. (1984). Some recent results in heuristic search theory. IEEE Trans. on PAMI, 6, 1, 1-13.

Shafer,G. and Logan,R. (1987). Implementing Dempster's rule for hierarchical evidence. Artificial Intelligence 33, 271-298.

knowledge-based systems in processing and interpretation of medical images

Aleven, V.A.W.M.M. (1988). A blackboard approach for the interpretation of Thallium-201 heart scans. Technical Report Delft University of Technology (in Dutch).

Backer,E., Gerbrands,J.J., Bloom,G., Reiber,J.H.C., Reijs,A.E.M., and Herik,H.J. van den (1988). Developments towards an expertsystem for the quantitative analysis of Thallium-201 scintigrams. Proc. IPMI'87, 293-306.

Bonadonna,F., Coppin,G., Rubino,A. and Valli,G. (1987). Computational methods in diagnostics of cardio-vascular images. Elettro-tecnica, 74, 3, 263-273.

Catros,J.Y. and Mischler, D. (1988). An artificial intelligence approach for medical picture analysis. Pattern Recognition Letters 8,2, 123-130. Dellepiane,S., Serpico,S.B. and Vernazze,G. (1986). Approximate reasoning and knowledge in NMR image understanding. Proc. 8ICPR, 943-946.

Dellepiane,S., Regazzoni, Serpico,S.B. and Vernazze,G. (1988). Extension of IBIS for 3D organ recognition in NMR multislices. Pattern Recognition Letters 8,2, 65-72.

Elion,J.L. and Nissen,S.E. (1987). A knowledge-based image processing system for the interpretation of coronary arteriograms. Proc. SPIE, 767, pf. 2, 428-432.Ellam,S.V., and Maisey,M.N. (1986).A Knowledge-based system to assist in medical image interpretation: design and evaluation methodology. Proc. 6th Annual Conf. of the Brittish Comp. Society Specialist Group on Expert Systems, 89-98.

Ezquerra,N.F., Zerbi,M., Peifer,J., West,M., Garcia,E.V., Depuey,E.G. and Hise,H.L. (1987). Knowledge-based cardiac image interpretation and display. Proc. Southcon/87, 4, 1, 1-7.

Ezquerra,N.F., Garcia,E.V., Depuey,E.G. and Robbibs,W.L. (1986). Development of an expert system for interpreting medical images. Proc. IEEE Conf. SMC, 1, 205-210.

Garcia,E.V., Briggs,W., Ezquerra,N.F. and Hise,H.L. (1987). An Expert System for interpreting Thallium-201 Tomographic images. Proc. IBM ACIS University Conf., 373-379.

Kumar,R. and Srihari,S.N. (1985). An expert system for the interpretation of cranial CT scan images. Proc. Conf. AIAA Expert Systems, 548-555.

Lebel,O. (1988). ARCHI: An expert system for biological objects recognition. Pattern Recognition Letters, 8,2, 131-139.

Menhardt,W. and Schmidt,K.H. (1988). Computer vision on magnetic resonance images. Pattern Recognition Letters, 8,2, 73-85.

Oosterlinck,A., Suetens,P., Wu,O. and Baird,M. (1987). Pattern Recognition and Expert image analysis systems in biomedical image processing. (Int. Symp. Pattern Recognition and Acoustical Imaging) Proc. SPIE, 768, 44-52.

Sagerer,G. (1988). Automatic interpretation of medical image sequences. Pattern Recognition Letters, 8,2, 87-102.

Smets,C., Verbeeck,G., Suetens,P. and Oosterlinck,A. (1988). A knowledge-based system for the delineation of blood vessels on subtraction angiograms. Pattern Recognition Letters, 8,2, 113-121. Suetens,P., Cleyenbreugel,J. van, Fierens,F., Smets,C. and Oosterlinck, A. (1987). An expert system for blood vessel segmentation on subtraction angiograms. Proc. SPIE, 767, pf. 2, 454-459.

Tascini,G. (1985). Knowledge-based system for decision making in medical image understanding. Proc. Melecon, 1, 239-242.

Todd-Pokropek,A. (1987). Medical imaging, Computer Bulletin, Vol. 3, pt. 3, 5-7.

PICTURE ARCHIVING AND COMMUNICATIONS SYSTEMS: PROGRESS AND CURRENT PROBLEMS

G.Q. Maguire Jr.

Columbia University in the City of New York
United States of America

ABSTRACT

The need to provide more interactive clinical consultations as well as retrieve images quickly (preventing image loss) and control the flow of image information has promoted the idea of incorporating a local area network (LAN) into imaging departments. This need was fueled by the observation that the full capabilities of digital image-producing modalities were rarely utilized routinely and by the proliferation of imaging techniques and clinical specialists participating in the generation and interpretation of them. This development is basically office automation for workers who deal with images in addition to text.

1. WHAT IS A PACS?

Picture Archiving and Communication Systems and the acronym PACS became known as a distinct endeavor due to the Society of Photo-Optical Engineers (SPIE) conference: *First International Conference and Workshop on Picture Archiving and Communication Systems (PACS) for Medical Applications (PACS I)*, held in January, 1982 at Newport Beach, California.

The first generation PACS emphasis was primarily to provide a filmless imaging department. That is, the goal was an electronic replacement for film and the clients were chiefly radiologists and pathologists. First generation systems are presently being developed and marketed commercially. Many of the concerns of first generation systems are related to the storage and communication of the large volumes of data in existing modes of radiology (or pathology, ...). To provide an initial feeling for the amount of data we will assume for a 700 bed university hospital (based on NYU Medical Center data) that the estimated data volume for a 10 year period of time is (Zeleznik, 1983):

Structured data 10^6 items (approximately 54 bytes of patient and study information) resulting in 5.4×10^7 bytes;

Textual data 4×10^6 documents resulting in 2.4×10^9 bytes; and

Image data 5 x 10^7 images resulting in 10^{14} bytes of data. For more information on the number of images see the proceedings of SPIE PACS I, PACS II, ...

Since this initial effort, a second generation of PACS is beginning to be defined; its goal is to regard images as just another data type in an office automation system and its clients are anyone who deals with images. The emphasis in second generation PACS is a system which integrates image generation, display and analysis, storage (both short term and archival), production of hardcopy, communication between all of the subsystems and between systems, and management of both the images and the system. This is very similar to the development of the General Motors Manufacturing Automation Protocol (MAP) which tries to tie all manufacturing systems together. However, in medical imaging there is no major ''consumer'' with the size of General Motors. Thus, although the number of medical imaging procedures is enormous, they are performed by a very large number of independent organizations (much like the automobile industry used to be). The result is that PACS progress has been much slower than that of MAP. An interesting aspect of second generation systems is that most of the things which the 1982 PACS conference specifically excluded as being covered by earlier conferences (e.g., image acquisition, image display, image processing, computerized patient record keeping, and management information system) are components of second generation systems! Thus just as MAP has caused the development of the Technical and Office Protocol (TOP) to interconnect the front office operations associated with manufacturing to the manufacturing systems, a development of second generation PACS will be an analogous system to allow Hospital Information Systems (HIS) and PACS to talk to each other. As we will see later this interaction is essential. Second generation systems are currently a topic of research and are quite far from commercial development.

2. BASIC COMPONENTS OF A PACS

A PACS consists of a number of components:

Input components	digital image sources, digitizers (for input of data produced by analog sources), digital media (tape, floppies, etc.)
Output components	viewing stations, hardcopy devices, and digital media (for output)
Storage	file servers, local disk drives, archiving systems
Communication	local area networks (LANs), wide area networks, busses, and interfaces
Control	protocols used in the interaction between the subsystems, network management systems,

interaction with the hospital (or other management) information system, and user commands. The minimal set of operations for a PACS includes: data archiving (including images and text), data retrieval, interactive data display, image processing/analysis, data output, personal workspace environment, electronic mail, user help, and data security. For a more detailed introduction to PACS system architecture see (Maguire, 1986).

Initial PACS systems were centralized, then some things were distributed, and future systems will be fully distributed systems. This evolution from centralized systems to distributed systems has been a direct result of the changes in the cost/performance of the equipment and personnel involved. The same basic functions are in the distributed systems as in centralized image processing architectures but with the internal bus replaced by networks.

The objects which exist in the PACS environment include images (pixels and associated data), parametric models, control information, text, remote procedure calls, and processes. The increasing use of (and development of) digital models and atlases of human and animal anatomy will force PACS to deal with not only pixel and voxel data but with other representations, such as Non-Uniform Rational B-splines (NURBS).

3. INITIAL PACS IDEAS AND SYSTEMS

For initial PACS designs see (Maguire, 1982a) and other papers in the same volume (SPIE PACS I). For a review of the major efforts prior to 1986 and an introduction to local area networking see (Noz, 1986).

Given the transfer rates of most archiving devices (approx. 200 kbytes/second for current optical disks), the time required to mount optical disk platters or other media (in the range of 5 to 30 seconds) and the probability of subsequent references after the initial reference, there is a need for buffers between the archive and the producers/consumers of images. This suggests a standard hierarchical storage system with caching such as has been extensively studied in computer systems. However, while objects in both situations have a high *locality of reference* within objects we do not yet understand the reference properties for images in a PACS. Therefore, we do not know which images will be referenced nor their order and time of reference.

4. STANDARDS

An important aspect of getting systems (such as PACS) widely adopted is either there must be a de facto standard (generally produced by the dominant entity in a field) or there has to be a national or international standard. Without standards a company who buys equipment risks a large financial investment and faces the possibility of having a system which they can not afford to maintain or expand. The desirable characteristics of standards for PACS are described in (Baxter, 1982a). One characteristic which this earlier paper does not emphasize enough is the the need for extensibility (adaptability). With the wide spread use of electronic mail and other distributed services we must begin to explore the use of these services to allow our standards to evolve via electronic mediation rather than via the much longer delay of committees.

4.1. AAPM Standard Format for Digital Image Exchange

In 1981, the American Association of Physicists in Medicine (AAPM) ''Task Force on Digital Image Data Exchange of the AAPM Science Council'' adopted the image format (Baxter, 1982b) which had been developed (jointly between New York University and the University of Utah) after several years of discussion with many medical imaging equipment manufacturers. This format was intended as a means of exchanging data between various imaging systems. It was not intended to be the format in which a specific manufacturer must store data, but rather was designed to be the universal intermediate representation for exchanging data. Therefore, for the effort of writing one program to read and write AAPM formatted tapes, N different systems could interchange data for an effort proportional to N, rather than the N^2 which was the norm.

The fundamental idea behind this format was the decoupling of the image data (viewed as an N dimensional array of values, each of which are B bytes in size) from the non-image data associated with this image. The non-image data was stored as a set of Key-Value Pairs (much like LISP properties and values - or the name-type-value of Selfridge (Selfridge, 1979)). These key-value pairs consisted of ASCII text for the key and value, which were separated by a ":=". The key-value pairs were kept in one file and the image data in another file. Any line in the directory (or ''header'') which did not contain a ":=" was considered to be a comment. The use of ASCII for the keys and values meant that the directory file could simply be printed on a terminal or line printer to determine any of the associated information - thus the user did not have to write a sophisticated program or deal with a complex tape format. There were a set of required keys for each image:

1. Image number - which identified the Nth image file
2. Bytes per pixel - the number of bytes required to store a pixel
3. Number representation - the number system in which to interpreted the binary pixel data (Two's complement integer, IEEE floating point, ...)

4. Number of dimensions - the number of dimensions in the image array
5. Size of dimension 1 .. Size of dimension N - the size of each dimension in the image array

A tape had a directory file as the first file on the tape. This directory consisted of the image headers for all the images on the tape and a set of keys which described this tape:

1. Tape Standard (whose initial value was 1.00) - this identifies which version of the standard had been used to generate this tape
2. Directory header - description of the tape
3. Institution - name of the institution which produced the tape
4. Department - name of the department which produced the tape
5. Date created - <Day>,<Month>,<Year> as decimal integers

A number of other keys were defined such as ''Date written'', ''Writer'', ''Grid <n> units'' (where <n> was the dimension number), ''Patient name'', ''Patient number'', and ''Exam type''.

The tapes were specified as being written using the applicable ANSI standards (X3.22-1973, X3.39-1973, X3.40-1976, and X3.27-1969) such that they could be read or written on standard 1/2 inch magnetic tape units.

For further information see (Maguire, 1982b, Baxter, 1982a, Baxter, 1982b).

4.1.1. Good

This format has been extensively used by two communities: (1) the radiation therapy research community and (2) the seismic mapping community. The first group used this format to allow the exchange of data for cross comparison of radiotherapy packages and sharing data. See (Goitein, 1987) for a description of the current keys used by the U.S. National Cancer Institute funded: Particle Intercomparison Project and High Energy Photon External beam Treatment Planning Project. The seismic community adopted this format because they wanted to use the 3D vibrating mirror display system developed at the University of Utah and the AAPM format was the only format which this system accepted!

4.1.2. Bad

It took from 1978 till 1981 for the AAPM task force to come to an agreement on the image format. This was done in a climate where manufacturers formats were proprietary information and to get these specifications required signing a non-disclosure agreement. I believe that this has hindered the development and spreading of programs to do the conversion from the vendor specific to/from the AAPM format as users might feel that they were violating their non-disclosure agreement if they distributed such conversion programs while vendors saw **NO** commercial advantage to making such program available.

Once the task force adopted this format it disbanded and there was no body to coordinate the creation of new keys - in retrospect this was a mistake.

4.2. ACR-NEMA Standard number 300-1985

This standard is the result of a joint effort by the American College of Radiology (ACR) and the National Electrical Manufacturers Association (NEMA).

4.2.1. Good parts

The major contribution of ACR-NEMA Standard is the definition of a logical format for images and an initial set of attributes for these images.

4.2.2. Bad parts

Some of the major failings of this standard are a result of the groups which developed this standard [i.e. ACR and the Diagnostic Imaging and Therapy Systems Division of NEMA] and their desire to do something quickly (in time for one of the meetings of the Radiological Society of North America (RSNA)). The exclusion of the non-U.S. manufacturers and all scientists and engineers who were not part of either ACR or NEMA led to lots of bad feelings, a tremendous loss in expertise, and probably less acceptance for the standard that if it had been developed by a broader body. A major component which was missing was expertise in modern protocol design; thus there did not exist a formal definition of the protocol prior to its adoption. Hence the lack of timeouts from certain states and other poorly specified features were only discovered when people tried to actually build interfaces. In addition, these interfaces were often not interoperable (i.e., the interfaces from two manufacturers would not talk to each other but could talk to other interfaces from the same manufacturer).

The development of a new physical and link level interface is VERY hard to justify economically as unless there is a very large volume of sales it is difficult to justify the roughly $1M cost of developing a VLSI implementation and without a VLSI implementation the cost for a single interface is quite high (i.e. $4-5K).

The ACR/NEMA Standard 300 image format is restricted to predefined attributes and user defined extensions. This mirrors the fixed formats commonly used by many vendors for storing images on magnetic tape or disk. In addition, many formats have restricted images to only two dimensions rather than allow higher (or lower) dimensionality.

The development of a PACS standard database for attributes of images which provides for easy extension by users is essential for future PACS developments and has not yet been addressed by ACR/NEMA. This database should be available (by a server) to developers and users via electronic mail.

ACR-NEMA has refused to separate the higher level protocols (including the image format) from the lower levels. Thus the emergence of protocols such as SPI (Standard Product Interface aka Siemens Philips Interface) (Greinacher, 1987).

4.2.3. Products

Matrix Instruments, Orangeburg, NY, has produced a 6U sized VME bus card which contains a MC68020 processor and memory along with drivers for the parallel ACR/NEMA interface. They call this card the ''ACR/NEMA Standard Interface Module (ANSIM)''. This card is really designed for Matrix and other medical OEMs to gain experience with the ACR/NEMA interface - rather than being an interface card which a manufacturer would simply configure into their system.

3M and University of North Carolina (UNC) have done experiments with an implementation of the ACR/NEMA interface - see Hemminger (Hemminger, 1986) for more information.

4.3. CART Standard Database

Computer Assisted Radiation Therapy (CART) (cart, 1984) is a nordic research and development program on computer aided radiation therapy. ''The ultimate vision in the CART program is an integrated information system in a radiotherapy clinic, where all the relevant data are easily transferred between the different functions (steps) in the radiotherapy procedure. Incorporated in this system are also the relevant tools for handling the data, which consist of both image- and non-image data.'' (Lamm, 1987)

The CART standard project ''was initiated with the aim of defining parameters and data sets for the integrated communication network as described in the CART prestudy report.'' (Moeller, 1987). The CART network can be thought of as a logical network composed of 2 networks carrying image data (Diagnostic images and Therapeutic images) and 5 networks carrying alphanumeric data (Patient Administrative data, Laboratory Results, Treatment Reference data, Treatment Documentation data, and Radiation Reference data). It was thought that traffic for the five alphanumeric logical networks would be carried on a normal office network, while the two image networks were to use some type of high speed network. In order to deal with all of the data in a consistent fashion a database was created at Uppsala which describes every variable giving its nomenclature (English name and its name in each of the nordic languages (along with English synonyms)), a definition (which includes references to the appropriate international standard or to the literature), specifications (presentation format, coordinate system, units, etc.), classification into one of several groups (administrative parameter, patient parameter, basic physical concept, etc.), and where this parameter will be provided to and from (i.e., which node in the CART model assigns a value to this parameter and which node(s) use the value of this parameter).

Thus sets of parameters form the data which a node provides or consumes - i.e. this specifies the data which must be transferred between two nodes. Therefore we can see that the CART standard is really an ISO OSI specification at the application and presentation level (i.e. layers 7 and 6).

Approximately 400 parameters have currently been defined in the database and it is estimated that there are perhaps 200 more parameters which are necessary (Moeller, 1987).

For more information see (CART, 1986, Moeller, 1985, Hyoedynmaa, 1987).

In keeping with the idea of extensibility described earlier: ''The CART Standard Database ought to be renamed to *CART Reference Database.''* [from a panel discussion chaired by W. Schneider, see CART Newsletter, No. 9, Nov. 1987, pp. 27-29]

4.3.1. Good

Very few PACS efforts have made appropriate provisions for the associated data which is attached to images. A notable exception has been the CART standard database (DB).

The most outstanding aspect of the CART standard database is that it exists as an electronic database (at UDAC - the Uppsala University Datacenter) which users can query electronically. Thus if a user needs to know about a particular parameter they can query the database and find (1) yes it is defined - in which case they know who entered the definition, along with the coordinate system, units, literature reference, etc. OR (2) it is not defined and they can provide a definition which will subsequently be available to anyone else who queries the database.

4.3.2. Bad

This database has not (yet) been used outside of the CART project members. In addition, the CART project has officially ended and thus there is no financial support for the continued maintenance of the database or extending this database to include NON-radiotherapy parameters.

A brief note about the end of the CART project: The project was born with a finite lifetime and was sponsored by the nordic industrial fund (Nordisk Industrifond) from June 1985 to June 1987. The funds goal was to increase the level of competence and interest within the nordic region and to encourage commercial development following the end of the project. This project has spawned several commercial developments and new companies.

4.4. More standards

As is often said: ''the best part of standards is that there are so many to choose from''.

Recently Pixar has proposed that others adopt their 3D image format standard (this is an outgrowth of the Pixar - Philips collaboration with a number of universities).

There are also a number of standards coming from the graphic community: PHIGS and PHIGS+, X window (with 3D extensions), ...

ANSI has just created a new committee (X3.H3.8) to define a standard Application Programming Interface (API) for Electronic Imaging.

5. NEW IDEAS

With the emergence of fiber optic and other optical busses we have seen the evolution toward treating the network as a bus. This evolution can be see in the following comparison of bandwidths:

- electrical busses - ~128 Mbytes/second;
- (Present) local area networks - with signaling rates of 100 Mbits/second using fiber-optic networks such as ANSI X3T9.5 Fiber Distributed Data Interface (FDDI); and
- (Future) fiber-optic networks - being viewed as both LANs and micro-busses and are expected to have data rates of 500 to 1000 Mbits/second.

5.1. New protocols and networks (including XTP and FDDI)

One of the things which we have learned from communication systems (and disk subsystems) is that performance of controllers and low overhead protocols will be very important. Presently, the processing requirements of protocols used for networks has limited their effective data transfer rate to very low rates. The overhead which the communication protocol imposes has lead to proposals for simpler protocols for bulk transfer of data, such as *NETBLT: A Bulk Data Transfer Protocol* proposed in RFC-998 (Clark, 1987).

With standard transport level protocols, it is not uncommon to have a transfer rate at the transport level of 1.8 Mbit/second over a 10 Mbit/second network (Strauss, 1987). Chesson (Chesson, 1987) shows that for a 10 Mbit/s ethernet only a very small fraction (1 Mbit/s) is available for data transfer because of the overhead due to protocols.

Recently the Protocol Engine Inc. (Chesson, 1987) effort seeks to combine the advantages of the low bit error rate (10^{-15}) and high data rate (the signalling rates of 100 Mbit/sec. to 1 Gbit/sec)of fiber optics with the low cost and high performance of VLSI implementations of a simple protocol optimized for transferring large amounts of data (XTP Protocol Definition, Revision 3.0) (Clark, 1987). This effort seeks to provide 10 MBytes/sec. of (FDDI) network data directly to the receivers random access memory (RAM). This will make the network performance comparable to SMD disks and parallel head transfer disks. The XTP protocol is a transfer layer protocol (unification of layers 3 and 4).

5.2. Fast channels and networks

The ANSI High-Speed Channel (HSC) being considered by ANSI X3T9.3 defines a 100 MegaBYTE/sec and a 200 MegaBYTE/sec (i.e. 1.6 Gbps). The draft version of the standard goes out for voting next month (October 1988). Los Alamos National Laboratory has been exploring this channel over the last two years (starting from 50Mbps). The list of companies participating in HSC includes: Cray, ETA, Amdahl, Aptec, Convex, DEC, FPS, Thinking Machines,... (for a total of 85 companies). One commercial company which is moving toward HSC is Ultra Network Technology, San Jose, CA - the contact is James Perdue (408) 922-0100. Ultra has built network frame buffers for Apple which talk to the UltraNet hubs at 100 MegaBytes/sec which Apple and the Air Force Weapons Lab have been using to connect to their Crays. Don Tolmie at LANL is a contact for more info on HSC (505) 667-5502. For more information see the news item entitled ''Los Alamos High-Speed Channel Will Spur 1 Gigabits/s Networks'' (Batacan, 1988).

5.2.1. A fast network

Scientific Computing Systems, San Diego, CA, has introduced the VectorNet with a peak transfer rate of 1.4 Gbits/s (178 Megabytes/sec). Their network interface processor consists of a MC68020 processor and a direct memory access (DMA) controller on a VME bus card. The network uses token-ring access protocols and can use either fiber or copper cable to join up to 254 nodes. They also have FDDI and Ethernet gateways. A contact is Stephen Campbell at (619) 546-1212.

5.2.2. Fast disk subsystems

While the data rates of future fiber-optic networks may help by increasing the transfer rate of the network there still remain some problems, such as the fact that disk transfer rates have increased very slowly, although capacities have increased more rapidly. One method, first proposed by Michele Kim of IBM T. J. Watson Research Lab in 1983, is the use of bit parallel access to sets of drives (with redundancy). Therefore, lots of inexpensive standard disk drives combined with electronics for error correction and detection and de-skewing would produce the performance of a real-time disk at a much lower cost. Commercial implementations of such disk systems are just beginning to appear.

Maximum Strategy Inc. has developed the (Strategy 2) disk array controller which provides up to 32 GB of storage which can be accessed at a sustained rate of 20 MB/s with burst rates of up to 40 MB/s. The controller has built-in redundancy for fault tolerance.

5.2.3. Multiplexing

Will Hicks, president of Applied Fiber Optics, has described (at the 7th FiberTour conference, 7 June 1988, Boston, MA) an all optical transcontinental broadband telecommunications systems using both time-division multiplexing and frequency-division multiplexing (which he refers to as ''distributed optical multiplexing'') as well as all-optical amplifiers.

5.2.4. Modulation rates

Robert Olshansky, et al., of GTE Laboratories has developed a semiconductor laser which can be modulated at 22 GHz. They have combined these high-speed lasers, fiber optics, and microwave electronics to create a system which runs at 20GHz. For more details see the April 4, 1988 issue of Fiber Optics News.

5.3. Fiber disks

The basic idea is to build an optical ring network which is run at very high data rates and with the size of the network equal to the size of the buffer. Then a user could simply wait for the data that they were interested in to go by! This is reminiscent of the mercury delay lines used in the 1950s. See (Maguire, 1988) for a description of this basic idea.

5.4. Backup

''Digital Paper'' - high density recording film developed by ICI Electronics, in the U.K. The base material is a 25 µm thick polyester film called Melinex. The film is coated with a IR sensitive dye which operates with the same 780 nm or 830 nm solid-state lasers used in optical disk drives. This is a write once optical media where data is encoded as a 1 µm pit in the film.

This film is being used by:

- Creo Products (Burnaby, B.C., Canada) to make an Optical

Tape recorder which uses a 12 inch reel of tape (880m long and 35 mm wide). The drive has a transfer rate of 3 MBytes/sec and a capacity of 1 TByte presently (increasing to ~3 TB in the future). The present drive has a 60 sec. maximum search time and an average 28 second access time. The price is $200,000 for the reader/writer unit and the expected storage costs $0.005/MB for offline storage and $5/MB for on-line storage.

- Bernoulli Optical Systems Corp. (Boulder, Colorado, U.S.) is using 5 1/4 inch disks of the same film to make flexible optical disks with a transfer rate of 1.5 MB/s and a capacity of 1 GB/disk. The avg. access time is 40 ms and the expect cost is $0.05/MB for off-line storage. The price for the drive has not been announced yet.

The performance of both of these systems is better than current jukebox systems which use 5 1/4 inch or 12 inch rigid optical disks due to their slow transfer rates (0.25 - 1 MB/s) and costs of $0.10-0.25/MB offline and $3-$30 on-line.

See the April 1988 issue of *Photonics Spectra* for the initial announcement and (Fisher, 1988) for more details.

5.5. Erasable optical media

Magneto-optic disks have a storage capacity of 300Mbits/in.^{2} (46.5Mbits/cm^2) compared with Winchester magnetic disks 43Mbits/in.^{2} (6.7Mbits/cm^2). Thus a 5 1/4 inch drive has a capacity of 600 Mbytes. Although there has been concern about the lifetime of the media, use of amorphous rare-earth transition metals has increased the number of write-erase cycles to over 10 million and accelerated lifetime testing suggests lifetimes of at least 10 years. For further information see (Freese, 1988a, Freese, 1988b).

Peter Vogelgesang and Judith Hartmann of the Information and Imaging Technology Sector Laboratory of 3M (St. Paul, Minnesota) have been working on an erasable optical tape using helical scanning technology. They are concentrating on tape as opposed to disk technology due to the limited surface areas of disks and manageable disk diameter. The result is that building a disk jukebox is an expensive way to increase surface area, while the area of a tape is limited by the thickness of the tape and the size of spool upon which it is wound. Their prototype uses an 8 inch (diameter) reel of tape with a 4 inch core diameter and a 11 inch (diameter) scanning drum rotating at 900 rpm. With a linear bit density of 12,500 bits per inch they produce a 6.47 Mbits/s transfer rate. A 2000 foot reel of tape can record data at this full transfer rate for 74 hours! The total capacity of the tape is 1.72×10^{12} bits (or 215 GBytes).

See (Vogelgesang, 1988a, Vogelgesang, 1988b) for more information.

5.6. Memory vs. networks

One model for how to deal with very large high speed networks is the memory network (MEMNET) model proposed by David J. Farber. Rather than have the user think in terms of a network, the model simply treats all the resources on the network as being in one address space. Thus memory accesses provide the transfer of information rather than system calls or message passing. The result is the elimination of the need for complex layered protocols and the possibility of truly high performance by eliminating unnecessary overhead. With the advent of optical interconnects within processor and the move toward optical interconnects at the chip level, except for delay it does not matter if the connection is 10mm or 10m or 10km or 10,000km. This idea is similar to that used in the MULTICS file system or in the Apollo DOMAIN single level address space. For more information see the recent University of Delaware Ph.D. dissertation of Gary Delp.

6. GENERATING PACS CONFIGURATIONS

One of the most important questions which must be asked when designing a PACS is ''Who is the user?''. What PACS is for and what characteristics a PACS must have, depends on who you ask.

Since the requirements for a PACS is a function of who sets them, current designs and simulations, such as those of Massar (massar, 1985), primarily emphasize the requirements of physicians. For example, prefetching algorithms (such as that of Bakker et al. (Bakker, 1986)) are being designed to provide images to a human, primarily a diagnostician. The PACS requirements of those interested in image processing may be different. Thus the requirements for a program which involves retrospective analysis of a series of studies and/or patients, may require a much different approach. These requirements might be typified by:

- Image processing tasks might benefit by migrating the process performing the analysis nearer to the archive.
- Images retrieved for analysis purposes, might **not** necessarily have to be preserved in the on-line buffer for ''current'' patients.
- It would be beneficial to assign ''background'' priority to these analyses so that the processing and retrievals can be done when the interactive demand is low. Romney et al. (Romney, 1986) have studied the number of images generated as a function of the time of day for interactive use. Their study shows large variations in the demands for and production of images.

We have been trying to develop an expert system which given a set of requests will generate a PACS configuration which can fulfill these requirements. A complete description of this system will appear in Tim Lee's - M. S. thesis ''Using a Simple Expert System to Generate PACS Configurations''.

The expert system controls a simulator to determine if the behavior of a system with the proposed configuration can handle the specified requests. The input specifications include:

- event trace of actual requests for images - based on hospital/radiology information systems (HIS/RIS),
- product descriptions such as the specifications for network interfaces, magnetic disks, optical disks, automatic tape systems,
- desired availability of equipment,
- ''mean time between failures'' of the various components, and
- number and type of objects is constrained due to the amount of money that an organization can afford.

6.1. Why build software vs. using a human expert

The reasons we have chosen to build an expert system rather than rely on human experts include: (1) we believe that PACS configurations will be quite dynamic due to equipment changes from month to month and failure throughout the day - thus requiring a lot of effort by a human expert; (2) many organizations will have PACS systems but there are few experts, thus most will lack an expert in residence; (3) we want to explicitly codify the rules for configuring a PACS (if possible) and (4) to explore multiple configurations, by examining: fault tolerance, sensitivity to changes, and to explore technology alternatives.

In order to use this system to plan for growth and for designing systems which are robust we have adopted a method similar to impulse testing of physical systems. Sets of rules which generate emergency cases are used to stress the system. Emergencies can stem from any one of the following: requests for scans of an emergency patient, emergency retrieval, buffer disk failure, network failure, workstation failure, failure of acquisition node, archive failure, buffer full, archive media full, and archive full.

One of the things we have learned is that for configurations which are non-trivial the large amount of data which the simulator produces overwhelms the user. Thus Norman Chin has developed a program using the X-window (graphics) system to display system event traces.

6.2. Status

The expert system is currently being developed using simulated trace data. The initial configuration and the interfacing to the simulator is complete, while the analysis of a configuration and the modifications to the configuration sections of the expert are just beginning. The next step in this project will be to use real data from the BAZIS group to compare our results against their simulations for the same configuration. We also plan to use the expert to evaluate the effects of technological changes on the system components to determine the desirable characteristics of PACS components, using a sensitivity analysis.

See (Lee, 1987) for more details.

7. CONCLUSION

While implementation of PACS is occuring slowly, the technology base is changing very rapidly, thus changing fundamental design decisions. We can not afford to ignore these changes nor can we afford to develop systems/networks/disks/... just for the medical imaging market. Thus we need to adapt the technology being developed by others to the special needs of medical imaging. Unfortunately as was discovered in the U.S. Defense Advanced Research Projects Agency Internet (ARPANET) most of the applications which it was designed for turned out to be only a small part of what it actually has been used for. The analagous change in users behavior with the introduction of office automation should be expected as PACS systems are introduced. Therefore, the challenge is to develop a system which will support todays needs and can evolve in a cost effective manner to serve tomorrows needs as well.

ACKNOWLEDGMENTS

This research was partially supported by equipment grants from the AT&T Foundation and Hewlett-Packard Corporation and U. S. National Science Foundation grant CDR-84-21402.

REFERENCES

Bakker, A., Didden, H., de Valk, J.P.J. and Massar, A.D.A. *Considerations on an algorithm for activation of images in a multi-layer storage within a PACS.* S.P.I.E., February, 1986.

Batacan, P. Los Alamos High-Speed Channel Will Spur 1 Gigabits/s Networks. *Computers in Physics*, September/October 1988, *2, (5)*, 12-13.

Baxter, B.S.; Hitchner, L.E.; and Maguire Jr., G.Q. *Characteristics of a protocol for the exchange of digital image information*, pages 273-277. Society of Photo-Optical Engineers, January, 1982.

Baxter, B.S.; Hitchner, L.E.; and Maguire Jr., G.Q. A Standard Format for Digital Image Exchange. American Association of Physicists in Medicine Series, Report number 10. American Institute of Physics, N.Y., N.Y., 1982. 11 pages.

Dahlin, H. and Hamren, M. CART - Computer Aided Radio Therapy - Presentation of an Integrated Information System in Radiotherapy. Medical Computer Physics/UDAC. Akademiska sjukhuset, S-751 85 Uppsala, Sweden, 1984. 73 pages.

Moeller, T. CART NEWS: CART standard database.

Chesson, G. The Protocol Engine Project. *UNIX Review*, September 1987, *5*, *(9)*, 70-77.

Clark, D.D., Lambert, M.L. and Zhang, L. NETBLT: A Bulk Data Transfer Protocol. Network Working Group, Request for Comments: 998.

Fisher, M.J. Digital Paper Promises Cost, Storage Gains for Optical Media. *Datamation*, May 1988, , .

Freese, R.P. Optical disks become erasable. *IEEE Spectrum*, February 1988, *25*, *(2)*, 41-45.

Freese, R.P. Optical disks become erasable. *OE Reports*, June 1988, *(54)*, 4.

Goitein, M. Specifications for tape format for exchange of planning information amongst the members of the Particle Intercomparison Project and the High Energy Photon External beam Treatment Planning Project. In Sandra Zink (Ed.), *Evaluation of Treatment Planning for Particle Beam Radiotherapy*, Bethesda, MD: U.S. National Cancer Institute, 1987.

Greinacher, C.F.C. Systemarchitektur und Baunsteine eines digitalen Bildinformationssystems fuer die Radiologie. In Lemke, H.U., Rhodes. M.L., Jaffee, C.C. and Felix, R. (Ed.), *Computer Assisted Radiology*, Berlin, West Germany: Springer-Verlag, 1987.

Hemminger, B.M., Thompson, B.G., Pizer, S.M. and Stancil, P. *The ongoing implementation of a prototype medical communications systme at the University of North Carolina*. S.P.I.E., February, 1986.

Hyoedynmaa, S. CART Standard Database - A Development Resource for Industry. *CART Newsletter*, November 1987, *(9)*, 25-26.

Lamm, I.-L. Images and Radiotherapy. *CART Newsletter*, November 1987, *(9)*, 3-5.

Lee, T.M., Maguire Jr., G.Q., Liu, M. and Noz, M.E. Using a Simple Expert System to Generate PACS Configurations. In Lemke, H.U., Rhodes. M.L., Jaffee, C.C. and Felix, R. (Ed.), *Computer Assisted Radiology*, Berlin, West Germany: Springer-Verlag, 1987.

Maguire Jr., G.Q.; Zeleznik, M.P.; Horii, S.C.; Schimpf, J.H.; and Noz, M.E. *Image processing requirements in hospitals and an integrated systems approach*, pages 206-213. Society of Photo-Optical Engineers, January, 1982.

Maguire Jr., G.Q.; Baxter, B.S.; and Hitchner, L.E. *An AAPM standard magnetic tape format for digital image exchange*, pages 284-293. Society of Photo-Optical Engineers, 1982.

Maguire, G.Q. Jr. *NATO ASI Series, Series F: Computer and Systems Sciences*. Volume F19: Looking Back at PACS attempts - What has Changed since PACS I. In K.H. Hoehne (Ed.), *Pictorial Information Systems in Medicine*, Berlin: Springer-Verlag, 1986.

Maguire Jr., G.Q., Noz, M.E., Bakker, A., Bijl, K., Didden, H., and de Valk, J.P.J. Introduction to PACS for Those Interested in Image Processing. In C.N. de Graaf and M.A. Viergever (Ed.), *Information Processing in Medical Imaging*, New York, NY: Plenum Publishers, April, 1988. Invited Paper.

Massar, A.D.A. Simulation of a Communication System for Medical Images. In *Laboratorium voor schakelterchieu en techniek dor informatieverwerkende machines,* Univeristy of Delft, The Netherlands: Master's Thesis, 1985.

Moeller, T. Definition of information in radiotherapy. *Medical and Biological Engineering and Computing,* August 1985, *23,* , 945-948. Proceedings of the XIV International Conference on Medical and Biological Engineering and VII International Conference on Medical Physics, Espoo, Finland.

Moeller, T. and Hyoedynmaa, S. CART Standard. *CART Newsletter,* August 1987, *(6),* 9-10.

Noz, M.E., Maguire Jr, G.Q. and Erdman, W.A. Local Area Networks in an Imaging Environment. *CRC Critical Reviews in Medical Informatics,* April 1986, *1, (1),* 81-133. Invited Review Paper.

ter Haar Romney, B.M., Meijwaard, J., ten Hertog, A. de Graaf, C.N. and van Rijk, P. P. *Radiological information flow between departments and out-clinics in the Utrecht University Hospital in the Netherlands.* S.P.I.E., February, 1986.

Selfridge, P.G. *A Flexible Data Structure for Accessory Image Information.* Technical Report TR-45, University of Rochester, May 1979.

Strauss, P. OSI throughput performance: Breakthrough or bottleneck? *Data Communications,* May 1987, *16, (5),* 53-56.

Vogelgesang, P. and Hartmann, J. *Erasable optical tape feasibility study.* SPIE, Bellingham, WA, January, 1988. ISBN 0-89252-934-2.

Vogelgesang, P. and Hartmann, J. Erasable optical tape feasibility study. *OE Reports,* June 1988, *(54),* 5.

Zeleznik, M.P.; Maguire Jr., G.Q.; and Baxter, B.S. *PACS data base design,* pages 287-295. Society of Photo-Optical Engineers, May, 1983.

EVALUATION OF MEDICAL IMAGES

Charles E. Metz

Department of Radiology
The University of Chicago
United States of America

ABSTRACT

From a practical point of view, image quality in medicine must be defined in terms of the decisons that physicians can make by reading the images. Image-based decisions concerning the actual state of an object or patient are evaluated most meaningfully by ROC analysis, which has been used to quantify detection performance in sensory psychophysics since the early 1960s and more recently has been applied to evaluate diagnostic systems. After surveying briefly some considerations that motivate the use of ROC analysis in medical image evauation, this manuscript addresses various practical issues of experimental design and data analysis, including the selection of cases and observers, the need for standards of truth, reading-order effects, data collection, curve fitting, statistical testing, and possible generalizations of conventional ROC methodology.

1. INTRODUCTION

Before conducting a study that seeks to measure the quality of images, one must decide upon the sense in which image quality is to be evaluated. In the past, image quality usually was judged in physical terms, such as spatial resolution and noise level, or by some "figure of merit" based on physical measurements. From a practical standpoint, however, an imaging procedure is "good" only to the extent that it serves as a useful tool when it is applied to a particular task. In this sense, the quality of an imaging procedure in diagnostic medicine depends upon the quality of the medical decisions that can be made with it. Unfortunately, the medical decisions that an imaging procedure allows usually cannot be predicted from knowledge of the physical properties of the images, because the process of human visual perception is poorly understood. Thus, diagnostic image quality must be evaluated directly in terms of the medical decisions that are made from the images. But this is possible only if one can quantify the "goodness" of medical decisions in a meaningful yet practical way.

In the broadest sense, the value of a particular diagnostic imaging procedure depends upon the benefit of that procedure to society as a whole; not only the risks and financial costs of the procedure but also the availability of therapies and alternative diagnostic techniques must be taken into account. While this very general perspective encompasses the ultimate concern, it clearly does not provide an immediately practical basis for making measurements of image quality. Therefore, in the foreseeable future, attention must focused on a narrower definition of medical image quality.

Knowledge of each patient's state of health or disease is not a sufficient condition for effective medical care, but it is almost always a necessary condition for such care. Therefore, the quality of a medical imaging procedure can be defined and measured meaningfully in terms of the extent to which image-based diagnoses agree with the actual state of health or disease of each patient.

For simplicity, the evaluation of visual detectability usually is restricted to situations in which possible truth can be divided into two states (e.g., "abnormal" vs. "normal," "disease X present" vs. "disease X absent," etc.) and in which two corresponding decisions can be made. The adequacy and limitations of this restriction are discussed elsewhere (Metz, 1986a). Often the two states are indicated by the abstract terms "positive" and "negative" to denote a defined (possibly composite) truth state and its compliment. In studies that seek to measure medical image quality, the two states of truth can be chosen by the designer of the experiment to represent alternative diagnoses, the presence and absence of some diagnostically relevant image feature, or the presence and absence of an idealized geometric object. Images obtained from actual clinical cases, phantoms, or computer simulation can be used in the experiment, depending on the compromise between realism and convenience that is considered appropriate.

Various indices have been employed in the past to quantify decision performance in the two-state task (which we refer to here as "detection," noting that this definition includes two-state discrimination as well). Among these indices, the fraction of actually positive cases called "positive" by the observer is clearly inadequate when used alone, because it provides no information concerning the fraction of actually negative cases that might falsely be called positive. In passing, one should note that this is an important limitation of "contrast-detail" diagrams measured with a Burger phantom, in which simple visual signals with various sizes and contrasts are known by the observer to be present in a regular array (Loo et al, 1983). Also inadequate is the overall percentage of correct readings in a mixture of actually positive and actually negative images (traditionally called "diagnostic accuracy" in the medical literature), in part because the numerical value of this index depends strongly on the prevalence of actually positive cases and does not reveal the balance of false-positive and false-negative errors, which can have very different clinical consequences (Metz, 1978).

Superficially, the pair of indices called "sensitivity" and "specificity" in the medical literature (Lusted, 1978; Weinstein and Fineberg, 1980) seems more satisfactory, but it suffers from the important limitation that the numerical values depend in an uncontrollable way upon the "critical confidence level," or "decision criterion," that each observer adopts to distinguish "positive" image readings from "negative" readings (Metz, 1978).

2. THE ROC CURVE

The limitations mentioned above are overcome by measuring all of the combinations of sensitivity and specificity that an imaging procedure provides in a particular detection task as the critical confidence level is varied over all possible settings, and by plotting one of these indices against the other in a unit square. More commonly, "True Positive Fraction" (TPF), which is equivalent to sensitivity, is plotted against "False Positive Fraction" (FPF), which is equal to 1.0 - specificity, thereby producing a "Receiver (or Relative) Operating Characteristic" (ROC) curve (Green and Swets, 1966; Metz, 1978; Swets and Pickett, 1982). ROC curves rise from the lower-left corner of the unit square; bend to the right with decreasing slope; and finally enter the upper-right corner of the square. High ROC curves represent better detection performance than low ones, because with appropriate settings of the decision criterion a larger TPF can be obtained on a higher ROC for any given FPF, or a smaller FPF can be obtained for a given TPF. Therefore, if ROC curves of interest do not intersect each other, detection performance can be summarized by the "area index" A_z, which represents the area under an ROC inside the unit square (Swets and Pickett, 1982; Metz, 1986a). Alternatively, ROC curves can be compared in terms of the TPF values they provide at a particular FPF of practical interest (Swets and Pickett, 1982; McNeil and Hanley, 1984).

3. SELECTION OF CASES AND OBSERVERS FOR AN EVALUATION STUDY

The detectability of a lesion in an image obviously depends not only upon the physicial properties of the imaging system, but also upon the size, contrast, etc. of the lesion in question and upon the characteristics of non-lesion structures that are (or may be) present in the images. More generally, the ability of an observer to discriminate visually between two classes of objects or "cases" depends upon the subtlety of the differences between the two classes of cases in question. Therefore the ROC curve that is measured with a set of images -- and so, according to our practical definition of image quality, the <u>quality</u> of those images -- depends not only upon the physical properties of the images, but also upon the particular actually positive cases (phantoms, patients, etc.) and actually negative cases from which the images were made. This dependence may seem confusing and undesirable at first, but in fact it is consistent with the common observation that different imaging systems can excel in depicting different classes of scenes and in different tasks. Similarly, image quality as we define it here depends upon the skill of the observer who reads the images, but this dependence also is appropriate, because different imaging systems may be best when used by observers with different training or experience. Ideally, all observers who perform a particular image-reading task would possess the highest level of skill that is humanly attainable; in reality, however, images are read by observers with diverse skills.

Because the results of an image-evaluation study generally depends upon the cases and observers that the study employs, care must be taken to ensure that those cases and observers are selected appropriately. It is important to note that sampling issues must be addressed in any evaluation study, and that ROC methodology is no more demanding in this regard than other methods of analysis that provide less meaningful descriptions of detection performance (Metz, 1986a).

In designing a visual-detection experiment for the evaluation of image quality, one must decide first whether an absolute measure of the detectability of some class of diagnostic features or some particular disease is desired, or whether the goal is more simply to rank alternative imaging procedures (Metz, 1988). As noted in following sections, the sampling issues that must be confronted in these two kinds of experiments can be quite different.

3.1 Absolute measurements of detection performance

Reliable absolute measurements of disease detectability in a defined patient population are often extremely difficult to obtain, in part because the sample of patients included in the study must accurately reflect the population of patients at large about which conclusions are to be drawn, and many sources of potential bias must be taken into account (Begg and McNeil, 1988). For example, an experiment that attempts to measure the absolute detectability of lung nodules by chest radiography must ensure that the distribution of nodule sizes is the same in the study sample as in the defined population of interest, because the detectability of a nodule depends on its size. Formal "stratified sampling" techniques (Kendall and Stewart, 1976) have not yet been used formally in medical imaging but may prove useful for reliable absolute measurements of system performance. In the detection of lung nodules, for example, these techniques can help to ensure that an appropriate distribution of nodule sizes is used in a study (Metz, 1988).

The absolute detectability obtained in medical imaging depends not only on the difficulty of the cases, but also on the experience and skill of the observers (e.g., radiologists) who read the images; experienced mammographers have been shown to perform better than general radiologists in using xeromammograms to discriminate between malignant and benign breast lesions (Getty et al, 1988), for example. Therefore, if reliable absolute measurements of detectability are to be obtained from a medical imaging study, the relevant population of observers must be defined, and the sample of observers employed in the study must accurately represent that population. Similarly, to measure accurately the absolute detectability of disease that would be obtained in routine clinical practice, the conditions (e.g., time, ambient light level, etc.) under which images are read in the study must represent those that would be used in clinical practice.

3.2 System-ranking studies

Studies that seek only to rank systems are often much more straightforward. Sampling considerations still require attention in the study design, but now the only requirement is that these factors do not affect the ranking of the systems; their effects on absolute detectability are no longer of primary concern. The key need becomes one of ensuring that a system which would provide superior diagnostic performance in its real-world application is found better in the study, and that two systems providing equivalent performance in the real world are found equivalent by the study.

An individual faced with the task of designing this second kind of study has greater freedom in choosing the cases it will employ. A basic question that often arises in this task concerns the most appropriate level of case difficulty. Common sense suggests correctly that the cases shouldn't all be extremely easy, because then even a very poor imaging system would perform well. But should the cases cover a broad spectrum of difficulty to represent cases at large; should they be restricted to "subtle" cases; or should they be so challenging that even the best system finds them quite difficult? The answer depends in part on the statistical properties of ROC indices.

If the difficulty of cases in a particular diagnostic task depends primarily on only one factor, such as lesion size, and if we may assume that a system which is better in detecting small lesions will be better in detecting large lesions also, then the question at hand is reduced to one of choosing the best lesion size, or range thereof. If a single lesion size is to be employed in a system-ranking study, it should be chosen to maximize the difference between the expected values of the ROC index used to summarize performance, relative to the uncertainty in that difference -- in other words, to maximize the statistical power of the experiment. No formal guidance is available as yet concerning this choice, but a reasonable rule of thumb when the ROC area index, A_z, is used seems to be that the <u>average</u> of the A_z values of the two modalities should lie near 0.75 or 0.80 (Metz, 1988). For example, if lesion size is the primary determinant of detectability with each system, then a pilot experiment involving the two systems and several lesion sizes should be performed to determine approximately the lesion size that achieves this <u>average</u> level of performance. A range of lesion sizes distributed around this optimal size may be employed in the definitive experiment if that is considered desirable for "realism," but one must recognize that each lesion with a size substantially different from the optimal value will make a smaller contribution to statistical power.

Unfortunately, the task of selecting appropriate cases may not be so simple even for a system-ranking experiment if the detectability of the disease of interest depends strongly on two or more distinct factors. In mammography, for example, microcalcifications are detected better with special-purpose screen/film systems than with any current digital radiographic system due to the superior spatial resolution that screen/film radiography provides, whereas large, low-contrast masses of soft tissue may be detected better with digital radiography, which allows image contrast to be manipulated by the radiologist. Thus, the relative "quality" of mammographic images obtained with screen/film radiography and with digital radiography may depend on whether the primary task is considered to be detection of microcalcifications or detection of soft-tissue masses. Furthermore, since clusters of microcalcifications are a strong indicator of some but not all malignant breast lesions, whereas many soft-tissue masses in the breast are benign, the relative "quality" of screen/film radiography and digital radiography for the task of detecting malignant breast lesions may depend in a complex way upon the characteristics of the particular patient population to which the systems will be applied. In this situation, the designer of a system-ranking experiment must choose between two courses of action. The first option is to measure image quality separately for the detection of microcalcifications and for the detection of soft-tissue lesions, thereby producing separate detection tasks in which single factors may dominate. Alternatively, if a measurement of the overall detectability of malignant breast lesions is desired, the experimenter must attempt to select a sample of cases with not only a representative balance among non-lesion cases, benign-lesion cases, and malignant-lesion cases, but also with microcalcifications present in appropriate individual fractions of these three kinds of cases.

3.3 Establishing truth

Essentially all objective evaluation techniques measure the agreement between the results of a test and some external standard of truth; therefore, the true state of each case in an objective evaluation study must be known. In evaluation studies that employ phantom images or computer-simulated images, the experimenter knows the actual state of each case because he/she controls it. The results of such studies may be reliable, in the sense that they can be reproduced, but often they are unrealistic, in the sense that the phantoms or computer simulations from which the study's images were made do not adequately represent the complexity of real-world diagnostic tasks. Therefore, even the most carefully performed phantom and computer-simulation studies may leave doubt concerning whether similar results would have been obtained with real patients and real images.

Unfortunately, the establishment of diagnostic truth in clinical studies is usually difficult, both in principle (because "truth" is ultimately a philosophical matter) and in practice (because great effort may be required to determine the actual state of health or disease in a particular patient at a particular point in time "beyond a reasonable doubt"). But despite these dificulties, carefully designed clinical ROC studies can be done, and useful conclusions can be drawn from them; see, for example, reports listed by Swets and Pickett (1982) and by Metz (1986a).

Important issues that must be confronted in establishing truth in clinical evaluation studies have been reviewed elsewhere (Ransohoff and Feinstein, 1978; Swets and Pickett, 1982; Metz, 1986a; Begg and McNeil, 1988). Particular attention must be focused on biases that may be caused by the omission of clinical cases for which truth is particularly difficult to establish (Ransohoff and Feinstein, 1978; Gray et al, 1984).

4. READING-ORDER EFFECTS

When two or more images of a particular patient are read by the same observer, the image read last will tend to be interpreted more accurately than the image read first if <u>any</u> relevant information is retained by the observer from a reading of one image of the patient to the next. Thus, if all of the images of a patient sample made with modality "A" are read before any of the images of the same patient sample made with modality "B," the results of the study are potentially biased in favor of modality "B." Biases of this kind, called "reading-order effects," must be avoided in evaluation studies that use a single case sample to compare two or more imaging sysytems.

There are two methods by which biases due to reading-order effects can be reduced or eliminiated; the common theme is to vary the order in which the modalities' images are read so that reading-order effects tend to cancel out. In the first method, the order of the modalities is varied across observers; thus, the ranking of the modalities remains potentially biased for individual observers, but any such bias tends to cancel across observers. The second method attempts to negate the bias <u>within</u> each observer's results by breaking the case sample into several subsets and having each observer read the modalities in a different order for each subset. An example of the latter approach is provided elswhere (Metz, 1988).

5. DATA COLLECTION

5.1 ROC studies

In visual detection experiments, ROC curves can be measured by having an observer reread a set images several times in different reading sessions, using a different critical confidence level in each session to distinguish "positive" image readings from "negative" readings; then one point on the ROC can be estimated from each set of readings. However, this method is extremely inefficient, requiring M readings of each image to estimate M points on each ROC curve. In practice, visual-detection ROCs are almost always measured by obtaining "confidence-rating" data that represents the observer's relative confidence that each image was produced from an actually positive case. With these graded judgements reported on a K-category scale, K-1 points on an ROC can be estimated from a single reading of each image (Green and Swets, 1966; Metz, 1978, 1979, 1986a). Usually five or six categories are employed, thereby yielding estimates of four or five points on the ROC in addition to the lower-left and upper-right corners of the unit square, into which every conventional ROC must pass. A larger number of confidence-rating categories would be desirable in principle, but in practice many observers find it difficult to grade subjective judgements on finer scales.

Discrete confidence-rating categories usually are assigned verbal labels such as "definitely or almost definitely positive," "probably positive," "possibly positive," etc. to ensure that each observer uses the categories in a strictly ordered fashion. The way in which these labels may be interpreted by an observer does not bias the measured ROC as long as the observer's relative confidence in the two states of truth varies monotonically across the categories (Metz, 1986a). However, for statistical efficiency and to guard against "degenerate" data sets (Metz, 1988), it is desirable for the confidence ratings to yield ROC points that are more-or-less uniformly spread along the curve. When approximately half of the images in an experiment are actually positive, this is accomplished if the observer uses the categories with roughly equal frequencies.

Clinical images can be read either with or without case-history information, and with or without the results of other tests that would precede the imaging test in real-world applications, depending on the goals of the evaluation study (Metz, 1986a). Similarly, clinical images can be read either sequentially in controlled reading sessions or one at a time in the course of daily clinical practice. However, when the confidence-rating method is used in a controlled experiment, a brief training session should be held immediately before each image-reading session in which data will be acquired. In each of these training sessions, the observer should be shown a spectrum of cases that represents the range of decision difficulty that will be presented during data acquisition, and he/she should be motivated to use the confidence-rating categories in a way that will produce a roughly uniform spread of points along the ROC curve. Also, the observer must be made to understand that he/she should strive to use the rating scale in a constant way during data acquisition, because variation in use of the scale degrades decision performance (Goodenough and Metz, 1977; MacMillan and Kaplan, 1985). The cases that are presented during the training session should be typical of, but must not include, those that will be read during data acquisition.

5.2 The two-alternative forced-choice paradigm

As noted earlier, the index A_z represents the area under an ROC curve inside the unit square and often provides a useful summary of detection performance. The area index A_z can be interpreted as the average value of sensitivity on the corresponding ROC curve if the system's specificity is selected randomly between 0.0 and 1.0, or -- equivalently -- as the average value of specificity if sensitivity is selected randomly between 0.0 and 1.0 (Metz, 1986a). Perhaps surprisingly, the area index can be measured directly in a "two-alternative forced-choice" (2-AFC) experiment (Green and Swets, 1966) without measurement of -- or explicit reference to -- the ROC. The observer in a 2-AFC experiment views independent pairs of images together. One image in each pair is always actually positive, whereas the other is always actually negative, and the observer is required to state which image is which.

The chief disadvantage of 2-AFC experiments lies in the fact that the trading relationship between sensitivity and specificity (i.e., the ROC curve) is never determined. Therefore, the result of a 2-AFC experiment cannot be used in a higher-order efficacy analysis that involves costs and benefits (McNeil et al, 1975; Metz et al, 1975; Metz, 1978; Swets and Pickett, 1982), where a particular compromise between sensitivity and specificity must be considered. Further, the apparent efficiency of the 2-AFC method is somewhat deceptive, because greater statistical precision in the A_z index can be gained with a given number of images if confidence-rating data are obtained from the observer (Hanley and McNeil, 1982).

6. ROC CURVE FITTING

After a set of rating data has been obtained, one is faced with the task of fitting a continuous ROC curve to those data. Objective curve-fitting methods require that some functional form be assumed for the ROC curve. Many functional forms have been proposed (Egan, 1975; Swets, 1986a), but the "binormal" form has been used most widely in medical imaging. According to the binormal model, which includes two adjustable parameters, each ROC is assumed to have the functional form that would be produced by two "normal" (i.e., Gaussian) decision-variable distributions with generally different means and standard deviations (Green and Swets, 1966; Egan, 1975; Swets, 1979). It is important to note that any monotonic transformation of a decision variable changes the distributions that underlie the ROC but not the ROC itself; therefore, the binormal assumption concerns only the functional form of the ROC curve, which always can be examined empirically, and not the form of the underlying distributions, which cannot be determined in many applications of ROC analysis (Metz, 1986a).

Empirically, the binormal form has been found to provide satisfactory fits to ROC data generated in a very broad variety of situations (Swets, 1986b), and it has the convenient property that all possible ROC curves are transformed into straight lines if they are plotted on "normal-deviate" axes (Green and Swets, 1966; Swets, 1979). Plots of the same binormal ROCs on both conventional and normal-deviate axes are shown elsewhere (Metz, 1986a).

The two adjustable parameters of a binormal ROC can be taken to be the y-intercept and the slope of the straight line that represents the ROC when it is plotted on normal-deviate axes. These two parameters, usually denoted by "a" and "b", can be interpreted in terms of an effective pair of underlying Gaussian distributions as the distance between the means of the two distributions and the standard deviation of the "actually negative" distribution, respectively, with both expressed in units of the standard deviation of the "actually positive" distribution. Alternative ways of parameterizing a binormal ROC curve have been described by Swets (1979).

With the binormal model, the task of curve fitting becomes one of choosing numerical values for the {a,b} parameter pair to best represent the measured data. Least-squares methods are not appropriate because the assumptions underlying those methods are not valid for ROC data; instead, maximum-likelihood estimation should be used (Metz, 1986b). Maximum likelihood algorithms for ROC analysis are readily available (Dorfman and Alf, 1969; Grey and Morgan, 1972; Metz, et al, 1984, 1985; Swets and Pickett, 1982) and provide not only estimates of the parameters of the best-fit ROC curve, but also estimates of the uncertainties in those parameters (Metz, 1986b).

The chief limitation of the binormal model and curve-fitting algorithms based on it is the fact that ROCs with bizarre shapes sometimes are fit to data from experiments that involve a small number of readings if the observer uses the rating scale in a poorly-distributed way. In certain limiting cases, data from such experiments are fit exactly by a binormal model that involves one or more infinite parameter values, thereby making it impossible for iterative curve-fitting algorithms such as RSCORE II (Swets and Pickett, 1982) to converge. A modified version of RSCORE II called ROCFIT (Metz et al, 1985) identifies these "degenerate" data sets (Metz, 1988) before iteration is begun and determines the exact-fit binormal ROC directly. In this way, ROCFIT is able to avoid situations in which an iterative procedure fails mysteriously to converge, but the form of these exact-fit binormal ROCs is clearly inappropriate (Metz, 1988). The only really safe way to deal with degenerate data sets at present is to identify and discard them. Any fundamental progress in dealing with this problem seems to depend on the future development of a maximum-likelihood curve-fitting algorithm that assumes the ROC to arise from a family of curves with monotonically decreasing slope (Metz, 1988). The historical success of the binormal model in fitting statistically reliable ROC data (Swets, 1986b) suggests that this new family of curves must closely approximate the binormal family of curves over the range of FPF and TPF in which confidence-rating data are usually collected.

When several sets of rating data are available for estimation of a single ROC curve (from repeated readings by the same or different observers), two approaches can be used to combine the data to obtain a single ROC estimate. In the first method, the ratings are pooled to produce a single data set, which is then fitted by a conventional maximum-likelihood algorithm. However, this procedure may bias the result, because data pooling can depress the fitted ROC curve, especially if the ratings are generated with substantially different category boundaries (Swets and Pickett, 1982; MacMillan and Kaplan, 1985), and it *always* yields overly optimistic estimates of curve reliability (Metz, 1988). The alternative approach is to calculate an averaged -- rather than pooled -- estimate of performance. This can be accomplished either by averaging the normal-deviate values corresponding to the various true-positive and false-positive probabilities associated with each rating category and then computing the maximum-likelihood curve fit, or by calculating the maximum-likelihood curve estimate for each data set individually and then averaging the resulting parameter values (Swets and Pickett, 1982). Both of these averaging methods have potential advantages and disadvantages. Swets and Pickett (1982) advocate the first averaging method when the observers have similar skills, but the second may be preferable when a heterogeneous observer population is studied and a "population averaged" ROC curve is sought. One should note that invalid estimates of precision are produced by all current maximum-likelihood curve-fitting algorithms when either of these averaging methods is used.

7. THE "CHI-SQUARE" GOODNESS-OF-FIT TEST

Dorfman's RSCORE II program for binormal ROC curve fitting (Swets and Pickett, 1982) calculates and prints a conventional "Chi-square" goodness-of-fit statistic to indicate the quality of the obtained fit. For confidence-rating data collected on a scale containing K categories, this statistic is distributed asymptotically as a Chi-square random variable with K-3 degrees of freedom ($\chi^2_{df=K-3}$) in the limit as the expected number of responses in each category becomes large for both actually negative and actually positive trials. Unfortunately, as noted elsewhere (Metz, 1988), the so-called "Chi-square" goodness-of-fit statistic approximates $\chi^2_{df=K-3}$ very poorly in most practical ROC experiments: the fit is rejected much too often.

A modified form of RSCORE II, called ROCFIT (Metz et al, 1985), prints the "Chi-square" goodness-of-fit statistic if and only if the expected number of ratings in each category is at least 5 for actually positive trials and for actually negative trials -- a condition rarely met in practice. We recommend that users of RSCORE II ignore its goodness-of-fit statistic unless every entry in the 2 × K "data matrix" is at least 5.

8. STATISTICAL TESTS FOR DIFFERENCES BETWEEN ROC CURVES

In this section we survey briefly a variety of issues that arise in testing the statistical significance of differences between ROC curves estimated from confidence-rating data. These issues have been considered in greater detail by McNeil and Hanley (1984) and by Metz (1986b, 1988).

Metz (1988) has pointed out that one must answer at least three distinct questions before selecting a statistical test:

-- In what sense should a difference between two ROC curves be quantified in the particular situation at hand?

-- Are the estimates of the ROC curves of interest statistically independent or are they potentially correlated?

and

-- Was each ROC curve of interest measured from one or from more than one reading of each image?

The answers to these questions determine the test (or tests) that can be used to evaluate the statistical significance of an apparent difference between measured ROCs.

8.1 Ways in which differences betweens ROCs can be quantified

The null hypothesis that one must use to evaluate the statistical significance of a difference between two ROCs is dictated by the way in which the difference is to be quantified. In general, the most appropriate null hypothesis depends on the way in which the two imaging systems will be employed in practice.

One approach, which can be addressed with a Chi-square test (Metz and Kronman, 1980; Metz et al, 1984; Metz, 1986b), considers the null hypothesis that two sets of rating data arose from exactly the same binormal ROC curve. This null hypothesis of identical ROCs may be unnecessarily strict in some evaluation studies, however. For example, in practice we may not care whether the two imaging systems yield exactly the same TPF at all possible FPFs, but instead we may wish only to compare the TPFs of the two systems at a particular FPF of clinical interest -- FPF = 0.10, say. Then the null hypothesis of "no difference" would correspond to the condition that the two ROCs have equal TPFs at FPF = 0.10, and it could be addressed by an appropriate univariate z-score test (McNeil and Hanley, 1984; Metz et al, 1984; Metz, 1986b). Alternatively, we might be willing to consider two systems equivalent in clinical practice if they yield the same average value of TPF for FPFs ranging from 0 to 1 -- in other words, if the two systems' ROC curves have the same area (A_z) beneath them when they are plotted on conventional axes (Metz, 1986a). This null hypothesis can be addressed by other univariate z-score tests (Swets and Pickett, 1982; Hanley and McNeil, 1982, 1983; McNeil and Hanley, 1984; Metz et al, 1984; Metz, 1986b).

The three null hypotheses mentioned here are not equivalent, because ROCs can cross. Thus, for example, the two _true_ ROCs in question may have equal areas beneath them yet be different curves, with different TPFs at all FPFs except one; or the true ROCs may have the same TPF at the FPF of interest, but be different curves and have different areas. Perhaps surprisingly, the results of the three statistical tests can disagree fairly frequently even when the ROCs are identical, and the null hypothesis of the "A_z" test and/or of the "TPF" test can be rejected even when the null hypothesis of identical ROCs is accepted (Metz, 1988)! Therefore, it is important to select the most pertinent null hypothesis and a corresponding test _before_ statistical testing is begun, and to then perform only that single test. If multiple tests are performed, conflicting conclusions may be drawn at best, and at worst a highly significant difference will be found after enough tests have been done -- even when the ROCs of the two systems are, in fact, identical.

With regard to the prior selection of a single test, one should note that the A_z ("area") test can be performed in several different ways (McNeil and Hanley, 1984), depending in part on whether the two ROCs in question are estimated (i) by using the "trapezoidal rule," with the discrete operating points obtained directly from the confidence-rating data connected by straight line segments to _approximate_ the area beneath the continuous ROC, or (ii) by fitting a continuous binormal ROC to the data and then calulating the _exact_ area under this ROC by means of a mathematical formula. Our experience with Monte Carlo simulation studies indicates that the results of these two approaches often agree but that the second method is somewhat more reliable.

8.2 Independent vs.correlated ROC estimates

If estimates of two ROC curves are statistically independent, then no matter how the difference between the ROCs is quantified, the statistical variability of the difference depends only on the variabilities of the individual ROC estimates themselves. Several relatively simple significance tests apply to this situation (Metz and Kronman, 1980; Hanley and McNeil, 1982; Metz, 1986b).

When the estimates of the two ROC curves are not independent but instead tend to vary above and below their means together -- for example, because a case sample that is atypically difficult for one imaging system tends to be atypically difficult for the other as well -- then the difference between the ROC estimates will vary less than it would if the estimates were independent. Thus, as intuition would suggest, the statistical power of a system-ranking study can be increased by applying both systems to the same case sample. Increased statistical power will be realized only if the covariance of the two ROC estimates is estimated by some means and then factored into the statistical test, however.

For comparisons in which each ROC is estimated from a single reading of each image, this can be accomplished by basing statistical tests on a "bivariate binormal" model (Metz et al, 1984; Metz, 1986b) that generalizes the conventional binormal model which underlies conventional ROC curve-fitting algorithms such as RSCORE II and ROCFIT. An alternative approach (Hanley and McNeil, 1983) can be used when the trapezoidal-rule approximation of the ROC area index is considered adequate.

8.3 Differences between ROCs estimated from multiple readings of each image

For simplicity, our discussion up to this point has focused on ROC curves measured from a single reading of each image. However, in many experiments each image from each modality is interpreted by more than one observer, and perhaps more than once by each observer. The advantage of multiple readings is fairly obvious: the effect of "between-reader" differences in skill is "averaged out" across the observer sample, and the effect of "within-reader" variation is reduced by each reader's replication. Much less obvious, unfortunately, are ways in which the variation-reducing effects of multiple readings can be taken fully into acccount in statistical tests for differences in modality performance.

Swets and Pickett (1982) have described a general model for variation in the difference between two area indices, A_z, measured with multiple readings of each image; it applies equally to any univariate difference between ROC indices. This model includes terms that represent the "case-sample" variation, "between-reader" variation, and "within-reader" variation mentioned above, together with correlation coefficients that account for the effects of matching readers and cases across modalities. The model shows clearly how the power of a statistical test for ROC differences is affected by the numbers of readers and readings in a multiple-reading experiment, and it prescribes a procedure for performing such a test. Swets' and Pickett's approach is the the most comprehensive one formulated to date, and in principle it is adequate for statistical analyses of data from most multiple-reading ROC experiments. The method is somewhat awkward, but it should be used whenever possible.

Of alternative techniques, the conventional "t-test for paired data" is clearly the simplest. This method can be employed whenever several readers interpret the images from both of two modalities. The test involves: (i) fitting an ROC individually to each reader's confidence-rating data for each modality, and extracting a univariate index (such as A_z) from each fitted curve; (ii) calculating, for each reader, the difference between the two modalities' indices; and (iii) using Student's "t" statistic to test the null hypothesis that the population mean of the resulting list of differences is equal to zero. This statistical test accounts fully for the effects of "between-reader" and "within-reader" variation, but its statistical power is governed strongly by the number of readers in the experiment (which determines the number of degrees of freedom of the t statistic), and it does not account for "case-sample" variation at all. The latter limitation may be acceptable with large random case samples and/or with small but carefully selected case samples that can be assumed to be representative, but with small random samples or small loosely selected samples it should be accepted only with great caution.

9. GENERALIZATIONS OF ROC ANALYSIS

As noted earlier, conventional ROC analysis is limited to situations in which possible truth can be divided into two states and two corresponding decisions are available to the observer. Several generalizations of conventional ROC analysis have been proposed to overcome this limitation.

"Location ROC" (LROC) analysis (Starr et al, 1975) measures the ability of an observer not only to decide between the presence or absence of a single abnormality in an image, but also to localize that abnormality when it is present. Swets and Pickett (1982) have suggested that LROC analysis can be used also to measure the ability of observers to detect and classify single abnormalities; the measured curve is then called a "joint" ROC. A generalization of the LROC approach called "Free-response operating characteristic" (FROC) analysis (IAEA, 1977; Bunch et al, 1977) applies to situations in which a particular kind of abnormality may be present at more than one location in an image and the observer is required to detect the presence of the abnormality at each of its locations. For experiments that seek to measure the detectability of lesions, FROC analysis can be more efficient than ROC or LROC analysis because it allows more than one lesion to be be included in each image, thereby reducing the number of images needed to estimate reliably the probability (at each decision criterion) of detecting a lesion when it is actually present.

Unfortunately, LROC analysis and FROC analysis share two substantial practical limitations: no formal curve-fitting procedures and no generally applicable statistical tests for differences in performance have been developed. Thus, at present, neither LROC nor FROC analysis can be recommended broadly. Hope does exist for overcoming these limitations, however (Metz, 1988). In particular, a method for fitting FROC curves on the basis of an ad hoc but apparently reasonable model was proposed recently by Chakraborty (1988); though not yet fully tested, this approach may prove useful.

ACKNOWLEDGEMENT

This work was supported by the U.S. Department of Energy under Grant DE-FG02-86ER60418.

REFERENCES

Begg, C.B., and McNeil, B.J. (1988). Assessment of radiologic tests: control of bias and other design considerations, Radiology 167, pp. 565-569.

Bunch, P.C., Hamilton, J.F., Sanderson, G.K., and Simmons, A.H. (1977). A free response approach to the measurement and characterization of radiographic observer performance. Proc. SPIE 127, pp. 124-135.

Chakraborty, D.P. (1988). Maximum likelihood analysis of free response receiver operating characteristic (FROC) data. Personal communication.

Dorfman, D.D., and Alf, E. (1969). Maximum likelihood estimation of parameters of signal detection theory and determination of confidence intervals -- rating method data, J. Math. Psychol. 6, pp. 487-496.

Egan, J.P. (1975). Signal detection theory and ROC analysis, Academic Press, New York.

Grey, D.R., and Morgan, B.J.T. (1972). Some aspects of ROC curve-fitting: normal and logistic models, J. Math. Psychol. 9, pp. 128-139.

Getty, D.J., Pickett, R.M., D'Orsi, C.J., and Swets, J.A. (1988). Enhanced interpretation of diagnostic images, Invest. Radiol. 23, pp. 240-252.

Goodenough, D.J., and Metz, C.E. (1977). Implications of a "noisy" observer to data processing techniques. In: Information processing in medical imaging, C. Raynaud and A.E. Todd-Pokropek (eds.), CEA, Orsay, pp. 400-419.

Gray, R., Begg, C.B., and Greenes, R.A. (1984). Construction of receiver operating characteristic curves when disease verification is subject to selection bias, Med. Decis. Making 4, pp. 151-164.

Green, D.M., and Swets, J.A. (1966). Signal detection theory and psychophysics, Wiley, New York. Reprint with corrections: (1974). Krieger, Huntington NY.

Hanley, J.A., and McNeil, B.J. (1982). The meaning and use of the area under a receiver operating characteristic (ROC) curve, Radiology 143, pp. 29-36.
Hanley, J.A., and McNeil, B.J. (1983). A method of comparing the areas under receiver operating characteristic curves derived from the same cases, Radiology 148, pp. 839-843.
International Atomic Energy Agency (1977). IAEA co-ordinated research programme on the intercomparison of computer-assisted scinigraphic techniques. In: Medical radionuclide imaging, IAEA, Vienna, 1, pp. 585-615.
Kendall, M., and Stewart, A. (1976). The advanced thoery of statistics, Hafner, New York, 3, pp. 182-243.
Lusted, L.B. (1978). General problems in medical decision making, with comments on ROC analysis, Seminars Nucl. Med. 8, pp. 299-306.
Loo, L.-N., Doi, K., Ishida, M., Metz, C.E., Chan, H.-P., Higashida, Y., and Kodera, Y. (1983). An empirical investigation of variability in contrast-detail diagram measurements, Proc. SPIE 419, pp. 68-76.
MacMillan, N.A., and Kaplan, H.L. (1985). Detection theory analysis of group data: estimating sensitivity from average hit and false alarm rates, Psychol. Bull. 98, pp. 185-199.
McNeil, B.J., Keeler, E., and Adelstein, S.J. (1975). Primer on certain elements of medical decision making, N. Engl. J. Med. 293, pp. 211-215.
McNeil, B.J., and Hanley, J.A. (1984). Statistical approaches to the analysis of receiver operating characteristic (ROC) curves, Med. Decis. Making 4, pp. 137-150.
Metz, C.E. (1978). Basic principles of ROC analysis, Seminars Nucl. Med. 8, pp. 283-298.
Metz, C.E. (1979). Applications of ROC analysis in diagnostic image evaluation. In: The physics of medical imaging: recording system measurements and techniques, A. Haus (ed.), Am. Inst. Phys., New York, pp. 546-572.
Metz, C.E. (1986a). ROC methodology in radiologic imaging, Invest. Radiol. 21, pp. 720-733.
Metz, C.E. (1986b). Statistical analysis of ROC data in evaluating diagnostic performance. In: Multiple regression analysis: applications in the health sciences, D. Herbert and R. Myers (eds.), Am. Inst. Phys., New York, pp. 365-384.
Metz, C.E. (1988). Some practical issues of experimental design and data analysis in radiological ROC syudies, Invest. Radiol. (in press).
Metz, C.E., and Kronman, H.B. (1980). Statistical significance tests for binormal ROC curves, J. Math. Psychol. 22, pp. 218-243.
Metz, C.E., Kronman, H.B., Wang, P.-L., and Shen, J.-H. (1985). ROCFIT: a modified maximum likelihood algorithm for estimating a binormal ROC curve from confidence-rating data, Dept. Radiology, Univ. Chicago, Chicago. Mini- and micro-computer versions of this and other software for ROC analysis are available from the author at no cost.
Metz, C.E., Starr, S.J., Lusted, L.B., and Rossmann, K. (1975). Progress in evaluation of human observer visual detection performance using the ROC curve approach. In: Information processing in medical imaging, C. Raynaud and A.E. Todd-Pokropek (eds.), CEA, Orsay, pp. 420-436.

Metz, C.E., Wang, P.-L., and Kronman, H.B. (1984). A new approach for testing the significance of differences between ROC curves measured from correlated data. In: Information processing in medical imaging, F. Deconinck (ed.), Nijhoff, The Hague, pp. 432-445.

Ransohoff, D.F., and Feinstein, A.R., (1978). Problems of spectrum and bias in evaluating the efficacy of diagnostic tests, N. Engl. J. Med. 299, pp. 926-930.

Starr, S.J., Metz, C.E., Lusted, L.B., and Goodenough, D.J. (1975). Visual detection and localization of radiographic images, Radiology 116, pp. 533-538.

Swets, J.A. (1979). ROC analysis applied to the evaluation of medical imaging techniques, Invest. Radiol. 14, pp. 109-121.

Swets, J.A. (1986a). Indices of discrimination or diagnostic accuracy: their ROCs and implied models, Psychol. Bull. 99, pp. 100-117.

Swets, J.A. (1986b). Form of empirical ROCs in discrimination and diagnostic tasks: implications for theory and measurementof performance, Psychol. Bull. 99, pp. 181-198.

Swets, J.A., and Pickett, R.M. (1982). Evaluation of diagnostic systems: methods from signal detection theory, Academic Press, New York.

Weinstein, M.C., and Feinberg, H.V. (1980). Clinical decision analysis, Saunders, Philadelphia.

Part 2. Theoretical Aspects

Section 2.1 3-D

A 3D MODEL OF THE GLOBAL DEFORMATION OF A NON-RIGID BODY

D. Friboulet, I.E. Magnin

URA CNRS 331, Laboratoire de Traitement du Signal et Ultrasons
France

ABSTRACT

We propose a three-dimensional (3D) deformation model for the quantification of the dynamical global deformation component of the human left ventricle (LV). Under the assumption that the LV behaves within each studied period of the cardiac cycle as an homogeneous non rigid body, its overall motion can be decomposed into :

- a pure rigid displacement component described by a translation and a rotation
- a pure deformation component associated with a dilatation operator.

When no temporal point to point correspondence is available, we show that the 3 motion components can be determined from the computed evolution of the LV center of gravity, and of the principal axes and moments of inertia.
The model is applied to a normal LV, from clinical data provided by Magnetic Resonance Imaging (MRI) acquisitions. Its adequacy is then quantified by comparing the predicted deformed LV shape to the actual one.

1. INTRODUCTION

The human heart left ventricle (LV) experiences a complex three-dimensional (3D) motion through the cardiac cycle. From this global motion, the deformation component contribution is the main factor to be quantified toward a reliable characterization of the heart function.

Experimental 3D studies using biplane cineradiography (Kim et al, 1985 - Meier et al, 1980 - Potel et al, 1984 - Walley et al, 1982) are based on the tracking of surgically radiopaque markers or of contrast opacified bifurcation points on the coronary artery tree, and clearly demonstrate the complexity of the LV behavior. Using the theoretical framework of linear elasticity and assuming a local linear homogeneous deformation, Meier et al (1980) and Kim et al (1985) described the in vivo dog heart local deformation by a local time-varying stretch and rotation tensor. The temporal evolution of these operators shows that the heart deformation is a spatially inhomogeneous and sequential phenomenon . Although very interesting, these local results do not lead to the definition of a global 3D deformation model of the beating heart. In another way, Potel et al (1984) derived a more global concept by computing a so-called center of contraction at each time interval along the cardial cycle of human subjects. They gave evidence that 91 % of the LV wall motion is directed toward that point. Unfortunately the position of this center of contraction cannot be analytically related to the LV geometry.Walley et al (1982) spread Meier's linear homogeneous assumption to the whole LV and demonstrated that such a model prediction correlates well with the actual wall motion.

Each of the various just mentionned approaches allows a quantification of a deformation component because they establish a temporal point to point correspondence (TPP) on the moving LV or heart walls.

Clinical studies based on ultrasound echocardiography, X-ray scanners or Magnetic Resonance Imaging (MRI), do not allow to build such a TPP correspondence through the cardiac cycle. Hence, the access to the 3D deformation components from these data is a theoretically impossible task except by introducing simplifying assumptions or a priori knowledge about the LV geometry or about the LV deformation itself. For example, the use of the finite element method (Vinson et al, 1982 and Chandran, 1984) is an attempt to exploit the myocardium mechanical properties. Because these properties are ill-defined (Vinson et al, 1982), and because the temporal range of the method is limited to the diastole, it does not seem to suit well the LV function assessment.Recently Azhari et al (1987) defined a 3D analytical descriptor to reduce the LV geometry and volume to a monodimensional signal. They successfully tested the hypothesis of isolateral and isolongitudinal contraction.

The present study introduces a purely geometrical approach leading to the estimation of the 3D global LV deformation. For that purpose, the LV cavity is considered as a 3D non-rigid body experiencing a 3D linear homogeneous transformation. The parameters of this model are computed from the LV cavity center of gravity, moments and axes of inertia. Hence, this model does not require any TPP correspondence through the cardiac cycle and can then be applied to 3D clinical data. The validity of the proposed model is quantified by comparing the LV predicted geometry to the actual LV geometry obtained from MRI data.

2. BASIC ASSUMPTIONS

As stated in section 1, the main problem toward the LV motion quantification is that clinical data (i.e. cross section or projection of the LV) do not allow to establish any direct TPP 3D correspondence of the LV wall.

2.1 Assumption A1

Then, the proposed method first introduces the only assumption, A1, emerging from experimental studies, that is general enough to be applied to clinical data. A1 states that :

> The motion of the whole LV considered as a non rigid body can be described at each small time interval by a translation and a linear homogeneous transformation.

The linear homogeneous transformation is mathematically expressed by its associated 3 x 3 matrix T. The polar decomposition theorem (Meier et al, 1980 - Walley et al, 1982) states that this linear operator can uniquely be decomposed into a rotation operator, R, and a dilatation (or stretch) operator, D, as follows

$$T = RD. \qquad (1)$$

The dilatation characterizes the non-rigid component, i.e the pure deformation component of the overall motion. Expressed in any coordinate system, D is a positive symmetric operator. The 3 mutually perpendicular eigenvectors of D give the so-called 3 principal directions (or axes) of dilatation, in which a segment experiences a pure length change (i.e : these directions remain unchanged by D as shown in Fig.1). The 3 associated eigenvalues provide the fractional length change

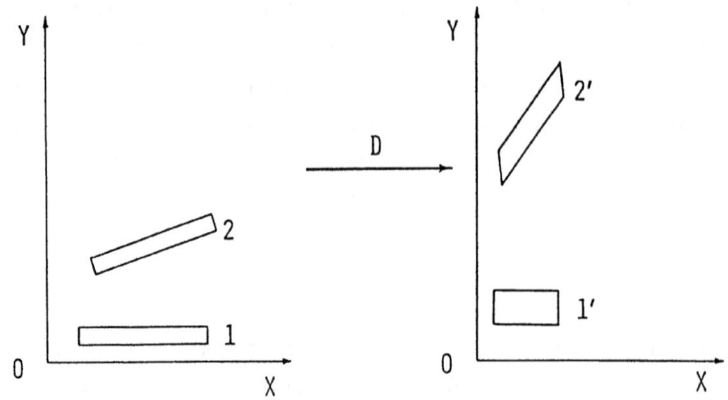

Fig.1. Schematic representation of a bidimensional dilatation effects.In this example the principal dilatation axes are Ox and Oy, with respective dilatation coefficients 0.5 and 2. The left side of the figure shows two rectangles, 1 and 2, the right side shows the two dilated rectangles , 1' and 2'. Because they are aligned with the dilatation axes, the sides of the rectangle 1 undergo a length change and not a directional change. On the contrary the sides of the rectangle 2 experience both length and direction changes.

along the principal dilatation axes, and we will call them "dilatation coefficients".
The two remaining parameters, - translation and rotation - are related to the pure rigid displacement component of the motion.

2.2 Assumption A2

The computation of the transformation T requires a second assumption, A2, (Friboulet et al, 1988) whose theoretical implications are detailed in section 3. Assumption A2 states that :

> The 3 principal directions of dilatation coincide with the 3 LV principal axes of inertia.

The 3 mutually perpendicular principal axes of inertia provide a practical tool for the quantification of the 3D LV geometry. The 3 associated principal moments of inertia give the weights of the 3 principal axes of inertia. These parameters correspond to the 3 eigenvectors and eigenvalues of the so-called matrix of inertia. For instance the 3 principal moments of inertia of a sphere are equal, because of the symmetry, and any 3 mutually perpendicular axes can be choosen as the principal axes of inertia. In the case of an ellipsoïd given by the equation

$$\frac{x^2}{a^2} + \frac{y^2}{b^2} + \frac{z^2}{c^2} = 1 \qquad a > b > c$$

the 3 axes of inertia are Ox, Oy, Oz and are clearly separated by their associated moments of inertia.
In the same way, the values taken by the 3 axes and moments of inertia of the LV versus time give a dynamical measure of its varying shape (see Fig.3 and 4).

3. MOTION PARAMETERS CHARACTERIZATION

In the following section, we consider a non-rigid homogeneous object experiencing a non-rigid motion which verifies the assumptions A1, A2. The derived expressions are related to a small time interval, Δt. Within Δt, the motion parameters (dilatation, rotation and translation) are supposed to be constant. Hence, the time variable will not appear in these expressions.

3.1 Notations

3.1.1 Motion parameters

$R_a = (O, i, j, k)$: arbitrary coordinate system
$T = (\tau_{ij})$; $i = 1, 3$; $j = 1, 3$: matrix of the linear transformation
R : Rotation matrix
D : Dilatation matrix

Without loss of generality, the origin of the current coordinate system is supposed to belong to the rotation axis. In the same way, we arbitrarily take the origin of the current coordinate system as the dilatation center (i.e the intersection of the 3 dilatation axes). Indeed, changing the position of the origin is equivalent to arbitrarily choose another translation vector.

VD_i ; $i = 1,3$: Vector corresponding to the ith principal axis of dilatation

δ_i ; $i = 1,3$: Scalar corresponding to the ith coefficient of dilatation. Since we deal with a physical deformation, we have $\delta_i > o$; $i = 1,3$. It is to be noted that $o < \delta_i < 1$ will denote shortening and $\delta_i > 1$ will be related to lengthening.

B : translation vector

$OX = X = (x_i)$; $i = 1,3$: Vector to a point of the space

$OX' = X' = (x'_i)$; $i = 1,3$: Vector to the transformed X point

F : global transformation

J : Jacobian of F

For any point we have, according to the assumption A1, (expressed in R_a) :

$$(R_a) \quad X' = F(X) = T\,X + B \tag{2}$$

From the polar decomposition theorem (1), it follows that

$$(R_a) \quad X' = RD\,X + B \tag{3}$$

3.1.2 Object parameters

S : object or domain limited by the surface of the object.

V : volume of S

G : center of gravity of S. We have

$$(R_a) \quad G = \frac{1}{V} \int_S X \, dv \tag{4}$$

$R_G = (G,i,j,k)$: coordinate system with origin at G

G' : Image of G by F, ie $G' = F(G)$

$MI = (m_{ij})$; $i = 1,3$; $j = 1,3$: Matrix of inertia of S computed in R_G. We have

$$(R_G) \quad m_{ij} = m_{ji} = \int_S x_i \; x_j \; dv \qquad (5)$$

VMI_i ; $i = 1,3$: Normed vector corresponding to the ith eigenvector of MI. (VMI_1, VMI_2, VMI_3) is choosen to be a right-hand system and gives the directions of the 3 principal axes of inertia of S.

μ_i, $i = 1,3$: eigenvalue associated with VMI_i. μ_i is the principal moment of inertia related to the plane perpendicular to VMI_i.

In the same way we define the transformed object S' and the associated computed parameters V*, G*, MI*, m_{ij}*, VMI_i* and μ_i*.

3.2 Jacobian of F

It is well known that the Jacobian of a transformation is related to the induced volumic variation, ie

$$J = \frac{V^*}{V} = \det(T) \qquad (6)$$

From the definition of T it follows that

$$J = \delta_1 \; \delta_2 \; \delta_3 \qquad (7)$$

3.3 Characterization and removal of the translation

The translation component can be eliminated by the knowledge of the centers of gravity of S and S' (Walley et al, 1982). Because the transformation is linear, G^* is also the image of G by F, and we can write

$$(R_a) \quad G^* = F(G) = G'$$

Applying the definition (2) in R_G gives

$$(RG) \quad GX' = T\;GX + B \qquad (8)$$

which obviously leads to

$$(R_G) \quad GG' = B \tag{9}$$

The relation (9) characterizes the translation. Moreover, the introduction of (9) in (8) gives

$$(R_G) \quad G'X' = T\ GX \tag{10}$$

The relation (10) allows to define the translation indexed coordinate system, R_B, in which the transformation F is now expressed as :

$$(R_B) \quad F(X) = X' = T\ X \tag{11}$$

3.4 Characterization of T.

The aim of this section is to show that the remaining motion parameters, (τ_{ij}), cannot be directly computed from the m_{ij} and $m_{kl}{}^*$ without the introduction of the assumption A2.
It can be easily shown that the coefficients of MI* can be expressed as a function of the coefficients of MI according to the following equations :

$$(R_B) \quad m_{kl}{}^* = J \sum_{i=1}^{3} \sum_{j=1}^{3} \tau_{ki}\ \tau_{lj}\ m_{ij} \tag{12}$$

$$k = 1,3 \quad ; \quad l = 1,3$$

If we note that (12) verifies the symmetry condition :

$$m_{kl}{}^* = J \sum_{i=1}^{3} \sum_{j=1}^{3} \tau_{ki}\ \tau_{lj}\ m_{ij} = J \sum_{j=1}^{3} \sum_{i=1}^{3} \tau_{lj}\ \tau_{ki}\ m_{ji} = m_{lk}{}^*$$

and that J can be computed from the relation (6), it follows that (12) is a non-linear system with 9 unknowns (τ_{ij} ; i = 1,3 ; j = 1,3) and 6 equations.

This system is under-determined. Adding constraints or a priori knowledge to (12) is necessary to restrain the solution field and find an acceptable physical solution. Once the coefficients τ_{ij} have been estimated, T can then be uniquely decomposed into R and D, according to the expressions derived in (Walley et al, 1982) :

$$D^2 = T^t\ T \qquad \text{(t is the transposition index)}$$
$$R = T\ D^{-1}$$

In this paper, a direct solution of (12) has been considered. The assumption A2 makes this direct resolution possible and, according to (Walley et al, 1982 and Azhari et al, 1987), meets the physical requirements of the phenomenon.Anyway, the solution derived from A1 could be regarded as the initial solution of a more adaptative resolution method. This approach will be considered in a future work.

3.5 Characterization of R :

A2 states that the principal directions of dilatation coincide with the principal axes of inertia of the moving object, ie formaly :

$$VMI_i = VD_i \quad ; \quad i=1,3 \qquad (13)$$

This relation implies that, by definition of D, the dilatation D does not affect the directions of the axes of inertia, which, in particular, remain mutually perpendicular. Hence, any change in the direction of the axes of inertia is only due to the rotation R, and we can write

$$VMI_i^* = R\ VMI_i \quad ; \ i = 1,3 \qquad (14)$$

Moreover, the expression (14) implies that we are able to establish a unique ordered correspondence between the VMI_i and the VMI_i^*. Section 3.7 details the computational problems related to this correspondence. In order to keep the discussion clear, such a correspondence is supposed to be effective and unambiguous.

If we define the 3 x 3 orthogonal matrices :

$$VM = (VMI_i \; ; \; i = 1,3) \text{ and } VM^* = (VMI_i^* \; ; \; i = 1,3)$$

we can rewrite (14) as :

$$R = VM^* \; (VM)^{-1} = VM^* \; (VM)^t \qquad (15)$$

The relation (15) uniquely characterizes the rotation R.

As we did in section 3.3 for the translation component, we can now remove the rotation from the global motion by indexing the directions of the coordinate system to the directions of the axes of inertia of the moving object.

Because of the orthogonality of the matrix R, the relation (11) can be expressed as

$$(R_B) \qquad R^t \; X' = D \; X$$

If X is expressed in the coordinate system (G, VMI_1, VMI_2, VMI_3) , R^t X' represents the coordinates of X' in the coordinate system (G^*, VMI_1^*, VMI_2^*, VMI_3^*). Hence, we can define the translation rotation indexed system, R_{BR}, in which the transformation F is now expressed as

$$(R_{BR}) \qquad F(X) = X' = D \; X. \qquad (16)$$

3.6 Characterization of D

In the new indexed coordinate system R_{BR}, and under the assumption A2, D, MI and MI* take a diagonal form and (12) can then be reduced to

$$(R_{BR}) \quad \begin{cases} m_{ij}^* = m_{ji}^* = 0 \quad \text{if } i \neq j \\ m_{ii}^* = \mu_i^* = J \; \delta_i^2 \; m_{ii} = J \; \delta_i^2 \; \mu_i \end{cases}$$

which leads to the three equations :

$$(R_{BR}) \qquad \mu_i^* = J \; \delta_i^2 \; \mu_i \; ; \; i = 1,3 \qquad (17)$$

Introducing the expression of the Jacobian (7) in (17) finally gives

$$(R_{BR}) \qquad J = \left[\prod_{i=1}^{3} \frac{\mu_i^*}{\mu_i} \right]^{1/5} \qquad \delta_i = \left[\frac{1}{J} \frac{\mu_i^*}{\mu_i} \right]^{1/2} \qquad (18)$$

Hence, the dilatation D is determined by :

-its principal direction of dilatation coinciding with the principal axes of inertia according to A2
-its dilatation coefficients provided by the equations (18)

3.7 Computational aspects :

Obviously, the characterization of R and D is based upon the correspondence we establish through the relations (15) and (18) between the two sets :

(VMI_i, μ_i) ; i = 1,3 for S and (VMI_i*, μ_i*) ; i = 1,3 for S'.

Because the VMI_i (VMI_i*) and the μ_i (μ_i*) are associated with the eigenvectors and eigenvalues of Mi (Mi*), this correspondence is not a priori unique and yields computational ambiguities.

Indeed, let us call EV_a, EV_b, EV_c the 3 unordered mutually perpendicular eigenvectors of MI, characterized by the 3 eigenvalues μ_a, μ_b, μ_c. The 3 vectors EV_a, EV_b and EV_c provide only an orientation (ie not a direction) information. Hence, building the righthand system RS : (VMI_1, VMI_2, VMI_3) is equivalent to choose one of the 24 righthand systems that we can create from the set (EV_a, $-EV_a$, EV_b, $-EV_b$, EV_c, $-EV_c$).We can arbitrarily take one of these 24 possibilities, for S and select a unique system RS.

The same problem arise for S' which can be associated with 24 distinct sets RS* : (VMI_1*, VMI_2*, VMI_3*).Consequently, we could form theoretically 24 correspondences (namely between the unique set RS and the 24 potential sets RS*) between S and S', leading to 24 potential solutions for R and D.

In the absence of any other a priori information about R and D, (ie because the axes and moments of inertia only provide a uncomplete description of an object), these 24 solutions would have to be tested to select the right one (or ones).

In the field of an experimental study, the introduction of physical constraints about the range of the rotation angles and the dilatation coefficients allows to select a unique solution. In the case of the LV motion, because the orientations related to VMI_1 and VMI_2 were found unreliable, (see the experimental results in section 4.4) the unique acceptable physical solution is obtained by :

- imposing a nul rotation around the axis associated with VMI_3
- among the two remaining choices left for VMI_3, selecting the solution leading to the minimum rotation angle.

4. EXPERIMENTAL

4.1 Data acquisition

The cardiac images were obtained on a 0,5 T Thomson-CGR Magniscan 500 MRI scanner, from a cycled multiple-slice spin-echo sequence. This sequence was synchronized with the electro-cardiogram signal. The selected imaging planes lied perpendicularly to the LV long axis, to minimize the partial volume effects. The in-plane resolution of the so-obtained 256 x 256 images was 1,64 mm.
The mean normal subject's heart rate was 70 beats/min. The acquisition sequence provided ten 10 mm thick adjacent sections at ten times of the cardiac cycle, with a temporal resolution of 72 ms. The time delay (TE) of the spin-echo was 26 ms.The LV contours were drawn by a cardiologist on the displayed images, by means of a trackball.

4.2 Experimental motion parameters estimation

At each scanning time, t_n, the center of gravity, G, and the matrix of inertia, MI, of the LV are estimated from the contours by using Simpson's rule. The 3 eigenvectors and eigenvalues of MI are calculated with a standard computation procedure. These parameters are then sorted to obtain small dilatation coefficients and small rotation angles, leading to the 3 ordered vectors VMI_1, VMI_2, VMI_3 and to the 3 associated moments of inertia, μ_1, μ_2, μ_3. The translation, the rotation and the dilatation of the LV between each time interval (t_n, t_{n+1}) are then determined from the equations (9), (15) and (18) derived in section 3.

4.3 Model validation method

The quantification of the capability of the proposed model to describe the LV behavior uses a surface comparison algorithm.From the LV contours, and at each scanning time t_n, the surface of the LV is generated through a triangulation procedure (Ekoulé et al, 1987).

First, without using the model, we measure the similarity of the surfaces associated with the LV at time t_n and t_{n+1} by performing the following steps (Fig.2) :

each triangular surface element (SEL) of the LV surface at time t_n is considered as the elementary distance measure area.

- the straight line passing through the center of gravity of each measure SEL is computed. This line is the measure line (Fig.2)
- The measure line intersects the surface of the LV at time t_{n+1} at a given point. The distance computed between that point and the center of gravity of the measure SEL is considered as the elementary distance measure. Arbitrarily, a negative sign will denote a distance measured toward the inner part of the surface and a positive one a distance measured toward the outer part of the surface.

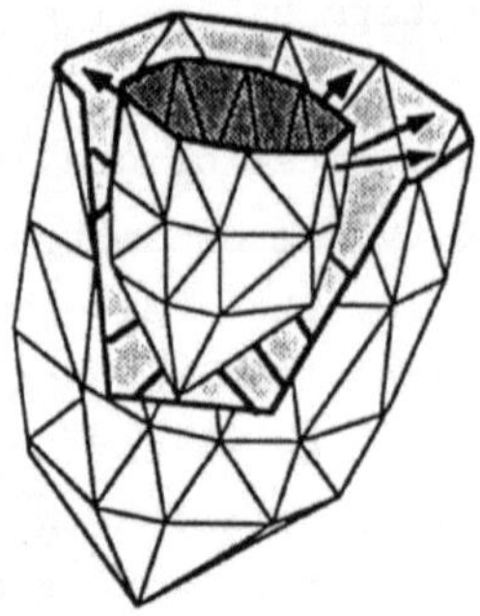

Fig.2. Schematic representation of the surface comparison method for a time interval situated in the dilatation phase of the cardiac cycle. The surface of the LV at time t_n is surrounded by the surface of the LV at time t_{n+1}. The arrows show the distance measure associated with some surface elements (SEL).

Let us call :

Nf : number of SEL of the LV surface at time t_n.

dr_i : elementary distance associated with the ith SEL for the comparison between the surfaces of the LV at time t_n and t_{n+1}.

The global similarity measure, E_r, between the surfaces at time t_n and t_{n+1} corresponds to the quadratic mean of the elementary distances :

$$E_r = \frac{1}{Nf} \sum_{i=1}^{Nf} dr_i^2$$

The model is then applied.From the experimental motion parameters and from the surface of the LV at time t_n, we can compute the predicted deformed and displaced LV surface at time t_{n+1}. This surface is compared to the actual LV surface at time t_{n+1} according to the above mentionned method. In the same way we define :

$$Ep = \frac{1}{Nf} \sum_{i=1}^{Nf} dp_i^2$$

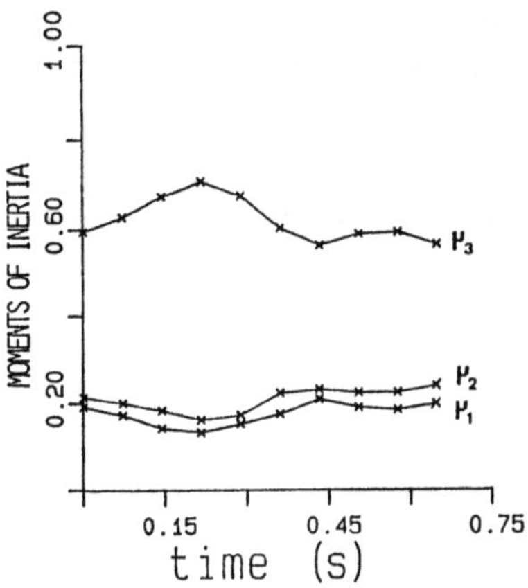

Fig.3. Evolution of the 3 principal moments of inertia μ_1, μ_2 and μ_3 in relative values through the cardiac cycle

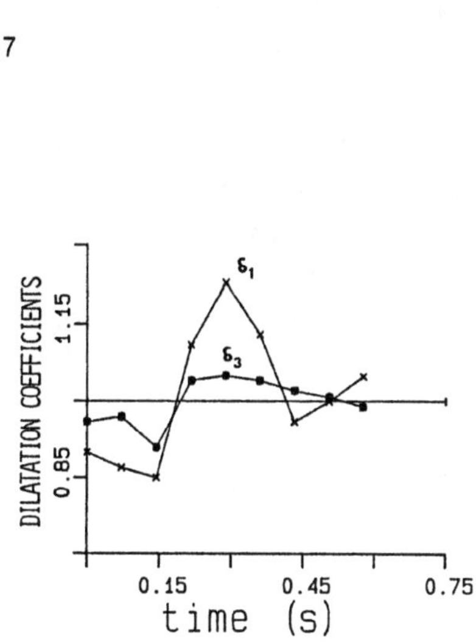

Fig.5 Evolution of the computed transverse (δ_1 : o) and longitudinal (δ_3 : x) dilatation coefficients.

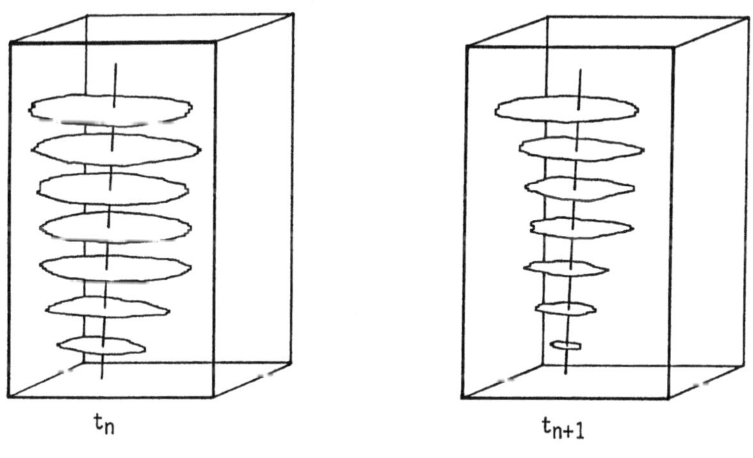

Fig.4. Projection of the LV contours together with the longitudinal axis of inertia VMI3 at two successive times in the contraction phase (systole) of the cardiac cycle.

where dp_i is the elementary distance associated with the ith SEL for the comparison between the predicted and actual deformed LV surfaces.

E_r is used as a reference measure (performed without the model) and we define the model performance index, MPI, between time t_n and t_{n+1}, as

$$MPI = \left[1 - \frac{E_p}{E_r}\right]^{1/2}$$

The MPI gives a measure of the relative similarity variation provided by the model between t_n and t_{n+1}. For instance, if the fit of the model to the LV motion is perfect, $E_p = 0$ and MPI =1.

4.4 Results

- Principal moments and axes of inertia

Fig. 3 shows the evolution (in relative values) of the 3 principal moments of inertia, μ_1, μ_2 and μ_3 along the cardiac cycle. At every time interval, $\mu 3$ takes large, well individualized values. For that reason the axis of inertia related to VMI_3 will be refered as the longitudinal axis and the plane containing VMI_1 and VMI_2 as the transverse plane (Fig.4).
The observed similarity between μ_1 and μ_2 implies that the LV has no well-defined privileged direction in the transverse plane. As a consequence, VMI_1 and VMI_2 were experimentally found not reliable and very sensitive to the contour tracing step.To overcome this difficulty and respect the small rotation angles constraint, we use the following model adaptation :

- μ_1 and μ_2 are both replaced by their joint mean value $(\mu_1+\mu_2)/2$.
- the rotation is computed only from the evolution of VMI_3.

The ultimate consequences of this adaptation is that the dilatation is homothetic in the transverse plane (i.e $\delta_1 = \delta_2$) and that VMI_3 is the third dilatation axis.Hence, δ_1 will be

refered as the transverse dilatation coefficient and δ_3 as the longitudinal dilatation coefficient axis.
Fig. 4 shows the orientation of the computed longitudinal axis, VMI_3, related to the LV geometry for 2 successive instants in the systolic part of the cardiac cycle.

- Dilatation coefficients.

Fig. 5 is a plot of the transversal and longitudinal dilatation coefficients, δ_1 and δ_3. The evolution of δ_1 allows to clearly distinguish the 3 phases of the cardiac cycle :

- the systole (or contraction phase), when $\delta_1 < 1$.
- the begining of the diastole,when $\delta_1 > 1$.
- the end of the diastole associated with slow motion, when $\delta 1$ is close to 1.

Fig. 5 implies that the transverse component of the dilatation plays a major part in the LV function during the systole and the begining of the diastole.

- Model performance index.

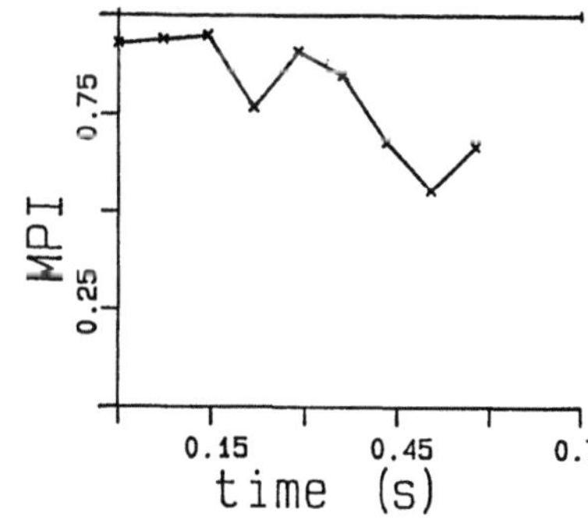

Fig.6a. Evolution of the Model Performance Index (MPI) versus time.

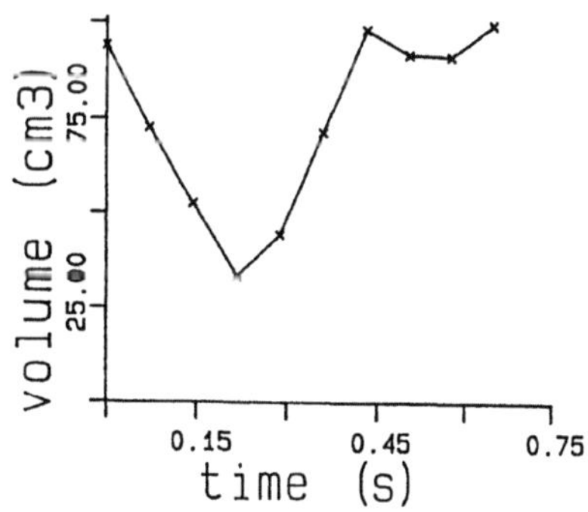

Fig.6b. Evolution of the LV volume versus time.

Fig. 6a and 6b respectively show the evolution of the MPI and of the LV volume through the cardiac cycle.The 4th point excepted, the MPI exhibits high and stable values (close to 90%) at the beginning of the curve and decreases then slowly to reach low levels (close to 60%). This means that the proposed linear model accurately describes the contraction phase (systole), associated with the three first segments of the volume curve (Fig. 6b). Its adequacy decreases progressively through the dilatation phase (beginning of diastole) and seems to be ill-adapted to the low-varying phase of the end diastole.

The observed discontinuity of the 4th point of the MPI plot could be explained by the coarse temporal sampling rate related to this particular instant of the cycle. Indeed, the sampling period being seemingly large (72 ms), we probably compare a still contracting LV (point 4, Fig. 6b) with an expanding LV (point 5, Fig. 6b).

We must recall that this interpretation is based on a single experimental study. It would be premature to raise general conclusions about the model behavior without further experiments.

5.CONCLUSION

We have described a method allowing to estimate the parameters of a particular case of non-rigid motion, precisely defined by the assumptions A1 and A2. The computation of these parameters (dilatation, rotation and tranlation) is derived from the evolution of the center of gravity, moments and axes of inertia of the moving object and thus does not involve a temporal point-to-point correspondence.
This model is aimed to study the behavior of the beating LV whose motion seems to meet the requirements of the assumptions A1 and A2.
The preliminary results (moments of inertia, dilatation coefficients) correlate well with previous experimental studies.
The adequacy of the model to the LV motion description is performed by comparing the model predicted deformed LV to the actual deformed LV. The derived Model Performance Index (MPI) allows to quantify the adequacy and, according to the A1 assumption, the degree of homogeneity of the LV behavior.

The potential interest of the method is demonstrated through the study of an actual normal LV. The values taken by the MPI seem to confirm the adequacy of this deformation to the physical studied phenomenon. These encouraging results lead us to apply the model to a set of normal and pathological data to get a quantitative information about the corresponding 3D LV deformation.

ACKNOWLEDGMENTS

The authors wish to thank Dr D. Revel and B. Dubuc for their help during MRI data acquisitions. We are grateful to Pr M. Amiel for advising and encouraging us.
This work was supported by a Région Rhône-Alpes Research and Technology grant.

REFERENCES

AZHARI H., SIDEMAN S., BEYAR R., GRENADIER E., DINNAR.U., 1987, An analytical descriptor of three-dimensional geometry : application to the analysis of the left ventricle shape and contraction. I.E.E.E. trans. on biomedical engineering, vol.BME-34, n° 5, pp. 345-355

CHANDRAN K.B., OLSHANSKY B., ATTARWALA Y.M.P., SKORTON D.J. , 1984, Finite element analysis of three-dimensional echocardiography data for the evaluation of diastolic left ventricular function, Automedica, vol. 5, pp 151-169

EKOULE A., PEYRIN F., ODET C., 1987, Description d'une procédure de triangulation entièrement automatique, Proceedings Cognitiva 87, pp 88-95

FRIBOULET D., EKOULE A., MAGNIN I., GOUTTE R., 1988, 3-D quantitative evaluation of the ventricular wall motion from MRI data, Proc. EUSIPCO-88, vol. 3, pp 1255-1258

KIM H.C., MIN B.G., LEE M.M.L., SEO J.D., LEE Y.W., HAN M.C., 1985, Estimation of local cardiac wall deformation and regional wall stresses from biplane coronary cineangiograms. I.E.E.E. trans. on biomedical engineering, vol. BME-32, n° 7, pp 503-512

MEIER G.D., BOVE A.A., SANTAMORE W.P., LYNCH P.R., 1980 b, Contractile function in canine right ventricle. Am. J. Physiol., vol. 239, pp. H794-H804

POTEL M.J., MAC KAY S.A., RUBIN J.M., AISEN A.M., SAYRE R.E., 1984, Three-dimensional left ventricular wall motion in man coordinate systems for representing wall movement direction, Invest. radiol., vol. 19, pp 499-509

VINSON C.A., GIBSON D.G., YETTRAM A.L., 1979, Analysis of left ventricular behaviour in diastole by means of finite element method., British heart journal, vol.41, pp 60-67

WALLEY K.R., GROVER M., RAFF G.L., BENGE J.W., HANNAFORD B., GLANTZ S.A. 1982, Left ventricular dynamic geometry in the intact and open chest dog., Circulation research, vol.50, pp 573-589

SIMULATION STUDIES FOR QUALITY ASSURANCE OF 3D-IMAGES FROM COMPUTED TOMOGRAMS

A. Pommert, M. Bomans, U. Tiede, K.H. Höhne

Institute of Mathematics and Computer Science in Medicine (IMDM)
University Hospital Eppendorf
Hamburg, F.R. of Germany

ABSTRACT

In recent years, techniques for direct display of three-dimensional (3D) gray level data from stacks of two-dimensional tomographic images have been developed. For clinical applications, it is important to ask whether the resulting images really show the "truth". This paper presents a simulation-based method which allows to measure and visualize the effects of different parameters and processing algorithms on the quality of 3D-imaging. First results demonstrate its usefulness for quality assurance.

1. INTRODUCTION

Radiological examinations are increasingly based on spatial sequences of cross-sectional images, as obtained in x-ray computed tomography (CT). In current practice, the three-dimensional (3D) structures represented in these images are usually reconstructed mentally, which is a difficult task even for an experienced observer.

In order to provide a more natural view of the objects under investigation, techniques for direct display of 3D-gray level data from sequences of tomographic images have been developed (Höhne et al. 1987). Unfortunately, the quality of the resulting 3D-images depends on many variables like slice aquisition parameters and processing algorithms used. As can be seen in fig. 1, a bad choice of parameters might even change the diagnostic meaning of an image. It is therefore important to assure that the 3D-images really show the "truth".

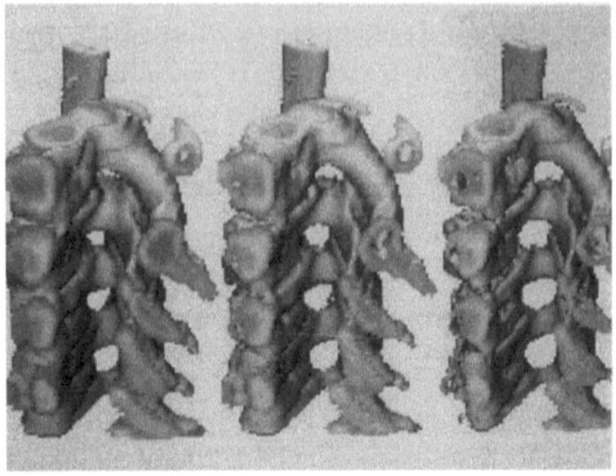

Fig. 1. Three 3D-images of a spine, segmented from the same CT data with different threshold values. Due to an unsuitably low threshold, the left representation shows a pathologic stenosis of the foramina.

As a means of investigating technical image quality, this paper presents a method which allows to measure and visualize the effects of different parameters and algorithms on the fidelity of 3D-imaging. Idea and first results will be discussed in the following sections.

2. METHOD

Basic idea of this method is to apply the usual 3D-viewing techniques to simulated tomographic sequences of artificial test objects. Unlike for real clinical data, the surface properties of these objects can be exactly calculated and thus be compared to the resulting 3D-images. The whole procedure consists of the following steps (investigated algorithms in brackets):

1. Simulated tomographic aquisition of a formally specified test object. Sampling is done in co-planar slices with adjustable distance and thickness (in the following exressed in units of pixel widths). We use the simplified assumption of uniform gray levels for object and environment, significantly different as in computed tomography. At the edges, the partial volume effect occurs.
2. Interpolation of the slices to form an isotropic gray level volume (linear interpolation, cubic splines).
3. Rotation of the volume to a specified view direction (trilinear interpolation).

4. Surface segmentation (thresholding, Marr-Hildreth operator).
5. Shading of the visible surfaces (z-buffer gradient method, gray level gradient method).
6. Comparison of the surfaces with the object specifications using the accuracy measures described in the next section.

Within this project, a tool was implemented in Pascal for calculating the simulated CT slices and performing the comparisons. Test objects may be any combinations of simple forms like box, ball, cone, cylinder, pyramid and gap. For this paper, a combination of pyramid and cone with a gap to simulate a fracture has been taken.

Processing steps 2 through 5 are identical to those performed for real clinical data (Höhne et al. 1987). The Marr-Hildreth operator used here is the three-dimensional extension of the Laplacian of a Gaussian (Bomans et al. 1987). The shading methods provide the surface normals of an object which are needed for surface rendering with the Phong illumination model (Bui-Tuong 1975). The z-buffer gradient method calculates these values from the depth image of the object, which is usually stored in the so-called z-buffer (see e.g. Chen et al. 1985).

In contrast, the gray level gradient method is based on the partial volume effect (Höhne and Bernstein 1986). It makes use of the correspondence between surface inclination and gray level gradients in the neighborhood of a surface voxel. The algorithm used here calculates the three components of a gradient G for a surface voxel at (i,j,k) directly from the gray levels g as

$$G_x = g_{(i+1,j,k)} - g_{(i-1,j,k)}$$
$$G_y = g_{(i,j+1,k)} - g_{(i,j-1,k)}$$
$$G_z = g_{(i,j,k+1)} - g_{(i,j,k-1)}$$

(Tiede et al. 1988). Normalisation of G gives the surface normal. Instead of 6 neighbors, the gradient can also be calculated with 26 neighbors weighted according to the Zucker-Hummel operator.

3. QUALITY MEASURES

For medical diagnostics, both size and shape (e.g. for tumors or deformations) as well as surface textures (e.g. for morphological structures) are essential features of an object. It is therefore important to consider both accuracy of surface localisation and surface inclination as quality measures for 3D-imaging.

A number of quantitative measures for evaluation of edge detectors in two-dimensional images, which are also based on comparison with an ideal reference object, are discussed in the literature (Peli and Malah 1982). The measure used here for 3D-data is the sum of the absolute volumes at all spots where segmented and ideal object do not match. From this, the average mean distance between segmented and ideal surface can be estimated as

$$M = \frac{\sum_i \Delta V_i}{F}$$

where ΔV_i is the absolute volume of a region i where ideal and segmented object do not match, and F is the total surface of the ideal object.

As a global measure, M describes the total error for the surface localisation of an object. The local scattering of this measure can be visualized in a volume deviation image, where the local volume deviations at the visible surfaces of an object are projected on the image plane. Thus the critical places of segmentation are easily recognized.

With a gradient deviation image, the local quality of surface inclination is presented. The surface normals calculated at the visible surfaces with one of the methods mentioned above are compared to the ideal surface normals, and the angle in between these vectors is displayed for every pixel (in the range of 0 to 90 degrees).

4. FIRST RESULTS

Variation of shading algorithms (thresholding segmentation): as can be seen in fig. 2, the gray level gradient method gives a much sharper image with less artefacts then the z-buffer gradient method. Furthermore, it shows more details like a gap which is only 0.5 voxels wide and thus below the spatial resolution of segmentation. In this sense, we might as well speak about a "fine segmentation". The corresponding gradient deviation images show that the z-buffer gradient method gives only a rough estimate of the real surface inclination, whereas the gray level gradient method comes very close to the ideal values.

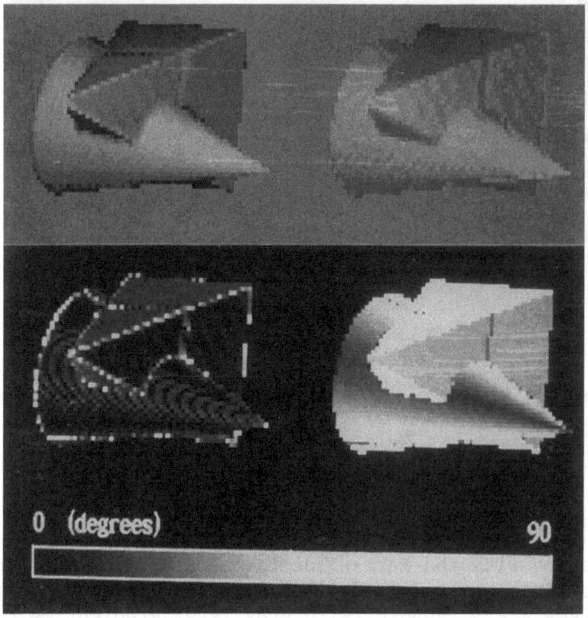

Fig. 2. 3D-images (top) and corresponding gradient deviation images (bottom) for different shading methods (all thresholding segmentation). The gray level gradient method (left) calculates the surface normals more accurately and thus results in better images than the z-buffer gradient method (right).

Variation of segmentation algorithms: As expected, thresholding minimizes the global measure M with a threshold right between gray levels of object and environment. Thin parts like peaks are wiped out with this method.

For small sigma, the Marr-Hildreth operator gives globally comparable results for M (test data are noise-free). As sigma is increased, strongly curved surfaces get "pumped up", as can be seen in fig. 3. If the gray level gradient method is used for shading, these local errors introduce strong artefacts because the surface is displaced from the gray level transitions.

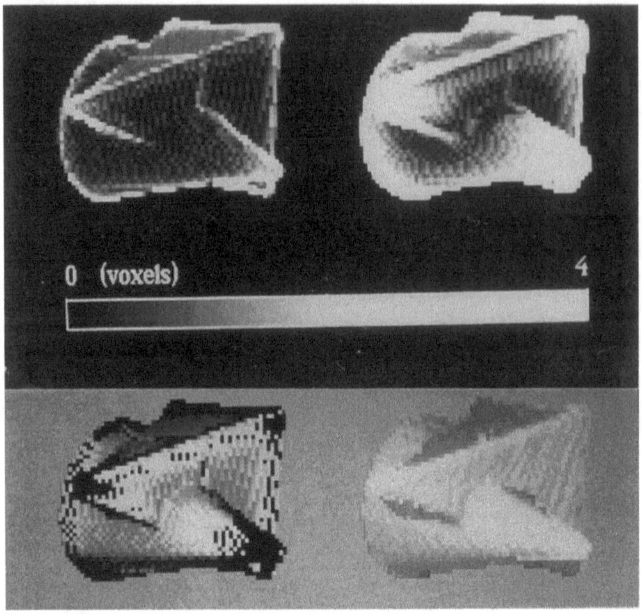

Fig. 3. Segmentation with the Marr-Hildreth operator. Top row: volume deviation images for sigma = 2 (left) and sigma = 4 (right). Bottom row: 3D-images for sigma = 2.5, shading with gray level gradient method (left) and z-buffer gradient method (right). Obviously, the gray level gradient method is not suitable here.

Variation of interpolation algorithms: for all slice distances investigated, (2-4), no advantadges of cubic splines over linear interpolation could be observed. Furthermore, splines introduce minimal "echoes" of strong intensity changes, which will be detected as surfaces by the Marr-Hildreth operator, thus creating artefacts. This again shows that algorithms for different processing steps have to be carefully coordinated.

Variation of slice aquisition parameters (thresholding, shading with the more sensible gray level gradient method): due to a loss of information, volume deviation goes up with increasing slice distance. However, for a given slice distance s, M can be minimized if slice thickness is also adjusted to s. If the thickness is increased further, M slowly goes up again.

As regards the gradient deviation, increase of slice thickness results in better display quality for regular surfaces. Quite surprisingly, gradient deviation is minimal for strongly overlapping slices (fig. 4).

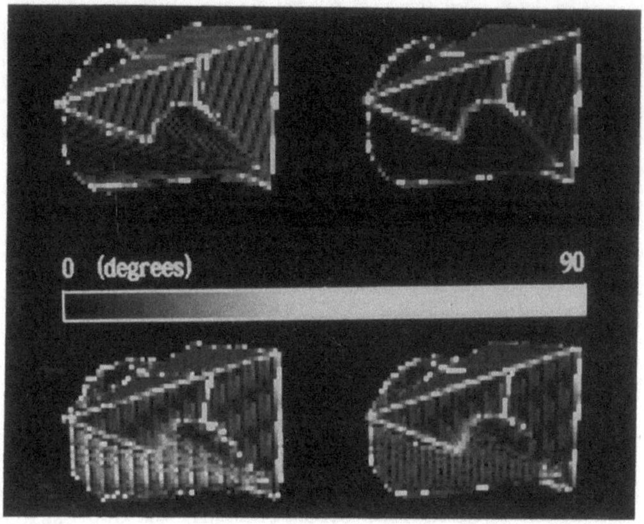

Fig. 4. Gradient deviation images for different slice aquisition parameters. Top: distance=1, bottom: distance=4, left: thickness=1, right: thickness=4. Gradient deviation is minimized for tight, overlapping slices.

5. CONCLUSIONS

A new method to measure and visualize the fidelity of 3D imaging has been developed and implemented. Based on simulation, it allows to investigate the effects of different aquisition parameters and processing algorithms within the complete processing context. As has been shown, this is a very important feature since proper coordination of different algorithms is essential for image quality.

An application of the results obtained so far to a real case is shown in fig. 5. For further investigations, especially in the field of segmentation, the model has to be refined in order to make it more realistic. Simulation of small structures (e.g. thin bones), variability of gray values within one tissue type and introduction of noise will be the next steps.

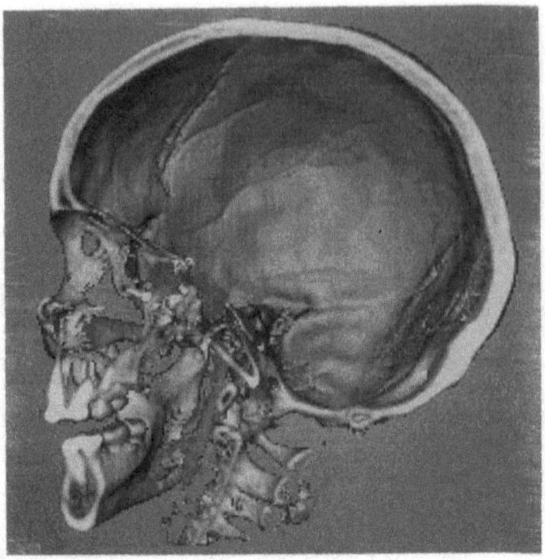

Fig. 5. 3D-image of a scull. Thresholding, gray level gradient shading, overlapping CT slices. Details like coronal suture and impressions of the meningeal artery are clearly visible.

ACKNOWLEDGEMENTS

The authors wish to thank the Werner Otto Stiftung, Hamburg, F.R. of Germany, for supporting this project.

REFERENCES

Bomans, M., Riemer, M., Tiede, U. and Höhne, K.H. (1987). 3-D Segmentation von Kernspin-Tomogrammen. In: Mustererkennung 1987, 9. DAGM-Symposium, Paulus, E. (ed.), Springer, Berlin, pp. 231-235.

Bui-Tuong, Phong (1975). Illumination for Computer Generated Images, Comm. ACM 18, pp. 311-317.

Chen, L.S., Herman, G.T., Reynolds, R.A. and Udupa, J.K. (1985). Surface Shading in the Cuberille Environment, IEEE Comput. Graphics Appl. 5, pp. 33-43.

Höhne, K.H. and Bernstein, R. (1986). Shading 3D-Images from CT using Gray-Level Gradients, IEEE Trans. Med. Imag. MI-5, pp. 45-47.

Höhne, K.H., Riemer, M. and Tiede, U. (1987). Viewing Operations for 3D-Tomographic Gray Level Data. In: Computer Assisted Radiology CAR '87, Lemke, H. (ed.), Springer, Berlin, pp. 599-609.

Peli, T. and Malah, D. (1982). A Study of Edge Detection Algorithms, Comput. Graphics Image Process. 20, pp. 1-21.

Tiede, U., Riemer, M., Bomans, M. and Höhne, K.H. (1988). Display Techniques for 3D-Tomographic Volume Data. In: Proc. NCGA '88, vol. 3, 188-197.

INTERACTIVE VOLUME RENDERING USING RAY-TRACING FOR 3-D MEDICAL IMAGING

Christian Barillot

Laboratoire "Signaux et Images en Médecine"
Université de Rennes-I, Rennes
France

ABSTRACT

The 3-D medical imaging over the last decade have got many evolutions to display volumetric medical data. These advances are summarized and compared introducing a new way for volumetrical display called "solid modelling". This paper focuses display methods towards image quality and interactivity. A new and efficient development of a ray-tracing algorithm featuring the volume rendering approach is explained and solutions to resolve interactivity problems using a special data structure are presented. This algorithm provides faster computation, convenient tools to interact with the original volume and several display modalities, allowing image quality and fast computation time to analyze the original data. A new volumetric visualization using voxel classification and transparency is presented.

1. INTRODUCTION

There are two familiar approaches to display 3-dimensional medical data. One may be called "surface modeling" (the most commonly used technique) and the second one called "solid modeling". In surface modeling, one is almost completely dedicated to displaying pre-extracted surfaces. A solid modeling approach knows as "volume rendering" deals directly with the original volume data set and is able to display shaded surfaces (interactively selected) or raw data values. Figure 1 shows comparisons between both methods.

Surface Modeling Methods

The earliest method is *Polygon Oriented*, it appeared with the arrival of triangulation algorithms with planar contours developed by Keppel (1975) and Fuchs et al. (1977). From the binary sequence of the source data, a set of contours representing the surface is extracted. The tiling of the surface is done by joining the different points belonging to the contours with triangles in order to optimize a criterion (volume, area of triangles, length of segments, ...) within a planar graph. This method gives a surface data base formed by polygons which is able to be displayed with standard computer graphics display algorithms.

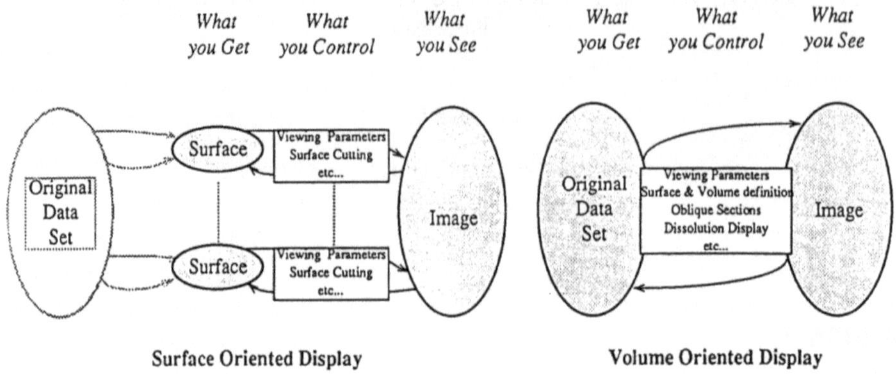

Figure 1: Comparison between *Surface Modeling* and *Solid Modeling* approaches

Another way to display the surfaces of anatomic structures is *Voxel Oriented*. Here, either binary volume or grey scale volume can be used. With either volume the data must be interpolated from the original sequence in order to obtain an homogeneous matrix of voxel in the three main axis of the space. Within the volume (binary or not), a surface tracking algorithm is executed to extract the components of the surface (voxels or face of voxels) in order to construct a surface database (Artzy et al., 1981 and Barillot et al., 1985). The display of this surface is done by rotating the surface elements to the point of view and processing visibility and shadowing as a function of the orientation of the surface at each surface element.

Solid Modeling Methods

Volume display techniques have appeared more recently and have the ability to display either surfaces with shading or part of the volume with the density values or both together. The main feature of these techniques is to display data directly from the grey scale volume. The selection of the data which will appear on the screen is done during the projection of the voxels as a function of different attributes such as density or gradient value (for surface segmentation), spatial coordinate (for volume cutting), volume set operation (union, intersection, difference of volume) or transparency for instance.

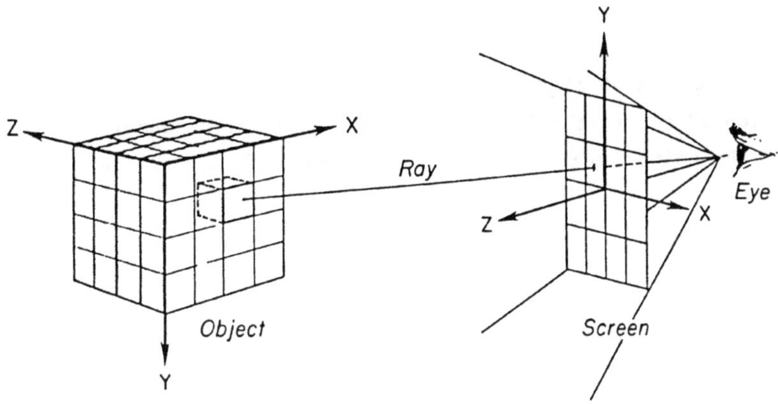

Figure 2: Ray-Tracing principal used for a Volumetric Display of a 3-D Object.

2. INTERACTIVE VOLUME RENDERING

One of the most efficient ways to implement a solid modeling method is to use a *Ray Tracing* algorithm (Höhne et al., 1987 and Schlusselberg et al., 1986). Unlike *Octree Encoding* (Meagher, 1982), you don't have to preprocess the volume in order to set up a new structure (hierarchical structure) on it. By using special implementations to compute the ray's intersection within the volume, the algorithm describes here is optimized to be fast (less than 10 seconds for one image) without compromising fidelity. With this interactive display, it is possible to redefine either a new point of view, a new volume or a new surface within the volume. This kind of interactivity allows the user to work not only with a particular surface but also to simultaneously see other information related to that surface.

2.1. Camera model

The main principal in a ray tracing algorithm is to define a viewing direction (of the object) and an eye position related to this direction. A screen is centered perpendicularly about this line (Z axis of the screen coordinate system). Then a ray is cast for each pixel of the screen from the eye position (infinite position for parallel projection) through the screen toward the volume (figure 2). All voxels that intersect the ray are used to determine the screen's displayed value. The same process is done for every pixel of the screen.

2.1.1 Ray Definition

A ray is simply a straight line in the 3-D space of the screen. The best definition of a straight line in 3-D is a parametric equation. A line through the point (x_0, y_0, z_0) and in the direction (A, B, C) is given by the equation:

$$\rho(t) : \begin{cases} x_t = x_0 + A \cdot t \\ y_t = y_0 + B \cdot t \\ z_t = z_0 + C \cdot t \end{cases} \tag{1}$$

So, for each ray the value (x_0, y_0, z_0) is given by the position of the pixel in the screen and the direction vector (A, B, C) is given by the direction between (x_0, y_0, z_0) and the eye position.
In the case of a parallel projection, these equations become very simple, a ray passing through the pixel (x_0, y_0) is simply defined as:

$$[x_t \; y_t \; z_t] = [x_0 \; y_0 \; 0] + [0 \; 0 \; 1] \cdot t \tag{2}$$

2.1.2 Geometrical Transformations

In order to compute the intersection between the ray and the volume, we need a geometrical transformation to rotate a ray defined in screen space to object space. The new position of the ray in object space is given by the formula:

$$[x'_t \; y'_t \; z'_t] = [x_t \; y_t \; z_t] \cdot [R] + \vec{T} \tag{3}$$

where

$[R]$ is the 3x3 rotation matrix function of the azimuth and the incidence of the viewing direction

$\vec{T}$ is the 3-D translation vector from the origin of the scene space to the center of the screen

By incrementing t in Equation (1) and rotating the result with (3), it is possible to find the position of the ray within the volume and to determine the intersection of the ray with the different voxels.

3. EDITING MODE AND MULTIPLE OBJECT DEFINITION

The visualization of an acceptable image is not the only goal of a 3-D medical imaging system, its main purpose is to allow the physician to work with his volumetric data base. Therefore, display quality is only one constraint of a 3-D medical imaging application; interaction

by changing the volume shape, doing measurements and linking different structures together must be also an aim for such a system.
In response to this, an object structure has been implemented wherein the object borders within the original volume are defined with a set of vertices associated with polygons. When the user wants to change the shape of an object to display, he can position these vertices, create or delete any vertex or polygon in order to construct the object shape desired. Another advantage of this technique is the ability to clip the original data set within the surrounding polygonal structure. This object structure gives the ability to store only a few amount of information to describe different part of the same original volume. These different objects (sub-parts of the same volume) can be displayed alone or together according to their own display parameters (sub-volume or surface definition).
When a polygon or part of it. intersects the original volume according to the current segmentation rules (e.g. a range of voxel density values), the pixels related to this clipping polygon are assigned with the real voxel density values. Hence it becomes relatively easy to do a digital dissection of the original data set according to the polygonal structure frame.

4. RAY-TRACING IMPLEMENTATION

Because the shape of an object to display is strictly defined by the polygonal structure, the computation of a ray throughout the volume will be particular. By using this structure, the entry points of rays within the volume are the projected points of the polygons on the screen. The ray equation is defined according to the polygonal structure projected on the screen and the associated polygon filling, each point filling a polygon becomes a ray entry point within the volume. To precompute the ray lengths, the hidden polygons are first projected to a z-buffer in order to record the object depth boundaries, the visible polygons are then projected defining an entry depth for each point or ray, the difference between these two depths or z values gives the ray length in the screen coordinate system.
The preprocess of a ray can be resumed as follow:

1. Projection of the polygonal structure on the screen.

2. Storage of the hidden polygon depth values on the z-buffer.

3. Computation of the visible polygons depth values during the polygon filling process and subtraction of these values with the ones previously stored on the z-buffer.

4. Casting of the rays on every polygon filling points according to their ray length.

The computation of a ray (ρ) throughout the volume is done according to Equations (2) and (3) with:

$$\rho(x_0, y_0, t) : [x_t\ y_t\ z_t] = [x_0\ y_0\ 0] + [0\ 0\ 1] \cdot t = [x_0\ y_0\ t] \tag{4}$$

Assuming that a point on the ray can be expressed as a function of the point before it on this ray; the ray equation using the chain coding becomes:

$$\rho(x_0, y_0, t) : \begin{cases} \rho(x_0, y_0, t_z) = \rho(x_0, y_0, t_{z-1}) + [r_{13}\ r_{23}\ r_{33}] \cdot \Delta t \\ \rho(x_0, y_0, t_0) : [x'_0\ y'_0\ z'_0] = [x_0\ y_0\ z_0] \cdot [R] + T \\ 0 \leq z \leq Z \end{cases} \tag{5}$$

where

(x_z, y_z, z_z)	is the ray position at the instant t_z according to the screen coordinate system
(x'_z, y'_z, z'_z)	is the ray position at the instant t_z according to the scene coordinate system
$R = \begin{bmatrix} r_{11} & r_{12} & r_{13} \\ r_{21} & r_{22} & r_{23} \\ r_{31} & r_{32} & r_{33} \end{bmatrix}$	is the rotation matrix
$T = [T_x\ T_y\ T_z]$	is the translation vector
$\Delta t = t_z - t_{z-1}$	is the step between two consecutive points on the ray
$Z = z_0 - z\_buff(x_0, y_0)$	is the length of the ray ($\rho(x_0, y_0, t)$)

Because the polygon filling algorithm works line per line (Rogers, 1985), two consecutive points on a line are consistently apart.
Finally, the ray equation becomes:

$$\rho(x_0, y_0, t) : \begin{cases} \rho(x_0, y_0, t_z) = \rho(x_0, y_0, t_{z-1}) + [r_{13}\ r_{23}\ r_{33}] \cdot \Delta t \ \ if\ 0 < z \leq Z \\ \rho(x_0, y_0, t_0) : \begin{cases} [x'_0\ y'_0\ z'_0] = [x_0\ y_0\ z_0] \cdot [R] + T \ \ if\ (x_0, y_0) \in L_e \\ [x'^{i}_0\ y'^{i}_0\ z'^{i}_0] = [x'^{i-1}_0\ y'^{i-1}_0\ z'^{i-1}_0] + [\Delta x\ \Delta y\ \Delta z] \\ otherwise \end{cases} \end{cases} \tag{6}$$

where

$\Delta x = r_{11} + r_{31}(z_0^i - z_0^{i-1})$
$\Delta y = r_{12} + r_{32}(z_0^i - z_0^{i-1})$
$\Delta z = r_{13} + r_{33}(z_0^i - z_0^{i-1})$
$(x_0^{i=0}, y_0^{i=0}) = (x_0, y_0) \in L_e$
L_e is the list of points (i.e. pixels) belonging to the polygon edges

5. DISPLAY MODALITIES IN VOLUME RENDERING

One of the advantages of this implementation is to interactively (in a few seconds) render an anatomical volume using several display modalities. Two families of display have been implemented: *Reflection oriented display* and *Transmission oriented display*.

5.1 Reflection oriented display

For all of these display types, voxel density values are used to specify surfaces within the original volume data set. For example, if the value of the voxel intersected by a ray belongs to a range of density values, this voxel is defined as a point on the surface. Four types of display functions have been implemented in this family:

1. Depth shading

   For this method, the value assigned to a pixel is a function of the distance between the screen and the intersected voxel on the surface.

2. Depth gradient shading

   This method is basically the same as the depth shading, but a post-processing on the resulting image is performed to enhance contrast using a depth gradient vector (Gordon and Reynolds, 1985).

   This depth vector behaves like a surface normal vector computed on a projected voxel. This set of vectors is used to process the final image according to the depth shading previously recorded and a derivation of the diffuse lighting model:

$$I = I_{depth} \cdot \cos\theta = I_{depth} \cdot (\vec{N}_{depth} \cdot \vec{L}) \qquad (7)$$

where :

I_{depth} is the value assigned by the depth shading.

θ is the angle between the light source $\vec{L}$ and the surface normal estimated vector $\vec{N}_{depth}$.

Assuming that on the screen coordinate system, the illumination vector becomes $\vec{L} = [0\,0\,1]$, after normalization of $\vec{N}_{depth}$, the lighting model becomes:

$$I = I_{depth} \cdot \left(\frac{1}{\| N_{depth} \|}\right)^{p} \quad \text{with } 0 \le p \le 1 \qquad (8)$$

where :

p is the weighting value reducing artifacts and smoothing the resulting image.

3. Gradient shading

To use the diffuse reflection lighting model without artifacts, a better normal vector associated with the pixel must be computed. Currently, the best way to do that is to compute the normal vector, for the voxel intersected by the ray, according to its real density value.

A natural way to compute a normal vector at a point (here a voxel) is to compute the gradient associated with this point. For a 3-D volume, this gradient can be computed using a 6, 18 or 26 neighborhood (Barillot et al., 1985 and Höhne and Bernstein, 1986). In practice, we have used both the 6 and 26 neighborhood. Once this gradient is computed, the illumination model can be written according to Lambert's law:

$$I = I_{max} \cdot \cos\theta = I_{max} \cdot (\vec{N} \cdot \vec{L}) \qquad (9)$$

where

$$\vec{N} = \vec{N_v} = \vec{\nabla_v} = [\delta_v^x \; \delta_v^y \; \delta_v^z]$$

4. Volumetric compositing display

One of the major problems in displaying a surface is to decide whether a given point belongs to this surface. With the previous options, this decision is a binary decision and hence produces artifacts. This display option does not assume that the surface has a constant density value. Instead, it assumes that the surface has a range of density values wherein the different voxel are together into a semi-transparent gel (Levoy, 1988).

The implementation of this method consists firstly to classified the voxels according to a range of voxel density values called "fuzzy range". This classification assigns to a "fuzzy voxel" (i.e. voxel belonging to this fuzzy range) a parameter α function of its amount of true-structure and none-structure. Assuming that at the top of the fuzzy range value, a voxel belongs to the true-structure (classification 1) and at the bottom of the same range, it belongs to the none-structure (classification 0); a voxel belonging to the fuzzy range will be classified according to its rank within these values (e.g. linear repartition between classification 0 and classification 1).

Nevertheless, this classification can be improved by adding surface information. Assuming that two voxèls with the same structure information (e.g. density value) should not have the same influence for the display whether there are on a surface or not. For that reason, the classification coefficient is weighted by a surface information (e.g. gradient intensity).

The merging of the voxel luminescences and opacities is performed using a transparency model. For two surfaces (one semi-transparent, one opaque) the resulting luminescence can be expressed as:

$$I_{tot} = I_{tsp} \cdot \alpha + (1 - \alpha) \cdot I_{opq} \tag{10}$$

where

I_{tsp}	is the transparent surface luminescence
I_{opq}	is the opaque surface luminescence
$\alpha : 0 \leq \alpha \leq 1$	is the transparent surface opacity coefficient

Extending this model to mix the information not only from two voxels but from a set of voxel along the ray, the volumetric compositing luminescence becomes:

$$I_{tot} = \sum_{n=0}^{N} \left(I_n (1 - \alpha_{n-1}) \prod_{m=n}^{N-1} \alpha_m \right) \tag{11}$$

where

α_{-1}	$= 0$
N	is number of voxel along the ray according to the fuzzy range
I_n	is the luminescence of the voxel v_n (n^{th} voxel along the ray)
α_n	is the opacity coefficient of the voxel v_n

This expression can be simplified by looking at the luminescence compositing C before and after the voxel v_n.

The resulting luminescence can be expressed as:

$$\begin{cases} I_{tot} = C_N \\ C_n = C_{n-1} \cdot \alpha_{n-1} + (1 - \alpha_{n-1}) \cdot I_n; \quad 0 < n \leq N \\ C_0 = 0 \end{cases} \tag{12}$$

where

C_n is the luminescence compositing after the voxel v_n

After defining the fuzzy range, the fuzzy volume computation process consists firstly to test for each voxel v_n along the ray, if its density belongs to the fuzzy range, secondly to compute its classification coefficient α_n and then to update the luminescence compositing C_n. The ray process stops when it goes outside the volume or when the ray reachs the voxel v_N according to the classification $\alpha_N = 1$.

5.2 Transmission oriented display

For this kind of display, there is no surfaces identification involved. A ray passes totally through the volume and the pixel value is computed as an integrated function. Two different display methods have been implemented in this family : *brightest pixel display* (the pixel value is the brightest density value along the ray) and the *weighted summation display* (average of the voxel density values along the ray).

6. CONCLUSION

The interactive approach for 3-D display of medical data presented here is a proof that today it becomes possible for a physician to have a direct use of a 3-D data base from different medical image modalities (C.T, N.M.R, D.S.R, P.E.T, 3-D ultrasound, ...). The improvement in the next few years will be provided by the introduction of upcoming machine specialized on 3-D medical imaging with larger memory, faster CPU and closest connection to the mainframe and/or workstation computer. The new generation of workstation could even make suitable the implementation of such techniques directly on these stations without involving on special and expensive hardware development.
The approach described in this paper is inclined to render an anatomical volume represented by a 3-D matrix. Regarding to other medical applications (in vivo or not), this approach would not be suitable, the next step would be to link this class of display methods with others in order to extend the range and the efficiency of medical applications like in neurosurgery, in cardiac imaging or in reconstructive surgery for instance (Barillot et al., 1988).

ACKNOWLEDGEMENTS

The author expresses appreciation to their colleagues in the Biodynamics Research Unit at the Mayo Clinic whose make this work possible. Special recognition is given to Drs Richard A. Robb, Christopher McEwan and William E. Higgins for their supports and comments. Image data was provided by the Mayo department of physiology and Biophysics and by Dr. C. McEwan from the Mayo section of plastic surgery.

REFERENCES

Artzy E., Frieder G., and Herman G.T. (1981) - The Theory, Design, Implementation and Evaluation of a Three-Dimensional Surface Detection Algorithm - Comp. Graph. & Im. Proc. (15) , pp.1-24.

Barillot C., Gibaud B., Luo L.M. and Scarabin J.M. (1985) - 3-D Representation of Anatomic Structures from CT Examinations Proc. SPIE (602), Biostereometrics'85, Cannes, Fr., pp.307-314.

Barillot C., Gibaud B., Lis O., Luo L.M., Bouliou A., Le Certen G., Collorec R. and Coatrieux J.L. (1988) - Computer Graphics in Medicine: A Survey - CRC Press, Crit. Rev. in Biom. Eng. (15), pp.269-307.

Fuchs H., Kedem Z.M. and Uselton S.P. (1977) - Optimal Surface Reconstruction from Planar Contours - Comm. of the ACM (20), pp.693-702.

Gordon D. and Reynolds R. A. (1985) - Image Space Shading of 3-D Objects - Comp. Vis. Graph. & Im. Proc. (29), pp.361-376.

Höhne K.H. and Bernstein R. (1986) - Shading 3-D Images from CT Using Grey-Level Gradients - IEEE Trans. on Med. Imag. (15), pp.45-47.

Höhne K.H., Riemer M. and Tiede U. (1987) - Viewing Operation for 3-D Tomographic Grey Level Data - Proc. of CAR'87, Springer-Verlag Ed., pp.599-609.

Keppel E. (1975) - Approximating Complex Surfaces by Triangulation of Contours Lines - IBM J. Res. Develop. (19), pp.2-11.

Levoy M. (1988) - Display of Surfaces from Volume Data - IEEE Comp. Graph. & Appl.,May 1988, pp.29-37.

Meagher D. (1982) - Geometric Modeling Using Octree Encoding - Comp. Grap. & Im. Proc. (19), pp.129-147.

Robb R.A. and Barillot C.(1988) - Interactive 3-D image display and analysis - SPIE proc. on Hybrid Image and Signal Proc. (939), Orlando, FL, pp.173-202.

Rogers D. F. (1985) - Procedural Elements for Computer Graphics - Mac Graw Hill Book Company, New-York.

Schlusselberg D. S., Smith W.K. and Woodward D.J. (1986) - Three-Dimensional Display of Medical Image Volumes - Proc. of 7th NCGA, Vol.III, Anaheim, CA, pp.114-123.

Part 2 Theoretical Aspects

Section 2.2 Reconstruction

DATA AUGMENTATION SCHEMES APPLIED TO IMAGE RESTORATION

Valen E. Johnson, Wing H. Wong, Xiaoping Hu, and Chin-Tu Chen

The University of Chicago
United States of America

ABSTRACT

Many medical imaging systems produce images that are degraded by statistical noise and blurring. This paper describes a physical model for the generation of images associated with these systems and presents a restoration algorithm that is designed to compensate for these sources of degradation. The restoration algorithm is based on the idea of Bayesian data augmentation and utilizes a Gibbs prior for the image. This prior incorporates two essential features necessary for the restoration of such images. First, boundary detection is included so that nonhomogeneous regions can be identified. Second, an expanded neighborhood system is proposed to permit the deconvolution of blurring effects. Additionally, a procedure for choosing parameters of the Gibbs prior is discussed.

1. INTRODUCTION

Recent advances in computer technology have permitted the development of highly complex imaging systems in a wide variety of fields. A common feature of many of these imaging systems is that they produce images representing observations of photons. As a result, the images often suffer from similar sources of degradation and are amenable to similar restoration techniques. The purpose of this paper is to present a physical model for image generation that incorporates two primary sources of degradation associated with such imaging systems, and to propose an algorithm designed to alleviate their effects. The physical model for image generation happens to correspond to several existing image reconstruction and restoration problems, but the primary motivation for its description is to provide a framework for constructing algorithms for use in more complicated imaging systems.

The first source of image degradation arises from variations in the number of photons emitted by the object under study. We assume that the number of counts emitted during a finite interval of time follows a Poisson distribution, and that it is the mean intensity level of the Poisson distribution that is of interest. Hence, one objective of the restoration algorithm is to reduce the effect of the variation of observed counts on the processed image. This can be accomplished by smoothing or averaging counts arising from nearby locations, but to do so indiscriminately results in a loss of spatial resolution and degradation of boundaries within the image. Smoothing should therefore be restricted to homogeneous regions, and so estimation of boundaries within the image is also required.

In addition to the Poisson noise inherent in the observed counts, many images also suffer from the loss of spatial resolution or sharpness. We will refer to this loss of spatial resolution as blurring. Without specifying the precise nature of the underlying mechanism, blurring can be broadly regarded as misdirection or scattering of photons prior to their registration as detected events. Thus, another objective of the restoration algorithm is to recover information that is lost as a result of blurring.

The basis for the restoration algorithm is the Bayesian data augmentation scheme proposed by Tanner and Wong (1987). The data augmentation scheme can be described as the Bayesian equivalent of the EM (Estimation-Maximization) algorithm and permits the imputation of data for unobserved quantities. It is ideal for this application because estimation of mean intensities and boundaries is more straightforward when a set of "unblurred" observations (i.e. sharp images with perfect spatial resolution) is available, and estimation of the effect of blurring is simple when an estimate of the underlying mean intensities is known.

2. PHYSICAL DESCRIPTION OF THE OBSERVED IMAGE

The two primary components of image generation are the object to be imaged and the sensor used to observe the object. We assume that the sensor is discretized into an $m \times n$ array of pixels $\Xi = \{\xi_{i,j}\}$, and that the object under study, Θ, emits photons according to a Poisson process. Ideally, photons emitted by the object Θ travel directly to the sensor without scatter or attenuation and register precisely on an ideal sensor†. Under these conditions, we assume that there is a partition $\{\theta_{i,j}\}$ of Θ which corresponds naturally to the discretization of the sensor Ξ in the sense that photons emitted from a partition element $\theta_{i,j}$ can be registered only at pixel $\xi_{i,j}$. We also assume that either all emitted photons have the same probability of detection or that the probabilities of detection are known *a priori*. Let $\lambda_{i,j}$ be the mean Poisson intensity of photons that register at $\xi_{i,j}$ under these conditions, and let $n_{i,j}$ be the number of photons actually detected at $\xi_{i,j}$ during the course of image generation. Also, let $\lambda = \{\lambda_{i,j}\}$ and $\mathbf{n} = \{n_{i,j}\}$ denote the corresponding arrays. The objective of the image restoration can therefore be regarded as estimation of the array λ.

In reality, photons are scattered prior to impinging on the detector, and realistic sensors suffer from loss of spatial resolution. To reflect these phenomenon, we assume that the photons registered under the ideal circumstances described above are randomly deflected and registered at surrounding pixels according to a known spread function. Specifically, assume that photons originating from $\theta_{i,j}$ register at pixel $\xi_{r,s}$ with probability $p_{r,s}^{i,j}$. Denote the number of photons actually observed at pixel $\xi_{r,s}$ by $y_{r,s}$. Typically, the number of partitioning elements $\theta_{i,j}$ for which $p_{r,s}^{i,j} > 0$ is a relatively small subset of Θ and will be denoted $B_{r,s}$. As a result of this model for image generation, the observed (blurred) counts $y_{r,s}$ follow a Poisson distribution with mean intensity

† By an ideal sensor we mean a sensor with perfect spatial resolution.

$$\sum_{i,j \in B_{r,s}} \lambda_{i,j} p_{r,s}^{i,j} \quad . \tag{1}$$

Note that estimates of λ must be based on the blurred, observed counts $\mathbf{y}$ rather than the unblurred, unobserved counts $\mathbf{n}$.

3. BAYESIAN DATA AUGMENTATION

The statistical framework for the image restoration algorithm is provided by the data augmentation scheme proposed by Tanner and Wong (1987). The algorithm proceeds by imputing values for latent (missing) data based on current estimates of the posterior distribution, estimating the unknown parameters based on the imputed values, and then imputing new values based on the updated posterior distribution. The algorithm proceeds iteratively in this manner until the estimated posterior reaches equilibrium.

By considering the unobserved counts $\mathbf{n} = \{n_{i,j}\}$ as latent data and letting $\mathbf{y} = \{y_{i,j}\}$ denote the observed counts, the posterior distribution $p(\lambda \mid \mathbf{y})$ may be expressed as

$$p(\lambda \mid \mathbf{y}) = \int_N p(\lambda \mid \mathbf{n}, \mathbf{y}) p(\mathbf{n} \mid \mathbf{y}) d\mathbf{n} \quad , \tag{2}$$

where $N = \{0,1, \cdots\}^{nm}$ is the state space of $\mathbf{n}$. This representation requires knowledge of the predictive density $p(\mathbf{n} \mid \mathbf{y})$, which can be represented in terms of the posterior density $p(\lambda \mid \mathbf{y})$ as

$$p(\mathbf{n} \mid \mathbf{y}) = \int_\Lambda p(\mathbf{n} \mid \lambda, \mathbf{y}) p(\lambda \mid \mathbf{y}) d\lambda \quad , \tag{3}$$

where $\Lambda = [0,\infty)^{mn}$ is the parameter space for λ. Hence, given a current approximation to $p(\lambda \mid \mathbf{y})$, a sample of size h, denoted $\mathbf{n}^1, \mathbf{n}^2, \ldots, \mathbf{n}^h$, may be drawn from the density $p(\mathbf{n} \mid \mathbf{y})$ by sampling from $p(\mathbf{n} \mid \lambda, \mathbf{y})$. This sample in turn can be used to update the current estimate of $p(\lambda \mid \mathbf{y})$ by equating

$$p(\lambda \mid \mathbf{y}) = \frac{1}{h} \sum_{i=1}^{h} p(\lambda \mid \mathbf{n}^i, \mathbf{y}) \quad , \tag{4}$$

and the algorithm proceeds iteratively.

Because of the high dimension of λ and $\mathbf{n}$, the value of h that we chose for the restoration algorithm is one. In this case, the posterior distribution is approximated by the distribution of successive samples of λ obtained once the algorithm has reached equilibrium.

Of course, implementing the data augmentation scheme requires specification of the conditional posterior density $p(\boldsymbol{\lambda} \mid \mathbf{n},\mathbf{y})$ and the conditional predictive density $p(\mathbf{n} \mid \boldsymbol{\lambda},\mathbf{y})$. Additionally, both densities must permit efficient sampling. Otherwise the computational burden for even moderate values of m and n would make the problem intractable.

The description of the physical mechanism that results in the observed counts $\mathbf{y}$ also implies a certain form of the predictive density $p(\mathbf{n} \mid \boldsymbol{\lambda},\mathbf{y})$. As demonstrated in the Appendix, if

$$q_{r,s}^{i,j} = \frac{\lambda_{r,s}\, p_{i,j}^{r,s}}{\sum_{m,n} \lambda_{m,n}\, p_{i,j}^{m,n}} , \tag{5}$$

then sampling from $p(\mathbf{n} \mid \boldsymbol{\lambda},\mathbf{y})$ is equivalent to allotting the counts $y_{i,j}$ to neighboring pixels $\xi_{r,s}$ according to the multinomial probabilities $q_{r,s}^{i,j}$.

From a Bayesian viewpoint, the posterior density is obtained as the product of the likelihood function and a prior density. The likelihood function based on $\mathbf{n}$ is well defined as the product of independent Poisson observations. The goal in introducing the prior is to reduce the Poisson noise in the image, possibly by averaging or smoothing within homogeneous regions. To accomplish this requires the identification of homogeneous regions, and so the locations of boundaries must also be estimated. For these and other reasons, an expansion of a model proposed by Geman and Geman (1984) was selected to define the prior density.

Geman and Geman suggested the use of Gibbs priors in image restoration because they possess several convenient properties. Besides ease of sampling, Gibbs priors can be specified in terms of local properties of the image, so that the difficult task of specifying a prior by its global characteristics need not be undertaken. Additionally, it is possible to include boundaries in the formulation of the prior, thus avoiding the problem of smoothing across edges. An application of a Gibbs prior in medical imaging that does not include boundary detection is provided in Geman and McClure (1985).

In order to describe Gibbs models further, several definitions concerning properties of graphs are needed. The description and notation provided below closely parallel that given in Geman and Geman, and the reader interested in more complete details should consult that source.

Consider the pixel sites $\xi_{i,j}$ to be elements of a graph Ξ. A ***neighborhood system*** on Ξ is defined to be any collection of subsets $\mathbf{G} = \{G_\xi,\ \xi \in \Xi\}$ such that $\xi \notin G_\xi$, and for all ξ_1 and ξ_2, $\xi_1 \in G_{\xi_2}$ if and only if $\xi_2 \in G_{\xi_1}$. Furthermore, a ***clique*** † is defined to be any subset C of Ξ in which each element of C is a neighbor of every other element. Denote the set of cliques by $\mathbf{C}$.

---

† This definition of clique corresponds to what is often called a complete set in graph theory.

Given a graph Ξ and a neighborhood system $\mathbf{G}$, a Gibbs distribution has a density function of the form

$$\pi(\lambda) = \frac{1}{Z} \exp\{-U(\lambda)\} \,, \tag{6}$$

where Z is called the ***partition function*** and is a constant independent of λ, and $U(\lambda)$ is termed the energy function. The energy function $U(\lambda)$ can be expressed as

$$U(\lambda) = \sum_{C \in \mathbf{C}} V_C(\lambda) \,, \tag{7}$$

where the functions $V_C(\lambda)$ are arbitrary functions depending only on the values of the sites in C. The functions $V_C(\lambda)$ are referred to as the ***potentials.***

An important characteristic of Gibbs distributions that relates to their sampling properties can be gleaned from the form of their densities. Since the densities can be factored according to cliques, it follows that the conditional distribution of a site $\xi_{i,j}$ is dependent only on sites that belong to cliques containing site $\xi_{i,j}$ and having nonzero potential.

To specify a Gibbs distribution it is therefore necessary to define a neighborhood system and a set of potentials. The partition function is determined so that the resulting density has unit integral.

In order to estimate the boundaries within the image, the graph Ξ is expanded to include ***line sites*** between each pair of neighboring pixels. Denoting the line sites δ, a portion of the expanded graph is illustrated below.

$$\begin{array}{ccccccc}
\lambda & \delta & \lambda & \delta & \lambda & \delta & \lambda \\
\delta & & \delta & & \delta & & \delta \\
\lambda & \delta & \lambda & \delta & \lambda & \delta & \lambda \\
\delta & & \delta & & \delta & & \delta \\
\lambda & \delta & \lambda & \delta & \lambda & \delta & \lambda
\end{array}$$

The line sites δ are assigned the value zero when the neighboring intensity sites are thought to lie in the same region, and take the value one otherwise.

3.1 Approximation of the likelihood function and specification of the Gibbs prior

The likelihood function that results from the product of independent Poisson observations is a product of gamma likelihoods. Comparing a gamma likelihood to the form of the Gibbs densities given above, it is clear that the gamma likelihood cannot readily be incorporated into the Gibbs formulation without sacrificing the convenient sampling properties associated with such models. However by reparameterizing and modelling the square root of the Poisson intensities, the likelihood function can be well approximated by a Gaussian likelihood (see, for example, Rao (1973), page 426).

Another reason for reparameterizing on the square root scale is that this is the variance stabilizing transformation for the gamma family of distributions. On this scale the variance of the transformed intensities is close to one fourth, regardless of the number of counts observed. The almost constant variance on this scale is quite important when specifying the prior since the amount of smoothing does not have to be adjusted according to the information contained in the likelihood.

Specifically, if $\gamma = \sqrt{\lambda}$, the contribution to the conditional likelihood for γ from site $\xi_{i,j}$ can be approximated by a Gaussian likelihood with mean

$$\mu_{i,j} = \sqrt{n_{i,j}+1} - \frac{1}{\sqrt{n_{i,j}+1}} \tag{8}$$

and variance 0.25. Correspondingly, the prior is also specified on the square root scale, and in the illustration given above for the expanded graph Ξ, the intensity sites λ should be replaced by the transformed variables γ. For the remainder of the discussion on the Gibbs prior, references to intensity sites refer to the transformed variables γ.

Perhaps the easiest way to specify a Gibbs prior is to begin with the desired form for the conditional distribution of the intensity sites, and then work backwards to determine the required cliques, potentials, and neighborhood system.

The purpose in introducing the prior was to permit averaging within homogeneous regions to alleviate the degradation caused by Poisson noise. With the transformed Gaussian likelihood, it is natural to specify the conditional distribution of an intensity site, given the values of all neighboring sites, to be Gaussian with mean equal to the weighted average of those neighboring sites not separated by an edge (line sites with value one). In terms of Gibbs priors, this is accomplished by the introduction of the following cliques

$$C_1 : \quad \gamma_r \ \delta \ \gamma_s \quad or \quad \begin{matrix} \gamma_r \\ \delta \\ \gamma_s \end{matrix} \tag{9}$$

and defining the potential of these cliques to be

$$V_{C_1} = \begin{cases} \frac{1}{2\sigma_1^2}(\gamma_r - \gamma_s)^2 & \text{if } \delta = 0 \\ 0 & \text{if } \delta = 1 \end{cases} , \tag{10}$$

where σ_1^2 is a smoothing parameter. This clique and corresponding potential permits local smoothing, but it also determines a specific form of the conditional distribution of the line sites. With only this clique, formation of boundaries is dependent on γ only through the difference in neighboring intensity sites. *This is inadequate because the blurring mechanism combined with the Poisson noise masks the effect of sharp edges within such a small neighborhood.* It is therefore necessary to expand the neighborhood system under consideration so that differences between intensity sites beyond the range of the blurring function are considered.

For definitiveness, assume that the illustration below represents the probability vector associated with the blurring function (note that this blurring function represents a point spread function with a full width at half maximum of four pixels). This blurring mechanism does not distribute a significant number of particles beyond a distance of three intensity sites in either the vertical or horizontal direction. By introducing the cliques and corresponding potentials given below, the effect of blurring in masking a boundary between the extreme intensity sites is almost eliminated. Hence when a sharp boundary is present, the squared difference between the extreme intensity sites is large and the formation of a boundary between the intensity sites is likely.

0.0027	0.0065	0.0103	0.0130	0.0103	0.0065	0.0027
0.0065	0.0156	0.0247	0.0313	0.0247	0.0156	0.0065
0.0103	0.0247	0.0392	0.0495	0.0392	0.0247	0.0103
0.0130	0.0313	0.0495	0.0625	0.0495	0.0313	0.0130
0.0103	0.0247	0.0392	0.0495	0.0392	0.0247	0.0103
0.0065	0.0156	0.0247	0.0313	0.0247	0.0156	0.0065
0.0027	0.0065	0.0103	0.0130	0.0103	0.0065	0.0027

$$C_2 : \quad \gamma_r \;\; \delta_1 \;\; \gamma \;\; \delta_2 \;\; \gamma \;\; \delta_3 \;\; \gamma_s \tag{11}$$

$$V_{C_2} = \begin{cases} \frac{1}{2\sigma_2^2}(\gamma_r - \gamma_s)^2 & \text{if all line sites are } 0 \\ a_1 & \text{if two line sites are } 0 \\ a_2 & \text{if one line site is } 0 \\ \infty & \text{if all line sites are } 1 \end{cases} \tag{12}$$

Here σ_2^2 is another smoothing parameter and the purpose of a_1 and a_2 is to compensate for the existence of two or more edges in the same neighborhood. Specification of a_1 and a_2 is discussed in a later section.

The two cliques described above provide the means by which local smoothing is included in the prior, but they fail to incorporate a sufficient amount of prior information about the distribution and local characteristics of line sites. Because line sites typically occur together to form boundaries, the processed image can be improved by including cliques and potentials that encourage the formation of continuous boundaries while discouraging the formation of isolated edges.

Part of this information can be included by the introduction of the clique pictured below.† A relatively large potential (small probability) is assigned to such cliques when only one positive line site occurs, thus discouraging the formation of isolated boundaries. Boundaries are naturally associated with the assignment of two positive line sites within such cliques, so the potential assigned to this configuration is smaller. The physical interpretation of three and four positive line sites is slightly less obvious, but follows along the same lines.

$$C_3: \quad \begin{matrix} \gamma & \delta & \gamma \\ \delta & & \delta \\ \gamma & \delta & \gamma \end{matrix} \tag{13}$$

$$V_{C_3} = \begin{cases} 0 & \text{if all line sites are } 0 \\ a_3 & \text{if one line site is } 1 \\ a_4 & 2 \quad '' \quad \text{are } 1 \\ a_5 & 3 \quad '' \quad 1 \\ a_6 & 4 \quad '' \quad 1 \end{cases} . \tag{14}$$

An additional feature which we wish to model is the presence of sharp edges between regions. With blurring, it is difficult to distinguish a sharp, step function type boundary from a soft, continuous boundary on the basis of only the observed counts. Since many of the images that we consider have sharp boundaries (e.g. the separation of white and gray matter in the brain), it is beneficial to include potentials that encourage their formation. The clique below and its corresponding potential accomplishes this by discouraging the formation of soft boundaries that have as their representation a series of parallel edges.

$$C_4: \quad \begin{matrix} \gamma & \delta & \gamma & \delta & \gamma \\ \gamma & \delta & \gamma & \delta & \gamma \end{matrix} \tag{15}$$

† The intensity sites pictured in the remaining cliques do not actually belong in the cliques— they are illustrated only for clarity.

$$V_{C_4} = \begin{cases} 0 & \textit{if at least one } \delta \textit{ is } 0 \\ \infty & \textit{otherwise} \end{cases} \tag{16}$$

A final consideration in specifying the prior involves the treatment of outliers. In a large array, the probability of observing a small number of pixels with unusually large deviations from their mean is quite likely. On the other hand, an isolated region in the image that is only one pixel in size is unlikely to be resolved. Hence, we include the following clique in order to eliminate the appearance of "isolated" pixels in the image.

$$C_5 : \quad \begin{array}{ccc} & \delta & \\ \delta & \gamma & \delta \\ & \delta & \end{array} \tag{17}$$

$$V_{C_5} = \begin{cases} \infty & \textit{if all line sites are } 1 \\ 0 & \textit{otherwise} \end{cases} \tag{18}$$

The neighborhood system required to support this collection of cliques is pictured below. The first illustration represents the neighborhood for a line site that falls horizontally between two intensity sites, while the neighborhood for a line site falling vertically between intensity sites would be represented by a 90° rotation of this neighborhood. The second illustration represents the neighborhood for an intensity site.

$$\begin{array}{ccccccccccc}
 & & & \delta & & \delta & & \delta & & & \\
 & & & & \delta & & \delta & & & & \\
\gamma & \delta & \gamma & \delta & \gamma & \delta & \gamma & \delta & \gamma & \delta & \gamma \\
 & & & & \delta & & \delta & & & & \\
 & & & \delta & & \delta & & \delta & & &
\end{array}$$

$$\begin{array}{ccccccccccccc}
 & & & & & & \gamma & & & & & & \\
 & & & & & & \delta & & & & & & \\
 & & & & & & \gamma & & & & & & \\
 & & & & & & \delta & & & & & & \\
 & & & & & & \gamma & & & & & & \\
 & & & & & & \delta & & & & & & \\
\gamma & \delta & \gamma & \delta & \gamma & \delta & \gamma & \delta & \gamma & \delta & \gamma & \delta & \gamma \\
 & & & & & & \delta & & & & & & \\
 & & & & & & \gamma & & & & & & \\
 & & & & & & \delta & & & & & & \\
 & & & & & & \gamma & & & & & & \\
 & & & & & & \delta & & & & & & \\
 & & & & & & \gamma & & & & & &
\end{array}$$

As a result of this model, the conditional distribution of an intensity site $\gamma_{i,j}$, given the values of all other sites, is Gaussian with mean given by

$$\frac{4\mu_{i,j} + \frac{1}{\sigma_1^2}\sum \gamma_{r,s} + \frac{1}{\sigma_2^2}\sum \gamma_{s,s}}{4 + \sum_{r,s} \frac{1}{\sigma_1^2} + \sum_{s,s} \frac{1}{\sigma_2^2}}, \tag{19}$$

where the first sum in both the numerator and denominator extends over all first order neighbors not separated from $\gamma_{i,j}$ by an edge, and the second over all third order horizontal and vertical neighbors not separated by an edge. Note that the conditional mean permits piecewise linear variation in the image surface, but does not precisely fit more curved surfaces.

3.2 Guidelines for Determining Line Site Potentials

Consider the following configuration in which a value for the line site δ is to be drawn from its conditional distribution. Line sites not pictured have the value 0.

$$\gamma_1 \quad \gamma_2 \quad \gamma_3 \; \delta \; \gamma_4 \quad \gamma_5 \quad \gamma_6 \tag{20}$$

In this configuration, the odds of forming a positive line site, or an edge, are given by

$$\frac{\exp\{-2a_3 - 3a_1\}}{\exp\{\frac{-1}{2\sigma_1^2}(\gamma_3-\gamma_4)^2 - \frac{1}{2\sigma_2^2}(\gamma_1-\gamma_4)^2 - \frac{1}{2\sigma_2^2}(\gamma_2-\gamma_5)^2 - \frac{1}{2\sigma_2^2}(\gamma_3-\gamma_6)^2\}} \tag{21}$$

If Δ represents the typical value of the squared difference of intensity sites within a homogeneous region, then the odds ratio for the formation of a boundary in a homogeneous region can be set to some arbitrary value, say $1{:}c^2$, and the value of a_1 and a_3 can be determined, by equating

$$\exp\{-2a_3 + \frac{\Delta}{2\sigma_1^2}\} = \frac{1}{c} \tag{22}$$

and

$$\exp\{-3a_1 + \frac{3\Delta}{2\sigma_2^2}\} = \frac{1}{c}, \tag{23}$$

for example. In the images that we have studied, typical values of Δ are between 0.25 and 0.5 and appear to be reasonably robust to the values of the line site potentials and the value of the scale parameters σ_1 and σ_2. Taking Δ to be 0.25, empirical results suggest that a good value for a is approximately 4, so the odds of forming a line site in the situation depicted above, assuming the region to be relatively homogeneous, is about 1 to 16. Of course, if the region is not homogeneous, the squared differences between line sites tends to be large and dominate the odds ratio, making the formation of edges more likely.

A similar procedure can be used to find reasonable values for the remaining line site potentials. It should be noted that the choice of the configurations used to determine other potentials, particularly those corresponding to neighborhoods in which numerous other line sites are present, is not as clear and is somewhat arbitrary. Nonetheless, the procedure seems to provide an effective means by which line site potentials can be chosen.

The configurations that we have found useful in implementing this procedure are illustrated below for the cases corresponding to a_4, a_5, a_6, and a_2. Possible values for these odds ratios are 1:3.5, 1:16, 1:4, and 1:5, respectively. Note that line sites that are currently estimated as edges are represented by 1's, and line sites with value 0 are omitted.

$$\begin{array}{ccccccc} & & & 1 & & & \\ \gamma & \gamma & \gamma & \delta & \gamma & \gamma & \gamma \\ & & & 1 & & & \end{array} \qquad \begin{array}{ccccccc} & & & 1 \; \; 1 & & & \\ \gamma & \gamma & \gamma & \delta & \gamma & \gamma & \gamma \\ & & & 1 & & & \end{array}$$

$$\begin{array}{ccccccc} & & & 1 & & & \\ & & & 1 \; \; 1 & & & \\ \gamma & \gamma & \gamma & \delta & \gamma & \gamma & \gamma \\ & & & 1 & & & \end{array} \qquad \begin{array}{cccccccc} & & & & & 1 & & \\ \gamma & \gamma & \gamma & \delta & \gamma & 1 & \gamma & \gamma \\ & & & & & 1 & & \end{array}$$

4. COMPUTATIONAL ASPECTS OF IMPLEMENTATION

Because the data augmentation scheme produces samples from the posterior distribution of the image, it is necessary either to form a point estimate from these samples or modify the algorithm so that its output includes a single "best" image. The former option leads naturally to the use of the posterior mean, while the latter suggests the use of the maximum *a posteriori* (MAP) estimate or an estimate obtained by the method of Iterated Conditional Modes (ICM, see Besag (1986)). Due to computational considerations, we use the ICM estimate.

The ICM method departs slightly from the formal framework of the data augmentation scheme as described above in that samples are not actually drawn from the predictive and posterior distributions. Instead, values that maximize the conditional likelihood at individual sites are iteratively selected until the algorithm converges. As a result the ICM method requires fewer iterations to converge and each iteration takes less time.

When the ICM method is employed, the algorithm more closely resembles the EM algorithm than the Bayesian data augmentation scheme in the sense that a point estimate rather than a posterior distribution is obtained. However, the data augmentation approach provides a useful conceptual framework for the introduction of prior information, and the point estimate can be viewed as a starting value for the data augmentation scheme in those circumstances in which an estimate of the posterior distribution is required. Additionally, the data augmentation scheme provides a potential mechanism for assessing the image variability.

The most problematic aspect in using the ICM method is the dependency of the estimated image on the sequence in which sites are visited. If line sites are visited systematically, say from left to right, boundaries tend to form to the left of actual boundaries. This occurs because blurring encourages the formation of edges several pixels away from their actual location, and once an edge is formed the prior discourages the formation of neighboring edges. Hence, if line sites to the left of a boundary are consistently visited before line sites to the right, the location of the estimated boundary is likely to be to the left of the actual boundary. This problem can be alleviated by updating line sites in a less regular pattern.

5. EXAMPLES

Figure 1 below illustrates the results of the restoration algorithm for an observation from the phantom pictured in Fig. 1a. The observed image is shown in Fig. 1b and was obtained by generating Poisson random variables with mean intensities given by Fig. 1a and blurring these counts according to the point spread function described in Section 3.1. Restorations are pictured in Fig. 1c and Fig. 1d. The dimension of the image is 128×128 pixels.

The restoration pictured in Fig. 1c was obtained after 40 iterations of the ICM method using the observed counts as the initial estimate for λ and $\sigma_1^2 = 2.71$. Limited experience indicates that reasonable restorations can be obtained when $\sigma_2^2/\sigma_1^2 = 0.6$, so for the remainder of this section take $\sigma_2^2 = 0.6\sigma_1^2$.

The image pictured in Fig. 1d was restored by using Fig. 1c as an initial estimate of λ, performing 40 ICM iterations with $\sigma_1^2 = 1.0$, then reducing σ_1^2 to 0.37 and performing an additional 40 iterations.

Several features of these restorations stand out. Perhaps the most obvious characteristic of the restorations, particularly in the restoration obtained with $\sigma_1^2 = 0.37$, is the smoothness of the image that results from the elimination of the Poisson noise. In this respect, the algorithm appears quite successful. Also, the effect of blurring has been alleviated and sharp boundaries have been established within the restored image.

A troublesome aspect of the restorations is the misleadingly precise representation of the location of boundaries. An effective way of representing this uncertainty is clearly needed and several possible alternatives are being investigated.

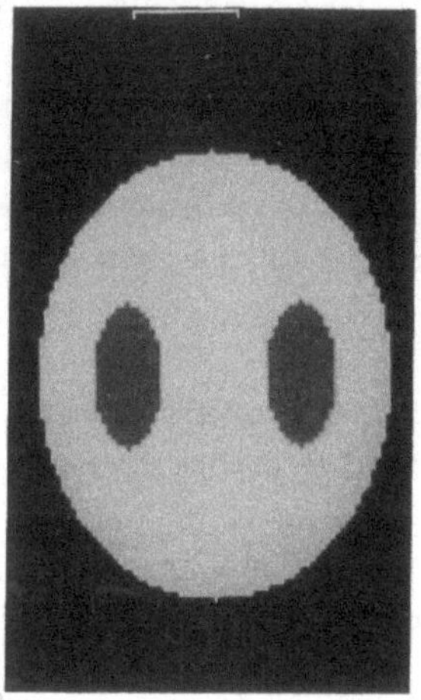

Figure 1a: Phantom of brain slice

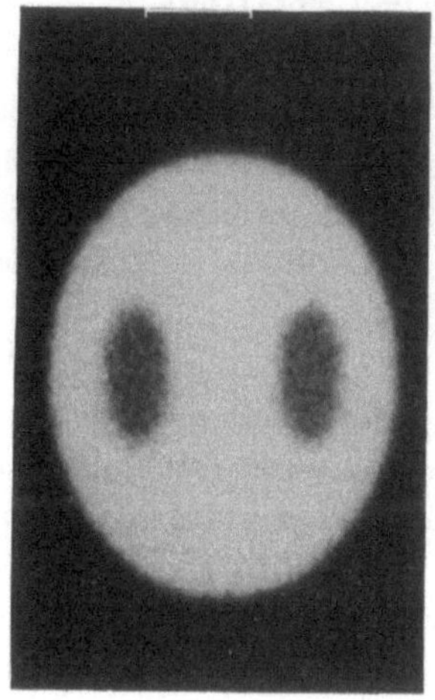

Figure 1b: Observed image

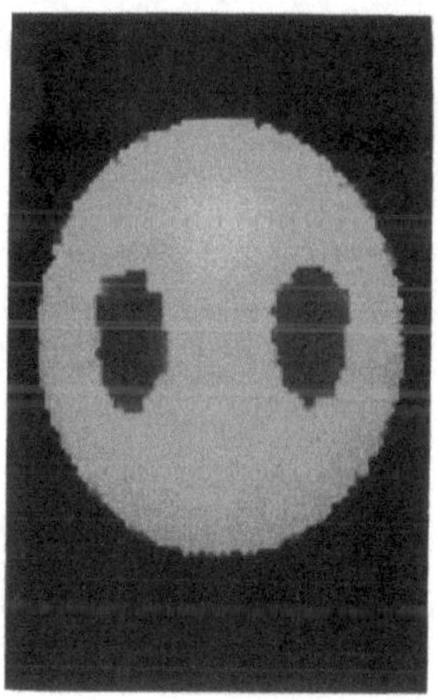

Figure 1c: Restoration with $\sigma_1^2 = 2.71$

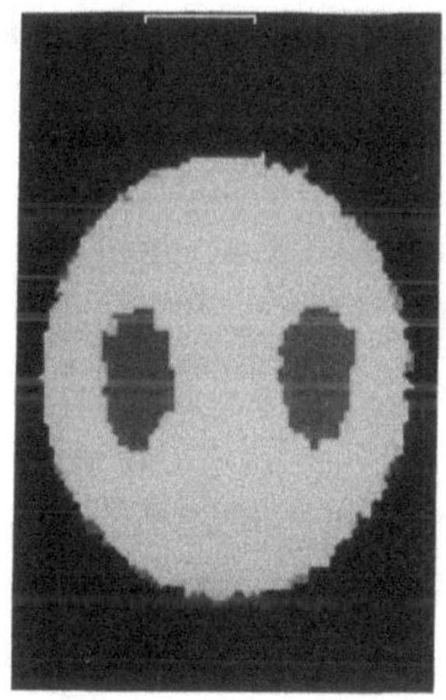

Figure 1d: Restoration with $\sigma_1^2 = 0.37$

Figure 1. Results of restoration algorithm. The brain backgound, the ventricle ellipses, and the small circular tumor have mean intensities of 400, 200, and 500, respectively.

6. CONCLUDING REMARKS

We have proposed a physical model for image generation that incorporates the primary sources of image degradation and that can be applied, with minor alterations, to a large number of imaging systems. Likewise, the proposed restoration algorithm can be applied to a variety of image restoration problems and has proven successful in alleviating image degradation in preliminary simulations.

Areas for further research include development of criteria for selecting smoothing parameters, refinement or expansion of the neighborhood system to more accurately describe prior image knowledge, and assessment of restoration uncertainty.

The algorithm is currently being adopted for application to positron emission tomography.

APPENDIX

The derivation of $p(\mathbf{n} \mid \lambda, \mathbf{y})$ is relatively straightforward. To simplify notation, treat $\mathbf{n}$, λ, and $\mathbf{y}$ as vectors rather than arrays, let $\{\theta_i\}$ represent the corresponding partition of Θ, and let p_i, $i=1, \ldots, k$ denote the probability that a count originating from θ_i is recorded at site 1. Also, let y_1^i denote the number of counts recorded at site 1 that originated in θ_i. Then

$$\mathbf{P}_\lambda\left[y_1^1=m_1, \ldots, y_1^k=m_k \,\middle|\, \sum_{i=1}^k y_1^i=y_1\right]=\frac{\mathbf{P}_\lambda\left[y_1^1=m_1, \ldots, y_1^k=m_k, \sum_{i=1}^k y_1^i=y_1\right]}{\mathbf{P}_\lambda\left[\sum_{i=1}^k y_1^i=y_1\right]} \quad (24)$$

The numerator can be written as

$$\mathbf{P}_\lambda\left[y_1^1=m_1, \ldots, y_1^k=m_k, \sum_{i=1}^k y_1^i=y_1\right]$$

$$=\sum_{n_1=m_1}^{\infty} \cdots \sum_{n_k=m_k}^{\infty} \mathbf{P}_\lambda\left[y_1^1=m_1, \ldots, y_1^k=m_k, \sum_{i=1}^k y_1^i=y_1 \,\middle|\, n_1, \ldots, n_k\right]\mathbf{P}_\lambda\left[n_1, \ldots, n_k\right] \quad (25)$$

$$=\sum_{n_1=m_1}^{\infty} \cdots \sum_{n_k=m_k}^{\infty}\left[\prod_{i=1}^k \frac{n_i!}{m_i!(n_i-m_i)!} p_i^{m_i}(1-p_i)^{n_i-m_i} \frac{e^{-\lambda_i}\lambda_i^{n_i}}{n_i!}\right], \text{ subject to } \sum m_i = y_1 . \quad (26)$$

$$= \sum_{n_1=m_1}^{\infty} \cdots \sum_{n_k=m_k}^{\infty} \left\{ \prod_{i=1}^{k} \frac{e^{-\lambda_i}(\lambda_i p_i)^{m_i} [\lambda_i(1-p_i)]^{n_i-m_i}}{m_i!\,(n_i-m_i)!} \right\} \tag{27}$$

But since

$$\sum_{n_i=m_i}^{\infty} \frac{[\lambda_i(1-p_i)]^{n_i-m_i}}{(n_i-m_i)!} = e^{\lambda_i-\lambda_i p_i} \quad , \tag{28}$$

Eq. (2) simplifies to

$$\prod_{i=1}^{k} \frac{e^{-\lambda_i p_i}(\lambda_i p_i)^{m_i}}{m_i!} \quad , \text{ where } \sum m_i = y_1 \,. \tag{29}$$

Letting $\sum \lambda_i p_i = c$, the denominator in Eq. (24) can be expressed

$$\mathbf{P}_\lambda\left[\sum_{i=1}^{k} y_1^i = y_1\right] = \sum_{m_1=0}^{y_1} \sum_{m_2=0}^{y_1-m_1} \cdots \sum_{m_k=y_1-m.}^{y_1-m.} \prod_{i=1}^{k-1} \frac{e^{-\lambda_i p_i}(\lambda_i p_i)^{m_i}}{m_i!} \frac{e^{-\lambda_k p_k}(\lambda_k p_k)^{y_1-m.}}{(y_1-m.)!} \tag{30}$$

where $m. = \sum_{i=1}^{k-1} y_1^i$. If the exponential terms are factored out of the sum, the remaining terms represent the multinomial expansion of $(\lambda_1 p_1+, \ldots, +\lambda_k p_k)^{y_1}$. Therefore,

$$\mathbf{P}_\lambda\left[\sum_{i=1}^{k} y_1^i = y_1\right] = e^{-c}\,(\lambda_1 p_1+, \ldots, +\lambda_k p_k)^{y_1} \quad , \tag{31}$$

and

$$\mathbf{P}_\lambda\left[y_1^1 = m_1, \ldots, y_1^k = m_k \,\middle|\, \sum_{i=1}^{k} y_1^i = y_1\right] = \frac{\prod_{i=1}^{k} \frac{(\lambda_i p_i)^{m_i}}{m_i!}}{(\lambda_1 p_1+, \ldots, +\lambda_k p_k)^{y_1}} \quad . \tag{32}$$

But since $\sum m_i = y_1$, this implies that

$$\mathbf{P}_{\lambda}\left[y_1^1=m_1,\ldots,y_1^k=m_k \,\middle|\, \sum_{i=1}^{k} y_1^i=y_1\right]=\prod_{i=1}^{k}\frac{\left[\frac{\lambda_i p_i}{(\lambda_1 p_1+,\ldots,+\lambda_k p_k)}\right]^{m_i}}{m_i!} \quad . \tag{33}$$

The last equation shows that the counts observed at site 1 can be distributed to the partitions $\{\theta_i\}$ according to a multinomial distribution with probability vector

$$\frac{\lambda_i p_i}{(\lambda_1 p_1+,\ldots,+\lambda_k p_k)} \quad . \tag{34}$$

Noting that $n_j = \sum_{i=1}^{k} y_i^j$, this demonstrates that a sample of $\mathbf{n}$ can be obtained by redistributing the counts $\mathbf{y}$ based on the current estimate of $\boldsymbol{\lambda}$.

ACKNOWLEDGMENTS

Computations for this work were performed using computer facilities supported in part by National Science Foundation Grants number DMS 87-03942 to the Department of Statistics at The University of Chicago and Department of Energy Grants DE-FG02-86ER60418 and DE-FG02-86ER60438 to the Franklin Mclean Memorial Research Institute at The University of Chicago. The work of Wing Wong was supported by a Guggenheim Fellowship and National Science Foundation Grants number DMS 86-01732.

REFERENCES

Besag, J. (1974). Spatial interaction and the statistical analysis of lattice systems, J. Royal Statist. Soc. series B 36, pp. 192-326.

Besag, J. (1986). On the statistical analysis of dirty pictures, J. Royal Statist. Soc. series B 48, pp. 259-302.

Geman, S., and Geman, D. (1984). Stochastic relaxation, Gibbs distributions, and the Bayesian restoration of images, IEEE Trans. Pattern anal. Machine Intell. 6, pp. 721-741.

Geman, S., and McClure, D.E. (1985). Bayesian image analysis: An application to single photon emission tomography. Proc. Amer. Statist. Assoc. Statistical Computing Section, pp. 12-18.

Rao, C.R. (1973). Linear statistical inference and its applications, second edition, John Wiley and Sons, New York.

Tanner, M., and Wong, W.H. (1987). Calculation of posterior distributions by data augmentation, J. Am. Stat. Assoc. 82, pp. 528-540.

Vardi, Y., Shepp, L.A., and Kaufman, L. (1985). A statistical model for positron emission tomography, J. Am. Stat. Assoc. 80, pp. 8-25.

THE CONCEPT OF CAUSALITY IN IMAGE RECONSTRUCTION

Jorge Llacer, Eugene Veklerov and Jorge Nunez[1]

Engineering Division
Lawrence Berkeley Laboratory
Berkeley, CA 94720 USA

ABSTRACT

Causal images in emission tomography are defined as those which could have generated the data by the statistical process that governs the physics of the measurement. The concept of causality was previously applied to deciding when to stop the MLE iterative procedure in PET. The present paper further explores the concept, indicates the difficulty of carrying out a correct hypothesis testing for causality, discusses the assumption needed to justify the tests proposed and discusses a possible methodology for a justification of that assumption. The paper also describes several methods that we have found to generate causal images and it shows that the set of causal images is rather large. This set includes images judged to be superior to the best maximum likelihood images, but is also includes unacceptable and noisy images. The paper concludes by proposing to use causality as a constraint in optimization problems.

1. INTRODUCTION

Over the past two years we have been involved in a detailed study of the behavior of solutions to the image reconstruction problem in Emission Tomography (ET) by the Maximum Likelihood Estimator (MLE) and Maximum Entropy (ME) methods, with data generated both by computer simulation and by the ECAT-III tomograph of UCLA. Initially, the study focused on explaining the reasons for image deterioration after a certain number of iterations in the MLE method and on finding a cure for this problem.

As we progressed into the study of this phenomenon, we became aware of the fact that the observed deterioration is a direct consequence of trying to find an image which would have the highest probability of having generated

1) J. Nunez is on leave from the Faculty of Physics, University of Barcelona, Spain.

the initial noisy data (which the MLE is designed to do), but, in an apparent contradiction, a true radioactive source equal to the image found could not have yielded the experimental data by a Poisson disintegration process in a statistical sense. This contradiction results from the simple fact that the likelihood criterion favors images whose forward projections are as close to the data as possible, whereas the Poisson assumption imposes a certain deviation between the two.

Veklerov and Llacer (1987) observed that the images recovered after iterations within a certain range could have generated the data as ascertained by a hypothesis testing procedure that, upon further investigation, turns out to be based on an assumption discussed below. That initial finding led them to formulating a stopping rule for the MLE algorithm. We consider that for a radiologist to have confidence in the results of an image reconstruction, it is necessary but not sufficient that the recovered image, if it were a true emission source, could have generated the data. This has led Llacer and Veklerov (1988) to formulating the concept of causality. In this paper we explore that concept in some detail, discuss the difficulty in carrying out a correct test of causality within the framework of hypothesis testing, describe the assumption that must hold for the present tests to have statistical meaning and indicate a methodology that may justify that assumption. We are encouraged in that endeavor by the fact that causal images include some of the visually best reconstructions that we have seen, both from simulated and from true PET data.

2. DEFINITIONS OF CAUSALITY

In this section we define causality at three different levels of strictness without regard to the possible methods by which the hypothesis tests indicated or implied can be carried out. We also show schematically the region in projection space that includes causal images. Throughout this paper we shall use the standard notation introduced by Shepp and Vardi (1982):

$n^*(d),(d=1,\ldots,D)$	- the projection data or the number of coincidences detected in tube d;
$\lambda(b),(b=1,\ldots,B)$	- the emission density;
$p(b,d)$	- the transition matrix;
$\lambda^*(d) = \sum_{b=1}^{B} \lambda(b)p(b,d)$	- the means or forward projections;

where B and D are the number of pixels and the number of projections, respectively. We shall consider specifically the case of emission tomography (PET or SPECT) in which disintegration data follow Poisson statistics.

Definition 1: The image $\lambda(1), \lambda(2),\ldots, \lambda(B)$ is said to be a causal image with respect to data, $n^*(1), n^*(2),\ldots, n^*(D)$, if and only if the statistical hypothesis that $n^*(1), n^*(2),\ldots, n^*(D)$ is a Poisson sample with the means $\lambda^*(1), \lambda^*(2),\ldots, \lambda^*(D)$, respectively, can be accepted (not rejected).

The rationale behind Def. 1 is as follows: Let us assume that $\lambda^*(1) = \lambda^*(2) = \ldots = \lambda^*(D) = 100$ and all $n^*(d)$ are between 99 and 101. Then we have to conclude that the corresponding image is not causal because the deviation between the data and the means is less than what is imposed by the Poisson assumption. If, on the other hand, all $n^*(d)$ are either 50 or 150 in the same example, then the corresponding image is not causal either because the deviation between the data and the means is greater than what is imposed by the Poisson assumption. Put another way, a causal image is an image such that the distance between the data and the corresponding forward projection is "just right", not too small and not too large.

Definition 1 specifies the fullest form of causality or strong causality. That form appears difficult to implement as a constraint in an optimization problem. For that reason we introduce a new causality definition by relaxing Def. 1; since the new definition of causality requires less than Def. 1, it is appropriate to call this new form of causality "weak causality".

Definition 2: The image $\lambda(1), \lambda(2),\ldots, \lambda(B)$ is said to be a weakly causal image with respect to data, $n^*(1), n^*(2),\ldots, n^*(D)$, if and only if the second moments of $n^*(1), n^*(2),\ldots, n^*(D)$ are consistent with the Poisson hypothesis, namely:

$$\sum_{d=1}^{D} \frac{[n^*(d) - \lambda^*(d)]^2}{\lambda^*(d)} \simeq D \tag{1}$$

Indeed, the expected value of the numerator of each term in Eq. (1) is the variance, while that of the denominator is the mean. Therefore, Eq. (1) must be satisfied if the Poisson hypothesis holds. However, the reverse statement is not true, because the equality of the variance and the mean is necessary but not sufficient for a distribution to be Poisson or, put another way, Def. 2 is less demanding than Def. 1.

It is possible to strengthen Def. 2 by requiring that higher moments of n*(1), n*(2),..., n*(D) be consistent with the Poisson hypothesis as well.

Definition 3: The image $\lambda(1), \lambda(2), \ldots, \lambda(B)$ is said to be a k-causal image with respect to data, n*(1), n*(2),..., n*(D), if and only if the first k moments of n*(1), n*(2),..., n*(D) are consistent with the Poisson hypothesis, namely:

$$\sum_{d=1}^{D} \frac{[n^*(d) - \lambda^*(d)]^i}{f_i[\lambda^*(d)]} \approx D \qquad (2)$$

where $i = 1, \ldots, k$ and $f_i(x)$ is the expression for the i-th moment of the Poisson distribution with the mean x.

The first moment meets this requirement automatically for any unbiased image. Here are the first several $f_i(x)$: $f_2(x) = x$, $f_3(x) = x$, $f_4(x) = 3x^2 + x$, $f_5(x) = 10x^2 + x$. Note that a 2-causal image is simply a weakly causal image and when $k \to \infty$, a k-causal image becomes strongly causal. The main advantage of Def. 3 over Def. 1 is that the former is easier to implement as a constraint of an optimization problem.

Geometrically, all weakly causal images belong to an area resembling an ellipsoid and surrounding the maximum likelihood point. Indeed, if the left-hand side of Eq. (1) must equal the right-hand side to the accuracy of ε, all points satisfying Eq. (1) form the space between two egg-shaped surfaces. We will call this space the causality shell. The k-causal and strongly causal images form a subset of the shell with gaps between them. Figure 1 shows an example of a shell as well as lines along which likelihood is constant, which are regular ellipsoids for large values of data.

3. THE TESTS FOR CAUSALITY

Let us outline the procedure for testing for strong causality described in detail by Veklerov and Llacer (1987). We will follow this outline with a critical discussion of its validity. We want to test the hypothesis that n*(1) is a realization drawn from a Poisson distribution with the mean $\lambda^*(1)$, n*(2) drawn from a Poisson distribution with the mean $\lambda^*(2)$, etc. A widely used method for testing such a hypothesis, known as Pearson's goodness of fit tests, see e.g. Canavos (1984), is to define several mutually

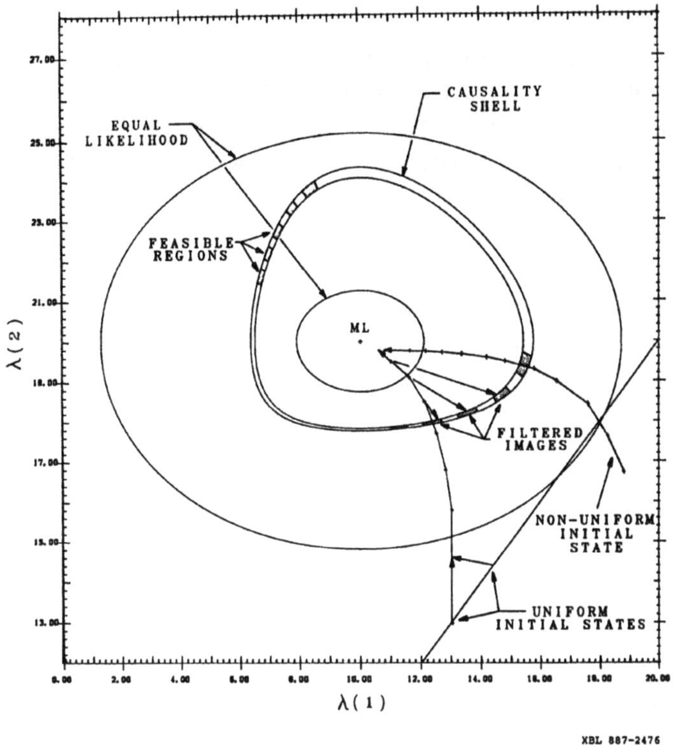

Fig. 1: Causality shell and ellipses along which likelihood is constant. MLE paths for different initial images. The results of filtering of "overdone" images after different MLE iterations are found to be causal.

exclusive and exhaustive classes, assign each realization to a class and then compare the observed numbers of realizations assigned to the classes with the expected numbers determined by the Poisson hypotheses. The discrepancy between the two is measured by the chi-square statistic. When the chi-square statistic exceeds a certain threshold, the hypothesis is rejected. Some technical difficulties arise in trying to adapt Pearson's procedures to our needs, but they are resolved in our previously referenced work.

A fundamental objection to the above procedure (pointed out to us by several workers) stems from the fact that the image which we are trying to test against has been generated by the same set of data that we are testing. Under those conditions, the question implied by testing for any of the definitions of causality given above cannot even be posed. Cramer (1946) proves that the chi square test, when certain parameters that describe the function against which we are testing are estimated from the sample, can only be used to test the hypothesis that the data are drawn from the defined probability distribution with some values of those parameters. In our case this implies that, with the parameters $\lambda^*(d)$ having been derived from the data $n^*(d)$, we can only test the hypothesis that the data were obtained from a Poisson distribution with some values of $\lambda^*(d)$ unknown to us.

This limited test seems totally unnecessary since we know that the data were really generated by a Poisson disintegration process. It follows that, by using the appropriate number of degrees of freedom, a hypothesis test comparing the parameters $\lambda^*(d)$ generated by the MLE method (at convergence) to the data $n^*(d)$ should never be rejected if, indeed, the MLE is a good method for estimating those parameters. We shall pursue this test in the future.

The hypothesis tests that can be derived from the above definitions of causality would have complete validity if the parameters that determine the Poisson process against which we are testing were completely and independently specified. In our case that implies that we have a complete and independent knowledge of the activity distribution that generated the data. If that were the case, a test would again be unnecessary. We know that the data would have been generated by such a source distribution. Indeed, the whole image reconstruction process would be unnecessary.

Because of our observations that images generated by the MLE (and other methods as described below), when started from a uniform image field and stopped according to the rule derived from strong causality hypothesis testing, are visually good, with a useful compromise between edge sharpness and low noise in regions of high activity, we have to believe that our concept of causality, although not fitting correctly in the framework of hypothesis testing, has a definite value. Results similar to ours have also been obtained by Hebert et al (1988) in Single Photon Emission Tomography (SPECT). Many other researchers (mostly in the field of Astronomy) have come up with variations of the same tests which they proposed to use as constraints or necessary conditions for image reconstruction, or as an analysis of residues: Skilling and Bryan (1984), Ables (1974), Gull and Daniel (1978), Narayan and Nityanda (1986), Reiter and Pfleiderer (1986) and Tarantola (1987).

The assumption that we believe can lead to a justification of our hypothesis testing is related to the case of a completely specified image, as indicated above, and is the following:

> We are given an image and its projection $\lambda^*(d)$. We do not know how it was obtained. We are also given a set of data $n^*(d)$ and we are asked to test the hypothesis that that image, as if completely specified, could have generated the data. We would then proceed with testing the hypothesis in the methods developed or implied for Definitions 1 through 3.

The above assumption would be justified in practice if an analysis of the distribution of the hypothesis testing function H of Veklerov and Llacer (1987) (for Def. 1) or of the chi square function values (for Def. 2) for reconstructions of a large number of different realizations of a fixed image source distribution were found to be chi square distributed with the expected number of degress of freedom. Such a finding would imply that the strict theory of hypothesis testing can be relaxed under some conditions. We are now beginning such a study.

4. ARE ALL CAUSAL IMAGES GOOD?

The causal images obtained by stopping the MLE algorithm at the right point by no means exhaust the set of causal images. We have conducted two series of experiments in order to get an idea of how different causal images can be from one another. We have used a mathematical brain-like phantom with 1 million counts shown in Fig. 5a.

The first series consisted of images generated by the same MLE algorithm but with drastically different initial images. We chose a four quadrant checkerboard image as an initial image with 1.95 and 0.05 times the average number of counts per pixel in the "hot" and "cool" quadrants, respectively.

It converged towards the same maximum likelihood image as in the case of a uniform initial image, but its MLE path crossed the shell at a different point, as is shown schematically in Fig. 1. Visually, a causal image obtained with the checkerboard image used as the initial one still retains the boundaries between the quadrants. Figure 2a shows the image after iteration 2, and 2b shows it after iteration 50, where the image is causal (H = 15.2). The quadrant boundaries gradually disappear as the iterative process progresses but they are still visible at iteration 300.

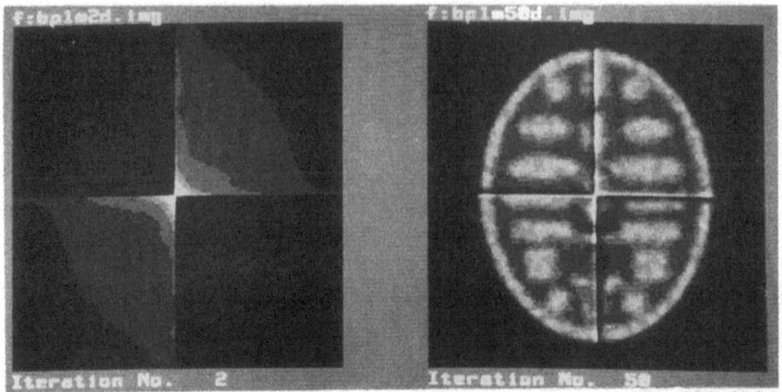

XBB 889-8516

Fig. 2: MLE reconstructions with the initial image being a 4 quadrant checkerboard. Image a) is after iteration 2; Image b) is after iteration 50; it is causal (H = 15.2).

The second series of experiments owes its origin to the idea put forth in Snyder et al (1987), whereby the use of a sieve and a resolution kernel is equivalent to post-filtering of the image generated by the regular MLE algorithm. Using that idea, we were able to produce a number of causal images as follows. We let the MLE process overshoot the causality shell. Thereafter, several "overdone" images were filtered by convolution with Gaussian distributions. By choosing the right standard deviations of the Gaussian distributions we were able to bring the images back to the shell, as is also shown schematically in Fig. 1.

It is interesting to note that it was necessary to use different values for the standard deviation to bring different MLE images back to the shell: the greater values were needed for images generated after more iterations. Figure 3 shows the value of the H function as a function of the standard deviation σ for several MLE images used in the convolution operation. The resulting causal images are shown in Fig. 4. Causal images obtained by filtering reconstructions with an increasing number of iterations show increasing noise. Image a) is a causal MLE image obtained by the stopping rule, image b) is an "overdone" MLE image after 50 iterations, images c), d), e) and f) are MLE images after 50, 100, 200 and 300 iterations, respectively, filtered to make them causal. The cleanest and sharpest image is image c). Furthermore, it is the "best" image we have ever seen for the data set at hand.

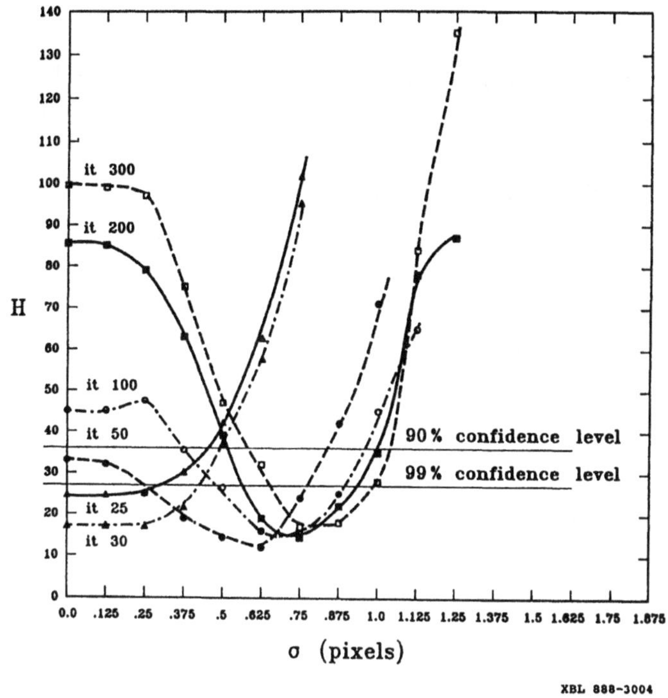

Fig. 3: Results of Gaussian filters with different standard deviations applied to MLE reconstructions after certain iteration numbers: 25, 30, 50, 100, 200 and 300. The hypothesis testing parameter H of the resulting images is shown as a function of the standard deviation. Different curves correspond to different iteration numbers.

5. MAXIMUM ENTROPY RECONSTRUCTION AND CAUSALITY

Another attempt to arrive at causal images was made by the use of maximum entropy techniques. The maximum entropy solution seems attractive because it yields the image with the lowest information content compatible with the data. Thus, the criterion will give an image which avoids any bias while satisfying the imposed constraints.

Following Frieden (1972) and Gullberg and Tsui (1987), we formulate the problem as follows:

Maximize:

$$-\sum_{b=1}^{B} \frac{\lambda(b)}{N_\lambda} \ln \frac{\lambda(b)}{N_\lambda} - \rho \sum_{d=1}^{D} \frac{n(d)}{N_n} \ln \frac{n(d)}{N_n} \tag{3}$$

where

$$N_\lambda = \sum_{b=1}^{B} \lambda(b); \quad N_n = \sum_{d=1}^{D} n(d); \ \lambda(b) \geq 0; \ n(d) \geq 0$$

subject to

$$\sum_{b=1}^{B} P(b,d)\lambda(b)+n(d)-n_m = n^*(d); \ d=1,\ldots, D \qquad (4)$$

The parameter ρ is the weight quantifying the relative importance of the entropy of the noise vs. that of the image. The true noise or the difference between the data and the forward projections can be represented as $n(d) - n_m$, where $n(d)$ is a biased noise term and n_m is a constant noise bias which insures that $n(d)$ is always positive.

Due to the Poisson nature of the data, we have chosen:

$$n_m = \max \ [2\sqrt{n^*(d)}] \ d = 1,\ldots, D$$

and

$$\frac{\sum_{d=1}^{D} (n(d) - n_m)}{D} = 0 \ ; \qquad N_n = \sum_{d=1}^{D} n(d) = D \ n_m$$

The constant ρ effectively controls the smoothness of the solution: the larger the constant, the less smooth image will result for a given noise. If the data contains very little noise, ρ can be made very high thus obtaining high sharpness. The bias n_m also controls the smoothness of the image. In this case, the large n_m, the smoother the image, with a background being higher.

Note that in this model, the noise $n(d)$ is a variable in the optimization problem and the solution gives unbiased estimates for both the image and the noise. However, the present model does not contain the information about the Poisson nature of the data.

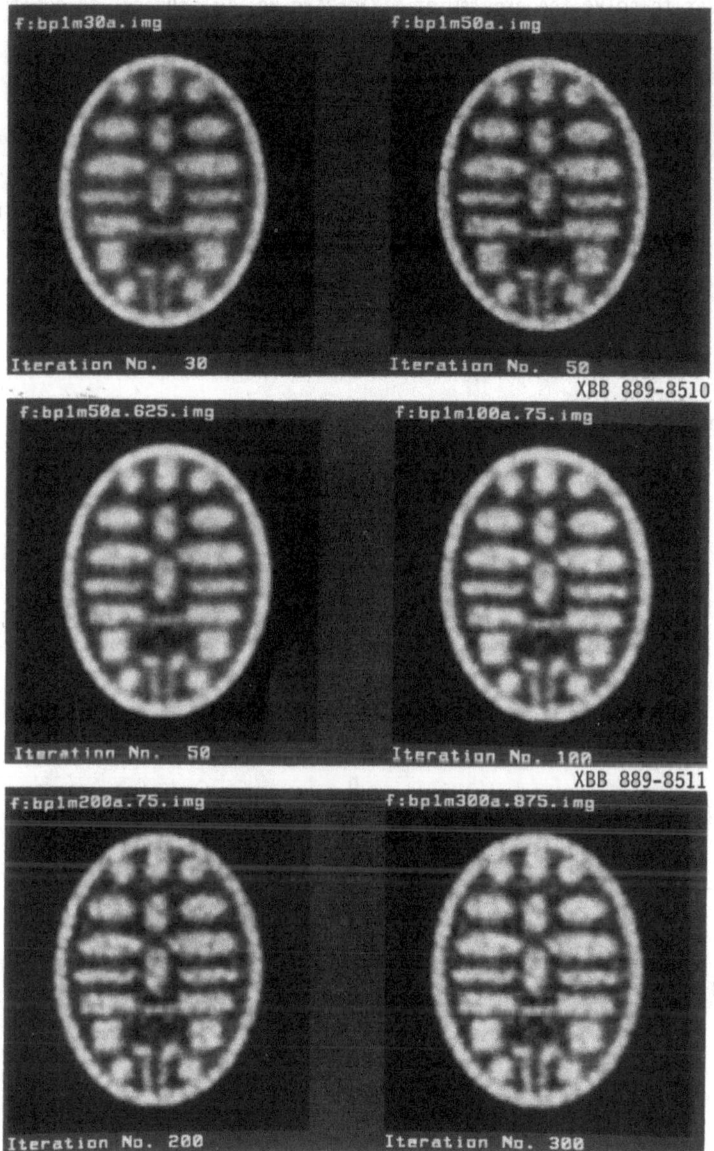

Fig. 4: Causal images obtained by the MLE or the MLE and subsequent filtering. a) MLE and the stopping rule (30 iterations, H = 17.5); b) MLE in the rising part of H (50 iterations, H = 30.25); c) is image b) filtered with sigma = 0.625 with resulting H = 13.2; d) MLE image after 100 iterations filtered with sigma = 0.75; e) MLE image after 200 iterations filtered with sigma = 0.75; f) MLE image after 300 iterations filtered with sigma = 0.875. Images d), e) and f) are all causal but with increasing noise.

In order to solve the problem of maximizing Eq. (3) subject to constraints in Eq. (4), we introduce the Lagrangian function and solve the unconstrained problem of maximizing:

$$L = -\sum_{b=1}^{B} \frac{\lambda(b)}{N_\lambda} \ln \frac{\lambda(b)}{N_\lambda} - \rho \sum_{d=1}^{D} \frac{n(d)}{N_n} \ln \frac{n(d)}{N_n}$$

$$- \sum_{d=1}^{D} \mu_d \, [n^*(d) - n(d) + n_m - \sum_{b=1}^{B} P(b,d)\lambda(b)] \, \frac{1}{N_\lambda}$$

where $\mu(d)$ $d = 1, \ldots, D$ are Lagrange multipliers.

The maximum of L is achieved when:

$$\frac{\partial L}{\partial \lambda(b)} = 0 \qquad b = 1, \ldots, B$$

$$\frac{\partial L}{\partial n(d)} = 0 \qquad d = 1, \ldots, D$$

$$\frac{\partial L}{\partial \mu(d)} = 0 \qquad d = 1, \ldots, D$$

which have the following solution:

$$\lambda(b) = N_\lambda e^{-1} \prod_{d=1}^{D} e^{P(b,d)\mu(d)}$$

$$n(d) = N_n e^{-1} e^{\mu(d) N_\lambda / \rho N_n} \tag{5}$$

where the Lagrange multiplier is determined by the system of D nonlinear equations:

$$N_\lambda e^{-1} \sum_{b=1}^{B} P(b,d) \prod_{m=1}^{D} e^{P(b,m)\mu(m)}$$

$$+ N_n e^{-1} e^{\mu(d) N_\lambda / \rho N_n} - n_m - n^*(d) = 0 \qquad d = 1, \ldots, D \tag{6}$$

The solution of the system as shown in Eq. (6), is derived using the non-linear Gauss-Seidel-Newton iterative algorithm found in Ortega and Rheinboldt (1970). As the iterative algorithm for obtaining the Lagrange multipliers progresses, we can substitute the multipliers in Eq. (5) and monitor the formation of the image as well as its various statistics.

Figure 5b shows the maximum entropy reconstruction of the same phantom as the one used in the previous reconstructions in the absence of noise at convergence after 10 iterations with $\rho = 50$. Note that the reconstruction is nearly perfect. This shows that the algorithm works very well in the case of little noise where it is possible to use large values of ρ thereby achieving high sharpness in the reconstructed image.

In the presence of noise, the behavior of the present version of the algorithm is not adequate due to the simplicity of the noise model. We have obtained promising results by adjusting ρ and smoothing the final image.

The images in Figure 5c through 5f are the ME reconstructions of the same phantom of 5a with noise (Poisson data), with $\rho = 8$. Image c) is generated after only one iteration in the calculation of the Lagrange multiplier and d) after twenty iterations at convergence. Image e) is obtained by filtering image d) by convolving it with the Gaussian distribution with $\sigma = 0.75$ pixels. Image f) is obtained by increasing the bias parameter n_m by 25% and filtering. Note that the structures of the phantom are clearly visible but we have found that none of the images are causal.

We have just recently concluded the first reconstruction experiments with a fully Bayesian reconstruction method, using an entropy prior distribution and the Likelihood criterion as the conditional probability of the data, given an image. An adjustable parameter which, in effect, changes the weight of one of the criteria with respect to the other and appears justifiable in terms of intensity jumps of the image and data has been used in the problem formulation. This work is still in progress but we can report that, by proper choice of the adjustable parameter, it is possible to obtain images that converge (in an iterative formulation of the reconstruction process) to stable causal images of a quality very similar to that of the MLE images of Fig. 4 c). We have, thus, another method to obtain causal images.

6. CAUSALITY-BASED ALGORITHMS

The experiments described in Sect. 4 have demonstrated that the set of causal images is rather large. It includes the best images we have ever seen, which are clearly superior to MLE images, but it also includes unacceptable images as well as noisy ones. This suggests that causality is a valuable concept in image reconstruction but, taken alone, it is not enough to guarantee good reconstructions.

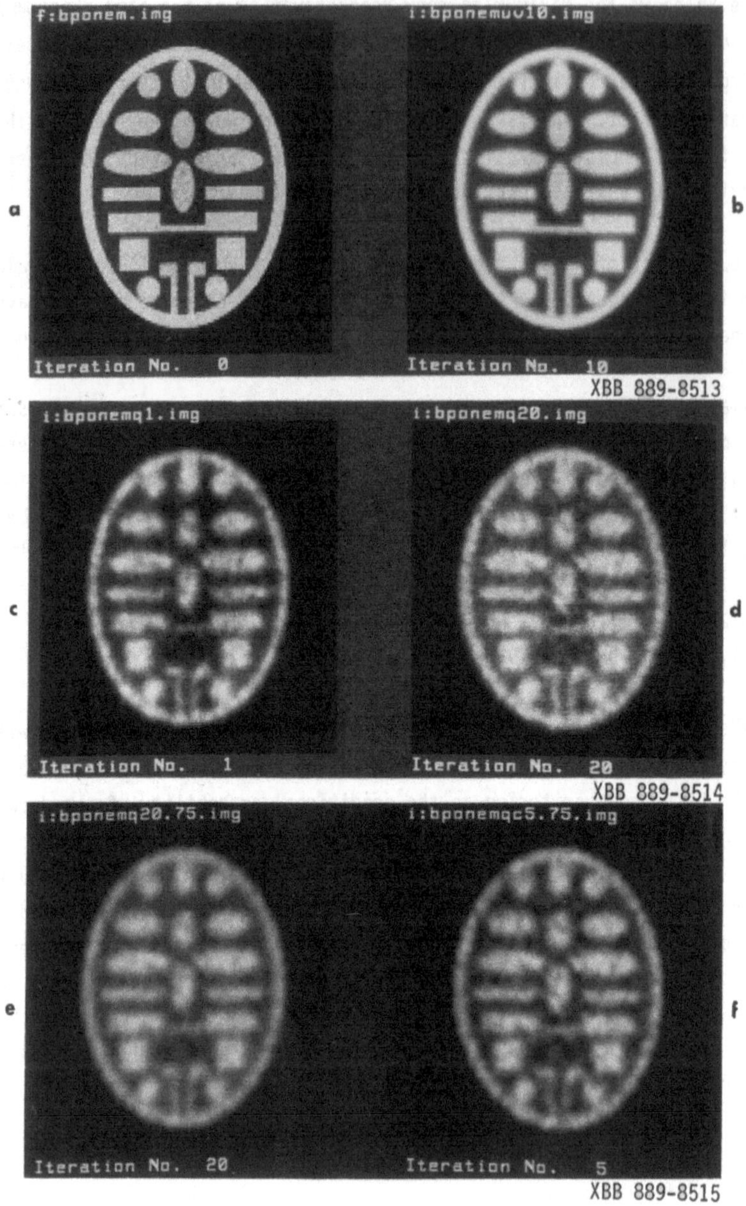

Fig. 5: ME images. a) source image with 1M counts; b) reconstruction with noiseless data; c) 1st iteration for obtaining Lagrange multipliers with H = 103; d) 20th iteration in Lagrange multipliers with very high H; e) 20 iterations filtered with sigma = 0.75 with H still too high (1142); f) obtained by increasing the bias parameter by 25% and filtering.

Therefore, it is promising to use causality as a constraint and try to optimize another criterion, such as entropy, while remaining on the shell. Then it may be more convenient, from the computation complexity point of view, to limit ourselves to weak or k-causality, that is:

Maximize:

$$- \sum_{b=1}^{B} \lambda(b) \ln \lambda(b)$$

subject to

$$\sum_{d=1}^{D} \frac{[n^*(d) - \lambda^*(d)]^i}{f_i[\lambda^*(d)]} \approx D$$

for $i = 1, \ldots, k$.

This approach has been proposed by Skilling and Bryan (1984) who, using our terminology, considered only weak causality ($k = 2$) and in the Gaussian rather than Poisson case.

ACKNOWLEDGMENTS

This work has benefited from discussions which J. Llacer has held with J. Skilling, S. Gull, D. Snyder and M. Miller. The authors are particularly grateful to Harrison Barrett for persistent and fair criticism of the work leading to the present paper, resulting in a much better understanding of the essential nature of the problem.

This work has been supported by a grant from the National Cancer Institute (CA-39501) and the U.S. Department of Energy under Contract No. DE-AC03-76SF00098. The work of J. Nunez has been supported in part by the Catalan Studies Program (U. of California and the Generalitat de Catalunya).

REFERENCES

Ables. J.G., (1974). Maximum entropy spectral analysis, Astron. Astrophys. Suppl., 15, pp. 383-393.
Cavanos, G., (1984). Applied probability and statistical methods, Little, Brown & Co.
Cramer, H., (1946). Mathematical Methods of Statistics, Princeton University Press.
Frieden, B.R., (1972). Restoring with maximum likelihood and maximum entropy, J. Opt. Soc. Am., 62, pp. 511-518.

Gull, S.F. and Daniell, G.J.,(1978). Image reconstruction from incomplete and noisy data, Nature, 272, pp. 686-690.
Gullberg, G.T. and Tsui, B.M.W., (1987). Maximum entropy reconstruction with constraints: iterative algorithms for solving the primal and dual programs, Proc. of Xth International Conference on Information Processing in Medical Imaging, Ultrecht, The Netherlands, pp. 181-199.
Hebert, T., Leahy, R. and Singh, M., (1988). Fast MLE for SPECT using an intermediate polar representation and a stopping criterion, IEEE Trans. Nucl. Sci., Feb. 1988.
Llacer, J. and Veklerov, E., (1988). The use of a stopping rule in iterative image reconstruction, LBL Report LBL-25314, presented at the 1988 AMS-IMS-SIAM Conference on Spatial Statistics and Imaging, to be publ.
Narayan, R. and Nityananda, R., (1986). Maximum entropy image restoration in astronomy, Ann. Astron. Astrophys., 24, pp. 127-170.
Ortega, J.M. and Rheinboldt, W.C., (1970). Iterative solutions of nonlinear equations in several variables, Academic Press, New York.
Reiter, J. and Pfleiderer, J., (1986). Improvement of MEM-deconvolution by an additional constraint, Astron. Astrophys., 166, pp. 381-392.
Shepp, L.A. and Vardi, Y., (1982). Maximum likelihood reconstruction for emission tomography, IEEE Trans. Med. Imaging, MI-1, pp. 113-122.
Skilling, J. and Bryan, R.K., (1984). Maximum entropy image reconstruction: general algorithm, Mon. Not. R. Astr. Soc, 211, pp. 111-124.
Snyder, D.L., Miller, M.I., Thomas, L.J. and Pollite, D.G., (1987). Noise and edge artifacts in maximum-likelihood reconstructions for emission tomography, IEEE Trans. Med Imaging, MI-6, pp. 228-238.
Tarantola, A., (1987). Inverse problem theory, Elsevier.
Veklerov, E. and Llacer, J., (1987)., Stopping rule for the MLE algorithm based on statistical hypothesis testing, IEEE Trans. Med. Imaging, MI-6, pp. 313-319.

ON THE RELATION BETWEEN ART, BLOCK-ART AND SIRT

Marcel C.A. van Dijke and Henk A. van der Vorst

Faculty of Technical Mathematics and Informatics,
Delft University of Technology,
Delft, The Netherlands

and

Max A. Viergever

Faculty of Medicine,
University of Utrecht,
Utrecht, The Netherlands

ABSTRACT

In this paper a theoretical comparison between ART, block-ART and SIRT is presented, by expanding the iteration matrices of the methods in the relaxation parameter. If ART, block-ART and SIRT are applied with the same relaxation parameter $\lambda \ll 1$, the three methods are practically equivalent. Such a small parameter is required for reasons of convergence for SIRT and to a lesser extent for block-ART. ART has the advantage that the relaxation parameter can be varied over a wider range. In problems with a large amount of noise, however, strong underrelaxation has to be applied to regularize the method. This confines the superiority of ART over SIRT and block-ART to systems of equations which are mildly inconsistent. A disadvantage of ART is that it is not suited for parallel implementation.

1. THE RECONSTRUCTION PROBLEM

The customary reconstruction procedures for most imaging modalities are based on discrete integral transforms, notably inverse Fourier and Radon transforms. The computational speed of these methods is higher than that of any iterative technique, especially since they may be implemented using the fast Fourier transform. Accordingly, iterative techniques are only used in particular cases where transform methods do not perform well: e.g., in severely incomplete data problems such as limited angle tomography, in imaging modalities where refraction plays a role (e.g. ultrasound, electrical impedance and microwave imaging), and in most forms of emission tomography.

In all of these cases, it has been proven successful to formulate the reconstruction problem in discrete terms (see e.g., Viergever,1988 for a survey). The problem then generally takes the form of a set of linear algebraic equations:

$$Ax = b \tag{1}$$

where x is the image vector, b is the data vector and A is the projection matrix.

The projection matrix A (MxN) is usually large, sparse, unstructured and rank deficient. Due to errors in the data acquisition, the system of equations is inconsistent: there is no exact solution to the problem. The problem of solving such a system is actually an optimization problem.

In this paper, we concentrate on iterative techniques, since direct techniques are too time-consuming and require too much memory allocation. For instance, implementing the reconstruction procedure as a multiplication of the Moore-Penrose inverse (stored on disk) with the data vector leads to larger reconstruction times than with the conventional iterative techniques.

We further confine ourselves to iterative methods converging to a least squares solution of the system $Ax = b$. If the noise in the data is additive, white, zero mean Gaussian, the least squares optimization criterion is equivalent to the maximum likelihood criterion (see Viergever,1988). The two main classes of least-squares iterative methods are Gauss-methods and projection methods. The latter class will be discussed only briefly.

The aim of the present paper is to provide a theoretical framework which enables a comparative study of the various Gauss-methods (ART, block-ART and SIRT).

The organization of the paper is as follows. ART, SIRT and block-ART will be discussed in sections 2,3 and 4 respectively. In particular, it will be shown that these methods are equivalent to Gauss-Seidel or Gauss-Jacobi methods. The expansion of the different iteration matrices in terms of the relaxation parameter will be described in section 5. Section 6 shows the relation between the three methods and section 7 will be devoted to the spectral response of the methods. A numerical experiment on a model problem will be given in section 8. A comparison between the Gauss-methods discussed in this paper and projection methods is presented in section 9 and the paper ends with the summary of conclusions in section 10.

2. ALGEBRAIC RECONSTRUCTION TECHNIQUE (ART)

The standard ART technique (Kaczmarz, 1937; Gordon, Bender and Herman,1970) for solving least squares problems iteratively, is a row-action method. The vector of unknowns is updated at each equation of the system in such a way that this one equation is satisfied exactly, after which the next equation is addressed. The ART algorithm is described by:

$$x^{k+1} = x^k + A_{i.}^{\dagger}(b_i - A_{i.}x^k); \quad i = (k \bmod M) + 1 \ ; k = 0,1,2,3,... \tag{2}$$

where $A_{i.}^{\dagger} = \dfrac{A_{i.}^T}{\|A_{i.}^T\|^2}$ is the Moore-Penrose or pseudo-inverse of row i of the system matrix ($A_{i.}$).

An equivalent formulation is

$$x^{k+1} = x^k + \frac{(b_i - A_{i.}x^k)}{\|A_{i.}^T\|^2} A_{i.}^T \tag{3}$$

We will assume, for convenience of notation, that the equations of (1) have been normalized to have unit norm and we introduce a constant relaxation parameter λ. One iteration step of ART can then be written as

$$\begin{aligned} x^{k+1} &= x^k + \lambda(b_i - A_{i.}x^k)A_{i.}^T \\ &= (I - \lambda A_{i.}^T A_{i.})x^k + \lambda b_i A_{i.}^T \\ &= \Omega_i x^k + \lambda b_i A_{i.}^T \end{aligned} \tag{4}$$

with $\Omega_i = I - \lambda A_{i.}^T A_{i.}$ (5)

We shall refer to Ω_i as an "ART-matrix".

The ART iteration cycles (involving all equations of the system) thus take the form

$$x^{(k+1)M} = Q_{art} x^{kM} + b_{art} \tag{6}$$

where $Q_{art} = \prod_{i=M}^{1} \Omega_i = \Omega_M \Omega_{M-1} \Omega_{M-2} \cdots \Omega_2 \Omega_1$

and $b_{art} = \sum_{t=1}^{M-1} \left(\prod_{s=M}^{t+1} \Omega_s \right) \lambda b_t A_{t.}^T + \lambda b_M A_{M.}^T$

The iteration matrix Q_{art} and the vector b_{art} are dependent on the order in which the equations are addressed. Therefore, we will use the notation $Q_{art_{1234}}$ to indicate that the equations are addressed in the order 1,2,3, and 4 respectively.

2.1 Convergence of ART

ART converges for relaxation parameters $0 < \lambda < 2$ since

$$\|Q_{art}\| = \|\prod_{i=1}^{M} \Omega_i\| \leqslant \prod_{i=1}^{M} \|I - \lambda A_{i.}^T A_{i.}\| < 1 \quad (\|A_{i.}^T A_{i.}\| = 1) \tag{7}$$

If the system of equations (1) is consistent, ART converges to a solution of this system. For right choices of the starting vector x^0, ($x^0 \in R(A^T)$) the iteration process converges to the minimum norm solution of (1). If the system is inconsistent, every sub-sequence, for instance the sequence of cycles through the system, converges, but not necessarily to the minimum norm least squares solution (Censor,1981). However, if strong underrelaxation is applied, the limits of the subcycles approach a (weighted) least squares solution of the system (1) (Censor et al., 1983). If the starting vector is chosen properly, the cycles converge to a weighted minimum norm least squares solution of this system.

2.2 ART as a Gauss method.

It can be shown (Bjorck and Elfving, 1979) that ART *without relaxation* applied to $Ax = b$ is equivalent to Gauss-Seidel on the system $AA^T y = b$; $x = A^T y$

The k'th minor step of Gauss-Seidel on $Ax = b$ can be written as:

$$x^{k+1} = x^k + \frac{(b_i - A_{i.} x^k)}{A_{ii}} e_i \quad ; i = (k \ mod \ M) + 1 \tag{8}$$

where e_i is the i'th basis vector. Consider Gauss-Seidel on the system $AA^T y = b$. The k'th minor step can be written as:

$$y^{k+1} = y^k + \frac{(b_i - (AA^T)_{i.} y^k)}{(AA^T)_{ii}} e_i \quad ; i = (k \ mod \ M) + 1$$

$$= y^k + \frac{(b_i - A_{i.} A^T y^k)}{\|A_{i.}^T\|^2} e_i \quad ; i = (k \ mod \ M) + 1 \tag{9}$$

Multiplying by A^T and substituting $x = A^T y$ in eq. 9, one obtains the iteration scheme for ART on $Ax = b$

$$x^{k+1} = x^k + \frac{(b_i - A_{i.}x^k)}{\|A_{i.}^T\|^2} A_{i.}^T \; ; i = (k \bmod M) + 1 \tag{10}$$

However, ART is relaxed in another way than Gauss-Seidel so that relaxed ART is not exactly the same as relaxed Gauss-Seidel (known as SOR).

3. SIMULTANEOUS ITERATIVE RECONSTRUCTION TECHNIQUE (SIRT)

SIRT methods distinguish themselves from ART methods in that they do not update the iterated vector after each equation, but after an entire sweep through all the equations, and thus, during one sweep, they use the same residual vector for each equation. Different members of the class of SIRT methods use different row- (R) and column (C) scalings. The iterations can be written as:

$$x^{k+1} = x^k + \lambda C A^T R (b - Ax^k) \tag{11}$$

The pre-scaling of the system matrix with the matrices R and C influences the distribution of the singular values of the matrix and the solution to which the process converges. The differences in e.g. rate of convergence and sensitivity to noise between different members of the class of SIRT methods are due to the different scalings and not to the method itself. All SIRT-type methods can be related to the method of Cimmino (1938) that has only a row scaling R: the rows are scaled to have unit norm (just like in ART).

$$\begin{aligned} x^{k+1} &= x^k + \lambda A^T R (b - Ax^k) \\ &= (I - \lambda A^T R A) x^k + \lambda A^T R b \end{aligned} \tag{12}$$

We scale the system beforehand such that $R = I$, as in section 2, and rewrite eq.(12) as

$$\begin{aligned} x^{k+1} &= Q_s x^k + b_s \\ \text{with} \quad Q_s &= I - \lambda A^T A \\ \text{and} \quad b_s &= \lambda A^T b \end{aligned} \tag{13}$$

3.1 Convergence of SIRT

The *general* SIRT algorithm can be rewritten as

$$x^{k+1} = (I - \lambda CA^T RA)x^k + \lambda CA^T Rb \tag{14}$$

$$= (I - \lambda W^T W)u^k + \lambda W^T f$$

where $W = R^{1/2}AC^{1/2}$, $u = C^{-1/2}x$ and $f = R^{1/2}b$

This is the Richardson iteration for the system

$$W^T Wu = W^T f \tag{15}$$

which converges to the minimum norm least squares solution of the scaled system $Wu = f$.

3.2 SIRT as a Gauss method

It can be shown that SIRT applied to Ax = b is equivalent to relaxed Gauss-Jacobi (JOR) on

$$AA^T y = b: \; x = A^T y \tag{16}$$

First note that the Gauss-Jacobi matrix of the system $AA^T \; y = b$, is $diag^{-1}(AA^T)$ which is equal to the relaxation matrix R we have chosen for SIRT.

The JOR iterations on this system can be written as:

$$y^{k+1} = y^k + \lambda R \; (b - AA^T \; y^k) \tag{17}$$

Multiplying by A^T and substituting $x = A^T y$, we get

$$x^{k+1} = x^k + \lambda A^T R(b - Ax^k)$$

$$= (I - \lambda A^T RA)x^k + \lambda A^T Rb \tag{18}$$

4. BLOCK-ART

The next iterative reconstruction technique we will consider is amidst ART and SIRT: it applies SIRT (Cimmino) to different groups of equations (blocks) of the system. To derive this block-ART algorithm, we split the system matrix A into K blocks B_i of L equations. (We have restricted ourselves to equal blocksizes for convenience of notation). Let b_i be the corresponding right-hand-side vectors and assume the rows of (1) have been normalized (see section 3).

The block-ART algorithm is given by:

$$x^{k+1} = x^k + \lambda B_i^T(b_i - B_i x^k) \quad \text{with } i = (k \bmod K) + 1 \tag{19}$$

$$= (I - \lambda B_i^T B_i) x^k + \lambda B_i^T b_i$$

$$= P_i x^k + \lambda B_i^T b_i \tag{20}$$

$$\text{with } P_i = I - \lambda B_i^T B_i \tag{21}$$

One iteration cycle (involving all equations) can be written as

$$x^{(k+1)K} = Q_{ba} x^{kK} + b_{ba} \tag{22}$$

$$\text{with } Q_{ba} = \prod_{t=K}^{1} P_t$$

$$\text{and } b_{ba} = \sum_{t=1}^{K-1} \prod_{s=K}^{t+1} (P_s)\ \lambda B_t^T b_t + \lambda B_K^T b_K$$

4.1 Convergence of block-ART

If the system of equations (1) is consistent, then the method converges to a solution of the system $Ax = b$. If in addition $x^0 \in R(A^T)$, the generated solution is the minimum norm solution $A^\dagger b$. If the system of equations is inconsistent, the sequence of subcycles converges, but not necessarily to the minimum norm least squares solution. If, however, block-ART is applied with strong underrelaxation, the subcycles converge to a weighted minimum norm least squares solution (Censor et.al., 1983).

4.2 Block-ART as a Gauss method

Since block-ART is just ART on the blocks of the system and ART was shown to be equivalent to Gauss-Seidel, block-ART is equivalent to block-Gauss-Seidel.

5. EXPANSION IN RELAXATION PARAMETER λ

In this section, the iteration matrices of the three iterative reconstruction methods ART, block-ART and SIRT are considered. They can all be written as a combination of products and sums of ART-matrices. In this way we obtain an expansion of the iteration matrices in the relaxation parameter λ.

We assume again that the rows of the system matrix A are normalized so that the dot product of the vectors $A_{i.}^T$ and $A_{j.}^T$ is equal to:

$$<A_{i.}^T, A_{j.}^T> = A_{i.} A_{j.}^T = \|A_{i.}^T\| . \|A_{j.}^T\| . \cos\beta(A_{i.}^T, A_{j.}^T) =$$

$$= \cos\beta(A_{i.}^T, A_{j.}^T) \equiv c_{ij} \tag{23}$$

We will write the different iteration matrices as combinations of ART-matrices, as defined in (5).

5.1 Algebraic Reconstruction Technique

The iteration matrix Q_{art} consists of a product of ART-matrices (see eq. 10):

$$Q_{art} = \prod_i \Omega_i$$

For example, the product of two ART-matrices can be written as:

$$Q_{art_{ij}} = \Omega_i \Omega_j = (I - \lambda A_{i.}^T A_{i.})(I - \lambda A_{j.}^T A_{j.})$$

$$= \Omega_i + \Omega_j - I + \lambda^2 c_{ij} A_{i.}^T A_{j.} \tag{24}$$

From eq. 24 we see that the iteration matrix of ART can be written as

$$Q_{art} = \prod_{i=M}^{1} \Omega_i = \sum_{i=1}^{M} \Omega_i - (M-1)I + 2^{nd} \text{ and higher order terms in } \lambda \quad (25)$$

$$= I - \lambda A^T A + 2^{nd} \text{ and higher order terms in } \lambda$$

5.2 Block-ART

The iteration matrix of block-ART is (see eq. 22):

$$Q_{ba} = \prod_{t=K}^{1} P_t = P_K P_{K-1} P_{K-2} \ldots P_2 P_1 \quad (26)$$

with $P_i = I - \lambda B_i^T B_i = \sum_j \Omega_j - (L-1)I$ (27)

where the summation j is over the rows in the block. (Note that B_i represents a block of A)

For instance, the iteration matrix for block-ART with two blocks of two rows can be written as:

$$Q_{ba_{ij,kl}} = (\Omega_i + \Omega_j - I)(\Omega_k + \Omega_l - I) \quad (28)$$

$$= I - \Omega_i - \Omega_j - \Omega_k - \Omega_l + \Omega_i \Omega_k + \Omega_i \Omega_l + \Omega_j \Omega_k + \Omega_j \Omega_l$$

$$= I - \Omega_i - \Omega_j - \Omega_k - \Omega_l + \Omega_i + \Omega_k - I + \lambda^2 c_{ik} A_{i.}^T A_{k.}$$

$$+ \Omega_i + \Omega_l - I + \lambda^2 c_{il} A_{i.}^T A_{l.}$$

$$+ \Omega_j + \Omega_k - I + \lambda^2 c_{jk} A_{j.}^T A_{k.}$$

$$+ \Omega_j + \Omega_l - I + \lambda^2 c_{jl} A_{j.}^T A_{l.}$$

$$Q_{ba_{ij,kl}} = \Omega_i + \Omega_j + \Omega_k + \Omega_l - 3I + \quad (29)$$

$$+ \lambda^2 \left\{ c_{ik} A_{i.}^T A_{k.} + c_{il} A_{i.}^T A_{l.} + c_{jk} A_{j.}^T A_{k.} + c_{jl} A_{j.}^T A_{l.} \right\}$$

5.3 SIRT

The expansion in the relaxation parameter λ of SIRT is easily found, since SIRT can be considered as block-ART with one block (K=1). Referring to the former section, the iteration matrix Q_s of SIRT can be decomposed as

$$Q_S = \sum_{i=1}^{M} \Omega_i - (M-1)I = I - \lambda \sum_{i=1}^{M} A_{i.}^T A_{i.} = I - \lambda A^T A \tag{30}$$

5.4 Summary

The iteration matrices of the three iterative techniques discussed in this paper can be written as a combination of ART-matrices in the following way:

$$Q_{art} = \prod_{i=1}^{M} \Omega_i = I - \lambda A^T A + 2^{nd} \text{ and higher order terms in } \lambda. \tag{31}$$

$$Q_{ba} = \prod_{i=1}^{K} \left(\sum_{j=iL}^{(i+1)L} \Omega_j - (L-1)I \right) = I - \lambda A^T A + 2^{nd} \text{ and higher order terms in } \lambda. \tag{32}$$

$$Q_s = \sum_{i=1}^{M} \Omega_i - (M-1)I = I - \lambda A^T A \tag{33}$$

6. EXAMPLE

We consider a simple example where we have two projections consisting of two parallel rays each (see Fig.1). By discretizing this problem we obtain a system of equations $Ax = b$.

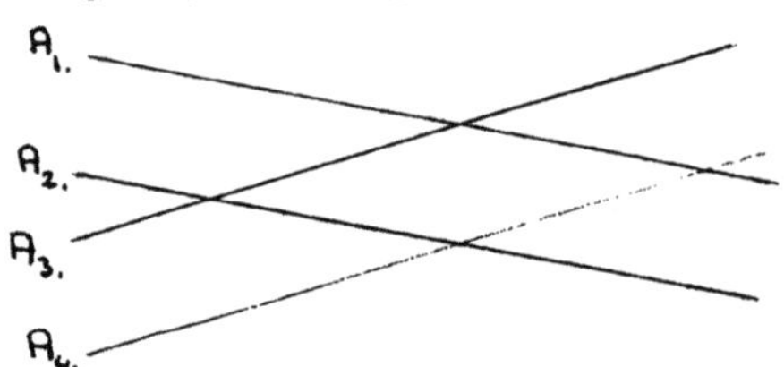

Fig.1. Simple model example: two projections of two parallel rays each.

For this specific example the iteration matrix for ART is:

$$Q_{art_{4321}} = \sum_{i=1}^{4} \Omega_i - 3I + \lambda^2 \left\{ c_{13} A_{i.}^T A_{3.} + \lambda^2 c_{14} A_{1.}^T A_{4.} + \lambda^2 c_{23} A_{2.}^T A_{3.} + \lambda^2 c_{24} A_{2.}^T A_{4.} \right\} \tag{34}$$

The iteration matrix for block-ART where the blocks coincide with the projections is equal to ART with the same ordering of the rows:

$$Q_{ba_{43,21}} = Q_{art_{4321}} \tag{35}$$

The iteration matrix for ART with a different formation of the blocks is

$$\begin{aligned} Q_{art_{4231}} = \sum_{i=1}^{4} \Omega_i - 3I &+ \lambda^2 \left\{ c_{13} A_{1.}^T A_{3.} + c_{14} A_{1.}^T A_{4.} + c_{23} A_{3.}^T A_{2.} + c_{24} A_{2.}^T A_{4.} \right\} \\ &- \lambda^3 \left\{ c_{13} c_{32} A_{1.}^T A_{2.} + c_{32} c_{24} A_{3.}^T A_{4.} + c_{24} c_{41} A_{2.}^T A_{1.} + c_{41} c_{13} A_{4.}^T A_{3.} \right\} \\ &+ \lambda^4 \left\{ c_{13} c_{32} c_{24} A_{1.}^T A_{4.} \right\} \end{aligned} \tag{36}$$

Block-ART with the same ordering of the rows has the following expansion:

$$Q_{ba_{42,31}} = \sum_{i=1}^{4} \Omega_i - 3I + \lambda^2 \left\{ c_{14} A_{1.}^T A_{4.} + \lambda^2 c_{23} A_{2.}^T A_{3.} \right\} \tag{37}$$

The expansion of the iteration matrix of SIRT is independent of the ordering of the rows and has no higher order terms in λ:

$$Q_s = \sum_{i=1}^{4} \Omega_i - 3I \tag{38}$$

6.1 Observations

The product of ART-matrices can be considered as an expansion in the relaxation parameter λ for all the methods considered. For ART, this expansion can also be obtained in another straightforward way. From the equivalence of unrelaxed ART and Gauss-Seidel, one can obtain the iteration matrix of relaxed ART:

$$Q_{art} = I - \lambda A^T (I + \lambda L)^{-1} A \tag{39}$$

where the matrix L is the lower triangular part of the matrix $A^T A$. The inverse matrix can be expanded in λ so that we obtain

$$Q_{art} = I - \lambda A^T (I - \lambda L + \lambda^2 L^2 - \lambda^3 L^3 + \ldots) A \qquad (40)$$

$$= I - \lambda A^T A + \lambda^2 A^T L A - \lambda^3 A^T L^2 A + \lambda^4 A^T L^3 A - \ldots$$

It is common knowledge that changing the order in which the equations are addressed in ART affects the convergence behaviour. In our development of the ART iteration matrix, the effect of addressing the equations in a different order is clear: other matrix products appear in the series expansion and other coefficients c_{ij} arise. The latter effect is probably the most important since this affects the number of non-zero terms in the development, as can be seen clearly from the model problem in this section.

In further investigations we hope to obtain from this expansion some more theoretical insight in the way to order the equations optimally or to form the blocks in block-ART optimally. From a *geometrical* point of view, it seems optimal to choose the ordering such that two consecutive equations are orthogonal (Youla, 1987). In our expansion, this ensures a minimum number of higher order terms, but note that SIRT has no higher order terms at all and yet converges poorly!

We have written the iteration matrices of both ART and block-ART as the iteration matrix $(I - \lambda A^T A)$ of SIRT plus some additional terms: the higher order terms in λ. For very small relaxation parameters, in some situations highly desirable (see section 7.3), these higher order terms vanish, and so ART, block-ART and SIRT are practically equivalent. This observation is in line with the fact that ART and block-ART with very small relaxation parameters (strong underrelaxation) applied to inconsistent systems converge to the minimum least squares solution of the (scaled) system (Censor et al., 1983).

The last observation is that block-ART with the usual formation of the blocks (corresponding to projections) (eq.35), is equal to ART with the same ordering of the rows (eq.34).

7. REGULARIZATION

If in the iterative reconstruction procedure the small singular values of the system matrix A appear too soon, the noise will blow up to an unacceptable level. The remedy to this problem is to regularize the reconstruction method. This can be achieved in various ways, e.g. by stopping the iteration procedure before it addresses the small singular values or by using a different optimization criterion as in Tikhonov regularization. In this section we confine ourselves to considering the regularizing effect of the methods themselves. Note that owing to the character of the reconstruction problem (being an inverse problem), the detailed information is in the small singular values, so that excessive regularization will destroy these (possibly important) details.

7.1 SIRT

Some theoretical results about the regularizing effect of SIRT have been given in Van der Sluis and Van der Vorst (1988). They describe the analytical relation between the iterated solution and the exact solution in the right-singular vectorspace in terms of the (relative) spectral response function. In the remainder of this subsection, a rough sketch of this work will be given.

The SIRT (Cimmino) algorithm can be rewritten as a Richardson iteration on the scaled matrix $W = R^{1/2}A$ (see eq. 14 with C = I):

$$x^{k+1} = (I - \lambda W^T W)x^k + \lambda W^T R^{1/2} b \tag{41}$$

or, if x^0 is the starting vector and $\hat{x}$ the minimum norm (least squares) solution to which the process converges:

$$x^k - \hat{x} = (I - \lambda W^T W)^k (x^0 - \hat{x}) \tag{42}$$

We compute the singular value decomposition (Golub and Van Loan,1983) of the scaled system matrix W: $W = U \Sigma V^T$, where U and V are orthonormal matrices containing the left- and right-singular vectors respectively and where Σ is the diagonal matrix containing the singular values σ_i of W.

To decompose the iterates x^k in the space spanned by the right-singular vectors v_i of A (right-singular vectorspace), we introduce the following relations:

$$\hat{x} = V\hat{z}, \; x^k = Vz^k$$

Multiplication by V^T and substitution into eq. (42) leads to:

$$V^T x^k - V^T \hat{x} = V^T (I - \lambda V \Sigma^2 V^T)^k (x^0 - \hat{x}) \tag{43}$$

$$z^k - \hat{z} = (I - \lambda \Sigma^2)^k (z^0 - \hat{z}) \tag{44}$$

If we now write this for each component in the right-singular vectorspace, we get

$$z_j^k - \hat{z}_j = (1 - \lambda \sigma_j^2)^k (z_j^0 - \hat{z}_j) \tag{45}$$

$$= \psi_k (\sigma_j^2)(z_j^0 - \hat{z}_j)$$

where $\psi_k(t) = (1 - \lambda t)^k$ is the relative response function. This function depicts how errors (in the right-singular vectorspace) are reduced after k iterations.

Figures 2 and 3 show the plots of the function $\psi_k(\sigma_j^2)$ after 1 iteration (k=1), 5 iterations (k=5), 10 iterations (k=10) and 50 iterations (k=50), for $\lambda = \lambda_{max}$ and $\lambda = \frac{1}{2}\lambda_{max}$, respectively.

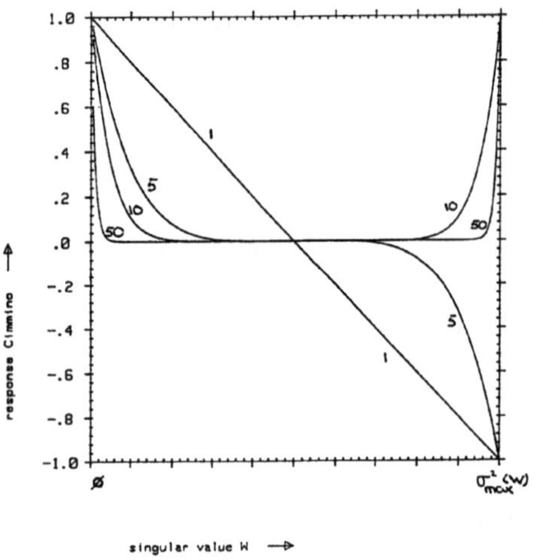

Fig.2
Relative response function of overrelaxed SIRT ($\lambda = \lambda_{max}$).

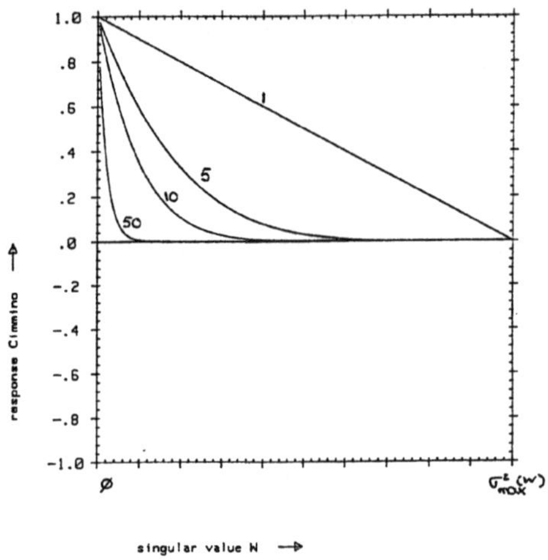

Fig.3
Relative response function of unrelaxed SIRT ($\lambda = \frac{1}{2}\lambda_{max}$).

7.2 ART and block-ART

For ART and block-ART, an analytical relation between the singular values of the (scaled) system matrix and the eigenvalues of the iteration matrix as described in the former section, is not (yet) known. (Note that the eigenvalues of the non-symmetric iteration matrix of ART are complex!) However, for a given model problem, the spectral response function can be determined experimentally. We take the solution to which the method converges and disturb this vector by adding all singular vectors to it. We now have a starting vector which has an error to the solution of 1 in each right-singular component.

$$x^0 = \hat{x} + \sum_i v_i \qquad (46)$$

We then compute the iterates x^k, determine for each k the iteration error $\hat{x} - x^k$, and decompose this error in the right-singular-vectorspace V of A. The components thus obtained give an indication how an error that is equally distributed over the right-singular vectorspace of A is reduced and we have experimentally obtained the relation

$$z_j^k - \hat{z}_j = \psi_k(\sigma_j^2)(z_j^k - z_j^0) \qquad (47)$$

7.3 Numerical comparison of the methods

We now discuss numerical calculations of the various response functions corresponding to a stylized geophysical reconstruction problem (Nolet, 1984). The size of the system matrix is 400 x 200 and its rank is 197.

To assess the value of the comparative study proposed in section 7.2, we applied unrelaxed SIRT to this model problem, because we theoretically know the response function of SIRT.

The result, shown in Fig.4, differs from the function in Fig.2, since we did not compute the response related to the scaled system matrix W, but to the original matrix A. This was done for reasons of comparibility, since other methods may scale differently. However, the regularizing effect of SIRT can be recognized in this figure.

Fig.5 shows the regularizing effect of strong underrelaxation applied to ART: the components corresponding to the small singular values of A converge very slowly, as compared to the convergence of the components corresponding to the larger singular values. For strong underrelaxation the ART and SIRT methods become similar, as is confirmed by Figs.4 and 5.

The relative response function of unrelaxed ART (Fig.6) indicates that ART converges considerably better than unrelaxed SIRT, however. A similar result is found when the optimal relaxation factor is taken for each method. From Fig.7 we see that block-ART has a convergence that can be compared with ART for the higher singular values of A. The lower singular values converge

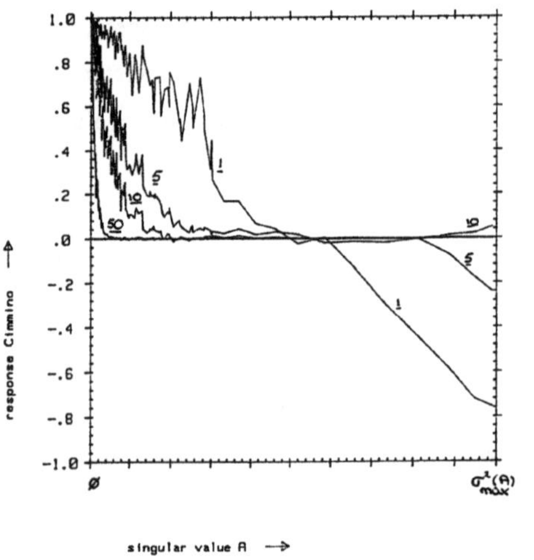

Fig.4
Relative response function of SIRT with $\lambda = \lambda_{max}$ =0.08 after 1, 5, 10 and 50 iterations respectively.

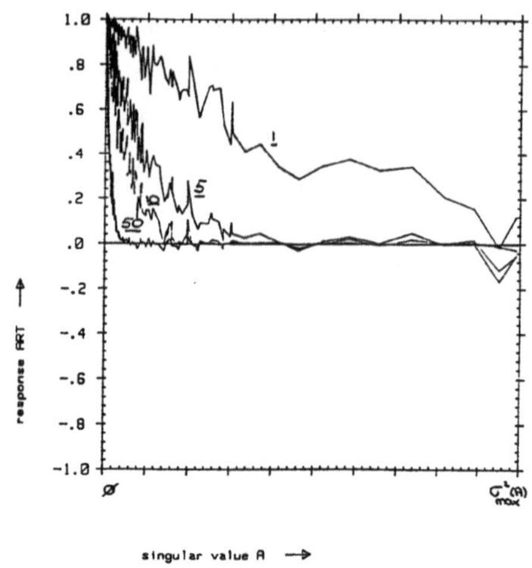

Fig.5
Relative response function of ART with strong underrelaxation (λ = 0.08) after 1, 5, 10 and 50 iterations respectively.

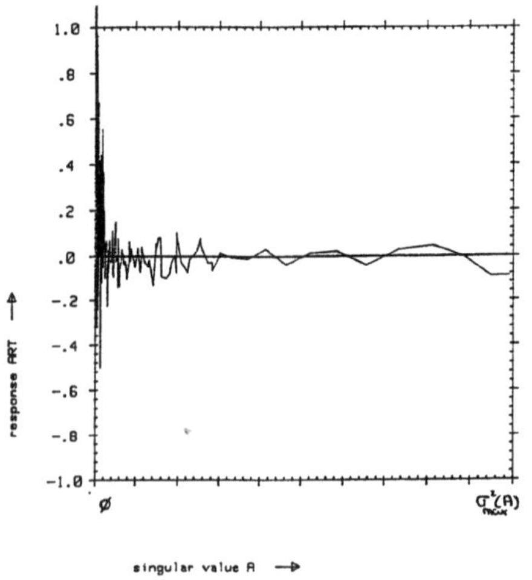

Fig.6
Relative response function of unrelaxed ART (λ=1.0) after 10 iterations.

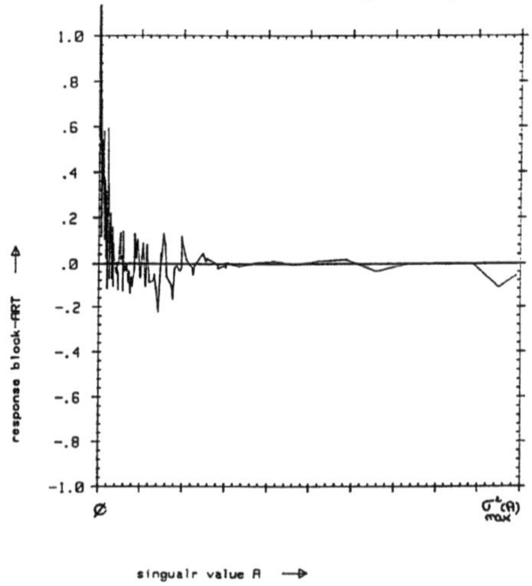

Fig.7
Relative response function of overrelaxed block-ART (λ = 0.4) after 10 iterations.

slower and therefore ART is to be preferred in systems that are nearly consistent, whereas in systems where there is a large amount of noise, unrelaxed block-ART will outperform ART without relaxation.

7.4 Conclusion

For regularizing ART, we thus need a small relaxation parameter λ. In cases where there is a large amount of noise, one has to make a trade-off. On the one hand we want fast convergence but on the other hand, the small singular values must not appear to soon. For very small λ, the higher order terms in the expansions in λ of the iteration matrices of both ART and block-ART vanish and therefore, the iteration matrices of the three methods considered become similar.

8. NUMERICAL EVALUATION OF CONVERGENCE BEHAVIOUR

For our numerical experiments, we used the same geophysical reconstruction problem as in section 7. In Fig.8 below, we give the L_2 distance between the iterate and the solution (mean squared error) of the different reconstruction techniques (with optimal parameters) of the consistent system, as a function of the number of iterations.

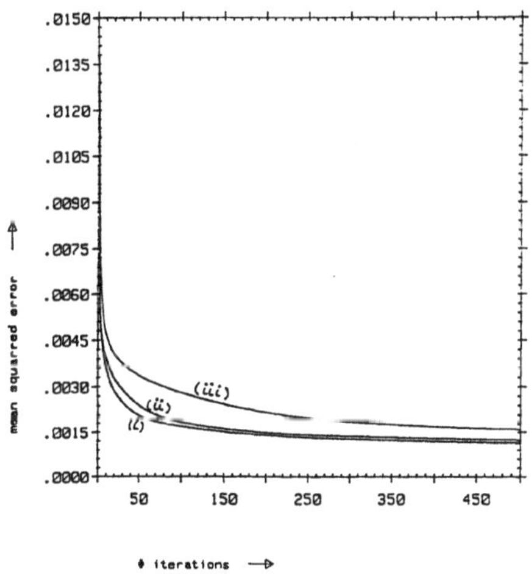

Fig.8
Mean Squared Error of the iterates of
(i) ART (λ = 1.0*),*
(ii) block-ART (λ = 0.4*) and*
(iii) SIRT (λ = 0.08*) as a function of the # iterations, for the consistent problem.*

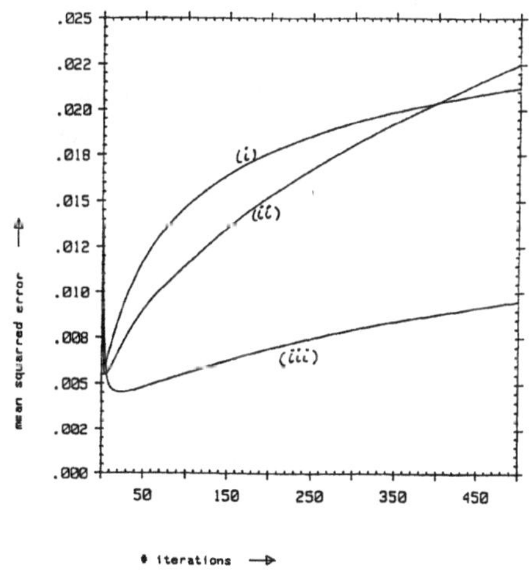

Fig.9
Mean Squared Error of the iterates of
(i) ART (λ = 1.0*)*
*(ii) block-ART (*λ = 0.4*) and*
*(iii) SIRT (*λ = 0.08*) as a function of the # iterations, for the inconsistent problem.*

From Fig.8 we can see that SIRT converges considerably slower than both ART and block-ART. The block-ART version we applied (5 blocks consisting of 4 projections) has a performance that can be compared to ART (with the same ordering of the equations). The advantage that the relaxation parameter for ART can be varied more than the relaxation parameter for block-ART does not lead to much better convergence for this consistent system.

In the next figure (Fig.9) we present L_2 distance of the iteraes of the different reconstruction techniques for the *inconsistent* system where the data were corrupted with 10% zero mean Gaussian noise, to the minimum norm least squares solution of the consistent model problem.

For a very small relaxation parameter, ART, block-ART and SIRT are practically equivalent. From Fig.9 we see that such choice of the relaxation parameter (the maximal relaxation for SIRT) outperforms ART and block-ART with larger relaxation parameters. Consequently, for reconstruction from noisy data, strong underrelaxation is mandatory.

9. PROJECTION METHODS

Projection methods, well known from the field of numerical linear algebra for solving a (sparse) set of linear equations are Lanczos, Conjugate Gradients (for both methods, see Golub and van Loan, 1983) and LSQR (Paige and Saunders, 1982). These three mathematically equivalent algorithms are optimal in the sense that they minimize the residual norm $\|b - Ax\|$ over a certain space (the so called Krylov space).

It has been shown recently (Van der Sluis and Van der Vorst, 1988) that Conjugate Gradients (CG) performs better than SIRT if CG is applied with the same scaling that is implicitly introduced in SIRT. The errors in the reconstruction process (both iteration and perturbation errors) for both methods are equally distributed over the spectrum, so that CG and SIRT produce similar images. However, the rate of convergence of CG is quadratic in the number of iterations, whereas the rate of convergence of SIRT is linear.

In this paper, SIRT has been compared with ART and block-ART. This has been done since ART, block-ART and SIRT appeared to have certain similarities. Since we know the relation between these methods, the position of Conjugate Gradients is clear from the earlier study.

We have shown that ART and block-ART with very small relaxation parameters (for regularization purposes) are similar to SIRT. Consequently, in this case, Conjugate Gradients is to be preferred over SIRT as well as over ART and block-ART. A comparison of convergence properties of ART and block-ART to CG under more general conditions is currently being studied.

10. SUMMARY OF CONCLUSIONS

The iteration matrices of ART and block-ART have been written as the iteration matrix of SIRT $(I - \lambda A^T A)$ plus the higher order terms in the series expansion in the relaxation parameter λ.

Since for the convergence of SIRT a small relaxation parameter is mandatory, this method is very similar to ART and block-ART for the same small value of λ . The higher order terms in the expansion in λ of both ART and block-ART vanish in this case.

We therefore conclude that SIRT is practically equivalent to ART with strong underrelaxation. The advantage of SIRT is that it is well suited for parallel implementation.

In the comparison between ART and block-ART we note that if in block-ART the blocks are formed by grouping disjoint equations (parallel rays), the method is equivalent to ART with the same ordering of the rows. If the blocks are formed by grouping L 'joint' equations (e.g., the rows all intersect one cell), the relaxation parameter is restricted to $0 < \lambda < \frac{2}{L}$. If the same small relaxation parameter is used for ART, the higher order terms of both ART and block-ART in the expansion are small and so both methods are quite similar to each other and also to SIRT. The major advantage of block-ART over ART is that it is suited for parallel implementation.

The overall advantage of ART is that the relaxation parameter can be varied over a wider range to obtain the optimum. This makes the convergence behaviour of ART superior to that of both block-ART and SIRT in problems that are only mildly inconsistent.

REFERENCES

Bjorck, A. and Elfving, T. (1979). Accelerated projection methods for computing pseudo-inverse solutions of systems of linear equations. BIT 19, 145-163.

Censor, Y. (1981). Row-action methods for huge and sparse systems and their applications, SIAM Rev. 23, 444-466.

Censor, Y., Eggermont, P.P.B., and Gordon, D. (1983). Strong underrelaxation in Kaczmarz's method for inconsistent systems. Numerische Mathematik 41, 83-92.

Cimmino, G, (1938) Calcolo approssimato per le soluzioni dei sistemi di equazioni lineari, La Ricerca Scientifica (Roma) XVI, serie II, Annno IX, Vol. 1, 326-333.

Dines,K.A. and Lytle,R.J. (1979). Computerized geophysical tomography. Proc IEEE 67, 1065-1073.

Eggermont,P.P.B., Herman,G.T., and Lent,A (1981). Iterative algorithms for large partioned linear systems with applications to image reconstruction. Lin. Alg. Appl. 40, 37-67.

Golub, G.H. and van Loan, C.F. (1983). Matrix Computations. North Oxford Academic.

Gordon, R., Bender, R. and Herman, G.T. (1970). Algebraic reconstruction techniques (ART) for three-dimensional electron microscopy and X-ray photography, J. Theor. Biol. 29, 471-481.

Herman, G.T. and Lent, A. (1976). Iterative reconstruction algorithms. Comput. Biol. Med. 6, 273-294.
Herman, G.T and Levkowitz, H. (1988). Initial performance of block-iterative reconstruction algorithms. In: Mathematics and Computer Science in Medical Imaging, M.A. Viergever and A. Todd-Pokropek (eds.), NATO ASI series, Vol 39, Springer-Verlag, Berlin, pp. 305 - 317.
Kaczmarz, S (1937). Angenaherte auflosung von systemen linearer gleichungen, Bull. Int. Acad. Pol. Sci. Lett. A., 355-357.
Nolet, G. (1985). Solving or resolving inadequate and noisy tomographic systems. J. Comp. Phys., 61, 463-482.
Paige, C.G. and Saunders, M.A. (1982). LSQR: an algorithm for sparse linear equations and sparse least squares. ACM Trans. Math. Softw. 8, 43-71.
Sluis, A. van der and Vorst, H.A. van der (1988). SIRT and CG type methods for the iterative solution of sparse linear least squares problems. Report of the Faculty of Technical Mathematics and Informatics no. 88-43 (submitted for publication).
Viergever, M.A. (1988). Introduction to discrete reconstruction methods in medical imaging. In: Mathematics and Computer Science in Medical Imaging, M.A. Viergever and A. Todd-Pokropek (eds.), Springer Verlag, Berlin, pp. 43 - 65.
Youla, D.C. (1987). Mathematical theory of image restoration by the method of convex projections. In: Image Recovery, Theory and Application., H. Stark (ed.), Academic Press, pp.29-77.

PRELIMINARY RESULTS FROM SIMULATIONS OF TOMOGRAPHIC IMAGING USING MULTIPLE-PINHOLE CODED APERTURES

J. N. Aarsvold,[1] H. H. Barrett,[1] T. A. Gooley,[1] T. J. Roney,[1] R. K. Rowe,[1] W. E. Smith[2]

[1] University of Arizona, [2] University of Rochester

USA

ABSTRACT

The use of multiple-pinhole coded apertures in single-photon emission computed tomographic (SPECT) systems makes possible SPECT geometries that image from multiple directions simultaneously and that have greater photon-collection efficiencies than conventional SPECT systems. However, the multiplexing of data that results from the use of such apertures is not well understood and, therefore, is not always optimally decoded. Presented in this paper are results from a simulation study of multiple-pinhole coded-aperture systems having pinholes that are uniformly spaced and of uniform width. Five apertures representing varying degrees of multiplexing are considered, and the results presented are in the form of pictorial displays that make possible qualitative comparisons of images from the five imaging systems. Shown are representations of the Hoffman brain phantom determined following singular value decompositions of the system matrices and reconstructions of simulated SPECT data using a Monte Carlo algorithm and prior information.

1. INTRODUCTION

Most single-photon emission computed tomographic (SPECT) imaging systems have two limitations imposed by their apertures or collimators. One is low photon-collection efficiency; the other is the small number of angles from which data may be collected simultaneously. One type of system that is not as restrictive as most conventional SPECT systems has multiple-pinhole coded apertures (Simpson and Barrett, 1980; Barrett and Swindell, 1981). Such apertures usually contain a relatively large number of pinholes systematically distributed about the object to be imaged. This implies that the geometrical photon-collection efficiency of such systems and the number of angles sampled simultaneously by such systems is significantly larger than that of systems having single-pinhole apertures or parallel-hole collimators. However, it also implies that radionuclide activity imaged through two or more pinholes is often recorded simultaneously by the same detector element. This type of simultaneous detection is referred to as multiplexing (see Fig. 1). If the multiplexed data can be completely decoded, the individual projections can be determined unambiguously; if not, some projections will not be uniquely specified. Thus, an understanding of the relationship between the increase in the number of measured projections by detection of more photons from more angles and the amount of ambiguity in the measurement caused by multiplexing is desired.

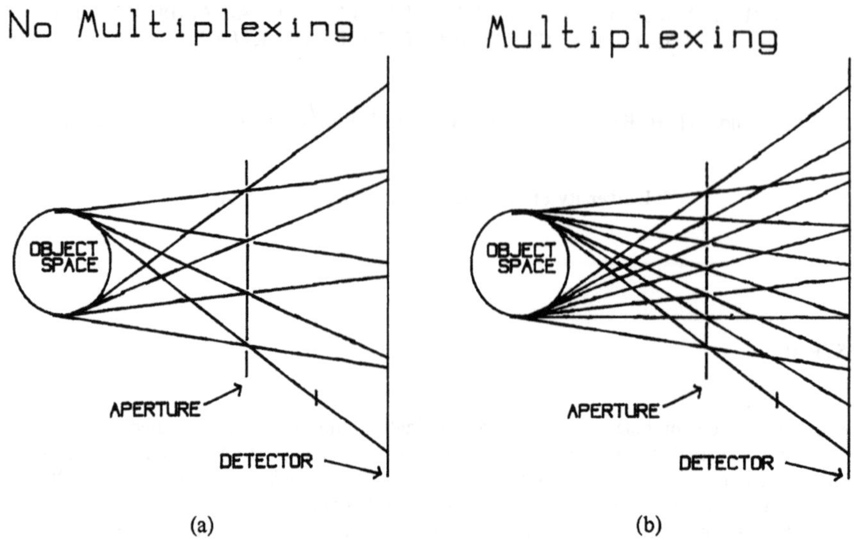

Fig. 1. Multiple-pinhole coded apertures. (a) Four uniformly spaced pinholes, no multiplexing. (b) Seven uniformly spaced pinholes, multiplexing.

Multiple-pinhole coded apertures can consist of complex arrangements of pinholes of different sizes. However, to date, even the properties of much simpler arrays used for tomographic imaging are not well understood. To perform qualitative and quantitative comparisons of various multiple-pinhole coded apertures, we have chosen to study systems having pinholes that are uniformly spaced and of uniform width. Of specific interest are the effects of the resultant multiplexing on the quality of reconstructions as the number of pinholes in an aperture is increased.

In this paper, we present preliminary results from simulations of the imaging of the Hoffman brain phantom (Data Spectrum Corporation, Chapel Hill, NC) using multiple-pinhole coded-aperture systems. These results are in the form of pictorial displays from which qualitative comparisons can be made. Included are representations of the phantom calculated using the singular vectors in the domains of the system matrices and reconstructions from simulated data. We begin with a description of the physical characteristics of the systems to be modelled, section 2, and an outline of the procedure used to obtain the simulated data, section 3. In section 4, we discuss the intrinsic imaging capabilities of each system as determined by the singular value decomposition (SVD) of each system matrix, and in section 5 we present reconstructions obtained using an iterative Monte Carlo algorithm and prior information. We conclude with a brief discussion, section 6.

2. MODELLED SYSTEMS

The SPECT geometry that was modelled for this study is the single-slice imager shown schematically in Fig. 2. The detector surface is configured as six sides of an octagon. Each side is 22 cm in length and is assumed to have 64 individual, uniform-width, 25%-efficient detector elements. Thus, the system records 384 measurements simultaneously over a range of 270°. The object space is 19.5 cm square, and reconstructions are represented on a 55 x 55 grid or 3025 pixels. Since the center of the object space coincides with the center of the detector-surface octagon, the perpendicular distance from the detector surface to the center of the object space is approximately 26.5 cm.

Of primary interest for this study is the multiple-pinhole aperture. As with the detector surface, the aperture surface is configured as six sides of an octagon. The centers of the two octagons are coincident and the aperture surface is 5 cm from the detector surface. We refer to the angular directions of the six sides of the aperture as the six views of the system.

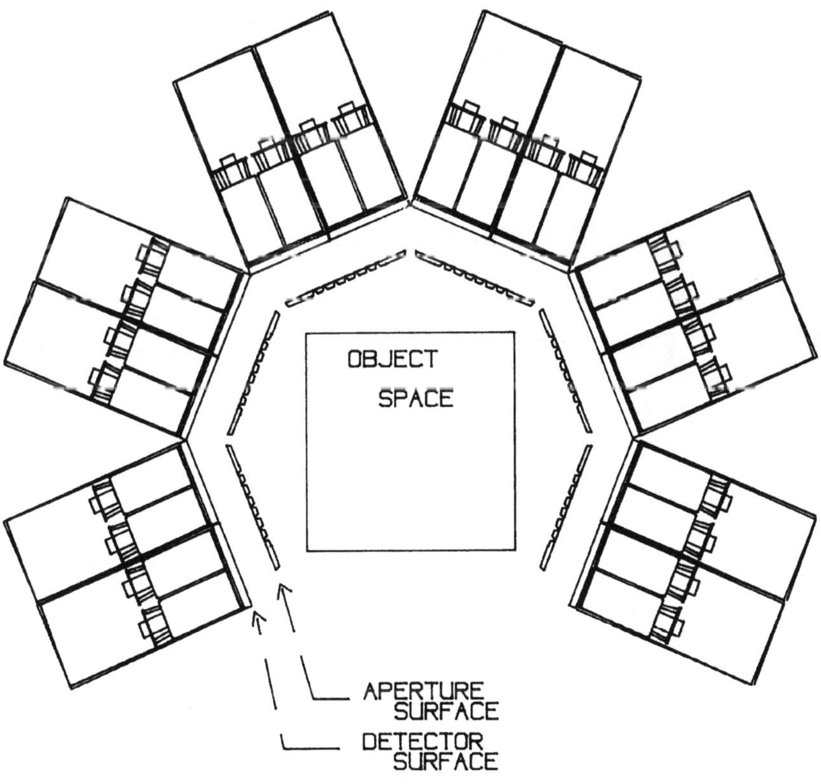

Fig. 2. Schematic of the hardware configuration modelled for the simulation.

Our study involved the use of five pinhole arrangements, of which three were coded apertures, i.e., they resulted in multiplexed image data. For all of the apertures, the pinhole widths were assumed to be 1 mm. Aperture 1 had a single pinhole per view, centered in each view. Aperture 2 had four uniformly spaced pinholes per view, spaced so as to utilize all detectors without multiplexing the data. The placement of the pinholes was such that if activity on the boundary of the object space was projected through the outer pinholes of each view, the activity would be incident on the outer detector elements of that view. (See Figs. 1 and 2.) This placement implied that the pinholes were approximately 4 cm apart and therefore the outer pinholes were 6 cm from the center of each view. Apertures 3, 4, and 5 had 7, 13, and 25 pinholes per view, respectively. The outer pinholes of each view of each of these apertures were positioned as those of aperture 2, and the remaining pinholes for each view were uniformly spaced between the two outer pinholes. Therefore, the apertures had uniform spacings between pinholes of 2 cm, 1 cm, and 0.5 cm, respectively, and resulted in image data multiplexed at sequentially increasing levels.

Figure 3 is a display of the image data for a single view for each of the five apertures. (The procedure used to create these data is outlined in Sec. 3.) Note that as the degree of multiplexing increases, the variation in the values of each data set decreases. Note, also, that aperture 1 does not sample the full 270°.

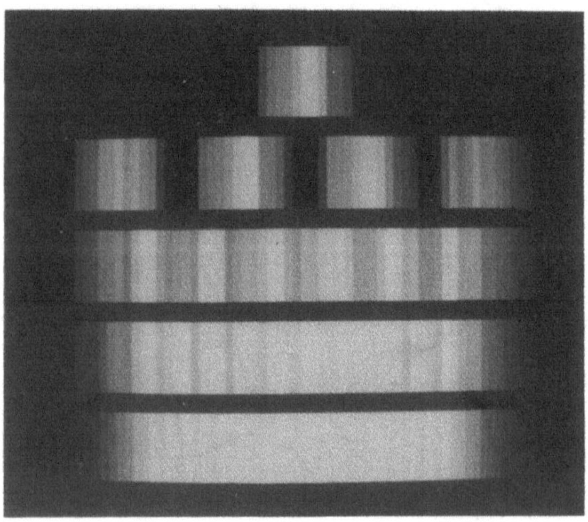

(a) 1 pinhole per view

(b) 4 pinholes per view

(c) 7 pinholes per view

(d) 13 pinholes per view

(e) 25 pinholes per view

Fig. 3. Simulated Hoffman brain phantom image data. Each horizontal stripe represents one view of the respective aperture. The data are one dimensional, but each set has been expanded vertically for clarity of display. Each set has been normalized to its respective maximum.

3. SIMULATIONS

Throughout the remainder of the paper a linear/statistical model of the imaging system will be assumed. This model is represented, using bold face type to designate vectors and matrices, by

$$\mathbf{g} = \mathbf{H}\mathbf{f} + \mathbf{n} , \tag{1}$$

where **g** represents the measured data; **H**, a linear approximation to the system point spread function (psf); **n**, zero-mean noise; and **f**, a discrete approximation to the object being imaged. For the systems under consideration, **H** is 384 x 3025 since **f** is 3025 x 1 and **g** is 384 x 1.

To perform the analysis and reconstructions reported in sections 4 and 5, it was necessary to simulate the collection of two forms of data. First, it was necessary to simulate the imaging of the phantom. Secondly, it was necessary to simulate experimental measurement of each system's point spread function, which is space-variant for the systems under consideration.

The collection of the image data **g** was simulated as follows. The Hoffman brain phantom was digitized on a 256 x 256 grid, and pixels were assigned values corresponding to the three activity levels within the phantom (of ratios 0:1:4). Each of these coarse pixels was further divided into fine pixels, and the point of origin of each photon was randomly selected from the appropriate set of fine pixels. Projection calculations then assumed the trajectory of a photon was from the center of a fine pixel to the detector surface. The point we wish to emphasize is that the simulation performed to obtain **Hf** was not done assuming a representation of the object on the 55 x 55 reconstruction grid, but done assuming a much finer representation, i.e., a closer approximation to the continuous object. Having obtained a value for each entry $(\mathbf{Hf})_i$ of the vector **Hf**, each entry $(\mathbf{g})_i$ of the vector **g** was then obtained by sampling from a Poisson distribution with mean $(\mathbf{Hf})_i$.

Each simulated image data set had an average of 60,000 cts/pinhole, e.g., the data set obtained from the system using aperture 3, seven pinholes per view, had approximately 2,500,000 counts. We did not model attenuation or scatter in the simulation.

Each column of a system matrix **H** represents the response of the system to a uniform source on one of the 3025 pixels of the reconstruction grid. Thus, the matrix **H** describes the system's psf and may be measured experimentally by stepping a "point" source, a pixel-size element of activity, to each reconstruction-grid pixel and collecting an image (Lefkopoulos et al, 1982; Rowe et al, 1988). This procedure was simulated by assuming that a 1 mCi, 3.55 mm square source was imaged for 1 min at each of the 3025 pixels. As in the simulation of the imaging of the phantom, each coarse pixel was subdivided into fine pixels and photons were assumed to emanate from the centers of the fine pixels.

4. SVD REPRESENTATIONS

In section 2, we described five imaging systems with different apertures that represent various degrees of multiplexing of data; and in section 3, we outlined a procedure to obtain, for each system, a matrix, **H**, that approximates the imaging properties of that system. In this section, we discuss a characterization of the intrinsic imaging capabilities of each system as determined using the SVD of $\mathbf{H}^T\mathbf{H}$ (where T denotes transpose). More detailed discussions of various aspects of this characterization have been presented elsewhere, for example, in Paxman, 1984, Barrett, 1988, and Defrise, 1988.

If $\mathbf{u}=\{\mathbf{u}_i \mid i=1,...,R\}$ denotes the set of singular vectors defined on the domain of **H** that are associated with the R nonzero singular values of $\mathbf{H}^T\mathbf{H}$, and $\mathbf{u}\dagger=\{\mathbf{u}_i \mid i=R+1,...,3025\}$ denotes the set of singular vectors, again defined on the domain of **H**, that are associated with the (3025-R) zero-valued singular values, then the set **u** spans the measurement space of **H** and the set $\mathbf{u}\dagger$ spans the null space of **H**. Note that R is the rank of **H** and that the domain of **H** is the 55 x 55 object space. An object **f** in the domain of **H** may then be represented as

$$\mathbf{f} = \mathbf{f}_u + \mathbf{f}_{u\dagger} = \sum_{i=1}^{R} (\mathbf{f},\mathbf{u}_i)\mathbf{u}_i + \sum_{i=R+1}^{3025} (\mathbf{f},\mathbf{u}_i)\mathbf{u}_i = \sum_{i=1}^{3025} (\mathbf{f},\mathbf{u}_i)\mathbf{u}_i \,, \tag{2}$$

where $(\mathbf{f},\mathbf{u}_i)$ denotes the inner product of **f** with $\mathbf{u}_i$; and it follows that

$$\mathbf{g} = \mathbf{Hf} + \mathbf{n} = \mathbf{H}(\mathbf{f}_u + \mathbf{f}_{u\dagger}) + \mathbf{n} = \mathbf{Hf}_u + \mathbf{Hf}_{u\dagger} + \mathbf{n} = \mathbf{Hf}_u + \mathbf{0} + \mathbf{n} = \mathbf{Hf}_u + \mathbf{n} \,. \tag{3}$$

Equations (2) and (3) indicate that under the assumptions of the model of Eq. (1), the intrinsic imaging capabilities of a given system for a specified object **f**, in the domain of **H**, may be characterized by

$$\mathbf{f}_u = \sum_{i=1}^{R} (\mathbf{f},\mathbf{u}_i)\mathbf{u}_i \,. \tag{4}$$

Though our calculations of this representation are not reconstructions from data, this representation is, in fact, equivalent to the best singular-vector reconstruction of noise-free data obtainable without the application of prior knowledge.

To determine these representations for the systems and the object under consideration, the Hoffman brain phantom, it is necessary to determine a useful 55 x 55 approximation to the phantom. The approximation that we have chosen is the grid of averages obtained by digitizing the Hoffman phantom on a 55 x 55 grid, the approximation that is the intended outcome of reconstructions. Throughout the remainder of the paper, **f** shall refer to this 55 x 55 version of the phantom.

Except for the matrix of the system using aperture 1, the system matrices were found to be essentially of rank 384. Therefore, each system adequately measures the components of f represented by the 384 singular vectors of its system matrix associated with the nonzero singular values of the system matrix. Figure 4 is a display of 18 of the singular vectors for the system matrix of the system using aperture 3, and Fig. 5 is a sequence of truncated versions of the representation in Eq. (4) calculated using the singular vectors for the system using aperture 3. Note in Fig. 4 that as the magnitude of the singular values decreases the "spatial frequency" of the singular vectors increases, i.e., the singular vectors contain finer details.

To close this section we present, in Fig. 6, the singular-vector representations calculated using Eq. (4) for each of the systems described in section 2. Note that the adverse effects from the multiplexing of apertures 3 and 4 do not seem to negate the increased information obtained from the increased photon-collection efficiency and finer angular sampling that results from the increase in the number of pinholes. However, the multiplexing occurring with aperture 5 does seem to cause a degradation of the representation image--using as the standard for comparison the representation associated with the full-view, no-multiplexing aperture, aperture 2.

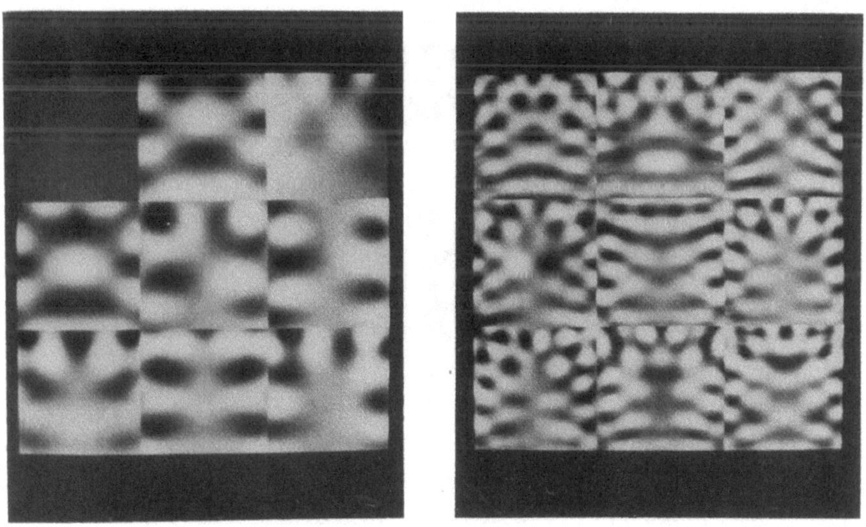

(a) (b)

Fig. 4. Some of the orthonormal singular vectors of the system matrix of the system using aperture 3 (see text). The scaling is such that dark is negative and light is positive. (a) The singular vectors associated with the singular values ranked 1-9 in magnitude. (b) The singular vectors associated with the singular values ranked 19-27.

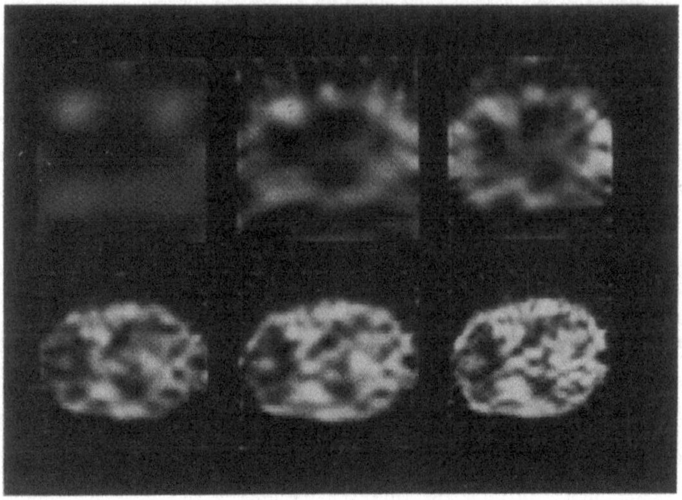

top: (a)-(c); bottom: (d)-(f)

Fig. 5. A pictorial display of the summation of Eq. (4) of the text, truncated at various values of N, N≤R. The singular vectors used are those for the system matrix of the system using aperture 3 (see Fig. 4). The object is **f** discussed in section 4. (a) N=16. (b) N=64. (c) N=128. (d) N=192. (e) N=256. (f) N=384. (All images involving the Hoffman brain phantom have a lower threshold of zero and an upper threshold that is equivalent to the maximum of the phantom.)

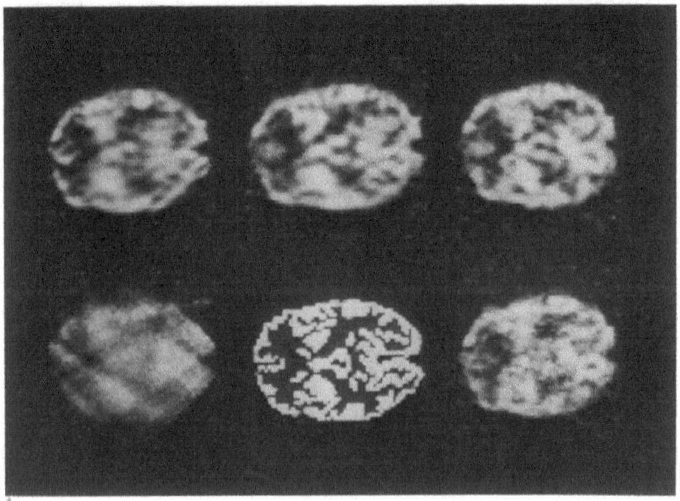

top: (a)-(c); bottom: (d)-(f)

Fig. 6. A comparison display of the summation of Eq. (4) of the text for each of the five systems discussed. (a) Aperture 2. (b) Aperture 3. (c) Aperture 4. (d) Aperture 1. (e) The 55 x 55 Hoffman phantom (see text). (f) Aperture 5. (See note of Fig. 5.)

5. RECONSTRUCTIONS

The characterization of the intrinsic imaging capabilities of the modelled systems described in the previous section suggest that some ambiguity in the measurement of individual projections can be countered by the measurement, even if not exact, of more projections. In this section, we present reconstructions using data from each of the five modelled systems. These reconstructions also indicate that modest amounts of multiplexing may be countered by added angular sampling. Further, they indicate that with the inclusion of prior knowledge, for example positivity, moderately multiplexed data can be significantly demultiplexed.

The reconstruction algorithm used is an iterative Monte Carlo algorithm that allows the incorporation of various types of constraints (Smith et al, 1983; van Laarhovan and Aarts, 1987). Four reconstructions were calculated for each set of data. Two reconstuctions were performed using the algorithm with an implementaion of the energy functional

$$E(\hat{\mathbf{f}}) = \sum_{i=1}^{384} (\mathbf{g} - \mathbf{H}\hat{\mathbf{f}})_i^2 , \qquad (5)$$

and two using the algorithm with an implementation of the energy functional

$$E(\hat{\mathbf{f}}) = \sum_{i=1}^{384} \frac{(\mathbf{g} - \mathbf{H}\hat{\mathbf{f}})_i^2}{(\mathbf{g})_i} , \qquad (6)$$

where $\hat{\mathbf{f}}$ is the estimate of f and $(\ )_i$ denotes the ith entry of the enclosed vector. The energy functional given by Eq. (5) is the familiar criterion of squared-error agreement with the data. The energy functional given in Eq. (6) if minimized over all possible $\hat{\mathbf{f}}$'s is equivalent to maximization of $p(\mathbf{g} \mid \mathbf{f}) = p(\mathbf{n})$, given the assumption that $p(\mathbf{n})$ is multivariate normal with means μ_i equal to zero and variances $\sigma_i^2 = (\overline{\mathbf{g}})_i \propto (\mathbf{g})_i$, for $i = 1,\ldots,384$.

Images of a typical progression of iterations of the algorithm is given in Fig. 7. Figure 8 is a matrix of reconstructions displaying the results of the four implementations of the algorithm from the data generated using aperture 3.

Finally, Figs. 9, 10, and 11 display reconstructions of data from the five modelled systems using the algorithm implemented with the energy functional of Eq. (5) without positivity, the energy functional of Eq. (5) with positivity, and the energy functional of Eq. (6) with positivity, respectively. The reconstructions resulting from the fourth implementation were all similar to the example given in Fig. 8, so have not been included. The images obtained using apertures 4 and 5 in Figs. 9, 10, and 11 indicate that improvement in reconstructions of multiplexed data are possible with a positivity constraint and the assumptions leading to the energy functional of Eq. (6).

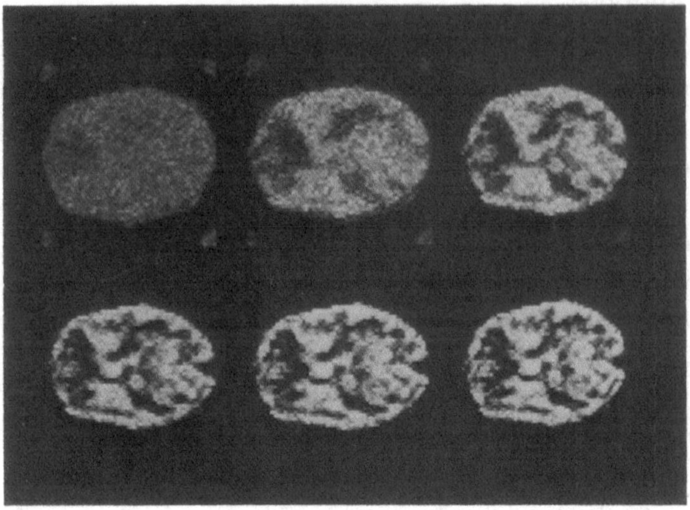

Fig. 7. A sequence of images representing intermediate iterations in the Monte Carlo reconstruction of the data from the system using aperture 3. The cost functional used was that of Eq. (6) with a positivity constraint. (See note of Fig. 5.)

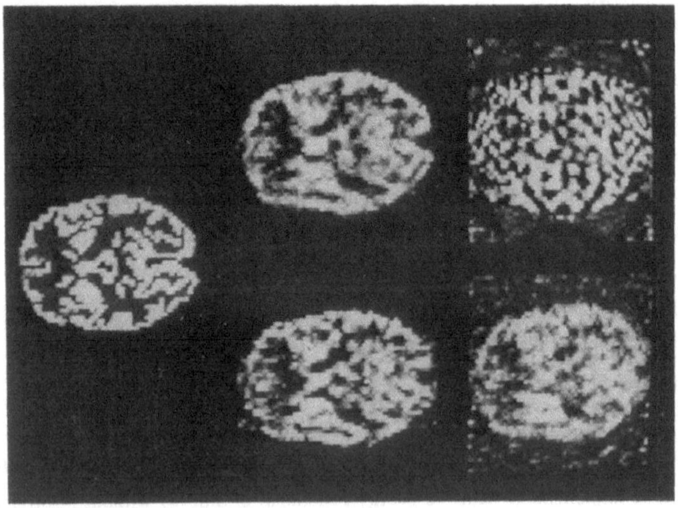

(a)

(b)-(c)

(d)-(e)

Fig. 8. Comparisons of different implementations of the reconstruction algorithm.
(a) The 55 x 55 version of the Hoffman phantom (see text).
(b) Reconstruction using the cost functional in Eq. (6) with a positivity constraint.
(c) Reconstruction using the cost functional in Eq. (6) without a positivity constraint.
(d) Reconstruction using the cost functional in Eq. (5) with a positivity constraint.
(e) Reconstruction using the cost functional in Eq. (5) without a positivity constraint.
(See note of Fig. 5.)

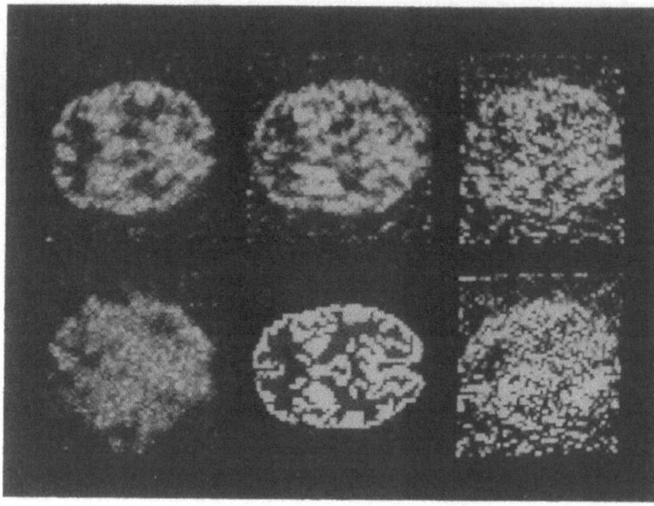

top: (a)-(c); bottom: (d)-(f)

Fig. 9. A comparison of reconstructions for data from each of the five systems discussed. The Monte Carlo algorithm was implemented using the cost functional in Eq. (5) without a positivity constraint. (a) Aperture 2. (b) Aperture 3. (c) Aperture 4. (d) Aperture 1. (e) The 55 x 55 version of the Hoffman phantom. (f) Aperture 5. (See note of Fig. 5.)

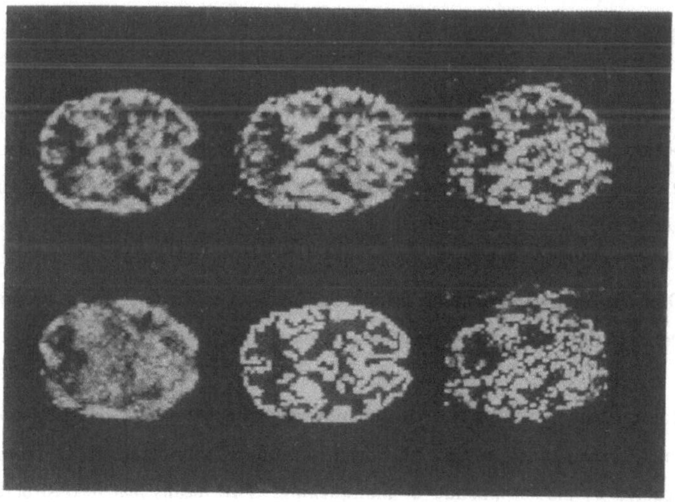

top: (a)-(c); bottom: (d)-(f)

Fig. 10. A comparison of reconstructions for data from each of the five systems discussed. The Monte Carlo algorithm was implemented using the cost functional in Eq. (5) with a positivity constraint. (a) Aperture 2. (b) Aperture 3. (c) Aperture 4. (d) Aperture 1. (e) The 55 x 55 version of the Hoffman phantom. (f) Aperture 5. (See note of Fig. 5.)

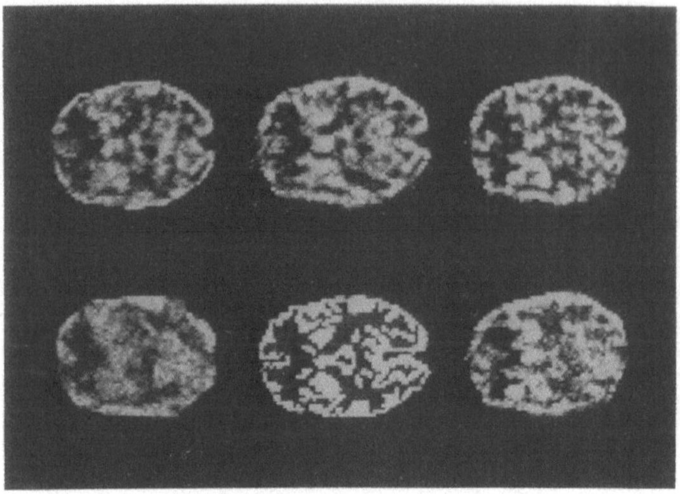

top: (a)-(c); bottom: (d)-(f)

Fig. 11. A comparison of reconstructions for data from each of the five systems discussed. The Monte Carlo algorithm was implemented using the cost functional in Eq. (6) with a positivity constraint. (a) Aperture 2. (b) Aperture 3. (c) Aperture 4. (d) Aperture 1. (e) The 55 x 55 version of the Hoffman phantom. (f) Aperture 5. (See note of Fig. 5.)

6. DISCUSSION

Using a model of a multiple-pinhole tomographic imager in which the only variable parameter is the number of pinholes, we have begun an analysis of the effects of varying degrees of multiplexing. Five multiple-pinhole apertures representing varying degrees of angular sampling and data multiplexing have been assumed in simulations, and the resulting data have been reconstructed. In addition, each simulated system has been approximated by a linear model and the SVD of the matrix representing the system has been calculated. These decompositions has been used to characterize some of the imaging properties of the systems.

Qualitative comparisons of images of reconstructions and singular-vector representations suggest that added angular sampling provides sufficient additional information to overcome the degradation caused by modest amounts of multiplexing. This means that reconstructions from the multiplexed data can result in images of higher quality than those from unmultiplexed data. Similar comparisons of images of reconstructions performed with and without a positivity constraint indicate that the positivity constraint alone is sufficient to perform a significant demultiplexing of some multiplexed data.

ACKNOWLEDGMENTS

This work was supported by the National Cancer Institute under Grant No. PO1 CA 23417. That support is gratefully acknowledged. The assistance of B. Moore and D. Spargur in the formation of the figures is gratefully acknowledged.

REFERENCES

Barrett, H.H., Swindell, W. (1981). Radiological imaging: Theory of image formation, detection, and processing, New York, Academic Press, Chap. 8.
Barrett, H.H. (1988). Image reconstruction and the solution of inverse problems in medical imaging. These proceedings.
Defrise, M. (1988). Regularization techniques in medical imaging. These proceedings.
van Laarhoven, P.J.M., Aarts, E.H.L. (1987). Simulated annealing: Theory and applications, D. Reidel Publishing Co., Dordrecht.
Lefkopoulos, D., Fonroget, J. Devaux, J.Y., Guilhem, J.B., Roucayrol, J.C., Guiraud, R. (1982). Quantitative 3D imaging with coded apertures by using SVD decomposition of the transmission matrix. In: Proceedings of the third world congress of nuclear medicine and biology. C. Raynaud (ed), Pergamon, Paris, pp. 503-506.
Paxman, R.G., Barrett, H.H., Smith, W.E., Milster, T.D. (1985). Image reconstruction from coded data: II. Code design, J. Opt. Soc. Am. 2, pp. 501-509.
Rowe, R.K., Aarsvold, J.N., Barrett, H.H., Chen, J., Hall, J.N., Landesman, A.L., Mar, L.S., Milster, T.D., Moore, B.A., Patton, D.D., Roney, T.J. (1988). The design and implementation of modular spect imaging systems. These proceedings.
Simpson, R.G., Barrett, H.H. (1980). Coded-aperture imaging, In Imaging in diagnostic medicine, S. Nudelman, D. Patton (eds.) Plenum, New York, Chap. 8.
Smith, W.E., Barrett, H.H., Paxman, R.G. (1983). Reconstruction of objects from coded images by simulated annealing, Opt. Lett. 8, pp. 199-201.

ASPECTS OF CLINICAL INFRA RED ABSORPTION IMAGING

S. Arridge, P. van der Zee, D.T. Delpy, M.Cope
Dept. of Medical Physics & BioEngineering
University College London
Shropshire House
11-20 Capper Street
LONDON WC1E 6JA
United Kingdom

ABSTRACT

This paper introduces the mathematical aspects of an imaging modality based upon the inversion of data from tissue transilluminated with infra red light. A Monte-Carlo model is discussed together with a simple analytic approach based on the so called P_1 or diffusion approximation to the Radiative Transfer Equation. Localised absorption is thought of as a perturbation on the homogeneous solution, and the expected resolution of the method is given in terms of the Singular Value Decomposition of the perturbation operator.

1. INTRODUCTION

Continuous monitoring of changes in the absorption of infra red light by tissues at multiple frequencies has become of increasing clinical interest over the last 10 years. The feasibility of detecting light transilluminating intact organs about 4-6 cm in diameter was shown originally by Jobsis (1977). Clinical applications of this monitoring technique include neonatal cerebral monitoring (Wyatt et. al. 1986), mammography (Watmough 1983) and muscle function (Tamura et. al. 1988). New developments in technology have allowed time resolved data, to be utilised, increasing the information content (Delpy et. al. 1988).

Together with these new techniques, the possibility of a new imaging modality has been suggested (Jarry et.al. 1984, Arridge et.al. 1985, Tamura et.al. 1987). This paper investigates some of the mathematical properties of the forward and inverse problems.

2. FORMULATION OF THE PROBLEM

The basic problem may be stated as follows, with reference to figure 1. The object to be imaged is a volume $\wp$ bounded by a surface $\mathfrak{S}$. An input light distribution $\underline{q}(\underline{r},t)$ is incident on the surface $\mathfrak{S}$, where $\underline{r}$ is a position vector, $\underline{\hat{s}}$ is a unit direction vector, and t is time. This input source distribution creates an angle dependent photon density $\phi(\underline{r},\underline{\hat{s}},t)$ in $\wp$. The flux measured is $\underline{\Gamma}(\underline{r},t) = -D(\underline{r})\underline{\nabla}\Phi(\underline{r},t)$ where $D(\underline{r})$ is some tensor to be specified and $\Phi(\underline{r},t) = \int_{\Omega}\phi(\underline{r},\underline{\hat{s}},t)d\Omega$, where the integration is over all directions. We have available a set of measurements of the output flux $\underline{\Gamma}_{\mathfrak{S}}$ at a

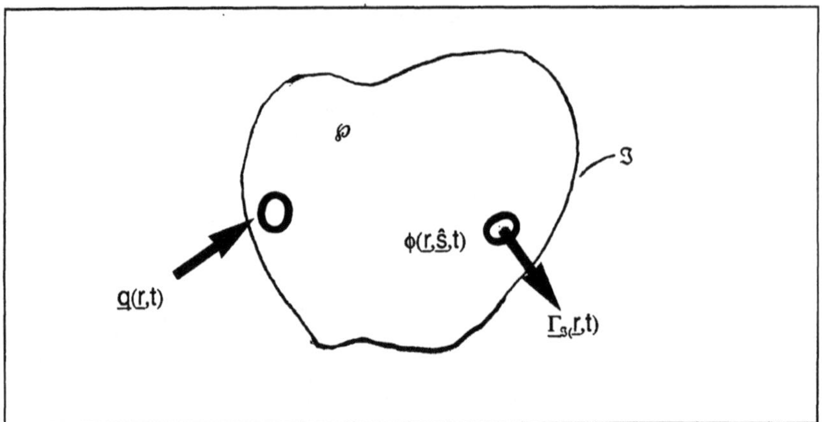

Figure 1: Generalised imaging schematic

set of measuring sites on the surface ℑ. What may be said about the distribution of absorption sites (the image of interest) within ℘?

To answer this it is necessary to have a model of photon propagation in tissue. If we suppose that the model is some general operator $\underline{\Gamma}_{\Im}$ = **L**[q], where **L** is spatially varying, we actually wish to find information about **L** given the measured $\underline{\Gamma}_{\Im}$ and known q. In the following we outline some approaches to the definition of the forward model, and a characterisation of its ill- conditioning by inspection of its Singular Value Decomposition. Results will be given for the non-time resolved case, but we also present a form for the time resolved case.

3. LIGHT PROPAGATION IN TISSUE

Tissue is an absorbing and highly scattering medium. The most general model that may be given is the theory of Radiative Transfer or Photon Transport Theory which has been extensively studied in Atmospheric Physics (Chandrasekhar 1950, Ishimaru 1978a, van der Hulst 1980), and in Neutron Transport (Case and Zweifel 1967).

Certain special cases have been investigated in detail although they are mostly limited to homogeneous media. The photon transport problem is termed by Case and Zweifel the one-speed transport equation, whose time independent form is:-

$$(\underline{\hat{s}}.\underline{\nabla} + (\mu_a + \mu_s))\phi(\underline{r},\underline{\hat{s}}) = (1/c)\underline{\hat{s}}.\underline{q}(\underline{r}) + \mu_s\int_{\Omega}\sigma(\underline{\hat{s}},\underline{\hat{s}}')\phi(\underline{r},\underline{\hat{s}}')d\Omega \qquad (1)$$

where $\sigma(\underline{\hat{s}},\underline{\hat{s}}')$ is the differential scattering cross section representing the probability of scattering from direction $\underline{\hat{s}}'$ into direction $\underline{s}$ and c is the speed of light.

What we would ideally require is a solution to this equation in the presence of a spatially varying absorption coefficient μ_a, the scattering coefficient μ_s and the normalised differential scattering cross-section $\sigma(\underline{\hat{s}},\underline{\hat{s}}')$. The image of interest in particular is then $\mu_a(\underline{r})$ although the other parameters may also hold clinical value; for example, the scattering coefficient of tissue will change with abnormal cell morphology.

3.1 Monte Carlo Modelling

The most sophisticated model of the forward problem is a Monte Carlo one where individual photons are followed through an object and arbitrarily complex boundary conditions and local absorption and scattering effects may in principle be included (Wilson and Adams 1983). Usually the simplification is made that $\sigma(\underline{\hat{s}},\underline{\hat{s}}')$ is not spatially variant, and further, that it is a function of $\underline{s}.\underline{s}'$ only, ie. a function of the cosine of the angle of scattering only. This is termed the phase function. Previous work has usually used a simple phase function described mathematically (van der Hulst 1980). However, we have incorporated a measured function from **in vitro** brain tissue (van der Zee and Delpy 1988a). This function is shown in figure 2. In general long computation times preclude the use of Monte Carlo techniques for a reasonable inversion algorithm.

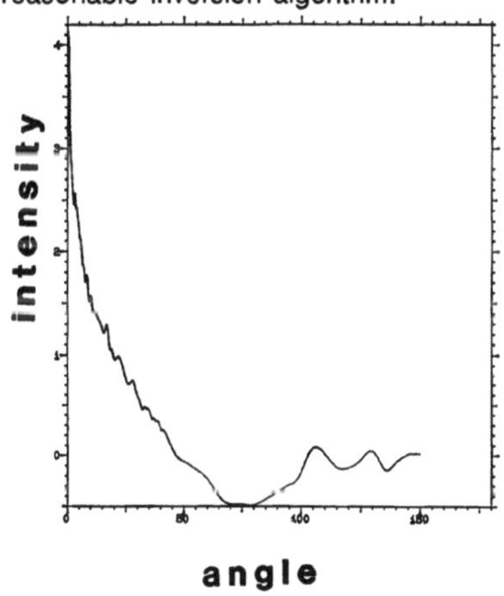

Figure 2: Phase function of in vitro brain tissue

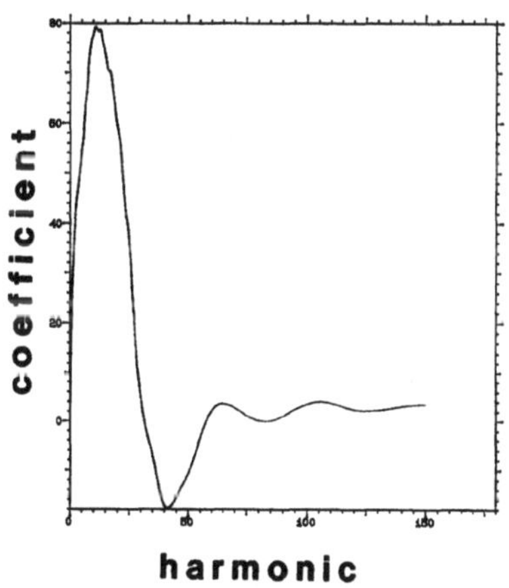

Figure 3: The phase function expanded in spherical harmonics

3.2 Analytic Approaches

The form of Eq. (1) is too complex to be solved by exact analytic techniques. However, some simple approaches have been made, and work done to show their validity within certain ranges of the optical parameters. (Reynolds et.al. 1976, Shimizu et.al. 1979, Groenhuis et.al. 1983).

The previously mentioned simplification of the differential scattering cross section to a phase function of $\underline{s}.\underline{s}'$ is again made. This permits Eq. (1) to be expanded in spherical harmonics (Lewis 1950, Bremmer 1964, Case and Zweifel 1967)

$$\phi(\underline{r},\hat{\underline{s}}) = \sum_{n=0}^{\infty}\sum_{m=-n}^{n} (2n + 1/4\pi)^{1/2}\, \phi_{n,m}(\underline{r})\, Y_{n,m}(\hat{\underline{s}}) \tag{2}$$

$$\underline{q}(\underline{r}) = \sum_{n=0}^{\infty}\sum_{m=-n}^{n} (2n + 1/4\pi)^{1/2}\, q_{n,m}(\underline{r})\, Y_{n,m}(\hat{\underline{s}}) \tag{3}$$

$$\begin{aligned}\sigma(\hat{\underline{s}},\hat{\underline{s}}') &= \sum_{n=0}^{\infty} (2n + 1/4\pi)\, \sigma_n\, P_n(\hat{\underline{s}}.\hat{\underline{s}}') \\ &= \sum_{n=0}^{\infty}\sum_{m=-n}^{n} \sigma_n(\underline{r})\, Y^*_{n,m}(\hat{\underline{s}}')\, Y_{n,m}(\hat{\underline{s}})\end{aligned} \tag{4}$$

where $P_n(\hat{\underline{s}}.\hat{\underline{s}}')$ and $Y_{n,m}(\hat{\underline{s}})$ are the Legendre and Associated Legendre polynomials respectively. The summation may be truncated if it assumed that the derivatives of ϕ vanish for n greater than some N. The result is a set of $(N+1)^2$ coupled partial differential equations known as the P_N approximation

Figure 3 shows the phase function of figure 2, expanded in spherical harmonics. It appears that the exact expansion requires a great many terms to realistically approximate the measured phase function. This stems from the highly forward peaked nature of the phase function. For example if the phase function were a delta function in the forward direction, the spherical harmonic expansion would "ring" infinitely (analogous to a Gibb's phenomenon), and this behaviour is observed in figure 3. However, using a limited number of terms can result in a function with negative lobes. This is non-physical because the phase function must represent a probability distribution. This has lead to the use of "synthetic" phase functions that are physically real (i.e. do not have negative probabilities in any direction), but which match the measured function in only a limited number of moments.

4. THE DIFFUSION OR P_1 APPROXIMATION

When N = 1, four coupled partial differential equations are required. However an elegant method described in Case and Zweifel reduces the four equations to a single diffusion type expression:

$$\underline{\nabla}.(D(\underline{r})\underline{\nabla}\Phi(\underline{r})) = \mu_a(\underline{r})\, c\, \Phi(\underline{r}) - (1/c)q_0(\underline{r}) \quad (5)$$

Here q_0 is the isotropic source term from the input distribution and $D(\underline{r})$ is given by:

$$D(r) = \frac{1}{3(\mu_a(\underline{r}) + (1-g)\mu_s(\underline{r}))} \quad (6)$$

i.e. D depends on the absorption coefficient, the scattering coefficient and g, the average cosine of the phase function.

If we make the assumption that scattering is a global characteristic and not spatially variant, then we have an elliptical partial differential equation for Φ and D that depends on a spatially varying μ_a, the image of interest.

4.1 Time Resolved Data

The use of ultrafast lasers and a picosecond resolution streak camera allows the possibility of time resolved measurement of the output flux. The time dependant form of Eq. (5) has been derived by Ishimaru (1978b) as

$$\{\nabla^2 - (3/c^2)\frac{\partial^2}{\partial t^2} - (1/D^2)\frac{\partial}{\partial t} - 3\mu_a(1-g)\mu_s\}\Phi = -[3/4\pi(1/c)\frac{\partial}{\partial t} + (1-g)\mu_s]q$$

and a similar analysis to that adopted here can be followed. Effectively we have increased the number of measurements allowing a less singular (more over-determined) problem. For the rest of this paper we consider only the time-independent form.

5. ANALOGY WITH IMPEDANCE TOMOGRAPHY

We note the close analogy with the problem of impedance tomography where, in the absence of current sources or sinks inside a volume, the potential is given by

$$\underline{\nabla}.(\sigma(\underline{r})\underline{\nabla}\Phi(\underline{r})) = 0 \quad (7)$$

In this case σ is the conductivity of tissue and Φ is the electric potential. The assumption usually made in this field, that σ is a scalar, is equivalent to the assumption in our problem that the effect of scattering may be replaced by an isotropic scatterer with its coefficient reduced by (1-g).

Breckon (1987) has shown the interesting relationship between (7) and the Schrodinger Equation. Analogously we can state that by using the same transformation:-

$$\Psi = D^{1/2}\Phi$$

$$\text{and} \quad \eta = (1/2)(\nabla^2\log(D)) + (1/4)(\underline{\nabla}\log(D).\underline{\nabla}\log(D)) \quad (8)$$

Then equation (5) above reduces to:-

$$(\nabla^2 - (\mu_a c/D))\Psi = \eta \tag{9}$$

A linear differential equation soluble in principle given the boundary conditions on $\mathfrak{S}$. We do not pursue this method further here.

6. PERTURBATION METHODS

General experimental conditions would necessitate the derivation of three-dimensional Point Spread Functions (PSFs) for point absorbers in general (inhomogeneous) distributions of absorbing and scattering media. Derivation via Monte-Carlo methods has attendant problems not only of lengthy computation times, but also because of the difficulty of normalising a delta-function which for Monte-Carlo purposes must have finite dimensions and strength. Figure 4 shows images of the perturbed photon flux for a narrow collimated laser input beam transilluminating a 10mm slab with a scattering coefficient of $2mm^{-1}$, absorption coefficient of $0.077mm^{-1}$, and average phase function cosine of 0.8, containing a 1mm radius sphere of high absorption ($20mm^{-1}$). These parameters were chosen to be representative of those found in tissue. The images were obtained by a Monte-Carlo simulation with the sphere at varying positions in the slab. For the larger peak the sphere is centred in the middle of the slab and on the axis of the input beam; for the smaller peak the sphere is still in the middle of the slab, but is laterally displaced 5mm from the input axis.

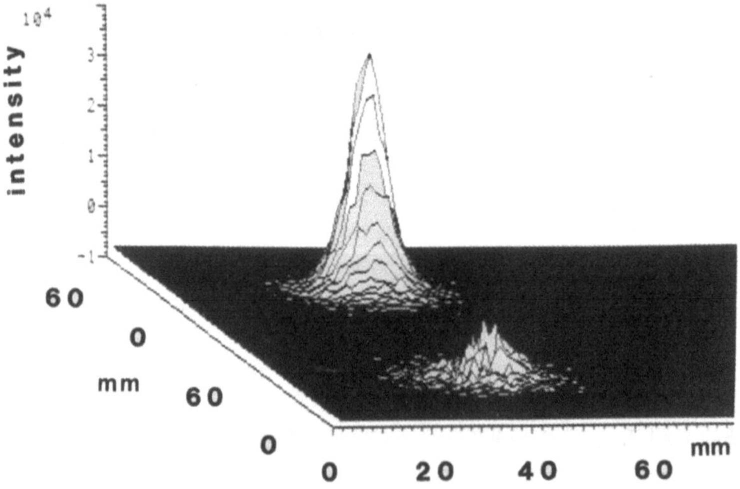

Figure 4: Monte-Carlo derived perturbation images of a 1mm diameter sphere of μ_a = $20mm^{-1}$ inside a 10mm thick slab of tissue with μ_a = $0.077mm^{-1}$ and μ_s = $2.0mm^{-1}$. See text for details.

In terms of the diffusion approximation, numerical techniques such as Finite Element Analysis could be employed for complex geometries. However because of the inevitable crudity of any images that might be reconstructed in infrared transillumination, a simple model is appropriate at least for a first approximation.

In a previous paper (Arridge 1985) we discussed the possibility of solving (5) for the homogeneous case ($\mu_a(\underline{r})$ = constant) and then treating variation in μ_a as a perturbation of the homogeneous solution. Using this method we derive

$$(\nabla^2 - (\mu_a c/D))\Phi = -(1/D)q_0(\underline{r}) \tag{10}$$

where q_0 is the isotropic source term from the input distribution.

Again numerical techniques would be necessary to satisfy the boundary conditions in general geometries, but some special cases may be solved analytically from Eq. (10) using Green's function methods. The boundary conditions used are of a mixed type:

$$G - \lambda \underline{n}.\underline{\nabla} G = 0 \tag{11}$$

where $\underline{n}$ is the unit vector normal to the boundary and λ is a constant dependent on the media. For the remainder of this discussion we consider the parallel slab problem, with the source $\underline{q}(\underline{r})$ consisting of a narrow collimated laser beam incident perpendicular to one surface, and the measurements taken on the other surface. For this problem the value of λ has been given by Reynolds et.al. (1976) as 0.2368/D. This geometry is not likely to be one used in a practical imaging system for the brain but it corresponds to an arrangement proposed by some workers for a system that uses a low light level image intensifier as a detection device, and is appropriate for diaphanography applications. Our purpose here is to find a mathematical limit for the resolution expected with such a system.

The boundary conditions of Eq. (11) then simplify to a one dimensional boundary condition and the Green's function may be found in a variety of ways. A detailed solution in the homogeneous case has been given by Reynolds et. al. (1980) and Groenhuis et.al. (1983), where a method of separation of variables was used. We have used a method of images (Arridge et.al. 1985) to derive a more tractable result. If we call the homogeneous solution $\Phi^{(0)}$, then the perturbed density is given by adding a term $\mu_a^{(1)}(\underline{r})\Phi^{(0)}(\underline{r})$ to the right in Eq. (10).

Then the equation of interest is

$$\Phi^{(1)}(\underline{r}) - \Phi^{(0)}(\underline{r}) = c\int_{\wp} (\mu_a^{(1)}(\underline{r}')/D^{(1)})\Phi^{(0)}(\underline{r}')G(\underline{r},\underline{r}')d^3\underline{r}' \tag{12}$$

In principle second and higher iterations may be formed. A simpler version of this expression was used in an earlier paper (Arridge 1985), but this derivation is presented here for the first time.

7. SINGULAR VALUE DECOMPOSITION OF THE PERTURBATION OPERATOR

We investigate the ill conditioning of Eq. (12) by the introduction of the Singular Value Decomposition (SVD). We note that Eq. (12) is equivalent to

$$\Delta\Phi = K[\mu_a^{(1)}/D^{(1)}] \tag{13}$$

Where K is the integral operator $c\int_{\wp}\Phi^{(0)}(\underline{r}')G(\underline{r},\underline{r}')d^3\underline{r}'$. The SVD method decomposes the kernel K into three matrices :-

$$K = V\Lambda U^T$$

where V is orthonormal in the space of the data, U is orthonormal in the space of the solution and U^T is the transpose of U, and Λ is the diagonal matrix of singular values. Then the columns of V are eigenvectors of KK^T and the columns of U are the eigenvectors of K^TK (Defrise 1988).

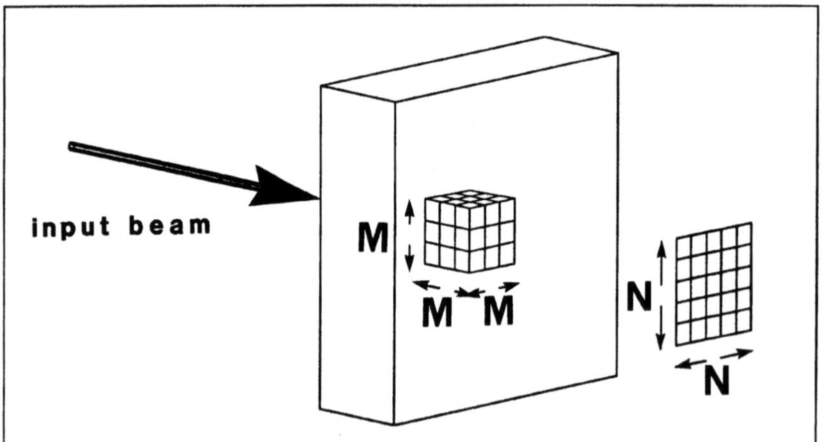

Figure 5 The relationship between object space defined on an M by M by M grid inside the slab, and data space defined by an N by N image.

7.1 Results

The use of so many simplifications needs careful justification. Our approach is to validate firstly against the more rigorous Monte-Carlo methods, and also experimentally against phantom studies. Experimental methods are described fully by Delpy et. al. (1988), and results from all three approaches (Monte-Carlo, analytic a

b

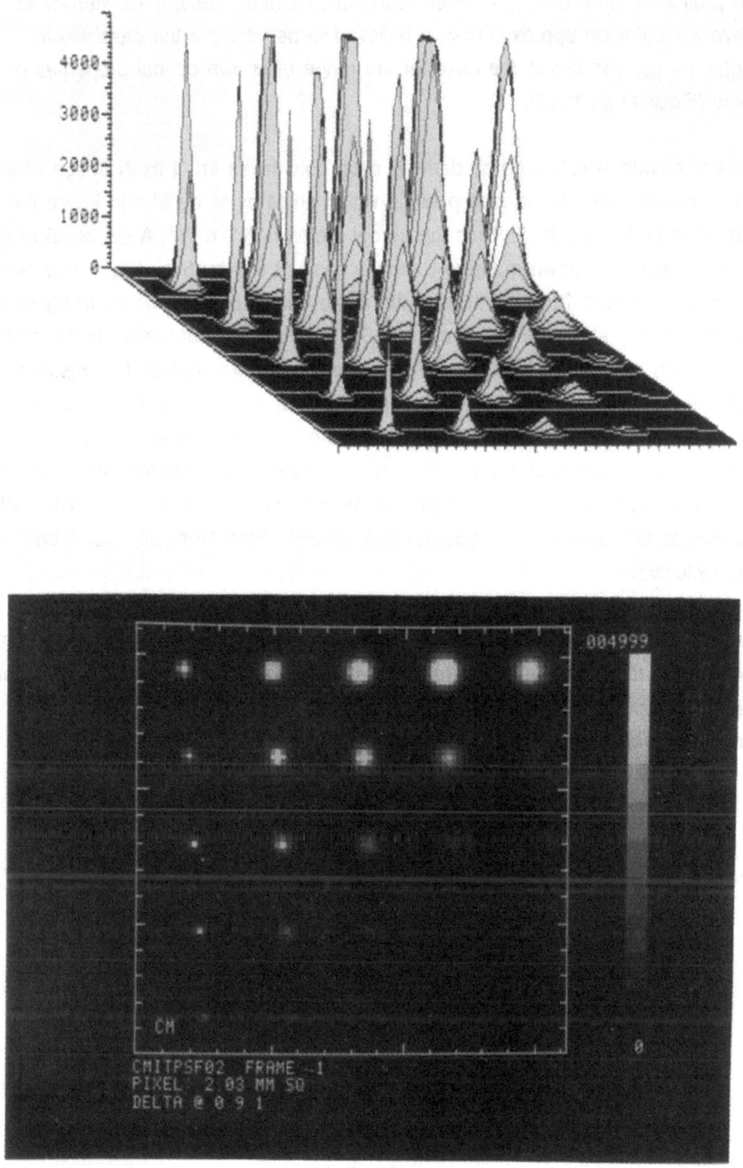

Figure 6: Images of the P_1 Green's functions displayed in greyscale and isometric format. The same optical parameters as in Figure 4 are used. The absorbers recede from the detector left to right, and from the input beam axis top to bottom.

and phantom) have been presented (van der Zee et.al. 1988b). As mentioned above the diffusion approximation is believed to be valid for the case where $(1-g)\mu_s >> \mu_a$, and this is the case for the range of known optical properties of tissue (Flock et.al. 1987).

The kernel was constructed by deriving each column as an N by N image of a delta function absorber at each position within an M by M by M grid inside the slab, as shown in figure 5. Thus the kernel matrix is N^2 by M^3. A set of 25 of these column images is shown in figure 6 for the case of a 5 by 5 by 5 grid and 16 by 16 images inside a 10 mm slab with the same optical parameters as in figure 4. The images are shown in greyscale and 3D format. The broadening of the peaks across each row shows how the Green's function spreads out as the absorber is positioned deeper into the slab. The delta functions are located at 2mm intervals from a depth of 1mm to 9 mm inclusive. As the absorber moves off axis the Green's function attenuates and shifts spatially, whilst retaining the same shape. This can be seen by comparing peaks in the same column of figure 6. The delta-functions in this direction are also at 2mm spacing, from 1mm off axis to 9mm off axis, inclusive.

Figure 7 shows the rate of fall in the log SV spectrum for a series of slabs from 10 to 40 mm in thickness, where here the scattering coefficient is $10mm^{-1}$. The rapid rate of fall represents the ill-conditioning and, as expected, it becomes increasingly worse as the slab increases in thickness.

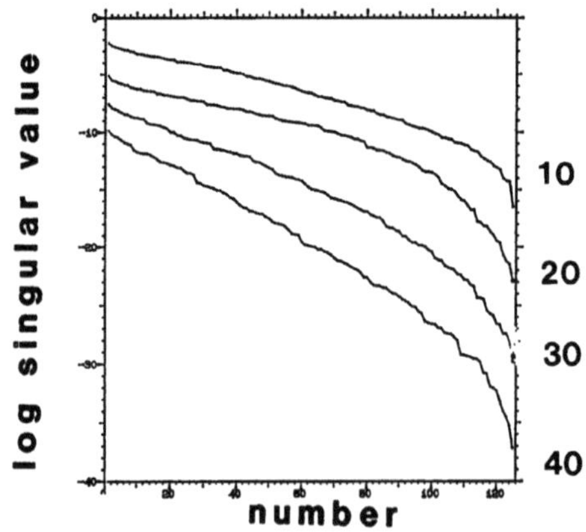

Figure 7: Log SV spectrum for tissue slabs of thickness 10mm, 20mm, 30mm, and 40mm, for $\mu_s = 10\ mm^{-1}$.

In a simulated reconstruction we show in figure 8 the inversion of a delta function absorber in two different slab thicknesses to which 1% noise has been added. In (a) the slab is 10mm thick and in (b) it is 40mm. In each case the delta function

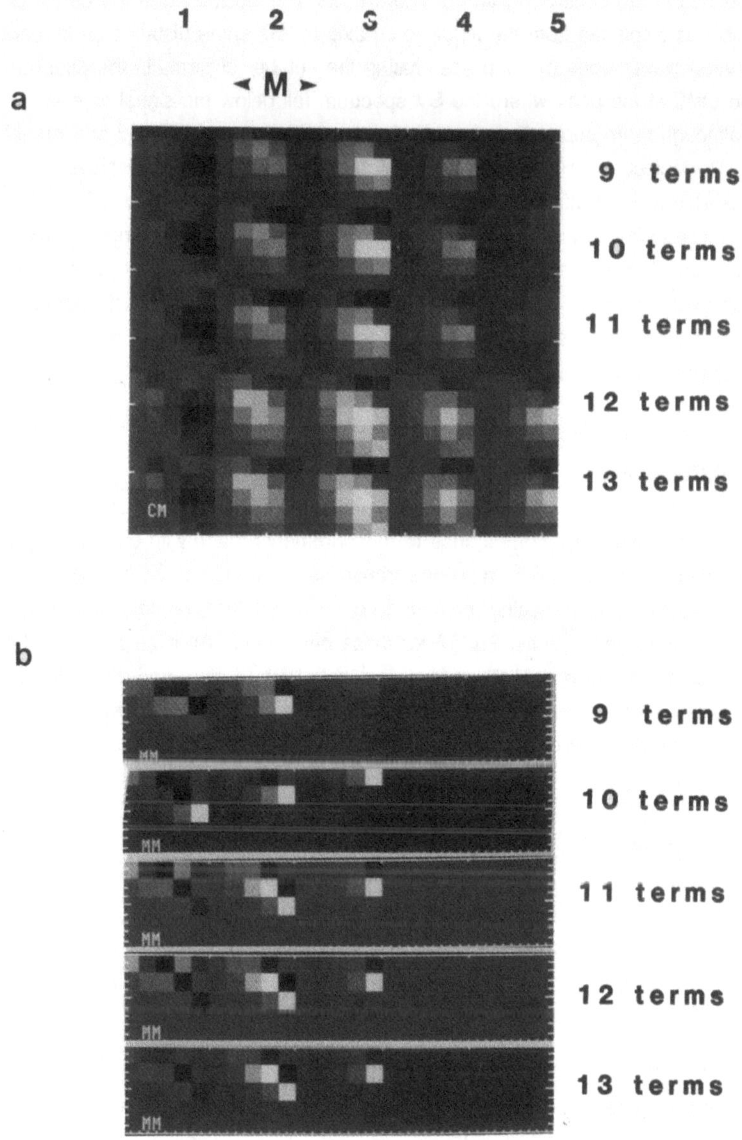

Figure 8: The reconstruction of a delta function absorber in the centre of a slab of tissue and displaced from the input beam axis by half the width of the slab. 1% noise has been added. The 3D object is represented as a 5 by 5 by 5 matrix, and displayed in five 5 by 5 slices, arranged side by side across each row. The true reconstruction would be a single pixel in the centre of this 25 by 5 array. Successive reconstructions obtained by adding progressively more terms to the generalised inverse are arranged in successive rows from top to bottom. In (a) the slab thickness is 10mm and in (b) it is 40mm. μ_s is $10mm^{-1}$ for both.

corresponds to the central function of figure 6; i.e. it is positioned in the centre of the slab and displaced from the input beam axis by the same distance as its depth. The inversion was done by simply truncating the number of terms in the pseudo-inverse of K, at the point where the SV spectrum fell below the signal to noise ratio, although more sophisticated regularizers may be used (Tikhonov and Arsenin 1977). The object is reconstructed on the 5 by 5 by 5 grid. This is displayed in figure 8 as a 25 by 5 matrix, where each vertical slice of the grid is a 5 by 5 matrix, placed side by side. Succesive approximations are arranged in rows from the reconstruction using 9 terms to that using 13, inclusive. In (a) the resolution appears good with 9 or 10 terms, but becomes noisy as more terms are added. In (b) under no circumstances was it possible to adequately restore the delta-function. Spurious peaks and noise appear however many terms are used.

8. CONCLUSIONS

The analysis of section 7 corresponds to measurements made with only one input site and measured output over all of the transmission surface of $\mathfrak{I}$. In order to more accurately assess imaging methods in other geometries, numerical techniques need to be employed to solve Eq. (5) for these geometries. An example would be a sphere with a set of optical fibres in a ring, able both to input and detect light. An imaging system of this type is being developed, and we hope to use such models as described here to predict the optimum arrangement of fibres for a realistic device.

ACKNOWLEDGEMENTS

This work was supported by the SERC, Wellcome Trust and Hamamatsu Photonics KK.

REFERENCES

Arridge, S.R., Cope, M., van der Zee, P., Hillson, P.J., and Delpy, D.T. (1985). Visualisation of the oxygenation state of brain and muscle in newborn infants by near infra-red transillumination. In: Information Processing in Medical Imaging, S.L. Bacharach (ed.), Martinus Nijhoff (Holland), pp. 155-176.

Breckon, W.R., Pidcock, M.K. (1987). Some mathematical aspects of electrical impedance tomography. In: Mathematics and Computer Science in Medical Imaging, M.A. Viergever and A. Todd-Pokropek (eds.), NATO ASI Series F, Vol. 39, pp. 351-362.

Bremmer, H. (1964). Random Volume Scattering, Radio Science Journal of Research, 680 (9), pp. 967-981.

Case, M.C., and Zweifel, P.F. (1967). Linear Transport Theory, Addison-Wesley, (New York).

Chandrasekhar, R. (1950). Radiation Transfer, Clarendon Press (Oxford).

Defrise, M. (1989). Regularization techniques in medical imaging, *These Proceedings*

Delpy, D.T., Cope, M., van der Zee, P., Arridge, S.R., Wray, S., and Wyatt, J. (1988 in press). Estimation of optical pathlength through tissue from direct time of flight measurement, Phys. in Med. & Biol.

Flock, S.T., Wilson, B.C., and Patterson, M.S. (1987). Total attenuation coefficients and scattering phase functions of tissues and phantom materials at 633nm, Medical Physics, 14 (5),835-841.
Groenhuis, J., Ferwerda, H.A., and Ten Bosch, J.J. (1983). Scattering and absorption of turbid materials determined from reflection measurement (Parts I and II), Applied Optics, 2216, pp. 2456-2467.
Jarry, G., Ghesquiere, S., Maarek, J.M., Debray, S., Bui-Mong-Hung, and Laurent, D. (1984). Imaging mammalian tissues and organs using laser collimated transillumination, J.Biomed. Eng., 6, pp 70-74.
Jobsis, F.F. (1977). Non invasive infra-red monitoring of cerebral and myocardial sufficiency and circulatory parameters, Science, 198, pp. 1264-1267.
Ishimaru, A. (1978a). Wave propagation and scattering in random media, Academic Press. (New York).
Ishimaru, A, (1978b). Diffusion of a pulse in densely distributed scatterers, J. Opt. Soc. Am., 68(8), pp 1045-1050.
Lewis, H.W. (1950). Multiple scattering in an infinite medium, Physical Review, 78(3), pp. 526-529.
Reynolds, L., Johnson, C., and Ishimaru, A., (1976). Diffuse reflectance from a finite blood medium : applications to the modelling of fibre optic catheters, Applied Optics, 15(9), pp 2059-2067.
Shimizu, K., Ishimaru, A., Reynolds, L., and Bruckner, A.P., (1979). Backscattering of a picosecond pulse from densely distributed scatterers, Applied Optics, 18(20), pp 3484-3488.
Tamura, M., Nomura, Y.,Hazeki, O. (1987) Laser tissue spectroscopy - near infrared CT. The review of Laser Engineering, 15 (8),74-82
Tamura, M., Hazeki, O., Nioka, S., Chance, B., and Smith D.S. (1988) The simultaneous measurements of tissue oxygen concentration and energy state by near-infrared and nuclear magnetic resonance spectroscopy. Adv. Exp. Med. Biol., 227, 359-364.
Tikhonov, A.N., and Arsenin, V.Y., (1977). Solutions of ill-posed problems, V.H.Winston and Sons (Washington D.C.).
van der Hulst (1980). Multiple light scattering, Academic Press. (New York).
van der Zee, P., Delpy, D.T. (1987). Simulation of the point spread functions for light in tissue, Adv. Exp. Med. & Biol., 215, pp. 179-192.
van der Zee, P., Delpy, D.T. (1988a). Computed point spread function for light in tissue using a measured volume scattering function, Adv. Exp. Med. Biol, 222, pp. 191-198.
van der Zee, P., Nalder, C., Delpy, D.T., Arridge, S.R., and Cope, M., (1988b) Optical characterization of tissue for near infra-red light. Lasers in ed. Sci. 3, Abstracts Suppl., 194.
Watmough, D.J., (1983). Transillumination of breast tissues : factors governing optical imaging of lesions. Radiology, 147, 89-92.
Wilson, B.C., and Adam G., (1983). A Monte-Carlo model for the absorbtion and flux distribution of light in tissue, Med.Phys., 10, pp 824-830.
Wyatt, J.S., Cope, M., Delpy, D.T., Wray, S., and Reynolds, E.O.R., (1986). Quantification of cerebral oxygenation and haemodynamics in sick newborn infants by near infra red spectrophotometry, Lancet, 8515, pp 1063-1066.

Part 2 Theoretical Aspects

Section 2.3 Perception

ON THE RELATIONSHIP BETWEEN PHYSICAL METRICS AND NUMERICAL OBSERVER STUDIES FOR THE EVALUATION OF IMAGE RECONSTRUCTION ALGORITHMS

G.T. Herman and K.T.D. Yeung

University of Pennsylvania
U. S. A.

ABSTRACT

Physical metrics, such as the root mean squared distance, are simple to calculate, but we do not have a great confidence in their medical relevance. We have a greater confidence in numerical observer studies, but they are more complicated to carry out. Suppose that a claim is made that a particular physical metric yields similar results to a particular numerical observer study. How can we evaluate the validity of such a claim? In this paper we introduce the notion of ''rank ordering similarity'' to measure the relationship (closeness, similarity) between algorithm performance measures. We use this notion to study the relationship between a particular physical metric (relative clipped error) and a particular numerical observer method (ROC on threshold). We also investigate the role of contrast in this context.

1. INTRODUCTION

An image reconstruction algorithm is supposed to recover medically relevant information that exists in a cross-section of the human body from measured integrals of x-ray attenuation (e.g., Herman, 1980). There is an enormous variety of such algorithms. The question arises: how well do they perform a specific medical task relative to each other?

An approach to answer this question with a high degree of confidence is that of ROC analysis (e.g., Green and Swets, 1966). In that approach the algorithms are compared based on the performance of human subjects using the reconstructed images to reach a diagnostic decision. The probability of reaching a true-positive decision is plotted against the probability of reaching a false-positive decision for each of the methods, and the relative merits of the methods is decided based on these plots. The problem with ROC studies is their complexity (and hence cost). A number of observers have to be used, each observer has to read many images, conditions for the study have to be carefully controlled, etc.

To overcome these difficulties, it has been suggested to replace the human observer by a numerical observer (e.g., Hanson, 1988). In this approach the diagnostic decision is made by a computer program which is supposed to emulate the behaviour of human observers. Our confidence in the results of such a study is tainted by the worry that the emulation of human behaviour by a computer program can never be perfect. However, considerable simplification is reached by using a numerical rather than a human observer.

An even simpler approach is by the use of physical metrics, such as the root mean squared distance between the reconstructed images and the (for the purpose of the experiment) known originals. For reasons of simplicity such metrics are most frequently reported in the literature, but in many cases with careful provisos regarding their medical relevance (Herman, 1980). However, if a physical metric could be shown to provide the same relative evaluation for different reconstruction algorithms as a more complicated procedure, it should be the evaluation method of choice (of the two) because of its computational simplicity. So the question arises: how do we compare the performance of the evaluators? This paper provides a first small step towards answering this question.

2. A SIMPLE STUDY

Our first work in this direction (Yeung and Herman, 1989) was very much influenced by the ideas propagated by Hanson (e.g., Hanson, 1988). The task we set ourselves is the detectability of low contrast signals in the presence of high contrast signals and their associated artifacts. We have taken a 127x127 pixel phantom which we subdivided into 14x14 square-shaped regions of 9x9 pixels each (see Fig. 1). Seventeen of these 196 regions were randomly selected to contain a circular-shaped low-contrast disc of density 0.1 (and diameter nine pixels). Nine of the regions were randomly selected to contain a circular-shaped high-contrast disc of density 1.0 (and diameter nine pixels). The rest of the regions (170 of them) were set to a uniform zero density. Nine such phantoms were generated, a typical one is shown in Fig. 1.

Three iterative reconstruction algorithms (EM, ART, and underrelaxed ART) were evaluated using the average relative clipped error (a physical metric) and the area under the ROC curve based on thresholding (a numerical observer method) for the nine phantoms; see (Yeung and Herman, 1989) for details. Fig. 2 shows the relative clipped error (plotted against iteration number) and Fig. 3 shows the area under the ROC curve (plotted against iteration number) for the three methods. Since a small relative error is ''good'' and a large area under the ROC curve is ''good'', one might say by looking at Figs. 2 and 3 that the relative performance of the three iterative methods is evaluated with similar results by the two performance measure methodologies.

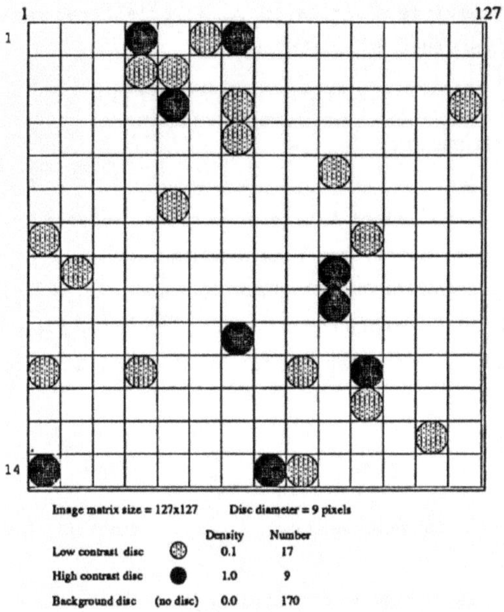

Figure 1 A schematic diagram illustrating a typical phantom.

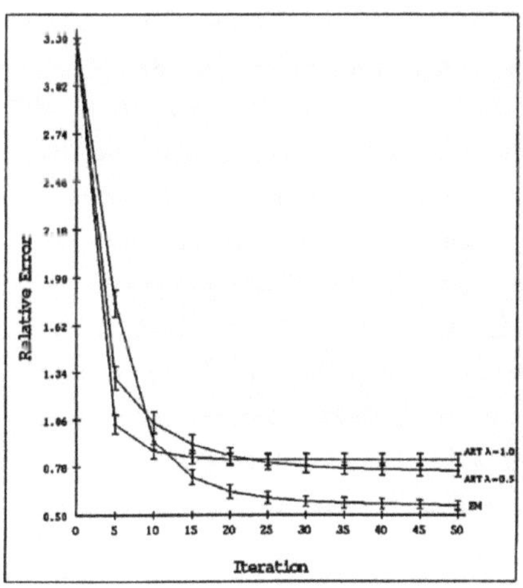

Figure 2 Relative error versus iteration.

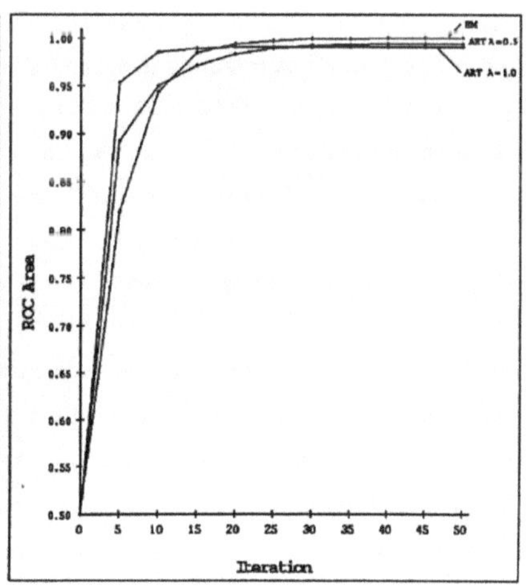

Figure 3 ROC area versus iteration.

How can we make this measure of similarity more precise? One way is by the following definition of ''rank ordering similarity''.

3. RANK ORDERING SIMILARITY

Let r and s be rank orderings of N reconstructions produced by two evaluation methods (or, for that matter, two rank orderings of anything else). More explicitly, by labelling the N reconstructions under study by integers between 1 and N, we use r_i and s_i to denote the rank orderings by the two evaluation methods of the i'th reconstruction. Then the similarity $S(r,s)$ between the two rank orderings is defined by

$$S(r,s) = 1 - \frac{1}{\left\lfloor \frac{N^2}{2} \right\rfloor} \sum_{i=1}^{N} |r_i - s_i|, \quad (1)$$

where $\left\lfloor \frac{N^2}{2} \right\rfloor$ denotes the largest integer not larger than $N^2/2$. (We assume $N \geq 2$.)

This measure of similarity has some desirable properties. Its value is 1 when the two rank orderings are the same and 0 when one is the reverse of the other. The behaviour of its mean is stable: considering all rank orderings equally likely, the expected values of $S(r,s)$ are $\frac{1}{3}$ for all odd N, and $\frac{1}{3}\left[1+2/N^2\right]$ for all even N; see (Yeung and Herman, 1989).

On the other hand, the measure also has some undesirable properties. For $N = 3$, and $r_i = i$ (for $1 \leq i \leq 3$), and s defined by $s_1 = 3$, $s_2 = 1$, $s_3 = 2$, we find that $S(r,s) = 0$, which is the same as for the s which is the reverse of r (i.e., $s_1 = 3$, $s_2 = 2$, $s_3 = 1$). This is not what our intuition tells us that a rank ordering similarity measure should do. For this reason we plan to study in the future some other rank ordering similarity measures. However, in this article we stick to the S defined by (1).

Using this measure we compared the rank orderings of 30 reconstructions produced by the simple study described in the last section. (The 30 reconstructions were those produced by EM, ART, and underrelaxed ART at the 5k'th iteration, where k=1,2,...,10.) Using r to denote the rank ordering of these reconstructions by the average relative clipped error and s to denote that by the area under the ROC curve based on thresholding, we find that $S(r,s)$ is 0.88. The significance of such high value of $S(r,s)$ can be appreciated from the fact that when we randomly generated 5 million rank ordering pairs, the value of $S(r,s)$ obtained was less than 0.75 in all cases. On the other hand, if r denotes the rank ordering produced by another physical metric (the average root mean squared clipped error), we get that $S(r,s)$ is only 0.51 (such a value or higher occurs with a probability of 1.5%). (A plot of the distribution of $S(r,s)$ for uniformly random r and s can be found in Yeung and Herman, 1989.) Thus our simple experiment seems to lead to the conjecture that we may be able to replace the numerical observer metric of the area under the ROC curve based on the thresholding by the physical metric of average relative clipped error, but not by the physical metric of average root mean squared clipped error.

To further investigate the potential validity of the above stated conjecture, as well as some related matters, we carried out a more complex study.

4. A MORE COMPLEX STUDY

In this study ten phantoms were used, each roughly representing the same human thorax cross-section ~~with some ''tumors''~~ at different places in the different phantoms. A more precise representation of the phantoms is given in Fig. 4. Outside the elliptical-shaped thorax the x-ray attenuation is assumed to be 0.0 cm^{-1} and inside it is assumed to be 0.2 cm^{-1} with the following exceptions. There are in the cross-section six pieces of bone (a ''vertebra'', the ''sternum'', and four ''ribs'') each of x-ray attenuation 0.4 cm^{-1} and ten ''tumors'' each of x-ray attenuation 0.205 cm^{-1}. The tumors are all circular in shape and of diameter 2 cm. They lie at ten of the twenty potential tumor locations shown in Fig. 4. The twenty potential tumor locations are arranged symmetrically in the left and right ''lungs''. For each of the phantoms, five out of the ten potential locations in the left ''lung'' were randomly selected to be actual tumor locations, the symmetric images of the other five are the tumor locations in the ''right'' lung. Since the ''bones'' are also left-right symmetric in the phantom, we expect left-right symmetries in the associated artifacts as well. The asymmetric tumor locations were decided upon so that we have an intrinsic control for each of the tumors. (The precise reasoning will become clearer after we discuss our definition of ''contrast''.)

We digitized the phantoms (see Herman, 1980, Section 4.1) in such a way that each pixel in the digitization is 1 cm^2 and each potential tumor location has its center at the vertex of four pixels (and is thus totally covered by those four pixels). A 57x57 digitization was used, which easily covers the ''thorax'' (whose major axis is 50 cm).

For each of the ten phantoms, we generated nine sets of projection data using SNARK77; see (Herman, 1980, Section 4.4). Parallel rays were used with a spacing of 1 cm between the rays. One data set was calculated using perfect integrals along the rays for 180 equally spaced directions. Two other data sets were

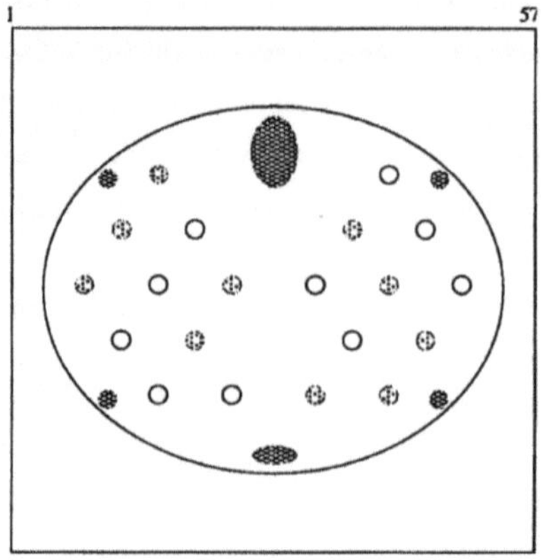

		density
	soft tissue	0.2
○	unoccupied tumor sites	0.2
◌	occupied tumor sites	0.205
●	bony structures	0.4

Figure 4 The thorax phantom.

obtained from the first one by taking subsets of it, each containing values for alternate sets of 90 equally spaced directions. The six further data sets were ''noisy'' versions of the three described above, assuming two different photon counts emitted at the source; see (Herman 1980, Section 4.4). We denote these nine different modes of data collection by PT, PE, PO, LT, LE, LO, HT, HE, HO, where T, E, and O refer to the total of all data, the subset with even directions, and the subset with odd directions, respectively, while P, L, and H refer to the perfect, low-noise, and high-noise measurements, respectively.

For each of the data sets, the ten phantoms were reconstructed using the convolution method with a generalized Hamming filter with $\alpha=0.8$, see (Herman 1980, Section 8.1). In contrast with the simple study described earlier, where we kept the data collection methodology fixed and varied the algorithms, here we use only one algorithm and compare the nine different data collection methodologies described in the previous paragraph. This allows us to compare the rank orderings provided by the measures with reasonable prior expectations: perfect data should yield reconstructions which are preferrable to those obtained from noisy data, the more data are collected, the better the results should be, etc.

The reconstructions were evaluated in a number of different ways, here we discuss the three which have to do with the conjecture stated at the end of the last section.

We let t_j and r_j denote the clipped values in the j'th pixel of the phantom and the reconstruction, respectively. The clipping was done by setting any value below 0.19 cm^{-1} to 0.19 cm^{-1} and any value above 0.22 cm^{-1} to 0.22 cm^{-1}. The relative clipped error is defined by the formula $\sum|t_j - r_j| / \sum|t_j|$ and the root mean squared clipped error is defined by the formula $\left[\sum(t_j - r_j)^2 / \sum(t_j - \bar{t})^2\right]^{\frac{1}{2}}$, where $\bar{t}$ is the mean of the t_j's. The physical metric associated with a particular mode of data collection is obtained by averaging the errors for the ten phantoms in the study.

The numerical observer measure has been obtained using the area under the ROC curves, which were obtained by varying a threshold and considering a two-pixel-by-two-pixel potential tumor location as having a tumor in it if the average value is above the threshold.

Using these definitions, we again found that the similarity between the rank orderings provided by the average relative clipped error and by the area under the ROC curve based on thresholding is high, it is 0.90. However, in this study the similarity between the rank orderings provided by the average root mean squared error and by the area under the ROC curve based on thresholding is even better; it is 0.95. So our more complex study confirms only half of the conjecture stated at the end of the last section.

5. CONTRAST AND SEPARABILITY

Based on the study described in the last section, we also considered some metrics which we thought might be more appropriate for correlating with human performance of tumor detectability. This is based on the idea of (local) contrast, which is a measure of strength of central signal compared with the background variation (noise). More precisely, we proceeded as follows.

Consider, Fig. 5. The central two-pixel-by-two-pixel area marked by **S** is one of the twenty potential tumor locations. We use s to denote the average value in those four pixels. We use a_i to denote the average value in those four pixels which are marked by $\mathbf{b_i}$ in Fig. 5 $(1 \leq i \leq 8)$. We use m and d to denote the mean and standard deviation of the a_i. Then we define the contrast c to be $(s - m)/d$.

In our experiments tumors have higher value than healthy tissue, so in a good reconstruction we would expect c to have a positive value at actual tumor locations and a zero value at potential tumor locations with no tumors. Thus one can use as a figure of merit the area under an ROC curve which is obtained by varying a threshold and considering a two-pixel-by-two-pixel potential tumor location as

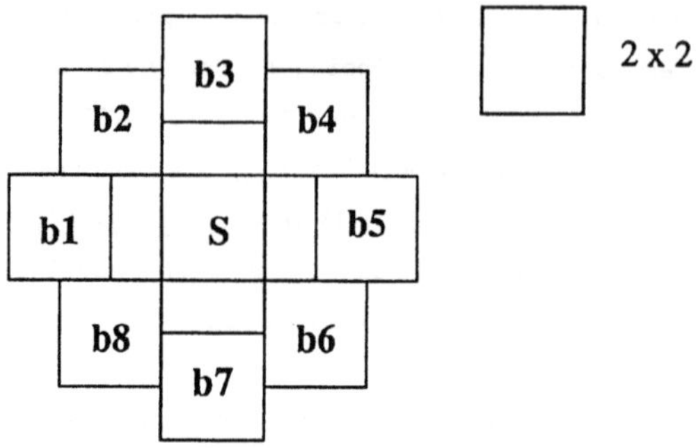

Figure 5 Spatial relationship between potential tumor location and its background neighbourhood.

having a tumor in it, if the contrast there is above the threshold. The similarity of rank orderings produced by this contrast based numerical observer measure and the one discussed in the previous section is high (0.875). However, we observed a to us surprising phenomenon: the performance of the numerical observer based on contrast is worse than that of the numerical observer based on average value for all nine methods of data collection. (For example, for the case of HO, the area under the ROC curve based on contrast is 0.922, while that based on average value is 0.971.)

A physical metric based on contrast can be introduced using the notion of separability from statistical pattern recognition theory (Fukunaga, 1972). This figure of merit is defined as follows. Let $\overline{c^t}$ and $\overline{c^n}$ denote the average values of contrast at potential tumor locations with and without tumors, respectively. Let v^t and v^n be the corresponding variances. We define the between class scatter, b, as $\left(\overline{c^t} - \overline{c^n}\right)^2$ and the within class scatter, w, as $\frac{1}{2}(v^t + v^n)$. Then the separability, J, is defined as b/w.

Using J as the figure of merit (the higher the better), the nine data collection methodologies were rank ordered as follows: PO, LT, HT, PT, LE, LO, PE, HE, HO. Some aspects of this are blatantly ridiculous; for example, for the total data sets, perfect data yield worse ranking than both low-noise and high-noise data.

Examining the source of this unexpected phenomenon, we found that it is due to the value of d (the standard deviation of the background) being very small at a single potential tumor location (the one near the center on each side of the ''lung'') with perfect data. This makes the c at that location very large (which is reasonable enough) and causes the within class scatter w to be also large, without a corresponding increase in b. This makes J small.

To demonstrate that we have indeed identified the source of the problem, we removed the offending potential tumor location and only used measurements derived from the other 9 locations. The resulting rank ordering was: PT, PE, PO, LT, HT, LE, LO, HE, HO; a perfectly reasonable rank ordering, whose similarity to that produced by the contrast based numerical observer study is 0.9. (As opposed to this, similarity of the rank ordering produced by J based on all ten potential tumor locations to that produced by the contrast based numerical observer study is only 0.65.)

It appears that separability, based on contrast, can have a very anomalous behaviour if we happen to hit on some unusual circumstances. This brings doubts upon its potential usefulness as a physical metric for the evaluation of image reconstruction algorithms. For a discussion of the behaviour of J in a somewhat different application area see (Fiete et al, 1987).

6. DISCUSSION

We have defined, in equation (1), a function $S(r,s)$ to measure the similarity between two rank orderings r and s. We have shown that this definition of rank ordering similarity can be effectively used to measure the relationship (closeness) between performance measures devised for the evaluation of image reconstruction algorithms.

Nevertheless, while we judge the idea of using some definition of rank ordering similarity in this context meritorious, we have some objections to the specific definition given by equation (1). One of this is stated in the discussion following the equation. Another one is to the hard-and-fast way rank-orderings have to be used in such a definition. There is nothing in the formula which allows us to distinguish between two identical rank orderings, but one in which the difference between the first and last item is statistically insignificant and another in which all differences can be shown to be significant. Further work will have to be done to remedy this situation.

We have specifically studied the relationship between physical metrics and the results of numerical observer studies. We have found evidence of strong similarity between the physical metric obtained by averaging the relative clipped error and the area under the ROC curve provided by the numerical observer based on thresholding the average value in potential tumor areas. Our initial experiments are more ambiguous regarding the physical metric obtained by averaging the root mean squared clipped error.

We have also looked at local contrast at the potential tumor areas as an alternative to averaging the values. Our initial results are disappointing. The numerical observer based on average outperformed the numerical observer based on contrast. The physical metric based on contrast separability gave us anomalous results on the original data set. Nevertheless, since local contrast is a much more appropriate model than average value of a human criterion for tumor detection, we intend to pursue research based on the notions of contrast and separability, hopefully correcting the problems which led to the observed anomalies.

ACKNOWLEDGEMENT

This work is supported by the National Institutes of Health under Grant HL28438.

REFERENCES

Fiete, R.D., Barrett, H.H., Smith, W.E., and Myers, K.J. (1987). Hotelling trace criterion and its correlation with human observer performance, Journal of the Optical Society of America A 4, pp. 945-953.

Fukunaga, K. (1972). Introduction to statistical pattern recognition, Academic Press, New York.

Green, D.M., and Swets, J.A. (1966). Signal detection theory and psychophysics, John Wiley & Sons, New York.

Hanson, K.M. (1988). Method to evaluate image-recovery algorithms based on task performance, Proceedings of the Society of Photo-Optical Instrumentation Engineers 914, pp. 336-342.

Herman, G.T. (1980). Image reconstruction from projections: The fundamentals of computerized tomography, Academic Press, New York.

Yeung, K.T.D., and Herman, G.T. (1989). Objective measures to evaluate the performance of reconstruction algorithms, Medical Imaging III: Proceedings of the SPIE 1092, to appear.

PSYCHOPHYSICAL STUDY OF DECONVOLUTION FOR LONG-TAILED POINT-SPREAD FUNCTIONS

J. P. Rolland[1], H. H. Barrett[1,2], G. W. Seeley[1,2]

[1] University of Arizona, Optical Sciences Center
Tucson, Arizona, United States

[2] University of Arizona, Department of Radiology
Tucson, Arizona, United States

ABSTRACT

The deconvolution of the tails of point-spread functions (PSFs) from simulated data has been examined through the use of psychophysical studies. Long-tailed PSFs arise frequently from scattering effects. They are composed of an unscattered component forming the core of the PSF, and of a scatter component referred to as the tail of the PSF. This type of PSF is commonly encountered, especially in medical imaging, and we have studied its effect on visual signal detection. The question is whether or not any improvement in performance can be gained by deconvolving the tail of the PSF. Experimental results show that the performance of the human observer is indeed improved by deconvolution.

1. INTRODUCTION

The performance of human observers for a specified detection task can be measured by psychophysical methods, as originally described in Green and Swets (1966). This paper treats a case of a simple two-hypothesis detection task, where the human observer is asked to detect a low-contrast signal on a smooth but noisy background. This type of image may simulate simple cases of typical nuclear-medicine images. The performance achieved may be described by a receiver operating characteristic (ROC) curve, as described by Metz (1978). The area under the ROC curve or the detectability index d′, as defined by Tanner and Birsdall (1958), may be used to quantify the performance.

Psychophysical studies are very useful for system assessment, but it is of great practical interest to be able to predict the outcome of psychophysical studies, especially for system design and optimization. In the case of white-noise images, many workers such as Burgess (1984) and Wagner and Brown (1985) have chosen the ideal observer as a standard of comparison for its ability to give good predictions of human observer performance. However, for images showing negative correlations in the noise, using the ideal observer as a standard of comparison results in low efficiencies for the human observer as suggested among others by Hanson (1979) and shown later on by Myers et al (1985). We expect to find another case of decrement in performance for images generated through long-tailed PSF imaging devices, and a natural question arises as whether or not deconvolution of the tail of the PSF leads to an improvement or a degradation of the human observer performance.

Point spread functions of this type may result from diverse physical processes. For example in nuclear medicine, they may occur due to scattering in the patient's body or to septal penetration when high-energy tracers are used, as described in Barrett and Swindell (1981). To determine the impact of deconvolving the tail of the PSF on visual performance, two psychophysical studies (with and without deconvolution) were carried out using computer-simulated images. Each study was composed of ten different experiments corresponding to ten different amounts of blur due to different spreads of the tail of the PSF.

Results show that the ideal observer greatly overestimates the human observer performance and that, for our well-defined detection problem, deconvolution does indeed improve the performance of the human observer.

2. THEORY

The nuclear imaging system used to simulate the image formation, is shown schematically in Figure (1a). The object considered in our study is either the superposition of a low-contrast signal on a smooth background distribution or the background distribution alone when no signal is present. The image-forming element, being for example a simple multibore collimator used for clinical studies, is characterized by a PSF $p_1(r)$, where $\mathbf{r}$ denotes a two-dimensional position vector, as being the sum of a sharply peaked core and an extended tail.

We choose, for mathematical convenience, to model the background, the signal, the core and the tail of the PSF as four Gaussian distributions. To approach some realistic cases, the width of the background distribution is taken to be the widest of the four Gaussians, while the extent of the tail, which is allowed to vary, must be large compared to the signal. To determine the effect of deconvolution, we process the noisy images through a deconvolving filter $p_2(\mathbf{r})$ as depicted in figure (1b). A different filter $p_2(r)$ is designed for each value of the tail width using Fourier transform theory.

To determine if we can predict the outcome of the psychophysical studies using the ideal observer approach, the detectability index, which is also the signal-to-noise ratio associated with the log-likelihood ratio, was evaluated according to signal detection theory expounded in Van trees (1968). In the case of long-tailed PSFs, the mean image, resulting from the convolution of the object with $p_1(r)$, presents a signal-dependent brightness level accross the image due to the convolution of the tail of the PSF with the signal. According to the brightness contrast effect well known in the field of perception, this brightness level that we refer to as a low-freqency level (because of its slow variation over the image) is ignored by the human observer when performing a detection task. On the other hand, this term is built into the ideal observer where it accounts for a substantial part of the detectability. This leads us to introduce a pseudo-ideal observer, referred to in the remainder of the paper as the no-low-frequency (nlf) ideal observer, characterized by a detectability index $d'_{ideal}(nlf)$ to account for the suppression of this low-frequency level. Another standard of comparison for human vision, known as the non-prewhitening ideal observer was also investigated as well as its combination with the nlf component. These standards of comparison were investigated to attempt to predict the performance of the human observer and to analyse the effect of linear deconvolution through efficiency measurements.

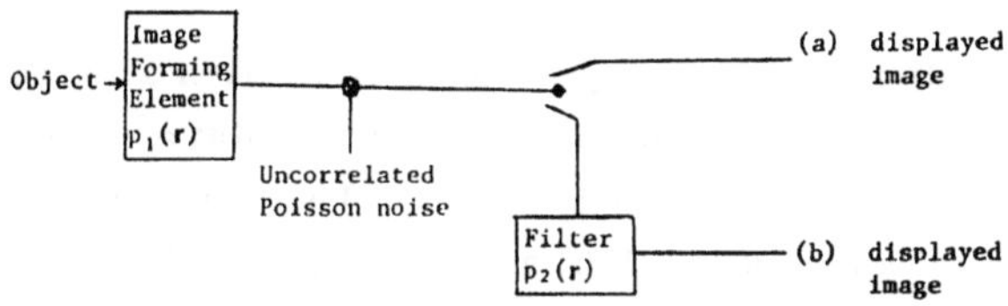

Fig. 1. (a) Basic scheme of nuclear-medicine imaging system. The object is convolved first with $p_1(r)$. Uncorrelated Poisson noise is then introduced to account for fluctuations in the detection process. (b) The noisy images are convolved with the deconvolution filter $p_2(r)$ before display.

3. METHODS

To evaluate the performance of the human observer, ten psychophysical experiments, corresponding to ten different values of the width of the tail PSF, were performed. Within each experiment fifty images were generated. Twenty-five images simulated a bright signal on a smooth background. The signal, which may represent a tumor, was known exactly and located in the center of the images. The signal was removed in the twenty-five other images. These two kinds of simulated objects were processed through the image forming element $p_1(r)$, and uncorrelated Poisson noise was introduced as depicted in figure (1a).

The parameters characterizing the signal, the background, the PSF, as well as the amount of noise, were adjusted such that small changes in detectability could be detected using ROC analysis. The diameters, at 1/e of the peak intensity, of the signal, background and core of the PSF were 8, 180 and 3 respectively. The width of the tail of the PSF was varied from about 52 to 105 pixels in diameter at 1/e of the peak intensity.

Images were displayed as 128 x 128 square arrays, and the contrast in each image was adjusted to cover 256 grey levels. One image covered 8 cm x 8 cm on the display. The light in the room was dimmed down to the same value for all the observers, and the contrast and brightness levels of the display monitor were constant over the study. Each subject viewed the five hundred images binocularly at a comfortable distance chosen by the observer. The observers were required to wear their usual correction glasses.

Six observers ran the studies, each composed of ten experiments. The observers first viewed sample images to ascertain the shape, size and location of the object. They did not have any information on the number of occurrences of the signal. The images were displayed in a random order. The ten experiments were also performed in a randomized order for each observer. The observer was presented a single stimulus on each trial. The stimulus was either the signal on a noisy background or a noisy background alone. The observer was instructed to rate his certainty on a six-point scale. The responses were recorded using a computer program written by Seeley et al. (1982).

Using the rating-scale data, an ROC curve is generated for each observer according to each experiment. Six points on a curve can be obtained by accumulating data across an observer. The curve is maximum-likelihood fitted to the data as described in Dorfman et al. (1969). The statistical efficiency η defined by Barlow (1978) as the squared ratio of the measured detectability d'_{meas} to d'_{ideal} is also derived as an evaluation of the performance of the human observer with respect to the ideal observer. By analogy with Barlow's definition, we can define a pseudo-efficiency η(nlf) as the ratio d'^2_{meas}/d'^2_{ideal}(nlf). A linear regression analysis was carried out on the data points resulting from the psychophysical studies, before and after deconvolution. Similarly, η(npw) is defined for the non-prewhitening ideal observer as the ratio d'^2_{meas}/d'^2_{ideal}(npw) and η(nlp) as d'^2_{meas}/d'^2_{ideal}(nlp) where nlp symbolizes the combination of the nlf component with the non-prewhitening ideal observer.

4. RESULTS

The range of detectability values and the average efficiencies over the ten psychophysical experiments, as well as the slopes, intercepts and correlation coefficients of the regression lines d'_{meas} versus $d'_{calculated}$ are given in tables 1a and 1b.

The average efficiencies η, accross the ten experiments, of 0.3% and 0.6% before and after deconvolution respectively, show that the ideal observer greatly overestimates the performance of the human observer. The values of d'_{ideal}(nlf), on the other hand, are in better agreement with the values of d'_{meas}, and the average efficiencies η(nlf) now take the values 10.6% and 19.5%. This seems to indicate that this pseudo-ideal observer gives a better prediction of human observer performance than the ideal observer itself. If we pursue the study further with the use of the non-prewhitening ideal observer and its combination with the nlf component, the average (over the ten experiments) efficiency of the human observer reaches 21% as shown in table 1a.

A comparison of the results before and after deconvolution shows the advantage of deconvolution by, first, the increase in efficiencies and, second, the increase of the slopes and intercepts of the regression lines that depict the relationship between the detectabilities measured through psychophysical studies and the detectabilities calculated using analytical standards of comparison for human vision. Figure 2 is an example of the effect of deconvolution performed on noisy images, for the case of the nlf ideal-observer.

Table 1a

Model / Index	Ideal	Ideal(nlf)	Ideal(npw)	Ideal(nlp)	Measured before Deconvolution	Measured after Deconvolution
Range of Detectability d' (min:max)	[22.38:34.24]	[2.32:6.80]	[2.41:7.65]	[2.26:6.42]	[0.41:3.04]	[0.65:3.59]
Average Efficiency η	0.3% / 0.6%	10.6% / 19.5%	0.3% / 16%	10.6% / 21.1%	Before	After

Table 1b

(Slope/Intercept) Correlation Coefficient	d'_{ideal}	$d'_{ideal}(nlf)$	$d'_{ideal}(npw)$	$d'_{ideal}(nlp)$
d'meas before deconvolution	(0.21/-4.50) 0.961	(0.57/-1.06) 0.976	(0.21/-4.50) 0.961	(0.57/-1.06) 0.976
d'meas after deconvolution	(0.24/-4.82) 0.977	(0.65/-0.88) 0.988	(0.55/-0.69) 0.988	(0.70/-0.98) 0.987

Table. 1. (a) Range of detectability values measured from psychophysical studies or calculated from analytical models over the ten experiments corresponding to different width of the tail PSF, and averaged efficiencies over the ten experiments, before and after deconvolution. (b) Slopes, intercepts and coefficient of correlation of the regression lines $d'_{measured}$ versus $d'_{calculated}$ for different standards of comparison for human vision.

5. CONCLUSION

This study shows that deconvolution of noisy images formed through a long-tailed PSF does indeed improve the performance of the human observer for the detection of a low-contrast disk signal on a noisy background. This is a very interesting result since the correlations in the noise, resulting from high-pass filtering, did not nullify its advantage. The study shows also that the the nlf ideal observer as well as the non-prewhitening ideal observer predicts efficiencies of 20% for the human observer performance.

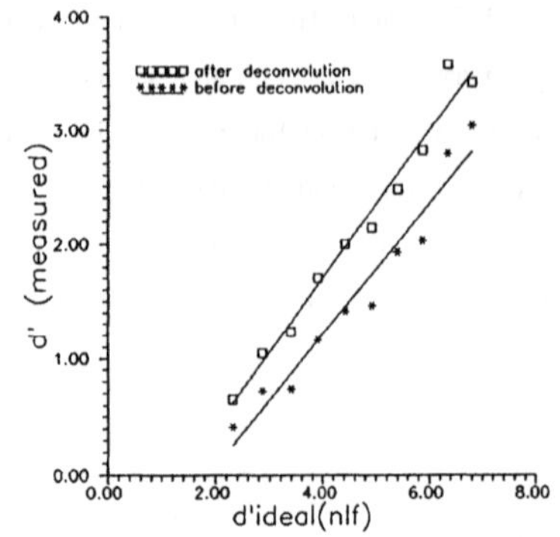

Fig.2. Measured detectabilities d'measured resulting from psychophysical experiments versus detectabilities d'ideal(nlf) calculated using the no-low-frequency ideal observer, before and after deconvolution. Each point on the regression lines corresponds to a different value of the width of the tail of the PSF that lies between 52 and 105 pixels in diameter at 1/e of the peak intensity.

ACKNOWLEDGEMENT

This research was supported by the National Cancer Institute through grant no. P01 CA 23417. The authors would like also to thank all the people who participated in the observer study for their patience and cooperation.

REFERENCES

Barlow, H.B. (1978). The efficiency of detecting changes in density of random dot patterns, Vision Res. 18, pp. 637-650.

Barrett, H.H., and Swindell, W. (1981). Radiological imaging: the theory of image formation, detection, and processing, Academic, New York, I and II.

Burgess, A.E. (1984). Statistical efficiency of perceptual decisions, Spie 454, pp. 18-26.

Dorfman, D.D., and Alf, E. (1969). Maximum likelihood estimation of parameters of signal detection theory and determination of confidence intervals-rating method data, J. Math. Psych 6, pp. 487-496.

Green, D.B., and Swets, J.A. (1966). Signal detection theory and psychophysics, Wiley, New York.

Hanson, K.M. (1979). Detectability in CT images, Med. Phys. 6, pp. 441-451.

Metz, C.E. (1978). Basic principles of ROC analysis, Seminars in Nuclear Medicine 4, pp. 283-298, III.

Myers, K.J., Barrett, H.H., Borgstrom, M.C., Patton, D.D., and Seeley, G.W. (1985). Effect of noise correlation on detectability of disk signals in medical imaging, J. Opt. Soc. Am. A2, pp. 1752-1759.

Seeley, G.W., Borgstrom, M.C., and Mazzeo, J. (1982). A general interactive computer program for running signal detection experiments, Behavior Res. Methods Instrum. 4, pp. 555-556.

Tanner, W.P., and Birdsall, T.G. (1958). Definition of d' and η as psychophysical measures, J. Acoust. Soc. Am. 30, pp. 922-928.

Van Trees, H.L. (1968). Detection, estimation, and modulation theory, Wiley, New York, I and II.

Wagner, R.F., and Brown, D.G. (1985). Unified SNR analysis of medical imaging systems, Phys. Med. Bio. 30 (6), pp. 489-518.

Part 2 Theoretical Aspects

Section 2.4 Image Processing

MATHEMATICAL MORPHOLOGY IN HIERARCHICAL IMAGE REPRESENTATION

O Ying-Lie and Alexander Toet

University Hospital Utrecht
Institute for Perception TNO
The Netherlands

ABSTRACT

Hierarchical structural image representations that are based on mathematical morphology are presented. Morphological operations specifically manipulate the shapes in the image, and only alter certain details without affecting the remaining image structure. A morphological hierarchical representation reflects the natural decomposition of shapes in the image.

1. INTRODUCTION

Image analysis requires a representation that reflects properties of shape. The representation of a shape by its natural components requires knowledge of the spatial structure of the shape. Shape patterns in images may occur at multiple scales. Complex large shapes can be represented as a union of simple shapes of smaller size. Thus, a hierarchical image representation may be obtained from a progressive decomposition of shapes.

A familiar hierarchical representation is the multiresolution representation. It allows analysing the image at a certain resolution level, thus avoiding unnecessary fine details. Typically, a multiresolution representation is an ordered sequence of images in which each image is a filtered version of its predecessor. The conventional multiresolution representations apply convolution filters to reduce the resolution. This results in an overall blurring of the image. In contrast, morphological operations specifically manipulate the shapes in the image. Moreover, morphological operations only alter certain details without affecting the remaining image structure.

2. HIERARCHICAL REPRESENTATION OF IMAGES

Image analysis requires a symbolic description of the structure of the image. The description supplies comprehensive information about the image for different purposes. The symbols used in the description should correspond to meaningful features in the image. Partial descriptions are subsets of the complete description and have the character of a logically filtered image.

An image representation consists of a list of symbols and a relation between these symbols. The symbols represent structural primitives of the image. The relation describes the structural composition of the image. A symbol may include attributes that depend on numerical measurements of the local structure on which the symbol is based.

In general, the hierarchical representation of an image is obtained by three consecutive or combined operations, namely:

application of filter operations to generate a sequence of images,

extraction of the image descriptions from the sequence of filtered images,

determination of the hierarchical relation from the sequence of image descriptions.

Hence, the algorithm is fully specified by the filter operator, the description method, and the hierarchical relation.

At each level, the filter operator removes unnecessary details in the image. Thereafter, the image representation is obtained by using a suitable description method. Finally, the hierarchical relation determines the connection between image representations at subsequent levels.

2.1 Filter operators

The sequence of images with progressive structural contents is generated by filtering. Each image in the sequence is obtained from its predecessor.

Filter operators are operators that are translation invariant over the image plane. These operators are compatible under change of scale of both the image plane and the grey values.

The image representation depends on the choice of the filter operator. Filtering may remove certain details in the image. The amount of removed details depends on the type and the size of the filter operator. Large operators may miss important details, while small operators leave extraneous details.

A decomposition of an image into a set of pattern primitives can be obtained by successive application of filters of certain types and sizes.

Linear filters such as convolution filters have the disadvantage of blurring the shapes in the image. These filters operate on frequencies in the image. As a result, important image features such as edges are smoothed. Nonlinear filters preserve image features better than linear filters. Examples of nonlinear filters are combinatorial filters such as majority filters, and order statistics filters such as local maximum, local minimum, and median filters.

2.2 Description methods

An image can be described by syntactic structural primitives. A fundamental issue in the structural description of patterns is the selection of the set of the pattern primitives in which the description is expressed. Both the identity and the attributes of these primitives should be relatively stable in the presence of phenomena which introduce small random variations in the measurements of the patterns.

Other than the usual greylevel representation, special features that specify shapes in the image can be extracted. The boundaries of shapes in the image can be described by connected lines and arcs.
A well known method for boundary estimation is that of thresholding. The description of an image at various threshold levels yields the multilevel representation. The structural description of the resulting level curves is induced by their nesting at different levels.
A unique description of a greylevel image can be given by level curves through critical points (Koenderink and van Doorn, 1979). These level lines decompose the greylevel image into a collection of juxtaposed and superimposed regions including a single extremum. Thus, an image can be described by the set of its greylevel extrema together with the containment relation of their corresponding regions (Pizer et al., 1985).
Other methods to extract object boundaries use greylevel statistics. Coherent regions can be extracted by aggregating elements with similar statistics. The structural description of region boundaries is deduced from their containment relation.
A natural shape representation by ribbon-like components is accomplished by the extraction of skeletal feature primitives (Gauch et al., 1987). The structural description of skeletal primitives is induced by their connectivity. Under certain circumstances, a shape can be reconstructed from its skeleton representation by means of an inverse skeleton transform.

2.3 Hierarchical relations

The hierarchical image representation is determined by the hierarchical relation between image representations at subsequent levels. The filter operator and the description method are chosen such that the image representations at subsequent levels have progressive structural contents.
In the identity hierarchical relation a correspondence is established between all points of successive hierarchical levels of the image.
Sampling is a natural hierarchical relation, resulting in the multiresolution representation mentioned earlier. A family of images with progressively decreasing structural content can be produced by progressively increasing the size of the filter operator. A hierarchical image representation can then be obtained by sampling the descriptions generated by operators of increasing sizes. Obviously, the sampling distance increases with the filter size.
Regarding the description method, the relation between subsequent hierarchical levels can be specified by following certain features or by finding the best link in a certain sense (Burt et al., 1981; Hong et al., 1982). A hierarchical image decomposition into a set of light and dark spots can be obtained by following the greylevel extrema in a family of images in which each is a slightly blurred version of its predecessor (Pizer et al., 1985).

3. MATHEMATICAL MORPHOLOGY

The theory of mathematical morphology is based on set algebra (Matheron, 1975; Serra, 1982). The basic operations on bivalued images can be extended to greylevel images (Serra, 1982, 1988; Sternberg, 1986). Morphological operations operate directly on the structure of shapes in the image. The change in these shapes is determined by the structuring element and the morphological operation.

Two-dimensional image shapes are represented as algebraic sets in Euclidian space **E**, which can be either the continuous space $\mathbf{R}^2$ or the discrete space $\mathbf{Z}^2$.

Each image shape is assumed to contain its boundary, and thus, can be represented by a closed subset of **E**. In addition, the structuring element is represented by a compact subset of **E**.

The notation employed in this paper is that of Matheron (1975) and Serra (1982). Other authors may use different notations.

The fundamental principles of mathematical morphology are (Serra, 1982):

invariance under translation $\phi(X_y) = (\phi X)_y,$ (1)

compatibility under change of scale $\phi(\lambda X) = \lambda(\phi X),$ (2)

local knowledge $\forall Z \ , \ \exists Y \subset Z \rightarrow \phi(X \cap Y) \cap Z = (\phi X) \cap Z,$ (3)

upper semicontinuity $X_n \downarrow X \leftrightarrow \phi X_n \downarrow \phi X,$ (4)

where ϕ denotes the morphological operator,

and X_y denotes the translation of X by y.

A morphological operator is increasing, that is

$$X \subseteq Y \leftrightarrow \phi X \subseteq \phi Y. \qquad (5)$$

3.1 Morphological operators

The primitives of morphological operations are erosion and dilation.

Erosion of a shape by a structuring element is the set of points such that all translates of the structuring element by the shape are fully contained in the original shape. Erosion is the intersection of translations of the shape by the structuring element.

Likewise, dilation of a shape by a structuring element is the set of points such that all the translates of the structuring element by the shape intersect the original shape. Dilation is the union of translations of the shape by the structuring element.

Consequently, the erosion $\epsilon(X,B)$ and the dilation $\delta(X,B)$ of a closed set X by a compact structuring element B are defined as

$$\epsilon(X,B) \equiv \{x : B_x \subseteq X\} = \bigcap_{b\in\widetilde{B}} X_b = \bigcap_{b\in B} X_{-b}, \tag{6}$$

$$\delta(X,B) \equiv \{x : B_x \cap X \neq \emptyset\} = \bigcup_{b\in\widetilde{B}} X_b = \bigcup_{b\in B} X_{-b}, \tag{7}$$

where the reflected set of B with respect to the origin is denoted by $\widetilde{B} = \{-b : b \in B\}$.

Morphological operators induce the inclusion relation. Erosion shrinks the shape (anti extensive), while dilation expands the shape (extensive),

$$\epsilon(X,B) \subseteq X \subseteq \delta(X,B). \tag{8}$$

Erosion commutes with intersection of sets, and dilation commutes with union of sets,

$$\epsilon((\cap X_n),B) = \cap(\epsilon(X_n,B)) \quad \text{and} \quad \delta((\cup X_n),B) = \cup(\epsilon(X_n,B)). \tag{9}$$

Opening is an erosion followed by a dilation. The opening of a shape by a structuring element is the union of translations completely contained in the shape.

Conversely, closing is a dilation followed by erosion. The closing of a shape by a structuring element is the complement of the translations of the reflection of the structuring element which are completely contained in the complement of the shape.

Hence, the opening $\alpha(X,B)$ and the closing $\kappa(X,B)$ of X by B are defined as

$$\alpha(X,B) \equiv \bigcup_{B_x \subseteq X} B_x = \delta(\epsilon(X,B),\breve{B}), \tag{10}$$

$$\kappa(X,B) \equiv \left(\bigcup_{B_x \subseteq X^c} B_x \right)^c = \epsilon(\delta(X,B),\widetilde{B}). \tag{11}$$

$$\tag{12}$$

Opening shrinks the shape (anti extensive), while closing expands the shape (extensive),

$$\alpha(X,B) \subseteq X \subseteq \kappa(X,B). \tag{13}$$

Furthermore, opening and closing are both idempotent operations

$$\alpha(\alpha(X,B),B) = \alpha(X,B) \quad \text{and} \quad \kappa(\kappa(X,B),B) = \kappa(X,B). \tag{14}$$

3.2 Greylevel morphology

Greylevel images are represented by 2-D functions whose domains are subsets of the continuous space $\mathbf{R}^2$ or the discrete space $\mathbf{Z}^2$, and whose ranges are subsets of $\mathbf{R}$.

A 2-D greylevel image function can be represented by a family of 2-D sets by thresholding the function at all possible grey values. The family of these sets is called the cross sections of the image function. The cross section of f at level t is given by

$$X_t = \{f(x) \geq t : x \in \mathbf{R}^2\} \quad , \quad t \in \mathbf{R}. \tag{15}$$

Furthermore, set inclusion of cross sections corresponds to an ordering of functions

$$X_t(f) \subseteq X_t(g) \quad , \quad \forall t \in \mathbf{R} \Longleftrightarrow f \leq g. \tag{16}$$

An upper semicontinuous function can be reconstructed exactly from all its cross sections using a supremum

$$f(x) = \sup\{X_t(f) : t \in \mathbf{R}\} \quad , \quad x \in \mathbf{R}^2. \tag{17}$$

The morphological operations on sets can be extended to greylevel functions. Set union and intersection of cross sections of two greylevel functions correspond, respectively, to pointwise supremum and infimum

$$X_t(f) \cup X_t(g) = X_t(\sup(f,g)) \quad \text{and} \quad X_t(f) \cap X_t(g) = X_t(\inf(f,g)). \tag{18}$$

For discretized greylevel functions, supremum and infimum, respectively, can be replaced by pointwise maximum and minimum.
The expressions for erosion and dilation can now be determined by modification of the expressions for set operations:

$$\epsilon(f,g)(x) = \min\{f(y) - g(y-x) : y \in \mathbf{B}\} \quad , \quad x \in \mathbf{Z}^2, \tag{19}$$

$$\delta(f,g)(x) = \max\{f(y) + g(y-x) : y \in \mathbf{B}\} \quad , \quad x \in \mathbf{Z}^2, \tag{20}$$

$$\tag{21}$$

where the domain $\mathbf{B}$ of g is a compact subset of $\mathbf{Z}^2$.

Morphological operations with bivalued structuring elements commute with thresholding. Hence, the operation reduces to performing morphological set operations on all the cross sections:

$$\phi(X_t(f)) = X_t(\phi(f)) \quad , \quad t \in \mathbf{R}. \tag{22}$$

Morphological operations on greylevel functions with bivalued structuring elements correspond with the nonlinear local maximum or local minimum filters

$$\epsilon(f,B)(x) = \min\{f(y) : y \in B_x\}, \tag{23}$$

$$\delta(f,B)(x) = \max\{f(y) : y \in B_x\}. \tag{24}$$

4. MORPHOLOGICAL HIERARCHICAL REPRESENTATION OF IMAGES

Image analysis frequently requires a hierarchical image representation that reflects the natural decomposition of shapes in the image. Morphological operations specifically manipulate the shapes in the image without affecting the remaining image structure. Therefore, mathematical morphology provides image representations that are directly related to shapes in the image.

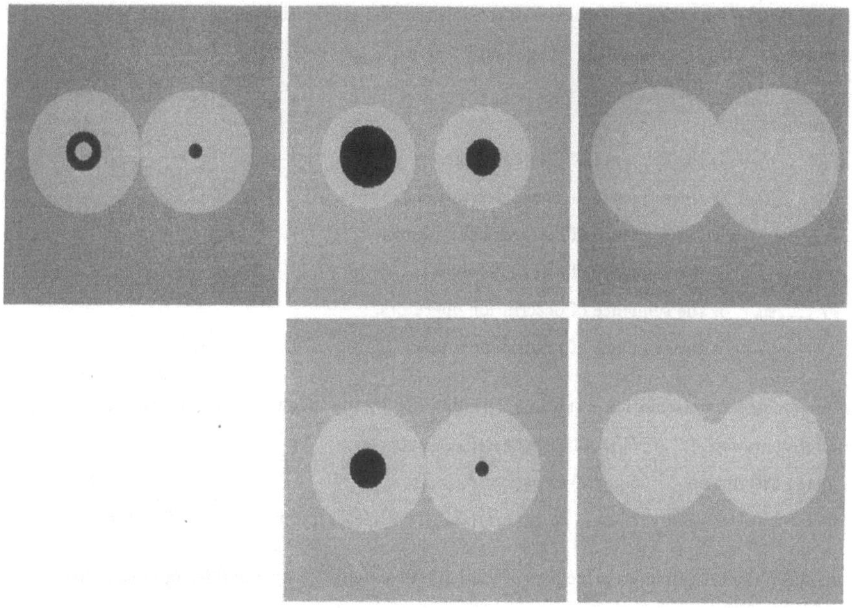

Fig. 1. The original image is shown on the upper left. This figure shows the effects of erosion (upper middle), dilation (upper right), opening (lower left) and closing (lower right) on the original image with a bivalued structuring element. The structuring element is a disk. Its size corresponds to the smallest detail in the original image.

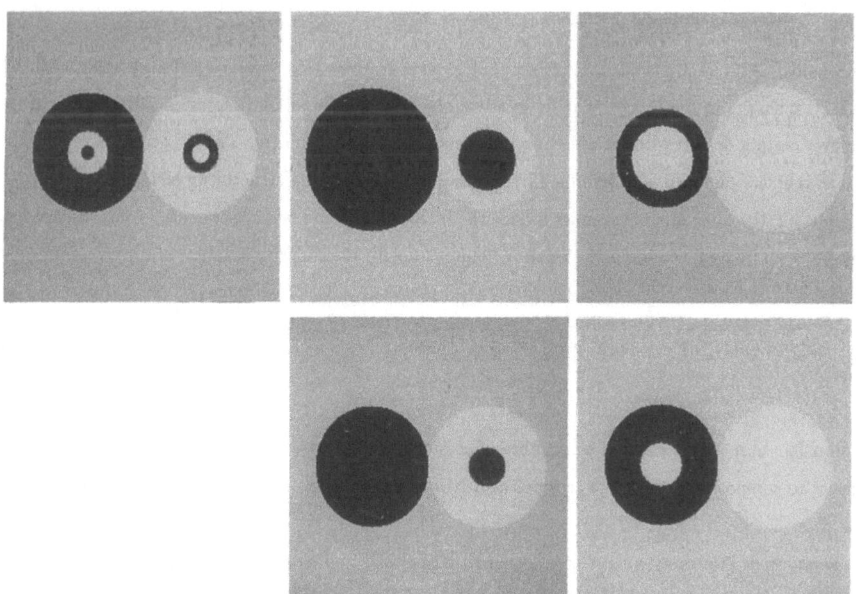

Fig. 2. As Fig.1. for the original image shown on the upper left.

Let $F(x,t)$ or, if appropriate, abbreviated to F denote the greylevel image, where $x \in \mathbf{Z}^2$ is the domain of the image and $t \in \mathbf{R}$ is the greylevel range.

In the sequel,

$F_0, F_1, \ldots, F_N$	is the sequence of filtered images,
$G_0, G_1, \ldots, G_N$	is the sequence of image descriptions,
$H_1, \ldots, H_N$	is the sequence of hierarchical relations,
$\varphi_1, \ldots, \varphi_N$	is the sequence of filter operators,
$\psi_0, \psi_1, \ldots, \psi_N$	is the sequence of description operators,
$\chi_1, \ldots, \chi_N$	is the sequence of relation operators.

The hierarchical representation of an image is obtained by the following three operations, namely:

the filter operation	$F_n(x,t) = \varphi_n(x,t)F_{n-1}(x,t)$	$n = 1, \ldots, N$,
the image description	$G_n(x,t) = \psi_n(x,t)F_n(x,t)$	$n = 0, \ldots, N$,
the hierarchical relation	$H_n(x) \;= \chi_n(x,t)\{G_{n-1}(x,t), G_n(x,t) : x \in H_{n-1}(x)\}$	$n = 1, \ldots, N$.

To maintain the fundamental principles of mathematical morphology, the filter operators must be morphological operators. The same consideration is also valid for the description operators.

4.1 Morphological filters

Morphological filters are sequences of morphological operations that have special properties with respect to shapes in the image. The main advantage of employing a morphological filter is that it removes certain details without affecting the remaining image structure. The shape and size of details that are removed is determined by the shape and size of the structuring element.
In addition, the filter operators must induce the inclusion relation on the sequence of filtered images. For greylevel images, the inclusion relation commutes with the ordering relation of grey values. Hence,

$$F_0 \supseteq F_1 \supseteq \cdots \supseteq F_N, \quad \text{or} \quad F_0 \subseteq F_1 \subseteq \cdots \subseteq F_N, \tag{25}$$

$$F_n(x,t) = \varphi_n(x,t)F_{n-1}(x,t) \quad n = 1, \ldots, N. \tag{26}$$

A filter operator may consist of a sequence of erosions or dilations. These operators are closed with respect to composition, therefore erosions or dilations with the same structuring element can be used as a filter.
An idempotent filter may consist of a sequence of openings and closings (Serra, 1988). The sizes of the structuring elements are chosen such that the inclusion relation holds.
Order statistics filters commute with thresholding. Therefore, these filters relate to morphological operations. In particular, opening and closing are respectively the lower and upper bounds of the median filter (Maragos, 1987).

4.2 Morphological description methods

A morphological description method extracts useful information of shapes in the image while preserving the fundamental principles of mathematical morphology and the inclusion relation induced by the filter operation. Accordingly,

$$G_0 \supseteq G_1 \supseteq \cdots \supseteq G_N, \quad \text{or} \quad G_0 \subseteq G_1 \subseteq \cdots \subseteq G_N, \tag{27}$$

$$G_n(x,t) = \psi_n(x,t)F_n(x,t) \quad n = 0,\ldots,N, \tag{28}$$

and in addition

$$G_n \subseteq F_n \quad n = 0,\ldots,N. \tag{29}$$

In case of a bivalued structuring element, the above mentioned properties also apply to summation and subtraction of images.

The multilevel description preserves the inclusion relation. For bivalued structuring elements it commutes with morphological operations.

Shapes in an image can be represented by a bivalued description. Using morphological operations, several of these descriptions can be obtained directly from the greylevel image without having to determine the boundaries of the shapes first.

A specific description method that contains condensed information about the size, orientation, and connectivity of a shape is the skeleton. The skeleton of both a bivalued or a greylevel image can be obtained directly by a sequence of morphological operations (Serra, 1982; Maragos and Schafer, 1986).

4.3 Morphological hierarchical relations

The hierarchical relation determines the relation between subsequent levels of image descriptions. More specifically, it links elements in the domains of successive descriptions. The links are restricted by previously established links and may depend on the grey values associated with these links.

The resulting hierarchical structure constituted by these links can be represented by a graph. In the graph, the nodes are elements of the image description that are associated by the hierarchical relation. Hence, the hierarchical structure preserves the morphological properties of the description method.

A special type of a hierarchical graph is the tree. The hierarchical structure of an image may be represented by several trees. The roots of these trees may be located on different hierarchical levels.

$$H_0 \supseteq H_1 \supseteq \cdots \supseteq H_N, \tag{30}$$

$$H_n(x) = \chi_n(x,t)\{G_{n-1}(x,t), G_n(x,t) : x \in H_{n-1}(x)\} \quad n = 1,\ldots,N, \tag{31}$$

and in addition

$$H_n \subseteq G_n \quad n = 1,\ldots,N. \tag{32}$$

The identity, the sampling, and most of the feature following relations preserve the morphological properties.

The sampling of a morphological filtered image can be related to the size of the structuring element of the filter operator. In accordance to the standard sampling theorem, the sampling distance must be smaller than half the size of the structuring element.

A sampled image can be reconstructed either by closing or dilation with a reconstructing element (Haralick et al., 1987). The shapes of the reconstructing element and the structuring element must be symmetric and equal. The size of the reconstructing element must be such that the dilation of the sample points fills the whole space. Under the above conditions, the accuracy of the reconstruction is bounded by the closing of its sampling as lower bound, and the dilation of its sampling as upper bound.

5. APPLICATIONS

Hierarchical image representation is a powerful tool for image analysis. The representation may consist of an ordered sequence of images with progressive structural contents. Global information can be used to impose constraints on local operations. Therefore, a hierarchical algorithm can be more efficient than operations performed on a single level.

A hierarchical decomposition of an image into a set of shape primitives can be obtained by successive application of morphological filters with structuring elements of various shapes and sizes. In the following, several useful hierarchical image representations are proposed.

5.1 A morphological multiresolution representation

A well known multiresolution image representation is the pyramid. Typically, a pyramid is an ordered sequence of images in which each image is a low-pass filtered and sampled copy of its predecessor. The purpose of the low-pass filter is to remove spatial frequencies which are higher than the Nyquist frequency of the sample spacing.

Low-pass filtering techniques are based on linear combinations of sinusoidal waveforms, a representation far removed from shape. Mathematical morphology operates directly on shapes. Thus, mathematical morphology provides a better basis for constructing pyramids than low-pass filters.

A multiresolution representation that is similar to the Gaussian pyramid (Burt, 1984) can be obtained by subsequently applying a morphological filter to remove details of the image. At each level, the morphological filter removes details of a size that is related to the size of the structuring element. Logically, the hierarchical relation is determined by sampling at intervals of half this size.

A multiresolution representation can be specified as follows:

morphological filter :	opening + closing - original	$\varphi_n = \alpha_n + \kappa_n - \iota_0$,
	with disks of increasing sizes,	
description method :	greylevel	$\psi_n = \iota_n$,
hierarchical relation :	sampling	$\chi_n = \sigma_n$, or
	identity	$\chi_n = \iota_n$.

Examples of multiresolution representations are given in Fig. 3 and Fig. 4. Fig. 3 shows two sampled multiresolution representations of bivalued images. A multiresolution representation of a greylevel image is shown in Fig. 4.

5.2 A morphological sieve

A band-pass or error pyramid is defined as the difference between low-pass filtered images at successive levels of a low-pass pyramid. Therefore, each image in the band-pass pyramid contains image features lost in one filter step to the next. In the band-pass pyramid features are segregated by size. The coarsest details are restricted to the top level and progressively smaller details are restricted to progressively lower pyramid levels.

A well known band-pass pyramid is the Laplacian pyramid (Burt, 1984; Burt and Adelson, 1983). Each level in the Laplacian pyramid is defined as the difference between two successive levels in the Gaussian pyramid.

The morphological sieve is analogous to the Laplacian pyramid. Each level represents the difference of two successive levels of a morphological multiresolution representation. Shapes are segregated by size with progressively coarser detail in each higher level. The representation acts like a sieve with progressively decreasing mesh size.

In contrast to the Laplacian pyramid, details represented in the morphological sieve are not deformed. This is due to the fact that morphological filters remove image details without adding greylevel bias. Thus, the pattern primitives in the morphological sieve representation have a well defined shape and size.

The morphological sieve is specified by:

morphological filter :	closing (opening)	$\varphi_n = \kappa_n \alpha_n$,
	with disks of increasing sizes,	
description method :	greylevel subtraction	$\psi_n = \iota_{n-1} - \iota_n$,
hierarchical relation :	sampling	$\chi_n = \sigma_n$.

Examples of sieves using the same original images as in the previous section are given in Fig. 5 and Fig. 6. Sampled sieves of bivalued images are shown in Fig. 5. Fig. 6 represents a sieve of a greylevel image.

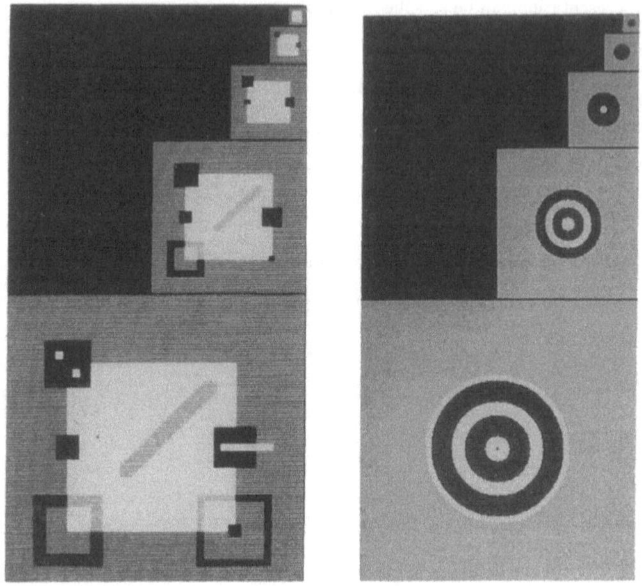

Fig. 3. Sampled multiresolution representations of bivalued images.
The original image is shown on the bottom. The filter operation is performed with disk shaped structuring elements. The size of the disks increases upwards.

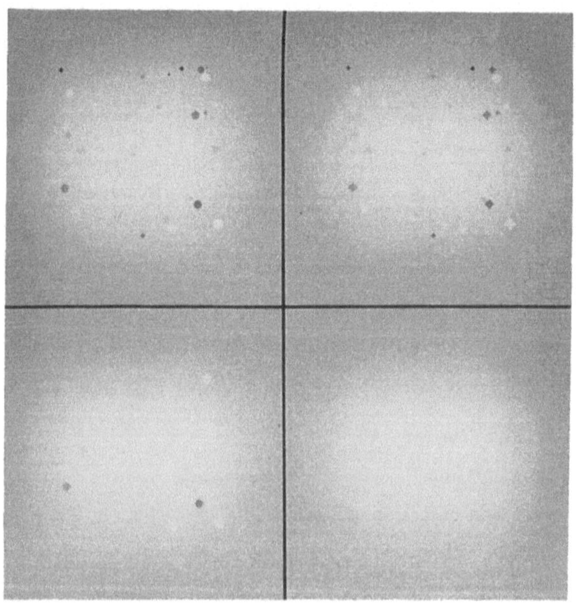

Fig. 4. A multiresolution representation of a greylevel image.
The original image is shown on the upper left. The filter operation is performed with disk shaped structuring elements of increasing size. The size of the disks increases from left to right and downwards.

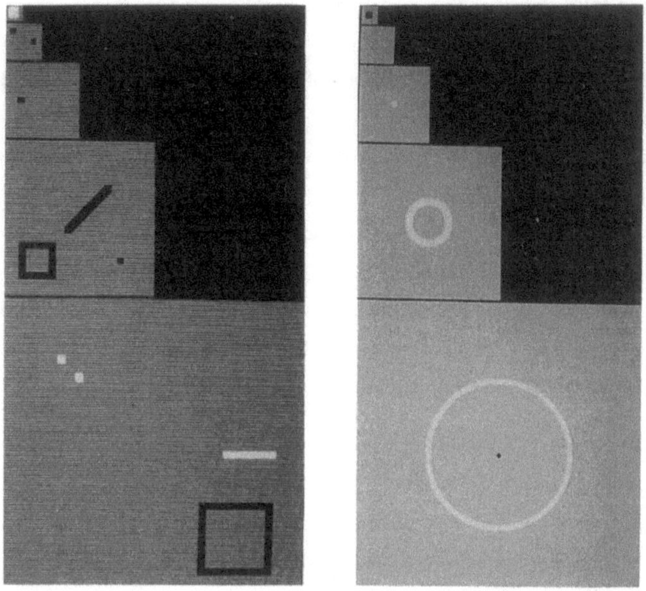

Fig. 5. Sampled sieves of bivalued images.
The original images are the same as in Fig. 3. The filter operation is performed with disk shaped structuring elements. The size of the disks increases upwards.

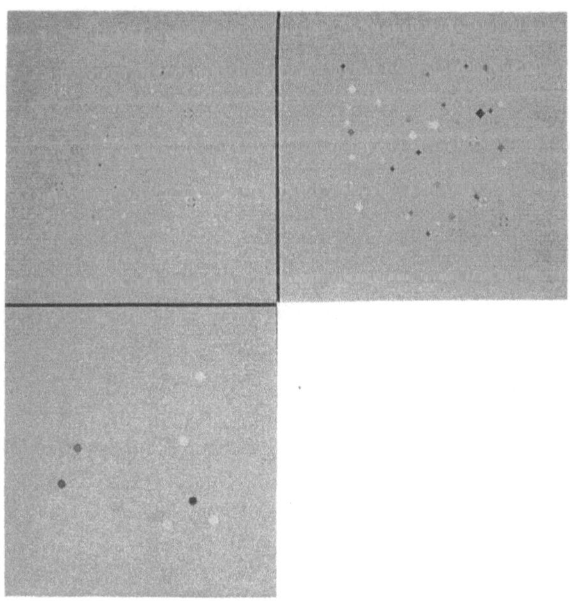

Fig. 6. A sieve of a greylevel image.
The original image is the same as in Fig. 4. The filter operation is performed with disk shaped structuring elements of increasing size. The size of the disks increases from left to right and downwards.

5.3 Application on an MRI image

In general, MRI images display fine anatomical details of the body. Both the spatial as well as the greylevel resolution of these images are quite high.
A shape enhancement representation can be obtained from the multiresolution representation by choosing an appropriate description method.
The subtraction of successive images in this representation results in a difference stack. From the difference stack, a set of bivalued masks are obtained by thresholding. Multiplication of the original image with these masks isolates those parts of the original image that conform to the size and shape of the structuring element and exceed a certain threshold contrast level. Subtracting the resulting stack of high contrast segments from the original image accentuates the low contrast areas of the image.

The shape enhancement algorithm comprises:

morphological filter :	opening + closing - original	$\varphi_n = \alpha_n + \kappa_n - \iota_0$,
	with small disks of increasing sizes,	
description method :	original - original ×	
	threshold of (greylevel subtraction)	$\psi_n = \iota_0 - \iota_0 \times X_t(\iota_{n-1} - \iota_n)$,
hierarchical relation :	identity	$\chi_n = \iota_n$.

Fig. 7 shows the results of the application of the shape enhancement algorithm on an MRI image.

The morphological sieve can also be used to generate a morphological gradient representation. Each level represents the difference of two successive levels of a morphological multiresolution representation. Subtraction of successive levels specifically accentuates details that are filtered between these levels.

A morphological gradient representation can be generated by:

morphological filter :	closing (opening)	$\varphi_n = \kappa_n \alpha_n$,
	with small disks of increasing sizes,	
description method :	greylevel subtraction	$\psi_n = \iota_{n-1} - \iota_n$,
hierarchical relation :	sampling	$\chi_n = \sigma_n$.

Fig. 8 illustrates the application of a morphological gradient representation on an MRI image.

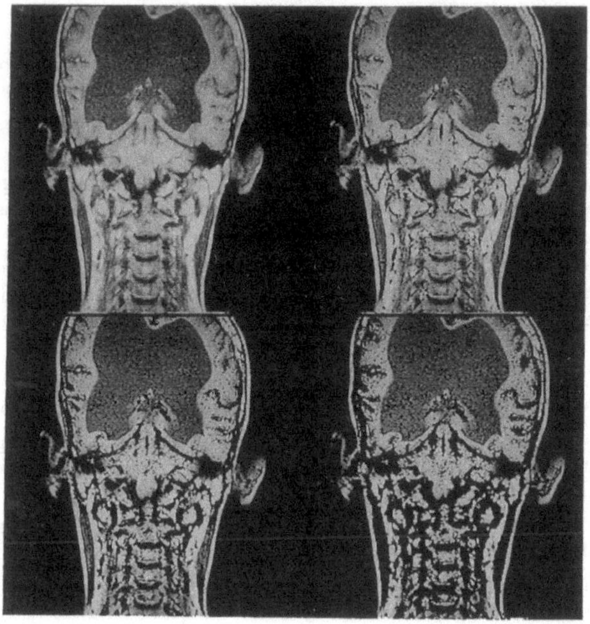

Fig. 7. Application of the shape enhancement algorithm on an MRI image. The original image is shown on the upper left. This example shows that the pronouncedness of the articulation of both the bone structure and the undulations of the cortex increases with the size of the structuring element.

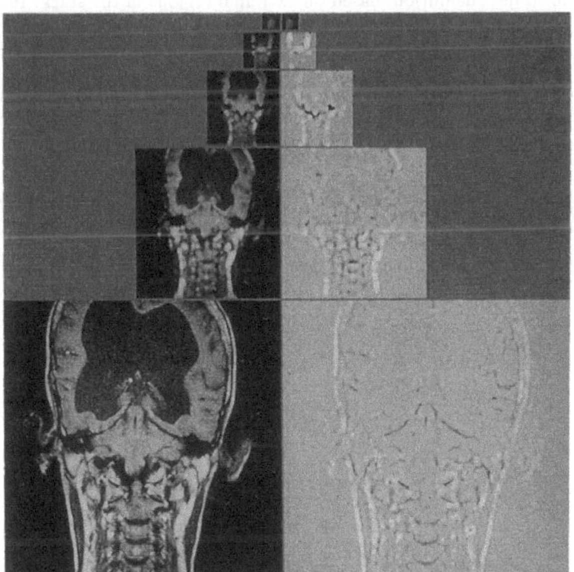

Fig. 8. Application of a morphological gradient representation on an MRI image. The original image is shown on the lower left. The left column represents the sampled results of the filter operations. The right column represents the morphological gradient.

REFERENCES

Burt, P.J. (1984). The pyramid as a structure for efficient computation. In: Multiresolution image processing and analysis, Rosenfeld, A. (ed.), Springer Verlag, Berlin, pp. 6-35.

Burt, P.J., Adelson, E.H. (1983). The Laplacian pyramid as a compact image code, IEEE Tr. on Comm., vol. COM-31, pp. 532-540.

Burt, P.J., Hong, T.H., Rosenfeld, A. (1981). Segmentation and estimation of image region properties through cooperative hierarchical computation, IEEE Trans. Systems, Man, and Cybernetics, vol. SMC-11, pp. 802-809.

Gauch, J.M., Oliver, W.R., Pizer, S.M. (1987). Multiresolution shape descriptions and their applications in medical imaging. In: Information Processing in Medical Imaging, Proc. of the 10th Conference, Graaf, C.N. de and Viergever, M.A. (eds.), Plenum Press, New York, pp. 131-149.

Haralick, R.M., Lin, C., Lee, J.S.C., Zhuang, X. (1987). Multiresolution morphology. In: Proc. 1st Int. Conf. Computer Vision, IEEE Computer Soc. Press, New York, pp. 516-520.

Hong, T.H., Narayanan, K.A. , Peleg, S., Rosenfeld, A., Silberberg, T. (1982). Image smoothing and segmentation by multiresolution pixel linking: further experiments and extension, IEEE Trans. Systems, Man, and Cybernetics, vol. SMC-12, pp. 611-622.

Koenderink, J.J., Doorn, A.J. van (1979). The structure of two-dimensional scalar fields with applications to vision, Biol. Cybern., vol. 33, pp. 151-158.

Maragos, P.A. (1987). Tutorial on advances in morphological image processing and analysis, Optical Eng., vol. 26, pp. 623-632.

Maragos, P.A., Schafer, R.W. (1986). Morphological skeleton representation and coding of binary images, IEEE Trans. Acoustics, Speech, and Signal Processing, vol. ASSP-34, pp. 1228-1244.

Matheron, G. (1975). Random sets and integral geometry, Wiley, New York.

Pizer, S.M., Koenderink, J.J., Lifshitz, L.M., Helmink, L., Kaasjager, A.D.J. (1985). An image description for object definition, based on extremal regions in the stack. In: Information processing in Medical Imaging, Proc. of the 9th Conference, Bacharach, S.L. (ed.), Martinus Nijhoff, The Hague, pp. 24-37.

Serra, J. (1982). Image analysis and mathematical morphology, Academic Press, New York.

Serra, J. (1988). Alternating sequential filters. In: Image analysis and mathematical morphology, vol. II: Advances in Mathematical Morphology, Serra, J. (ed.), Academic Press, New York, pp. 203-216.

Sternberg, S.R. (1986). Grayscale morphology, Comp. Vis. Graph. and Im. Proc., vol. 35, pp. 333-355.

FAULT-TOLERANT MEDICAL IMAGE INTERPRETATION

A. Sood and H. Wechsler
Department of Computer Science
George Mason University
Fairfax, Virginia 22030

ABSTRACT

Novel medical imaging techniques have been introduced over the last decade. They provide for the non-invasive study of internal structures and their dynamic behavior because of the large bandwidth characteristics of the new sensors. The processing and automatic interpretation of such images, however, lags far behind. We suggest herein a synergetic approach where novel techniques derived from artificial intelligence (AI), computer vision (CV) and neural networks (NN) could be integrated towards robust and automatic image interpretation. Such image interpretation would be relevant for the analysis of internal organs and/or tissue and to the understanding of time-varying (dynamic) imagery. Within the medical area it is important that such analysis be fault-tolerant (low sensitivity) to noisy data, occlusion/overlap, geometric distortions, and still be efficient.

1. INTRODUCTION

Computer Vision (CV) is concerned with the computational study of processes which produce useful descriptions from visual input. The visual processes we are considering can be thought of as information processing tasks, and they are analyzed at three levels, those of goal of computation, representation and algorithm, and implementation. Both representation and algorithms are crucial for understanding how a specific visual task can be solved and implemented. Another aspect that we emphasize throughout the paper is that of transformation and invariance. All mathematical derivation can be viewed simply as change of representation, making evident what was previously true but obscure. This view can be extended to all of problem solving--solving a problem then means representing it (transforming it) so as to make the solution evident. One of the characteristics sought to be derived via transformations is that of invariance. Vision is then a sequence of transformations, some of them carried out in parallel, whose goal is to capture some invariant aspect of the objects being imaged, as needed for recognition and interpretation.

Medical image processing (MIP) and interpretation requires the capability for fault-tolerant (FP) processing and a learning capability as well. Furthermore, there are many cues (like size, motion, surface, curvature) which need to be integrated in order to enhance the interpretation process. Finally, the interpretation process is greatly improved if models of the medical structures

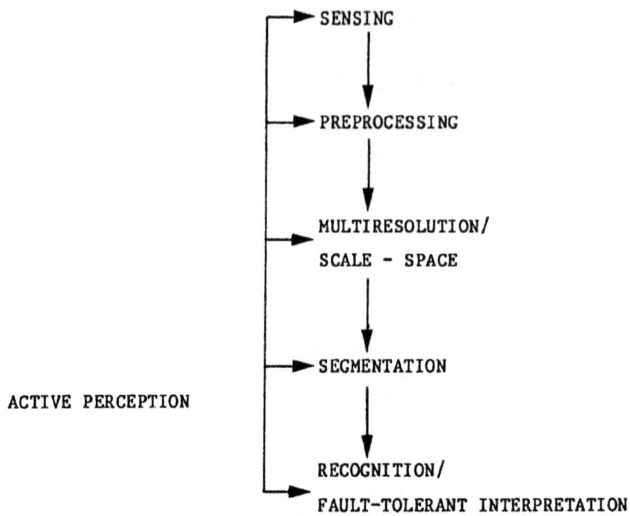

Figure 1. Visual Processing

being imaged are made available. Suggested solutions to the above requirements include distributed computation and self-organization as provided by neural network mechanisms, data fusion (DF) for multisensory data integration, and active perception (AP) for allocating computational resources and guiding the interpretation process in a model-based fashion.

The particular model for medical image processing which is considered herein is described in Fig. 1 and specific instantiations of its components are discussed in the next sections.

2. RESTORATION

Although automatic interpretation is the final goal for MIP, enhancement techniques which will lead to improved image quality and assist in obtaining relevant physical measurements are also needed.

One of the medical tasks that we considered (Acharya et al, 1987) was related to the artifacts (streaks-like) resulting from almost real-time data acquisition as in the case of CT (computer tomography) of the beating heart. Due to motion, the resulting data is blurred (think about photography of a moving car!) and artifacts can result. Novel techniques based on optimal estimation like the Kalman Filter (KF) were used to image the beating heart as if the data acquisition occurred at one instance in time.

The next technique, that of using the Gibbs (Boltzman) distribution was suggested by Geman and Geman (1984), is characteristic of optimization problems, and is used in Section 4.2. Here we discuss the generic model as it relates to image restoration and texture segmentation. The image data takes the form of a density function which can be thought of as behaving according to a Markov Random Field (MRF), i.e., the pixel value function is influenced by a finite neighbourhood-like memory. Assuming that what one observes is a degraded image D one seeks then to recover the original, restored image X. Assuming that the MRF behaves like a Gibbs distribution, one can write that the probability $P(\vec{X})$ of any of the possible restored images is given by

$$P(\vec{X}) = \frac{1}{Z} \exp[-E(\vec{X})/T] \quad (1)$$

where Z is a constant, $E(\vec{X})$ is an energy function, and T is known as the temperature of the system. One seeks the MAP (maximum a-posteriori estimate) which is equivalent to minimizing the energy function $E(\vec{X})$. The energy function is set such that minimal energy configurations $\vec{X}$ correspond to desirable restored images. The energy function is made up of the terms J_i (i = 1,2,3). The first term is a penalty-like component induced by introducing edge transitions. Non-homogeneities lead to raising the energy level but at the same time are useful to break the spatial coherence of the MRF. This might be needed for texture segmentation tasks. The second penalty term J_2 corresponds to lack of homogeneity (between regions defined by edge transitions allowed by the first term J_1. Finally, the third penalty term, J_3, accounts for lack of fidelity, i.e., the discrepancy between the degraded image D and the restored one. The MAP estimate is then obtained through a procedure like relaxation, where the specifics are implemented as simulated annealing. One lowers the temperature T of the system described by the Gibbs distribution and the corresponding physical system configuration is of lower energy. Eventually, the system reaches a stable state, where it settles in a (locally) minimal energy state corresponding to a restored image $\vec{X}$.

3. MULTI-RESOLUTION TECHNIQUES

Multi-resolution techniques have been extensively utilized, see for example the collection of papers in Rosenfeld (1984). The use of multi-resolution techniques in the context of medical imaging is discussed in a survey appearing in this volume (Pizer). Multi-resolution techniques have been useful in extracting information at different scales from an image. In this respect these techniques appear to have support from the human visual system. The use of multi-resolution techniques highlights the different spatial frequencies in the scene. Such techniques can be readily implemented in hardware using Pyramid architectures (Uhr, 1988). Computer systems based on such architectures are being built at Papia University, Italy, and at George Mason University, U.S.A.

Joint spatial/spatial-frequency (s/sf) representations were recently introduced to achieve high resolution. They indicate the frequency content in localized regions in the spatial domain. As such, those representations overcome the shortcomings of the traditional Fourier-based methods. The Wigner-distribution (WD) is one example of such (s/sf) representations, and it has been suggested for 2-D image processing by Jacobson and Wechsler (1988). The 2-D WD is defined as

$$W_f(x,y,u,v) = \iint_{-\infty}^{\infty} f\left(x + \frac{\alpha}{2}, y + \frac{\beta}{2}\right) f^* \left(x - \frac{\alpha}{2}, y - \frac{\beta}{2}\right)$$

$$\bullet \exp\left[-j(\alpha u + \beta v)\right] d\alpha d\beta \qquad (2)$$

We show next how to use multi-resolution techniques for developing methods for the measurement of the size of objects in the scene. We are interested in exploring this approach, because it is generally believed that multi-resolution techniques lead to more robust algorithms and fault tolerant processing. In fact, in our experiments we have found that our approach is insensitive to noise, and for this reason is particularly well suited to size measurements in the context of medical images. The details of the analytical development are available in Shah et al (1986) and Sood and Shah (1987). This section presents the highlights of the methodology.

Object size measurement from two-dimensional images can be treated as a sequence of one-dimensional problems. In a left to right line scan of an image, the left and right edges of an object will lead to positive and negative edges, respectively. Consequently, we model the problem of size evaluation as the problem of determining the width of a pulse. In passing it is noted that the height of the pulse is a measure of the contract variation in the image, we use scale space techniques for this measurement. Scale space filtering initially proposed by Witkin (1983), and Yuille and Poggio (1983) has shown that the scale space representation is unique. In the current application we have found that the scale space approach is insensitive to noise, and also enables the use of additional a priori information.

The overall strategy in our evaluation of the pulse width is shown in Figure 2. The input image is processed by a bank of Laplacian of Gaussian filters. Each filter has zero mean and a different variance. The zero crossings in each of the channels is determined, and the locus of the zero crossings leads to the scale space representation. The scale space representation of an isolated edge, pulse and staircase functions have the characteristics shown in Figure 3. It is emphasized that the three scale space representations have qualitatively different behavior.

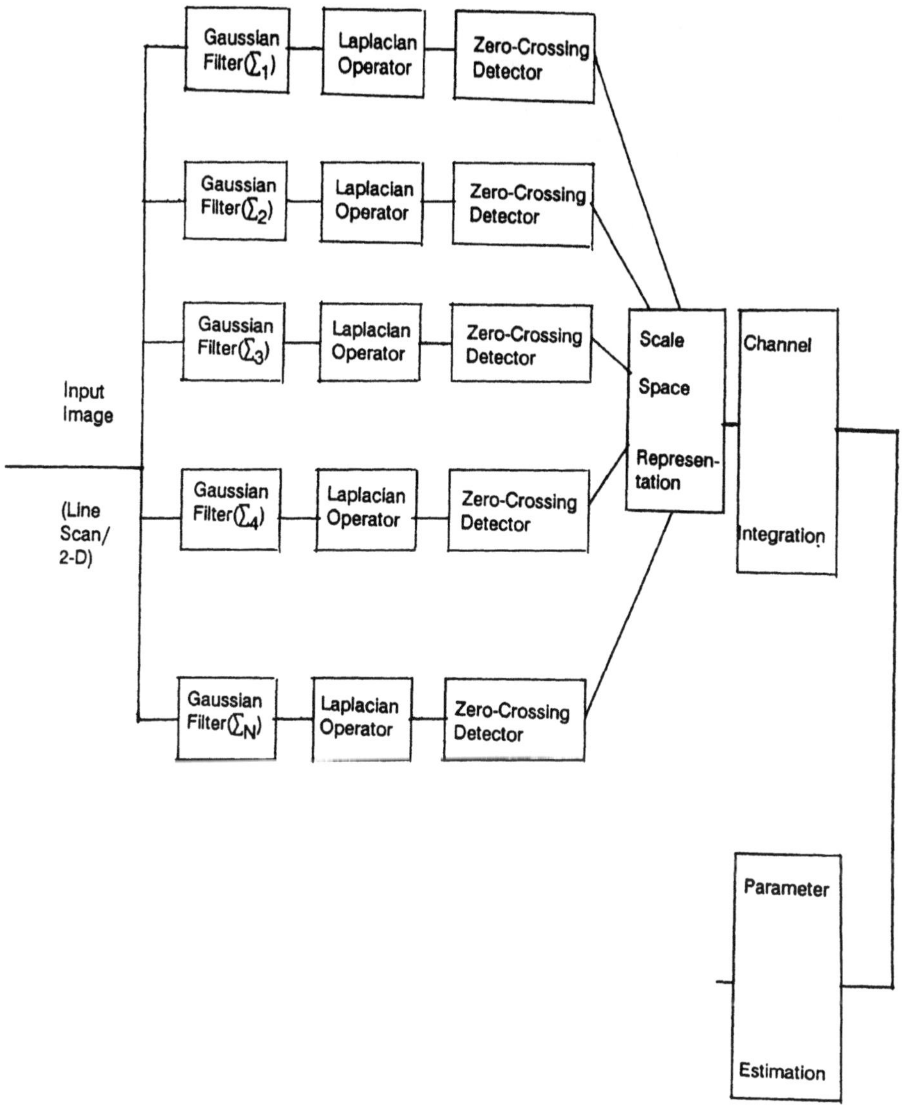

Figure 2. Overall Methodology for Scale Space Filtering

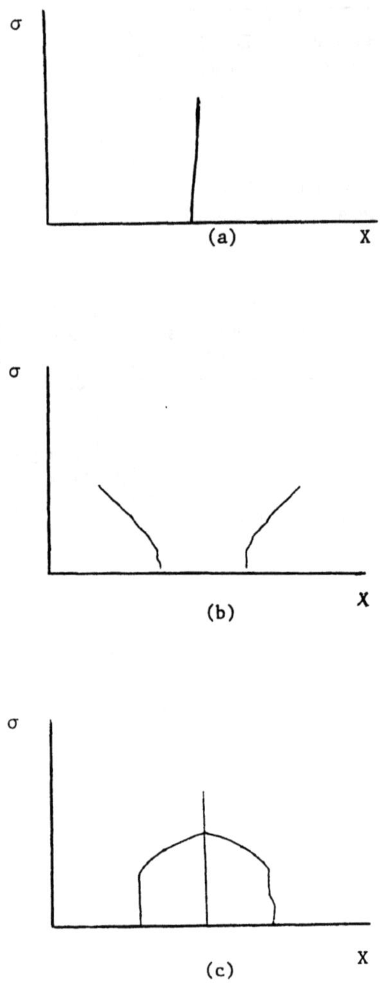

Figure 3. Scale Space Representations
(a) Isolated edge; (b) Pulse; (c) Staircase function

If we characterize the pulse as a combination of two step functions, then the pulse can be modeled as

$$f(x) = M[u(x - \omega_1) - u(x - \omega_2)], \tag{3}$$

where M is the pulse magnitude, $u(x)$ represents a step at $x = 0$, and $\omega_2 - \omega_1$ is the pulse width.

Denoting the Gaussian filter of standard deviation σ by $g^{\sigma}(x)$, then the output of a channel is

$$E_2(x,\sigma,\omega_1,\omega_2,M) = f(x) * \nabla^2 g^{\sigma}(x) = 0. \tag{4}$$

Substituting for $f(x)$ and integrating yields

$$E_2(x,\sigma,\omega_1,\omega_2,M) = M\left[\frac{x - \omega_1}{\sigma^2} g^{\sigma}(x - \omega_1) - \frac{x - \omega_2}{\sigma^2} g^{\sigma}(x - \omega_2)\right] = 0 \tag{5}$$

Now the problem reduces to an optimization problem: Given a data set $\{(x_i^k,\sigma_i)\}$, where x_i^k are the zero crossings in the ith channel with standard deviation of σ_i, find ω_1,ω_2 and M such that we minimize:

$$\sum_{i,k} [E_2(x_i^k,\sigma_i,\omega_1,\omega_2,M)]^2.$$

However, at the points of constant intensity the second derivative and E_2 are zero. So we need to exclude these points, by introducing an additional constraint on the first partial derivative, $E_1 = f(x) * \nabla g^{\sigma}(x)$.

The final formulation is:

Find ω_1,ω_2 and M to

$$\text{minimize} \sum_{i,k} [E_2(x_i^k,\sigma_i,\omega_1,\omega_2,M)]^2$$

$$\text{such that } [E_1(x_i^k,\sigma_i,\omega_1,\omega_2,M]^2 > 0$$

This problem was solved using Lagrange multipliers, and results are discussed below.

In many noisy cases the algorithm performed reasonably well, except in the cases where the signal to noise ratio was too low. Due to additive noise the zerocrossings shift substantially from their corresponding positions in the noise free case. At the shifted locations of zerocrossings the objective function is not necessarily close to zero. Therefore, the optimization program had problems in minimizing the function and in some cases the algorithm did not converge. The example in Figure 4 shows the case of pulse width = 25, and SNR = 2.5 . There is no extra contour due to noise, and only some of the zerocrossings are shifted. The algorithm computed a pulse width of 25.

4. SEGMENTATION OF RANGE IMAGES

Image segmentation has attracted extensive attention, and yet the segmentation of real images remains a difficult problem. We suggest that image segmentation algorithms can be made more reliable and robust if we employ multiple cues and/or multiple sensors. We have tried to work towards this end especially in the context of range images (depth maps). It is emphasized that range images provide 3-D information at each point in the scene. The results presented in this section have potential application for Computer Tomography (CT).

The focus of our research has been to characterize range images with the eventual objective of recognizing an object. As a first step towards characterizing the range image, it is necessary to segment the scene into regions of similar surface curvature. Thereafter using a set of primitive surfaces, we describe the different segments of the scene. This approach is similar to that adopted in Besl and Jain (1988), Al-Hujazi and Sood (1988).

A distinguishing characteristic of our research at George Mason's Computer Vision and Distributed Systems Research Laboratory is that we have developed a strategy to integrate multiple cues. Our strategy is based on the use of multi-resolution ideas and Markov Random Fields (MRF) to obtain reliable segmentation of the scene. The cues we currently use are based on the surface curvature and the edges in the scene. We plan to extend the algorithm and incorporate information regarding surface normals in the algorithm at a later date. The overall algorithm is shown in Figure 5, and can be divided into four steps:

1. Identify a seed region on the basis of surface characteristics.
2. Compute edge location and surface normal directions.
3. Use MRF to integrate the above information and obtain the range image segments.
4. Fit surfaces to the identified regions.

The second and fourth steps are based on standard techniques and for this reason we focus on the first and third steps.

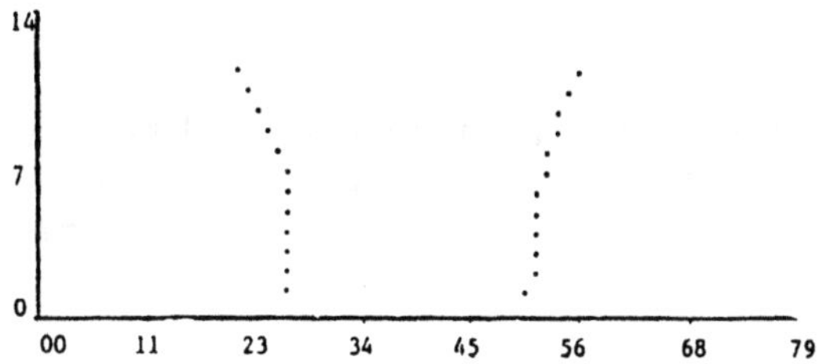

(a)

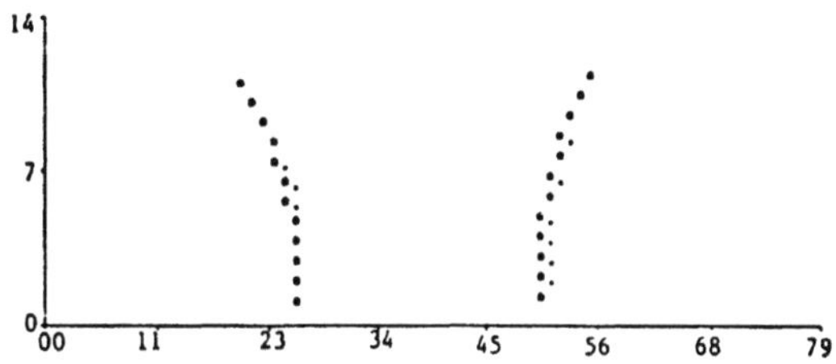

(b)

Figure 4. Pulse Width Measurement Example
(a) Scale-space of pulse with SNR = 2.5
(b) The results of fitting are shown by *

4.1 Seed Region

The mean (H) and Gaussian (K) curvature parameters can be utilized to characterize the surface curvature. In fact, just the sign of H and K are sufficient to classify the surface into one of eight primitive surfaces (Table 1).

H-K can be calculated using:

$$K = \frac{f_{xx}f_{yy} - f^2_{xy}}{(1 + f^2_x + f^2_y)/2}, \quad H = \frac{f_{xx} + f_{yy} + f_{xx}f^2_y + f_{yy}f^2_x - 2f_x f_y f_{xy}}{(1 + f^2_x + f^2_y)^{3/2}} \tag{6}$$

	K > 0	K = 0	K < 0
H < 0	Peak	Ridge	Saddle Ridge
H = 0	---	Flat	Minimal Surface
H > 0	Pit	Valley	Saddle Valley

Table 1. Surface type labels from surface curvature sign.

A major problem with the H-K-sign map is that in spite of preliminary smoothing H-K computation is very sensitive to noise. In the presence of noise H-K-sign map surface labels tend to connect the labels of neighborhood, but distinct, surface regions. On the other hand, the use of H and K allows us to reduce the number of levels from 256 (the original image) to 3 levels (-,0,+ for H and K) in the H-K sign map. In order to reduce the impact of noise, H and K are calculated in multi-scale fashion and only pixel location for which H-K signs are same at all scales are included in the seed region.

4.2 Region Growing

Our region growing strategy is based on the use of MRF. As noted in Section 2, the problem of obtaining the MAP estimate reduces to minimizing the energy function. In our current implementation the energy function consists of two parts--one part based on surface characteristics (V_s) based on the H-K sign map and the second based on the edges (V_e) obtained by using the Laplacian of Gaussian operator, i.e., the energy function is re-written as:

$$E(X) = z*(V_s) + V_e,$$

where z is zero if the computation is being undertaken at a pixel which is an edge point. V_s is the energy function evaluated in the neighbourhood of the pixel. In our experiments we have utilized the single pixel cliques. (Although higher order cliques should also be utilized the computational costs are very high and for this reason most researchers make similar assumptions.)

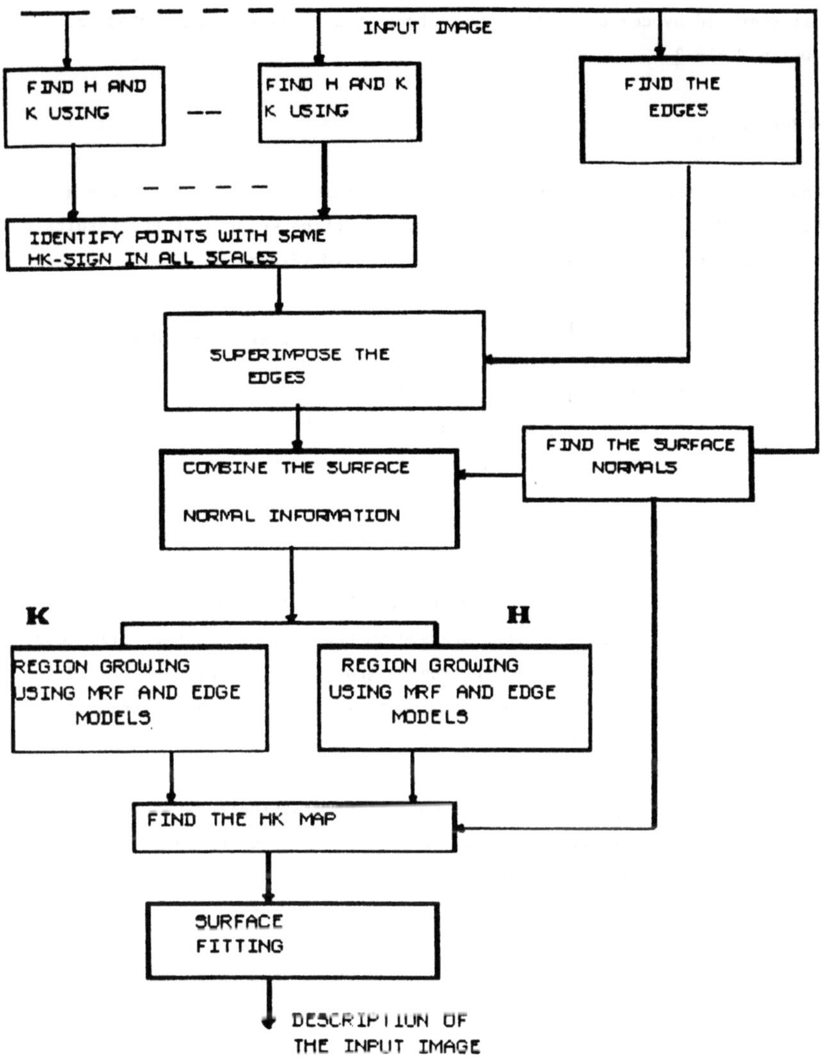

Figure 5. The Algorithm Flow Chart

V_e is computed by comparing the edge labels in a 4 pixel neighbourhood as shown in Table 2.

4.3 Results

We have experimented with synthetic and real images and the results are available in Al-Hujazzi and Sood (1988). In this section we present the results of processing a real image--a mug obtained from ERIM (Figure 6a).

A	0	0	0	0	0	0	0	0	1	1	1	1	1	1	1	1
B	0	0	0	0	1	1	1	1	0	0	0	0	1	1	1	1
C	0	0	1	1	0	0	1	1	0	0	1	1	0	0	1	1
D	0	1	0	1	0	1	0	1	0	1	0	1	0	1	0	1
V_e	0	1	1	0	1	0	0	1	1	0	0	1	0	1	1	0

* * where 0 no edge
C D 1 edge
* *
A B

Table 2. Energy Contribution (V_e) by the Edge.

Fig. (6d) and Fig. (6h) show the seed region obtained for H and K respectively. The range image is then processed in two different ways. Fig. (6e) and Fig. (6i) show the output of the region growing algorithm with the edge model. Fig. (6b) shows the final H-K-sign map. The segmentation results obtained were good with the exception of the handle of the coffee cup, which was not classified. This is because of the size of this region and the restriction in the algorithm on the number of pixels required for classification. In Fig. (6f) and Fig. (6g) the outputs of the region growing algorithm without the edge model are shown. Fig. (6c) shows their final H-K-sign map. In this case, the handle is classified as a planar region, also small regions of the cylindrical surfaces of the object are classified as planar. A comparative study of Fig. (6b) and Fig. (6c) illustrates that inclusion of the edge model leads to less misclassified points.

5. FAULT-TOLERANT IMAGE RECOGNITION

The challenge of the visual recognition problem stems from the fact that the projection of an object onto an image can be confounded by several dimensions of variability such as uncertain perspective, changing orientation and scale, sensor noise, occlusion, and non-uniform illumination. A vision system must not only be able to sense the identity of an object despite this variability, but must also be able to characterize such variability--because the variability inherently carries much of the valuable information about the world. Our goal is to derive the functional characteristics of image representations suitable for invariant recognition using a distributed associative memory (DAM). The main question is that of finding appropriate

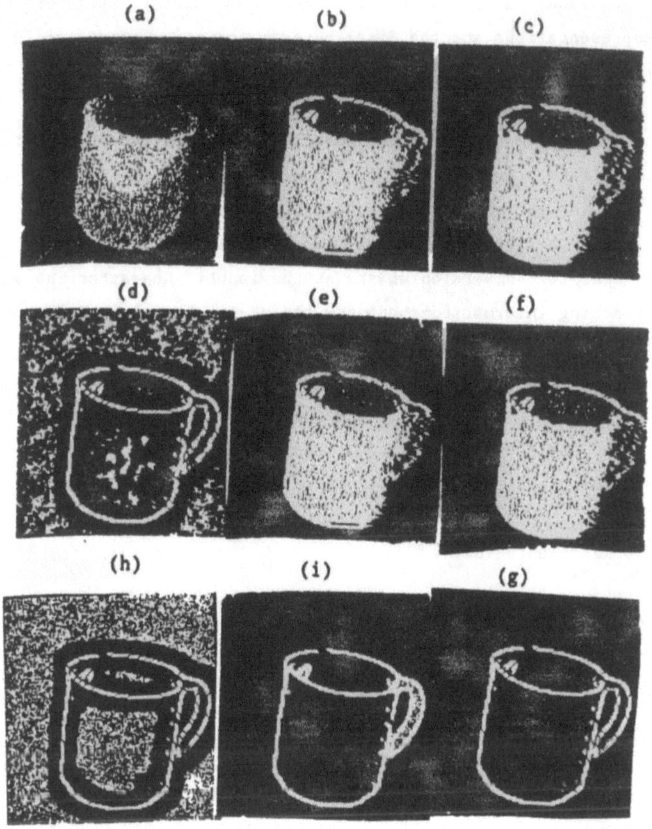

Figure 6. Results of Processing a Range Image--A Coffee Cup.

a) The original image.
b) The HK-map with edge model (Gray: Ridge surface; Black: Planar surface; White: Valley ridge).
c) Similar to (b) without edge model.
d) H seed region multi-scale output (Black: unclassified).
e) H region growing output with edge model (Gray: $H > 0$; Black: $H = 0$; White: $H < 0$; Handle: unclassified).
f) Similar to (e) without edge model (Handle classified as planar).
g) Similar to (i) without edge model (Black: planar).
h) K seed region multi-scale output (Black: unclassified).
i) K region growing output with edge model (Black: $K = 0$; Gray: unclassified).

Edges are superimposed on Fig. 6b through 6g.

transformations such that interactions between the internal structure of the resulting representations and the distributed associative memory yield invariant recognition.

We approach the problem of object recognition with three requirements: classification, reconstruction, and characterization. Classification implies the ability to distinguish objects that were previously encountered. Reconstruction is the process by which memorized images can be drawn from memory given a distorted version exists at the input. Characterization involves extracting information about how the object has changed from the way in which it was memorized. Our goal is to discuss a system (Wechsler and Zimmerman, 1988) which is able to recognize memorized 2-dimensional objects regardless of geometric distortions like changes in scale and orientation, and can characterize those transformations. The system also allows for noise and occlusion and is tolerant of memory faults.

5.1 Invariant Representation

We examine the various components used to produce the vectors which are associated in the distributed associative memory. The block diagram which describes the various functional units involved in obtaining an invariant image representation is shown in Figure 7. The image is complex-log mapped so that rotation and scale changes become translation in the transform domain. Along with the mapping, the image is also filtered by a space variant filter to reduce the effects of aliasing. The mapped image is then processed through a Laplacian in order to solve some problems associated with the conformal mapping. The Fourier transform of both the mapped image and the Laplacian processed image produce the four output vectors. The magnitude output vector $|\bullet|_1$ is invariant to linear transformations of the object in the input image. The phase output vector ϕ_2 contains information concerning the spatial properties of the object in the input image.

The first box of the block diagram given in Figure 7 consists of two components: Complex-log mapping and space variant filtering. Complex-log mapping transforms an image from rectangular coordinates to polar exponential coordinates. This transformation changes rotation and scale into translation. If the image is mapped onto a complex plane then each pixel (x,y) on the Cartesian plane can be described mathematically by $z = x + jy$. The complex-log mapped points w are described by

$$w = \ln(z) = \ln(|z|) + j\theta_z \tag{8}$$

where $|z| = (x^2 + y^2)^{\frac{1}{2}}$ and $\theta_z = \tan^{-1}(y/x)$.

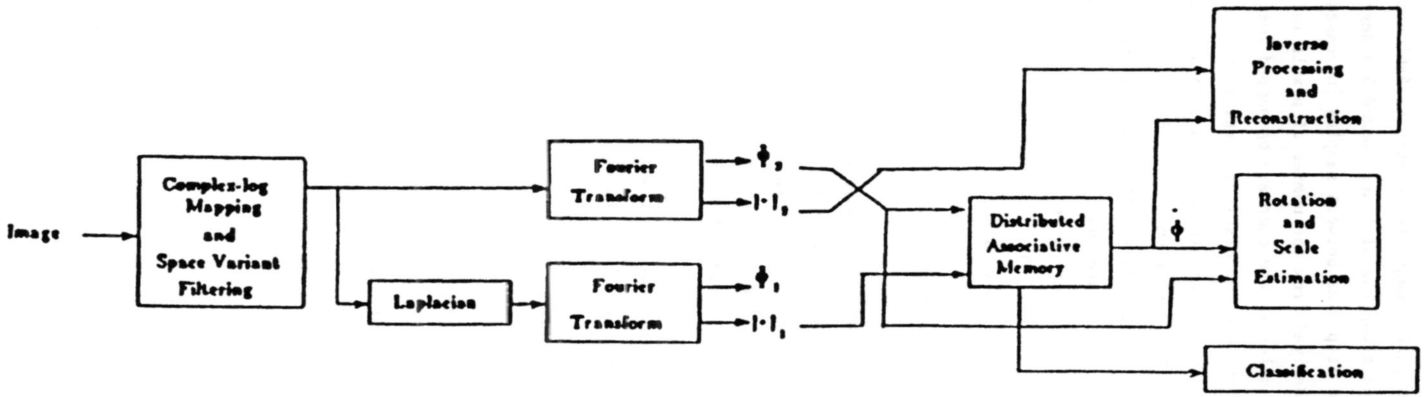

Figure 7. Block Diagram of the System.

Our system sampled 256x256 pixel images to construct 64x64 complex-log mapped images. Samples were taken along radial lines spaced 5.6 degrees apart. Along each radial line the step size between samples increased by powers of 1.08. These numbers are derived from the number of pixels in the original image and the number of samples in the complex-log mapped image. The nonlinear sampling can be split into two distinct parts along each radial line. Toward the center of the image the samples are dense enough that no anti-aliasing filter is needed. Samples taken at the edge of the image are large and an anti-aliasing filter is necessary. The image filtered in this manner has a circular region around the center which corresponds to an area of highest resolution. The size of this region is a function of the number of angular samples and radial samples. The filtering is done, at the same time as the sampling, by convolving truncated Bessel functions with the image in the space domain. The width of the Bessel functions main lobe is inversely proportional to the eccentricity of the sample point.

A problem associated with the complex-log mapping is sensitivity to center misalignment of the sampled image. Small shifts from the center cause dramatic distortions in the complex-log mapped image. Our system assumes that the object is centered in the image frame. Slight misalignments are considered noise. Large misalignments are considered as translations and could be accounted for by changing the gaze in such a way as to bring the object into the center of the frame. The decision about what to bring into the center of the frame is an active function and should be determined by the task.

The second box in the block diagram of Figure 7 is the Fourier transform. The Fourier transform of a 2-dimensional image f(x,y) is given by

$$F(u,v) = \int_{-\infty}^{\infty}\int_{\infty}^{\infty} f(x,y)e^{-j(ux + vy)}dx\, dy \tag{9}$$

and can be described by two 2-dimensional functions corresponding to the magnitude $|F(u,v)|$ and phase $\phi_F(u,v)$. The magnitude component of the Fourier transform which is invariant to translation, carries much of the contrast information of the image. The phase component of the Fourier transform carries information about how things are placed in an image. Translation of f(x,y) corresponds to the addition of a linear phase component. The complex-log mapping transforms rotation and scale into translation and the magnitude of the Fourier transform is invariant to those translations so that $|\bullet|_1$ will not change significantly with rotation and scale of the object in the image.

The Laplacian implemented is a difference-of-Gaussians (DOG) approximating the $\nabla^2 G$ function and is given by

$$\nabla^2 G = \frac{1}{\pi\sigma^4}\left[1 - r^2/2\sigma^2\right]e^{\left\{-r^2/2\sigma^2\right\}} \tag{10}$$

The result of convolving the Laplacian with an image can be viewed as a two step process. The image is blurred by a Gaussian kernel of a specified width σ, in a similar way to the approach described in Section 3. Then the isotropic second derivative of the blurred image is computed. The width of the Gaussian kernel is chosen such that the complex-log mapped image is visible--approximately 2 pixels in our experiments. The Laplacian sharpens the edges of the object in the image and sets any region that did not change much to zero. Below we describe the benefits from using the Laplacian.

The Laplacian eliminates the stretching problem encountered by the complex-log mapping due to changes in object size. When an object is expanded the complex-log mapped image will translate. The pixels vacated by this translation will be filled with more pixels sampled from the center of the scaled object. These new pixels will not be significantly different from the displaced pixels so the result looks like a stretching in the complex-log mapped image. The Laplacian of the complex-log mapped image will set the new pixels to zero because they do not significantly change from their surrounding pixels. The Laplacian eliminates high frequency spreading due to the finite structure of the discrete Fourier transform and enhances the differences between memorized objects by accentuating edges and de-emphasizing areas of little change.

5.2 Distributed Associative Memory (DAM)

The particular form of distributed associative memory that we deal with in this paper is a memory matrix which modifies the flow of information. Stimulus vectors are associated with response vectors and the result of this association is spread over the entire memory space. Distributing in this manner means that information about a small portion of the association can be found in a large area of the memory. New associations are placed over the older ones and are allowed to interact. This means that the size of the memory matrix stays the same regardless of the number of associations that have been memorized. Because the associations are allowed to interact with each other an implicit representation of structural relationships and contextual information can develop, and as a consequence a very rich level of interactions can be captured. There are few restrictions on what vectors can be associated. There can exist extensive indexing and cross-referencing in the memory. Distributed associative memory captures a distributed representation which is context dependent.

The *construction* stage assumes that there are n pairs of m-dimensional vectors that are to be associated by the distributed associative memory. This can be written as

$$M\vec{s}_i = \vec{r}_i \quad \text{for } i = 1,\ldots,n \tag{11}$$

where $\vec{s}_i$ denotes the ith stimulus and $\vec{r}_i$ denotes the ith corresponding response vector. We want to construct a memory matrix M such that when the kth stimulus vector $\vec{s}_k$ is projected onto the space defined by M the resulting projection will be the corresponding response vectore $\vec{r}_k$. More specifically, we want to solve the following equation:

$$MS = R \tag{12}$$

where $S = [\vec{s}_1|\vec{s}_2|\ldots|\vec{s}_n]$ and $R = [\vec{r}_1|\vec{r}_2|\ldots|\vec{r}_n]$. A unique solution for this equation does not necessarily exist for any arbitrary group of associations that might be chosen. Usually, the number of associations *n* is smaller than *m*, the length of the vector to be associated, so the system of equations is underconstrained. The constraint used to solve for a unique matrix M is that of minimizing the square error, $|MS - R|^2$, which results in the solution $M = RS^+$, where S^+ is known as the Moore-Penrose generalized inverse of S.

The *recall* operation projects an unknown stimulus vector $\vec{s}$ onto the memory space M. The resulting projection yields the response vector $\vec{r}$

$$\vec{r} = M\vec{s} \tag{13}$$

If the memorized stimulus vectors are independent and the unknown stimulus vector $\vec{s}$ is one of the memorized vectors $\vec{s}_k$, then the recalled vector will be the associated response vector $\vec{r}_k$. If the memorized stimulus vectors are dependent, then the vector recalled by one of the memorized stimulus vectors will contain the associated response vector and some *crosstalk* from the other stored response vectors.

The recall can be viewed as the weighted sum of the response vectors. The recall begins by assigning weights accordign to how well the unknown stimulus vector matches with the memorized stimulus vector using a linear least squares classifier. The response vectors are multiplied by the weights and summed together to build the recalled response vector. The recalled response vector is usually dominated by the memorized response vector that is closest to the unknown stimulus vector.

Fault tolerance is a byproduct of the distributed nature and error correcting capabilities of the distributed associative memory. By distributing the information, no single memory cell carries a significant portion of the information critical to the overall performance of the memory.

6. CONCLUSIONS

We have reviewed some novel computer vision techniques and discuss their relevance for robust image interpretation. Some of the techniques discussed are characteristic to parallel distributed processing (PDP) and active perception. Furthermore, we emphasized the significance of modeling for a successful interpretation where the dynamics of the perceptual activities are guided by top-down predictions.

REFERENCES

Acharya, R., Hefferman, P.B., Robb, R., and Wechsler, H. (1987), High-Speed 3-D Imaging of the beating heart using temporal estimation, Computer Vision, Graphics, and Image Processing, 39, 279-290.

Geman, S., and Geman, D. (1984), Stochastic relaxation, Gibbs distribution, and Bayesian restoration of images, IEEE Trans. on PAMI, Vol. 6, No. 6, 721-741.

Rosenfeld, A. (Ed.) (1984), Multiresolution Image Processing and Analysis, Springer-Verlag.

Pizer, S. (1989), Multiscale methods and the segmentation of medical images, this volume.

Uhr, L. (1988), Parallel Computer Vision, Academic Press.

Jacobson, L., and Wechsler, H. (1988), Joint spatial/spatial-frequency representations, Signal Processing, Vol. 14, No. 1, 95-102.

Shah, M., Sood, A. and Jain, R. (1986), Pulse and staircase edge models, Computer Vision, Graphics, and Image Processing, Vol. 34, 321-343.

Sood, A. and Shah, M. (1987), Scale space technique to finding primitives in images with application to road following, Proceedings SPIE--Applications of Artificial Intelligence V.

Witkin, A.P. (1983), Scale space filtering, Proceedings of IJCAI.

Yuille, A.L. and Poggio, T. (1983), Fingerprints theorems for zero crossings, MIT AI memo 730.

Besl, P. and Jain, R. (1988), Segmentation through variable-order surface fitting, IEEE Trans. on PAMI, Vol. 10, No. 2, 167-192.

Al-Hujaze, E. and Sood, A. (1988), Range data description based on multiple characteristics, Proceedings 1988 Goddard Conference on Space Applications of Artificial Intelligence.

Wechsler, H., and Zimmerman, L. (1988), 2-D Invariant object recognition using DAM, IEEE Trans. on PAMI, Vol. 10, 6 (in print).

SECOND MOMENT IMAGE PROCESSING (SMIP)

Jos Dechamps

Agfa-Gevaert, Systems Analysis, R&D Laboratories
Mortsel, Belgium

ABSTRACT

The class of SMIP-algorithms (Second Moment Image Processing) is presented, containing some forms of adaptive unsharp masking as well as approximations of adaptive histogram equalisation. SMIP-algorithms allow to drive the image-processing parameters by means of the mean and the standard deviation of the pixels in a continuously sliding window. The clipping technique as originally proposed by Pizer can be reformulated more generally for use in specific variants of this class. SMIP is especially suited for processing large images where the number of operations per pixel must be kept as small as possible. These algorithms lend themselves to fast software as well as hardware implementations.

1. INTRODUCTION

1.1 Unsharp masking

For X-ray chest images the most cited form of image-processing is undoubtedly unsharp masking (UM), which couples edge enhancement with a fast implementation. Let x represent a pixel and m the unsharp value, then the resulting pixel y is calculated as:

$$y = x + \alpha\,(x-m) \tag{1}$$

Mostly the unsharp image is obtained as a convolution with a rectangular flat kernel, leading to very fast implementations. The kernel must have the same size as the details one whishes to amplify. To avoid the overamplification of large differences (x-m), an S-shaped function T[z] is often used (normalised so that T'(0)=1):

$$y = x + \alpha\,T[x-m] \tag{2}$$

A second problem with UM is the excessive noise enhancement when a rather large amplification factor is used. This is mostly solved by an extra smoothing step afterwards.
Adaptivity appears in this algorithm in the form of different variants which mostly try to make the amplification factor dependent on some local statistical information such as x and/or m. However, neither the pixel value nor the average pixel value give a satisfactory indication of the image region since these values depend a lot on the irradiation dose (X-ray dose) used. The standard deviation gives a better indication since it is a relative measure instead of an absolute one.

1.2 Statistical differencing

The basic idea of these algorithms is to change the overall brightness and/or contrast of an image by changing the statistical properties of the gray-value distribution. This can be done globally or locally. A good survey is given by Cocklin et al (1983). For example, in order to obtain an image with a given mean m_0 and standard deviation s_0, one could use the formula:

$$y = m_0 + (s_0/s)(x-m) \tag{3}$$

wherein m is the calculated local (or global) mean, and s the calculated local (or global) standard deviation.
The same problem of overamplificating the differences (x-m) occurs for small values of s. A second problem concerns the time needed for calculating the local value of the standard deviation.

1.3 Adaptive histogram equalisation (AHE)

Another basic algorithm is global histogram equalisation, which leads to fast implementations, but is inadequate for chest radiology, as described by Johnson et al (1985). According to this technique the cumulative histogram is calculated, rescaled and used as a look-up table.
Adaptivity was first introduced here by Ketcham (1976) with the sliding-window algorithm (CAHE: Continuous Adaptive Histogram

Equalisation), calculating the local histogram for each window around the central pixel. Cocklin et al (1983) reports the principal problems with this algorithm: 1) the computational effort, 2) the uncertain diagnostic value, and 3) the overamplification of noise in regions with a very peaked local histogram (Sherrier et al, 1987).

This time-consuming procedure was improved by Pizer et al (1984, 1986, 1986) with the regionally adaptive histogram equalisation; this method divides the image in both directions in contextual zones or regions and determines the histogram for each region by subsampling the pixels. Smooth transitions between individual regions are obtained by means of an interpolation technique. This method is much faster, but unpredictable artefacts may be created at the borders of the contextual regions. Further improvements were made in CLAHE (Contrast Limited AHE), where Pizer proposes to limit the gradient of the cumulative histogram by clipping the local histograms.

2. SECOND MOMENT ALGORITHMS

2.1 First moment graphs (FMG)

The resemblance between the different algorithms can be illustrated graphically by representing the resulting pixel value y as a function of the original pixel value x and the unsharp value m in a 3D-graph:

$$y = F [x , m] \tag{4}$$

We shall refer to such a graph as a FMG. The basic UM-formula (1) is shown in Fig. 1. The resulting image y may contain values largely outside the original gray-level range. In practice, however, this effect is not so strong because the differences (x-m) remain reasonably small. For a 12-bit thorax X-ray, digitised from an Agfa-Gevaert RP1L-film, these values do not exceed the range [-200..+200], as shown in Fig. 2. Fig. 3 gives the (x-m) difference histogram for a thorax X-ray digitised from photostimulable phospor on 10 bits after range determination and show that these values are not larger than approximately 50, again 1/20th of the range.

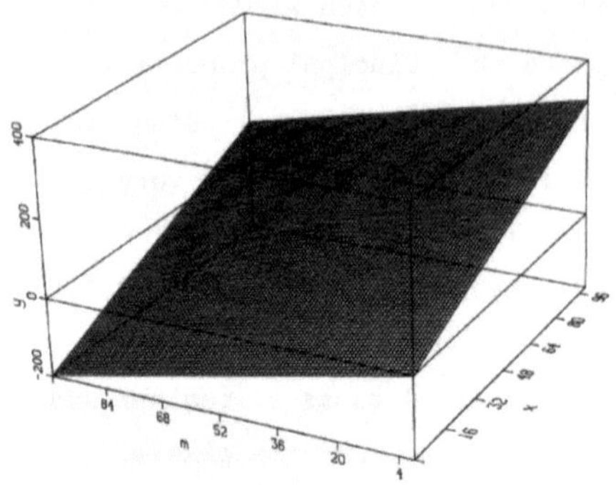

Fig. 1. UM-FMG: y=x+2(x-m)
with 0<x<100.

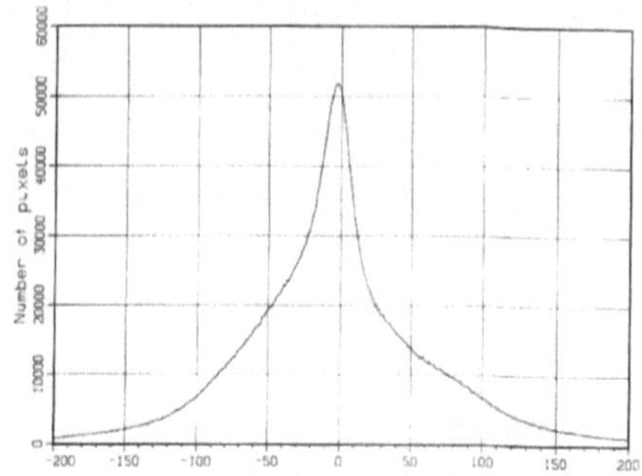

Fig. 2. Difference histogram
of thorax from RP1L-film

This means that the image-processing formula is essentially used around the bissectrix x=m. A narrow strip around this line determines the global result of the image-processing, while the zone outside this strip mostly gives rise to abnormal amplification of extreme values. This effect can be avoided by the S-shaped function T(z). Fig. 4 gives the FMG for T, and the corresponding unsharp masking FMG is shown in Fig. 5.

For the CAHE-algorithm, the local histogram for a given average value m is represented in Fig. 6. The window histogram has an average value m lying on the bissectrix, but has an arbitrary form.

By way of experiment we have tried to get an idea of the form of these local histograms. Because every pixel in the image has a corresponding local histogram of the window surrounding it, the average local histogram according to a given value of m was calculated for all values of m contained in the original image. Fig. 7 and 8 are the FMG's for an X-ray of a thorax, digitised with 10 bits from a photostimulable phosphor, seen from two different points of view. The cumulated histogram is represented in Fig. 9. For high values of m, some inaccuracies occur because of the limited number of local histograms having this average value and the plotting program itself. Fig. 10 gives the same FMG for an image digitised from Agfa-Gevaert RP1L-film with 12 bits. These average local histograms may be reasonably approximated by a normal distribution.

Also remark the resemblance between Fig. 9 and Fig. 4 and 5, indicating that AHE and UM are qualitatively similar.

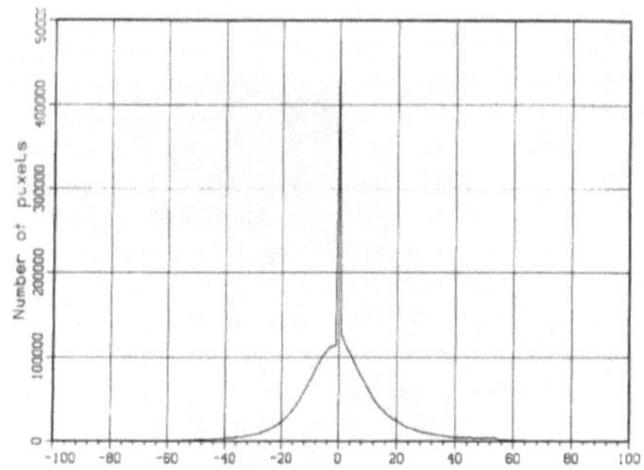

Fig. 3. Difference histogram of thorax from PSL

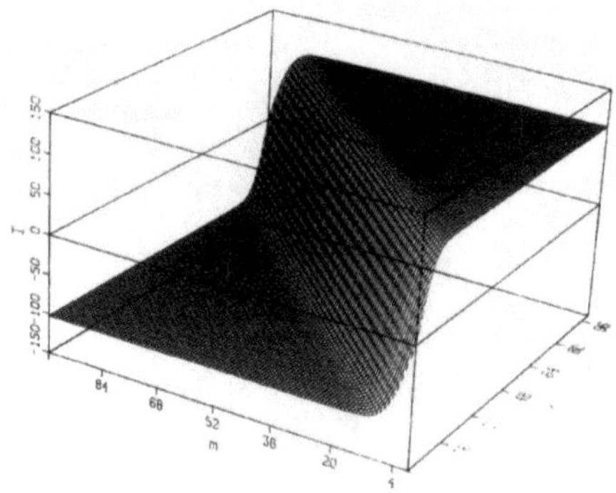

Fig. 4. FMG of possible S-shaped correction for UM

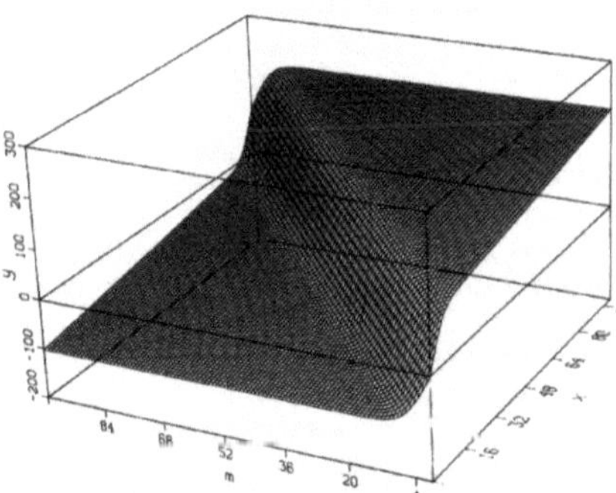

Fig. 5. UM-FMG with S-shaped correction shown in fig.4

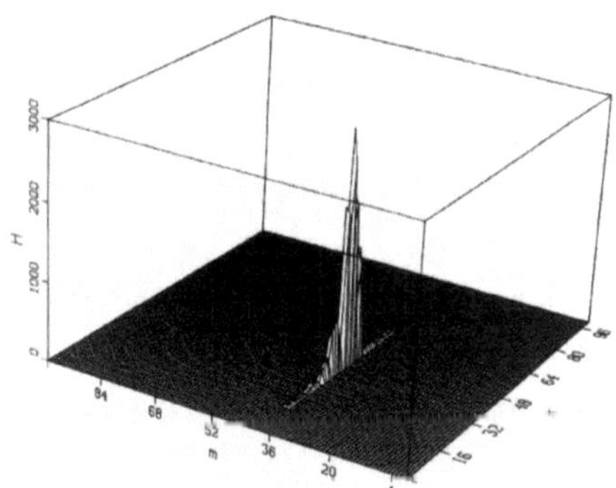

Fig. 6. Expected local histogram with mean value m

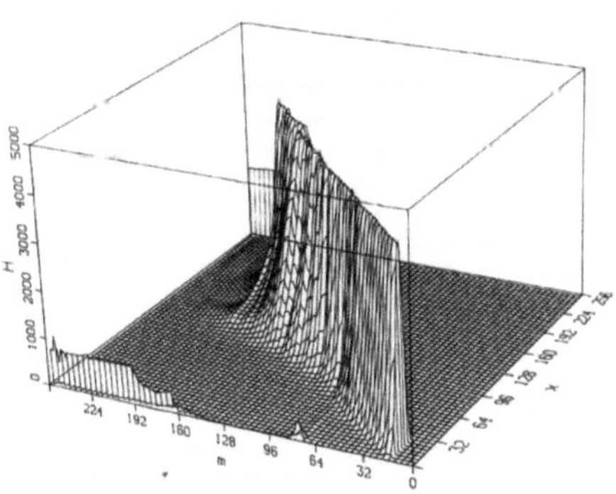

Fig. 7. Average local histogram of thorax from PSL

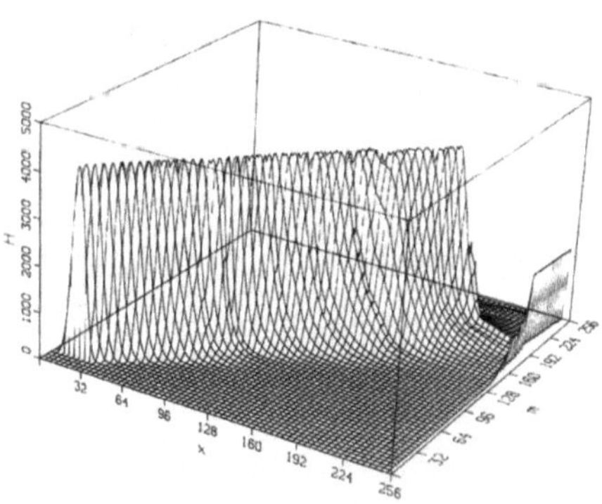

Fig. 8. Same as fig. 7, seen from another point of view

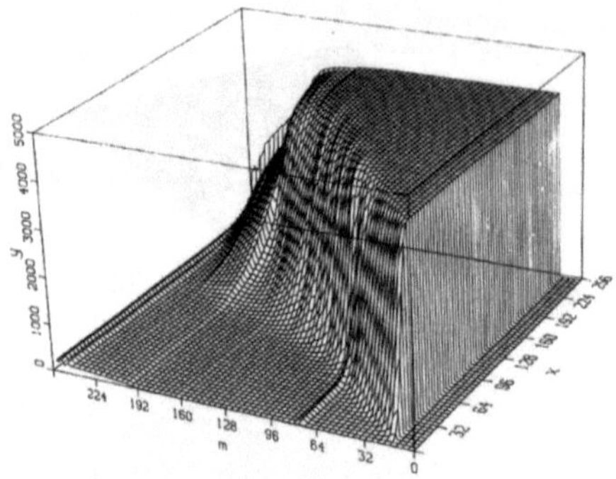

Fig. 9. FMG of cumulative average local histogram of PSL-thorax

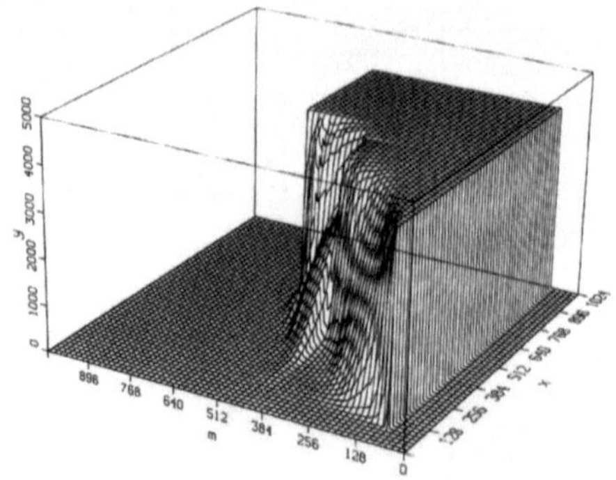

Fig. 10. FMG of cumulative average local histogram of thorax from RP1L-film.

2.2 Second moment adaptive histogram equalisation (SMAHE)

The time-consuming step of the CAHE-algorithm lies in the updating of the sliding-window histogram. This can be solved by approximating each local window histogram by a Gaussian so that AHE reduces to:

$$y = Y_{max} G[(x-m) / s] \tag{5}$$

wherein G is the standardised cumulative normal distribution, s the standard deviation, and m the unsharp image. In order to avoid overamplification in regions with a peaked histogram, the clipping technique as descibed by Pizer et al (1986) for the CLAHE-method can be used:

$$y = Y_{max} G[(x-m) / max(s,s_{min})] \tag{6}$$

where s_{min} is a given lower limit for s. We shall refer to both algorithms as SMAHE (Second-Moment AHE), because they both use the second central moment of the pixel distribution on the sliding window. The Gaussian assumption generates 2 new problems: most authors confirm that the computation of the standard deviation of a sliding window increases the computation time dramatically (Blume et al 1987, Cocklin et al

1982,1983, Pizer et al 1986). Cocklin et al (1983) summarizes ways to obtain an approximation for the standard deviation.
Another problem lies in the evaluation of the SMAHE-formula requiring one division per pixel. As will be explained below, SMIP allows to calculate the exact value of s and m simultanously with 13 additions and 4 bit-shifts per pixel, independently of the window size.

2.3. Relations between the different algorithms

2.3.1 Unsharp masking and the local histogram

So far we have assumed that T[z] is an S-shaped function with the properties that: T[0]=0 and T'[0]=1. It is possible to define a family of such functions by defining for an arbitrary value of s_0:

$$T[z] = g[z/s_0]\ s_0\ /g'[0] \tag{7}$$

with

$$g[u] = G[u] - 1/2 \tag{8}$$

a translated cumulative Gaussian curve through the origin (g[0]=0 ; g'[0] = G'[0] = $1/\sqrt{2\pi}$ = 0.39894.. =≈0.4). The above formula states that the S-shaped function used in UM serves as an estimation of the average cumulative local histogram for all windows, as explained in SMAHE.

2.3.2 Edge amplification and amplification factor

Some interesting relations with SMAHE can be deduced by calculating the edge amplification for a given value of m:

$$A[x-m] = y' = dy/dx \tag{9}$$

$$A[x-m] = 1 + \alpha\ T'[\ x-m\ ] \qquad \text{(UM)} \tag{10}$$

$$= 1 + \alpha\ g'[z/s_0]/g'[0] \tag{11}$$

By definition, the amplification factor is the value on the bissectrix:

$$A[0] = 1 + \alpha \qquad \text{(UM)} \qquad (12)$$

After the function g[u] has been introduced in the SMAHE-formula, the amplification factor becomes:

$$A[0] = Y_{max}\, g'[0] \,/\, s \qquad \text{(SMAHE)} \qquad (13)$$

2.3.3 Comparing the different algorithms

The different algorithms may be compared from different points of view. The table below compares them in the terminology of statistical differencing, giving the values for m and s used in the local window of the original image x, and their goal values in the resulting image y.

Algorithm	m(x)	s(x)	m(y)	s(y)	A[0]=s(y)/s(x)
UM	m	s_0	x	$s_0+\alpha.s_0$	$1+\alpha$=Cte
Stat.Diff.	m	s	m_0	s_0	s_0/s =Cte/s
SMAHE	m	s	$Y_{max}/2=m_0$	$Y_{max}\, g'[0]=s_0$	s_0/s =Cte/s

From the unsharp-masking point of view, these algorithms look very alike when written in simplified form:

$$y = x + C_1\, g[\,(x-m) \,/\, s_0\,] \qquad \text{(UM)} \qquad (14)$$

$$y = D_1 + C_2\, (x-m) \,/\, s \qquad \text{(statistical diff.)} \qquad (15)$$

$$y = D_2 + C_3\, g[\,(x-m) \,/\, s\,] \qquad \text{(SMAHE)} \qquad (16)$$

Because the frequency characteristics of UM are well known, analogous derivations can be made for SMAHE. Analogously, the notions used in UM can be translated in histogram terminology:

1) The S-shaped curve g used by UM is an approximation for the average cumulative histogram of all possible windows in the image.
2) On comparing UM with SMAHE, UM seems to make the assumption that the local histogram for each window has the same standard deviation; this value s_0 is the shape parameter often used for the function T.
3) The Gaussian assumption is situated in between the UM- and the CAHE-assumption. Indeed, UM assumes that the local histogram is the same curve for all window positions, with a variable first central moment m. SMAHE assumes varying first and second central moments. CAHE and CLAHE assume each local histogram may have an arbitrary form, i.e. with all central moments varying.

From the AHE point of view one can notice that:

1) The size of the sliding window in SMAHE corresponds with the kernel size of UM, so that the contrast-enhancement effect of SMAHE may be localised in the use of a larger window/kernel size than UM. When the kernel size of SMAHE is small, it will act as a Laplacian. When the kernel size of UM becomes larger, a correction corresponding with SMAHE is added. So, UM augments the visibility of higher frequency details, AHE locally augments the contrast in what might be regarded as low-frequency enhancement.
2) The use of a clip limit corresponds with a variable amplification factor with given upper limit. Clipping in CLAHE, for small values of the standard deviation s, corresponds to switching to the unsharp masking method, where this value is always kept fixed.

2.4 Second moment unsharp masking (SMUM)

It seems straightforward to propose an edge-enhancement formula adding the SMAHE correction to the original image.
In order to avoid the overamplification by small values of s, the clipping technique for the local histograms can also be applied:

$$y = x + \beta\, g[\,(x-m)\ /\ \max(s,s_{min})\,] \qquad (17)$$

It is possible to make β a function of x, m and/or s. In this way, a class of algorithms can be formulated, each one introducing new and alternative parameters. A lot of effort is still needed to find the 'best' formula with the right parameters for a specific task. The evaluation of the different variants by way of ROC is a very time-consuming operation. A more automatic procedure, as proposed by De Belder (1988), might be more appropriate and has been implemented at Agfa-Gevaert.

3. RESULTS

Fig. 12 and 13 represent SMAHE-results of the thorax shown in Fig. 11, with and without clipping. Fig. 14, 15 and 16 show some variants of SMUM, a little bit overamplified to make the effect visible. Fig. 17 and 18 show results from classical UM. The difference of the varying amplification factor of SMUM augments the visibity of details in the mediastinum zone.

4. IMAGE PROCESSING ON LARGE IMAGES

4.1 Problem description

Constructing useful image-processing techniques for large X-ray images (typical 2000x2500x10b) has the following implications:

1) The number of operations per pixel must be kept as small as possible. This goal is achieved by using only those operations leading to a fast implementation, of which a brief survey is given below. Each time the number of elementary instructions for every pixel, needed to execute the operation, is given:

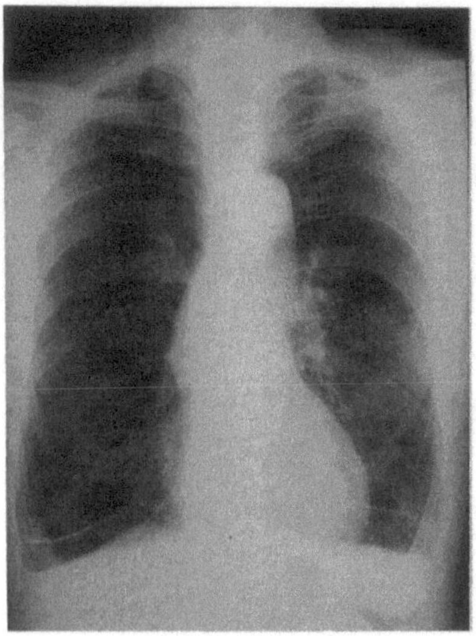

Fig. 11. Unprocessed image (12bit)

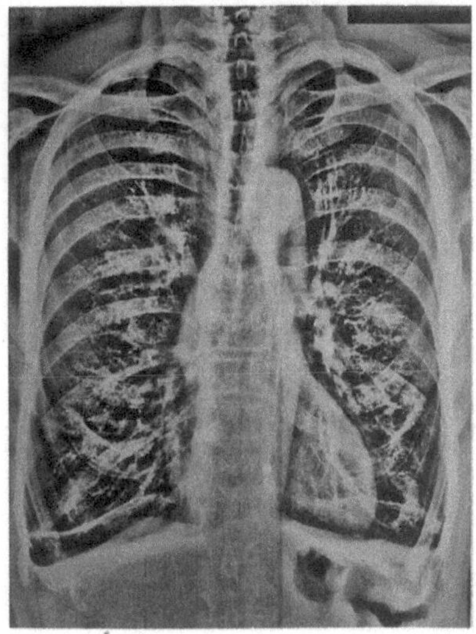

Fig. 12. SMAHE with S_{min}=0
(Kernel=128)

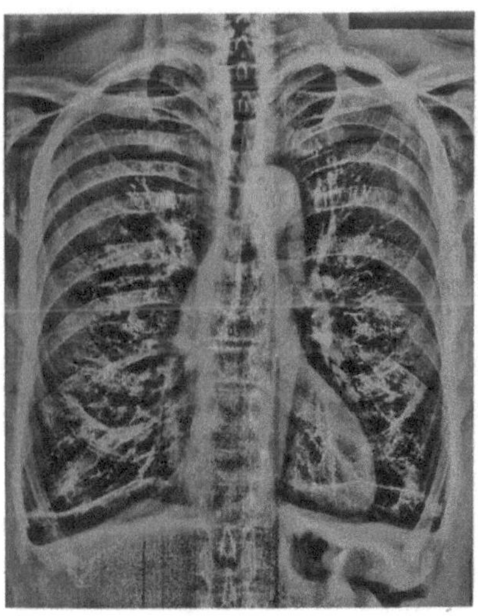

Fig. 13. SMAHE with S_{min}=60
(K=128)

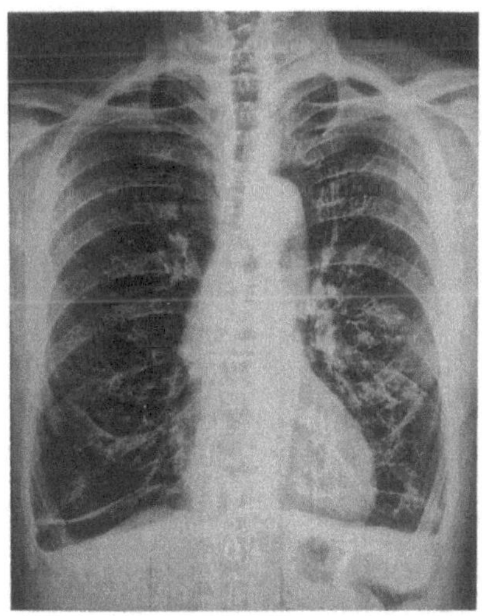

Fig. 14. SMUM
(K=128, S_{min}=10, β=500)

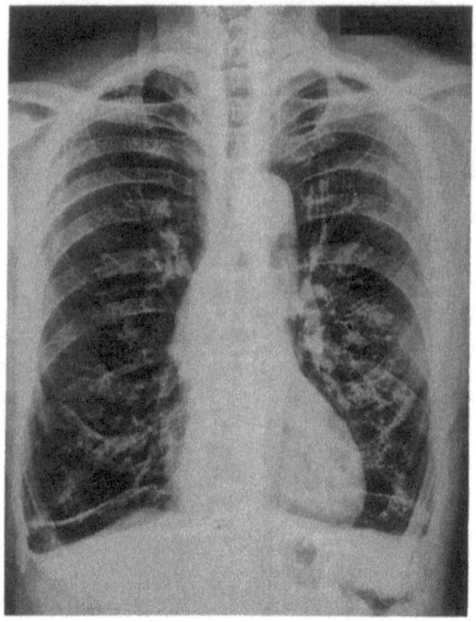

Fig. 15. SMUM

($K=128, S_{min}=30, \beta=500$)

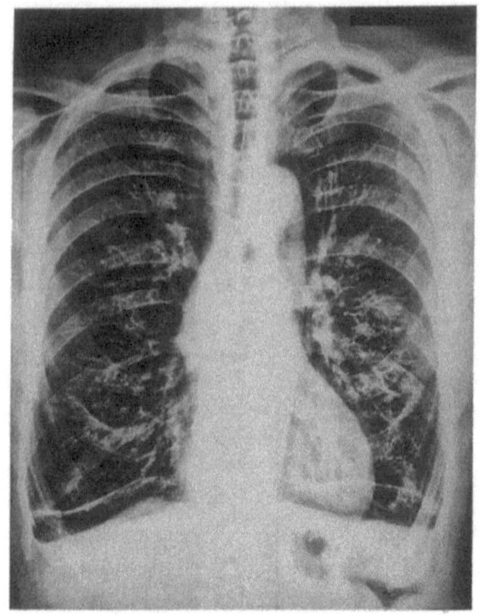

Fig. 16. SMUM

($K=128, S_{min}=30, \beta=750$)

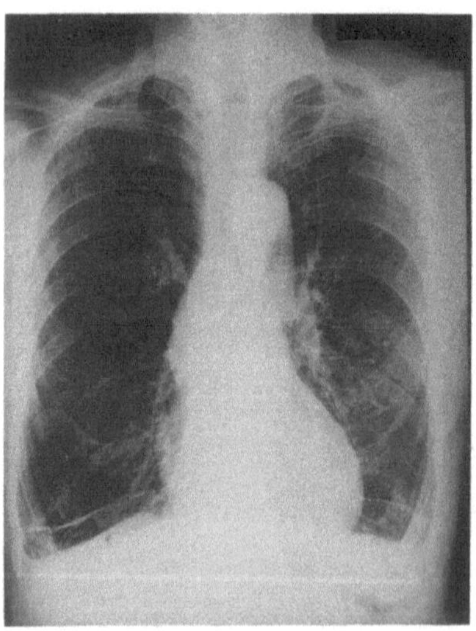

Fig. 17. UM ($K=128, \alpha=1, s_0=68$)

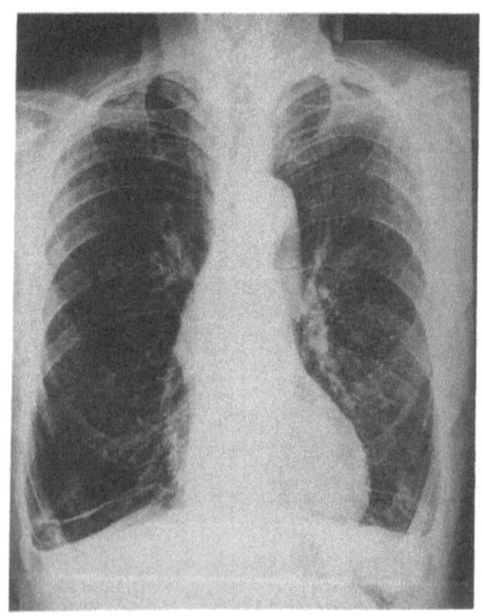

Fig. 18. UM ($K=128, \alpha=2, s_0=68$)

Operation	additions	bit-shifts
addition	1	0
subtraction	1	0
look-up tables 1D	1	0
2D	2	1
LUT-combination 1D	-1	0
multiplication by power of 2	0	1
multiplications by a const.(LUT)	1	0
division by a power of 2	0	1
divisions by constant (LUT)	1	0
power (LUT) (square, inverse..)	1	0
root (LUT) (square root,...)	1	1
convolution with flat-kernel	4	1
multiplication with variable $ab = [(a+b)^2-(a^2+b^2)]/2$	6	1
......	..	..

2) Another important factor is the relation between spatial resolution and pixel depth, which leads to a different optimal implementation method for large images. When an image-processing formula with a convolution is used for a large 10-bit image, it is favourable to calculate a FMG-LUT (4) even when this table has dimensions of 1024 by 1024.

The search for medical image-processing methods at Agfa-Gevaert was directed towards the maximal result that could be obtained, knowing that for large images large look-up tables might be made containing complex formulae, but whereby the number of operations for every pixel was to be kept as small as possible.

4.2 The double convolution

A fast implementation for the standard deviation of a sliding window can be obtained by a double convolution: the first one on the original image, the second one on the squares of the pixels found through a look-up table: $L[x]=x^2$. If we indicate the two different convolution results as:

$$S1 = x \otimes U \tag{18}$$

$$S2 = L[x] \otimes U \tag{19}$$

where U is a flat rectangular kernel with all elements equal to unity, then the standard deviation formula becomes:

$$s = \sqrt{S2/N - (S1/N)^2} \tag{20}$$

A fast implementation can be obtained when:

* N equals a power of 2 (or else with division-LUT)
* the squares are obtained with a look-up table
* the square root is also calculated by a look-up table constructed for example as follows:

$$\mathrm{SQRTLUT}\ [\ i\ ] = \mathrm{Trunc}\ (16\ \sqrt{i}) \qquad i=0..1023 \tag{21}$$

The square root from a number of 18 bits can then be approximated by using this LUT indexed with the 10 most significant bits. Other variants with shorter or longer look-up tables are possible and give different accuracies for small values of s: this is the range where the s-value in CLAHE is replaced by the fixed value s_{min}.

4.3 Implementation aspect of SMAHE

A fast implementation of the SMAHE-formula appears difficult because of the division by s. In a practical implementation, with 10-bit pixels a look-up table of 2048x512 can be constructed giving directly the result of SMAHE:

$$y = L[\ x-m\ ,\ s\ ] \tag{22}$$

By changing the look-up table we can switch from AHE- to CLAHE-variants as shown in Fig. 19 and 20. A further improvement can be obtained by combining this LUT with the square-root LUT.

4.4 Implementation aspect of SMUM

The evaluation of the SMUM-formula is completely analogous, except that an extra LUT is needed to rescale the resulting image:

$$y = c_1\ (x + \alpha\ g[\ (x-m)\ /\ s\ ]) + c_2 \tag{23}$$

Loss of accuracy over this LUT may be avoided by upshifting the UM-expression before the LUT. This implies that the 2D-LUT gives some extra bits accuracy. Rescaling then requires for 1 extra shift and 1 extra addition (rescale-LUT).

4.5 Computational effort

SMIP algorithms lead to very short programs, even when reflection is assumed for the processing of the borders, so that they easily fit in a cache of 256 bytes. The computational effort for the different algorithms is given in the table below, which shows the number of operations per pixel.

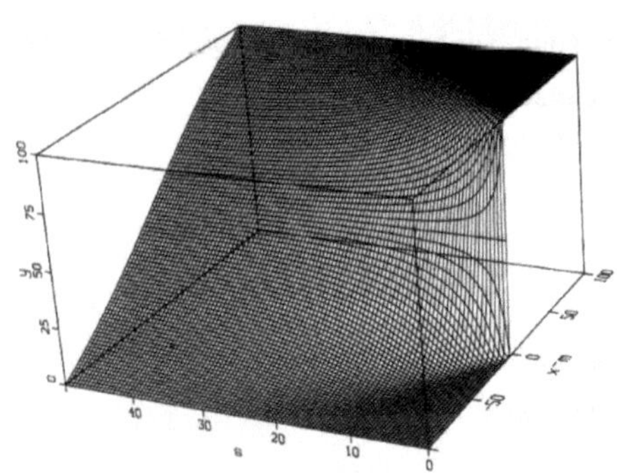

Fig. 19. SMAHE-LUT with s_{min}=0

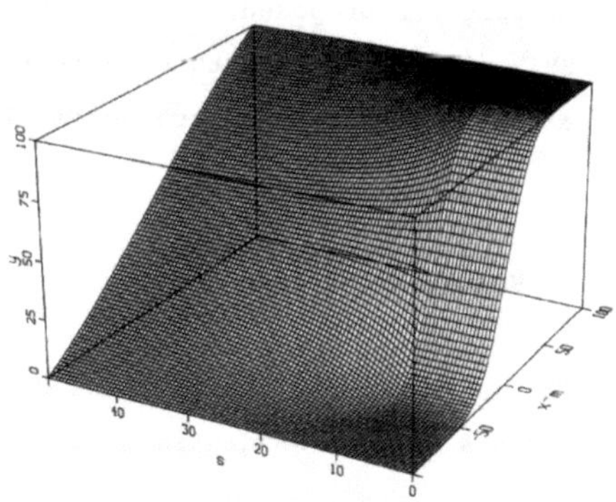

Fig. 20. SMAHE-LUT with s_{min}>0.

algorithm	additions	bit-shifts
Unsharp masking	5	2
Smoothing step	5	2
Total UM-method	10	4
SMAHE	15	5
Statistical differencing according to SMIP	15	5
SMUM	17	6

5. CONCLUSION

SMIP-algorithms allow to drive the image-processing parameters by means of the mean and the standard deviation of the pixels in a continuously sliding window. SMIP gives the possibility to propose new formulae for image-processing algorithms, that combines the advantages of both techniques: edge enhancement and contrast enhancement.
It was shown that CAHE can be fitted into SMIP, showing at the same time that CAHE and UM are very similar algorithms. This resemblance was shown mathematically as well as graphically. A new form of adaptive unsharp masking was proposed, making details in low contrast zones more visible. SMIP algorithms lead to very short and fast implementations, making these algorithms suited for use on small processors.

ACKNOWLEDGEMENT

The author would like to thank Mr M. De Belder for the many discussion and his helpful suggestions during these research activities.

REFERENCES

Austin J.D. and Pizer S., (1987). A Multiprocessor Adaptive Histogram Equalisation Machine, Information Processing in Medical Imaging, de Graaf C.N. and Viergever M.A., Plenum Press, New York and London, pp. 375-392.

Blume H.,(1986). Digital processing of diagnostic images, Medicamundi, Vol. 31, no.3, (4522 984 56701).

Blume H. and Kamiya K., (1987). Auto-ranging normalisation versus histogram modifications for automatic image-processing of digital radiographs, SPIE, Vol. 767, Medical Imaging 371.

Cocklin M.L., Kaye G., Kerr I.H. and Lams P.,(1982). Image Enhancement of Chest Radiographs Using Local Statistical Information, Comput. Soc Int. Symp. Med. Imaging Image Interpretation, Berlin.

Cocklin M.L., Gourlay A.R., Jackson P.H., Kaye G., Kerr I.H. and Lams P., (1983). Digital processing of chest radiographs, Image and Vision Computing, Vol.1, no.2.

De Belder M.,(1988). The evaluation of image-processing algorithms for use in medical imaging, Mathematics and Computer Science in Medical Imaging, Viergever M.A. and Todd-Pokropek A.E., Springer Verlag, Berlin-Heidelberg.

Johnson G.A., Danieley N. and Ravin C.E., (1985). Processing Alternatives for Digital Chest Imaging, Symposium on Digital Radiography, Radiologic Clinics of North America, Vol.23, no.2.

Ketcham D.J., Lowe R.W. and Weber J.W., (1984). Real-time image enhancement techniques, Seminar on Image Processing, Pacific Grove, Hughes Aircraft Company, California, pp. 1-6, SPIE/OSA Vol.74, pp. 120-124.

Pizer S.M., Zimmerman J.B. and Staab E.V., (1976). Adaptive grey level assignement in CT scan display, J. Comput. Assist. Tomogr., Vol.8, pp. 300-305.

Pizer S.M., Austin J.D., Perry J.R., Safrit H.D. and Zimmerman J.B., (1986). Adaptive histogram equalisation for automatic contrast enhancement of medical images, SPIE, Vol. 626, Medicine XIV/PACS IV.

Pizer S.M., Amburn E.P., Austin J.D., Cromartie R., Geselowitz A., Greer T., Romeny B.H., Zimmerman J.B. and Zuiderveld K., (1986). Adaptive histogram equalisation and its variations, Technical Report 86-013, Computer Science Department, University of North Carolina, 207A New West Hall 035A, Chapel Hill, NC 27514. To appear in Computer Vision, Graphics and Image Processing.

Sherrier R.H. and Johnson G.A., (1987). Regionally Adaptive Histogram Equalisation of the Chest, IEEE Transactions on Medical Imaging, Vol. MI-6, no.1.

Part 3. Applications

Section 3.1 Nuclear Medicine

APPLICATIONS OF ITERATIVE RECONSTRUCTION METHODS IN SPECT

B.M.W. Tsui, G.T. Gullberg*, H.B. Hu, J.G. Ballard,
D.R. Gilland, J.R. Perry, W.H. McCartney and T. Bernstein+

The University of North Carolina at Chapel Hill
USA
*The University of Utah
USA
+General Electric Medical Systems Group
USA

ABSTRACT

We have applied iterative reconstruction methods to compensate for the major image degrading effects of photon attenuation and geometric response in SPECT imaging. The compensation methods implement, in the projection and backprojection operations of the iterative reconstruction algorithms, an accurate model of photon attenuation in the patient's body and the geometric response of the collimated-detector system. To evaluate the corrective iterative reconstruction methods, data from a computer-generated phantom which simulated Tl-201 distribution in the thoracic region was used. Also, the techniques were assessed using data from a cardiac SPECT study with Tl-201. Our studies indicate that compensations for attenuation and detector response in SPECT are possible using iterative reconstruction techniques. The compensations are especially important when the attenuation coefficient distribution in the body region, such as the thorax, is non-uniform. The attenuation compensation scheme using the iterative maximum likelihood-EM (ML-EM) algorithm can provide reconstructed images with low noise amplification, and accurate quantitative information without distortions and artifacts. Also, compensation for detector response gives additional improvement in spatial resolution.

1. INTRODUCTION

The major factors which degrade the quality of SPECT images are photon attenuation, geometric detector response, and scatter radiation. Compensations for these factors are necessary to obtain the best SPECT image quality and accurate quantitative information. Most popular attenuation compensation methods used in SPECT assume that the attenuation distribution inside the body is uniform (Budinger and Gullberg, 1974; Kay and Keyes, 1975; Walters et al., 1976; Hsieh et al., 1976; Tretiak and

Delaney, 1978; Chang, 1978; Budinger et al., 1979; Tretiak and Metz, 1980; Gullberg and Budinger, 1981). This assumption is not true in the thoracic region where the various tissues have very different attenuation coefficients. Compensation for non-uniform attenuation distribution by theoretical analysis is difficult and the solution has not been derived (Gullberg, 1979). Detector response and scatter compensations are typically accomplished by deconvolution based on a single detector response function (King et al., 1984; Webb et al., 1985; Madsen and Park, 1985). However, because detector response is a function of the distance from the detector and scatter radiation depends on the configuration and geometry of the scatter medium between the radioactive source and the detector, a simple deconvolution scheme is incorrect.

Since analytical solutions for the correction of non-uniform attenuation and detector response in SPECT are difficult if not impossible to derive, an alternative approach is the use of iterative reconstruction algorithms (Gullberg et al., 1985; Tsui et al., 1988). In this paper, we describe the applications of iterative reconstruction methods in compensating for two of the image degrading effects in SPECT, namely, photon attenuation and detector response. The compensation methods implement an accurate model of photon attenuation and the geometric response of the detector in the projection and backprojection operations of the iterative reconstruction algorithms. Compensation of scatter radiation can be implemented in the same manner by treating it as a component of the total detector response. However, since knowledge of the exact scatter response function as a function of the configuration and geometry of the source distribution and scattering medium is required in the implementation but is unavailable, it was not included in our study.

To evaluate the corrective iterative reconstruction methods, we used data from a computer-generated phantom which simulated Tl-201 distribution in the thoracic region and from a clinical Tl-201 SPECT study. The computer-generated phantom allowed evaluation of the reconstruction algorithms in both the absence and presence of image noise. The compensation methods were also evaluated using clinical data.

2. ITERATIVE RECONSTRUCTION ALGORITHMS

The conventional filtered backprojection algorithm is based on analytical solutions of the reconstruction problem (Budinger et al., 1979). Iterative reconstruction algorithms rely on the ability to correct the pixel values of the estimate of the reconstructed image such that the projections from the estimation are as close as possible to the measured projection data. The correction procedure is repeated until the difference between the calculated and measured projections is less than a predetermined small value.

The distinctions between different iterative algorithms are based on the statistical criteria and correction sequences used. In the algebraic reconstruction technique (ART) in its additive and multiplicative forms (Gordon and Herman, 1974), a set of projection data are calculated from the current estimate of the reconstructed image and compared with the measured projection data. Correction factors are computed from the difference and applied to all pixels of the reconstructed image. These updated pixel values are then used in calculating the next projection. Although the ART algorithms are efficient, as a large amount of updating is incorporated during each iteration, they are very susceptible to noise (Herman et al., 1973; Budinger and Gullberg, 1974). The result is that the ART algorithms converge rapidly during earlier iterations but tend to diverge at higher iteration numbers if the data are noisy. This difficulty can be partially corrected by the use of a damping factor at the cost of a slower rate of convergence (Herman and Lent, 1976).

In the multiplicative ART (MART) algorithm, the (n+1)th estimate of the reconstructed image at pixel (i,j) is given by

$$x_{ij}^{n+1} = x_{ij}^{n}(p_{km}/p_{km}^{n}) \tag{1}$$

where x_{ij}^{n} is the nth estimate of the reconstructed image, p_{km} and p_{km}^{n} are the measured projection data and the projection data calculated from x_{ij}^{n} at projection angle θ_m and bin k, respectively. The calculated projection data are given by

$$p_{km}^{n} = \sum_{i,j} W_{ij}(k,m)\, x_{ij}^{n}, \tag{2}$$

where $W_{ij}(k,m)$ is the probability of emissions from pixel (i,j) being detected at projection angle θ_m and bin k.

The conjugate gradient (CG) algorithm is an additive technique which minimizes the chi-square function by the steepest descent technique. Here the (n+1)th estimate of the image is given by

$$x_{ij}^{n+1} = x_{ij}^{n} + a_n \Delta^n \tag{3}$$

where Δ^n is the direction of the nth step and a_n is the factor that determines the step length. Different from the gradient algorithm, the convergence rate of the CG algorithm is improved by making the step direction orthogonal to all previous steps (Huesman et al., 1977; Budinger et al., 1979).

The maximum likelihood (ML) approach incorporates the Poisson statistics in the measured data and maximizes the likelihood function of the projection data given the radioactivity distribution. By using an EM algorithm (Shepp and Vardi, 1982), an estimate of the reconstruction image can be obtained with a multiplicative algorithm where the the (n+1)th estimate of the reconstructed image at pixel (i,j) is given by

$$x_{ij}^{n+1} = \frac{x_{ij}^{n}}{\sum_{k,m} W_{ij}(k,m)} \sum_{k,m} \left[W_{ij}(k,m) \frac{p_{km}}{p_{km}^{n}} \right]. \tag{4}$$

The summations in Eq. (4) over (k,m) are the backprojection operations

$$b_{ij}^{n} = \sum_{k,m} W_{ij}(k,m)\, p_{km}^{n}. \tag{5}$$

Here the weighting factors $W_{ij}(k,m)$ represent the amount a projection bin will contribute to an image pixel in the backprojection operation.

Besides modeling the statistical noise accurately, the ML-EM algorithm possesses other important properties. If the initial estimate of the reconstructed image is positive, the EM algorithm automatically sets a non-negativity constraint. Also, the total number of counts in the reconstructed image at each iteration remains constant.

Different iterative algorithms exhibit different characteristics. In SPECT image reconstruction, important considerations are the convergence rate, noise propagation and processing time. To evaluate the different iterative algorithms, we used a computer-generated phantom consisting of four Gaussian-shaped hot spots with different sizes and superimposed on a uniform circular disk (Fig. 1). Simulation studies were performed using this phantom.

Figure 2 shows the noise-free reconstructed images obtained with the CG algorithm after the 3rd, 5th, 10th, 20th, 30th and 40th iterations. Similar reconstructed images can be obtained with the MART and the ML-EM algorithms. To compare the convergence rate, we calculated the mean-squared-error (MSE) after each iteration. The MSE after the nth iteration is defined as

$$MSE^n = \{\sum_{i,j}^{N} [x_{ij}{}^n - x_{ij}]^2/N\}^{1/2}, \tag{6}$$

where x_{ij} is the true image distribution and $x_{ij}{}^n$ is the estimate of the reconstructed image after the nth iteration, and N is the total number of pixels in the reconstruction image.

Figure 3 shows plots of the MSE as a function of the iteration number for the MART, CG and ML-EM algorithms in the noise-free case. The curves show that the iterative algorithms yield good estimates of the true image with low MSE values at high iteration numbers. Among the three algorithms, the MART gives the fastest and the ML-EM algorithm the slowest convergence rate

To evaluate the performance of the iterative algorithms in the presence of noise, Poisson noise fluctuations with standard deviation equal to 10% of the mean were added to the simulated projection data. Figures 4 (a), (b), and (c) shows the reconstructed images obtained with the MART, CG and the ML-EM algorithms, respectively, after the 4th, 6th, 10th, 20th, 30th, and 40th iterations. The MSE of the reconstructed images obtained with the three iterative algorithms as a function of iteration numbers are plotted in Fig. 5.

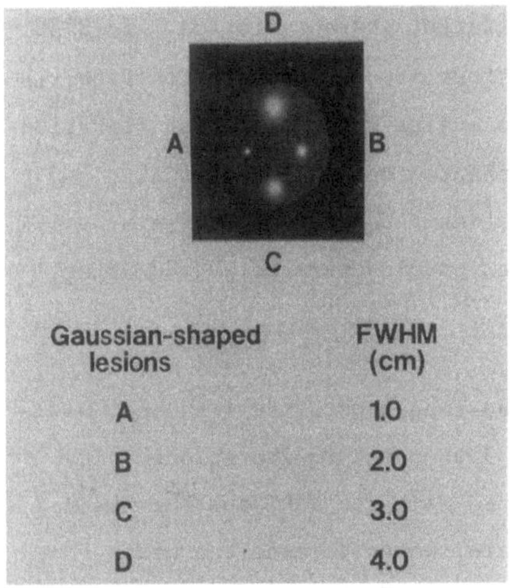

Fig. 1. Computer-generated phantom consisting of four Gaussian-shaped hot lesions superimposed on a uniform circular background. The FWHM of the lesions are 1cm, 2cm, 3cm, and 4cm. The object contrast is 1 for all lesions.

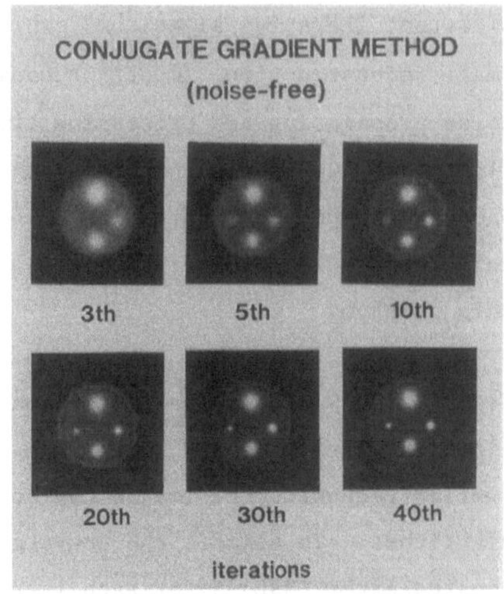

Fig. 2. Noise-free reconstructed images of the computer-generated phantom shown in Fig. 1. The images were obtained with the iterative CG algorithm after the 3th, 5th, 10th, 20th, 30th, and 40th iterations.

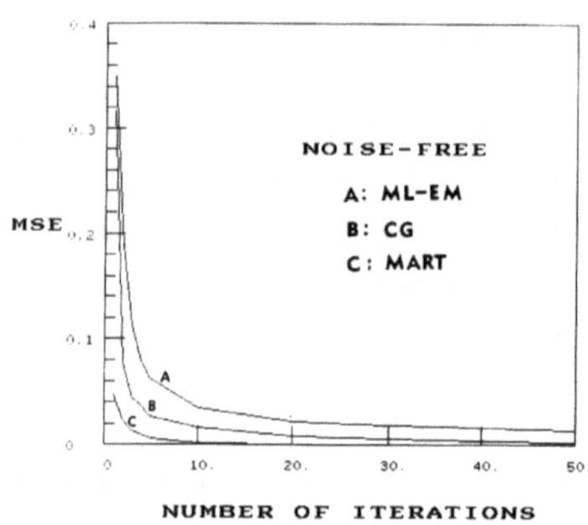

Fig. 3. Mean-Squares-Error (MSE) of the noise-free reconstructed images obtained from the computer-generated phantom shown in Fig. 1. The reconstructed images were generated from the iterative MART, CG and ML-EM algorithms plotted as a function of iteration number.

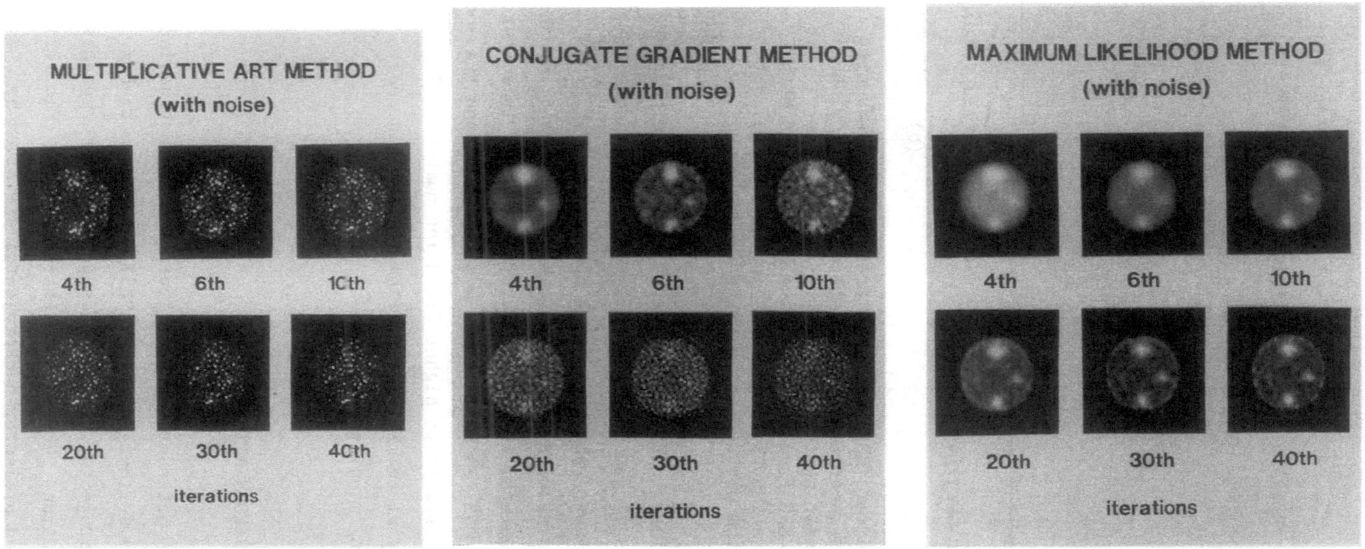

(a) (b) (c)

Fig. 4. Reconstructed images of the computer-simulated phantom shown in Fig. 1 with approximately 10% Poisson noise fluctuations added to the projection data. The images were obtained with the iterative (a) MART, (b) CG, and (c) ML-EM algorithms after the 4th, 6th, 10th, 20th, 30th, and 40th iterations.

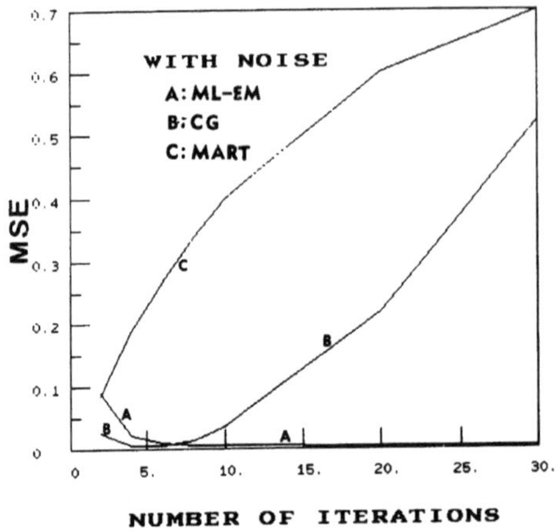

(a)

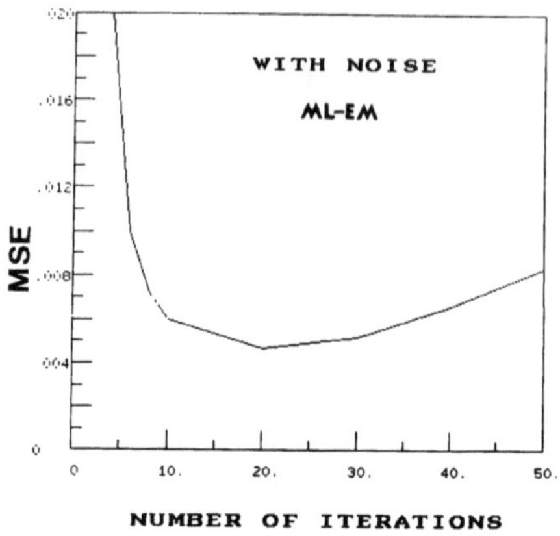

(b)

Fig. 5. Mean-squared-error (MSE) obtained from the reconstructed images of the computer-generated phantom shown in Fig. 1 with approximately 10% Poisson noise fluctuations added to the projection data. (a) MSE from the iterative MART, CG and ML-EM algorithms plotted as a function of iteration number. (b) MSE from the ML-EM algorithm plotted in an expanded scale.

Figures 4 and 5 show the fast initial convergence and rapid divergence at higher iterations of the MART and CG algorithms. Although the ML-EM algorithm provides the slowest rate of initial convergence, it gives the lowest noise amplification at high iteration numbers. Figure 5 (b) shows the MSE from the ML-EM algorithm at higher iteration numbers in an expanded scale. It shows that the MSE reaches a broad minimum at around the 20th iteration before diverging slowly at higher iteration numbers. The low noise propagation property of the ML-EM algorithm is the main reason for its choice in the following studies.

3. COMPENSATION FOR NON-UNIFORM ATTENUATION

The compensation of photon attenuation using iterative reconstruction techniques was proposed by Gullberg et al. (1985). As shown in Fig. 6, a ray-driven projector-backprojector can be designed which incorporates the exact attenuation distribution. Here the weighting factors (Eqs. (2) and (5)) used in the attenuated projector-backprojector are given by

$$W_{ij}(k,m) = \begin{cases} \dfrac{A_{ij}(k,m)}{\mu_{ij}} \{ 1 - \exp[-\mu_{ij} l_{ij}(k,m)] \}, & \mu_{ij} > 0 \\ l_{ij}(k,m)\, A_{ij}(k,m), & \mu_{ij} = 0 \end{cases} \tag{7}$$

where μ_{ij} is the attenuation coefficient of pixel (i,j) and $l_{ij}(k,m)$ is the projection ray length which is within pixel (i,j). The attenuation factor $A_{ij}(k,m)$ is the exponential of the line integral of the attenuation from the point b_{ij} (Fig. 6) to the detector:

$$A_{ij}(k,m) = \exp[- \int_{b_{ij}}^{\text{detector}} \mu(s_{km})\, ds_{km}], \tag{8}$$

where b_{ij} is the entrance to the pixel (i,j) and s_{km} is the length along the projection ray. The attenuated projector-backprojector incorporates the exact attenuation distribution and was used in the iterative algorithms to compensate for the attenuation effect in the SPECT reconstruction.

4. COMPENSATION OF DETECTOR RESPONSE

The detector response degrades as the distance from the detector increases for most collimated-detector systems used in nuclear medicine imaging. To model the distance-dependent detector response function, we use a fan of projection rays originating from each projection bin as shown in Fig. 7. The projection rays in each fan are weighted according to the shape of the distance-dependent detector response function to model the resolution degrading effect of the collimated-detector. Assuming that the detector response function is independent of the position of the projection bin and the projection angle, the weighting function is stored in a look up table to speed up the processing time during the iterative process. The fan-shaped projector-backprojector was used in the iterative algorithms to compensate for the detector response. Furthermore, each projection ray in the fan can incorporate the attenuation effect as described in the previous

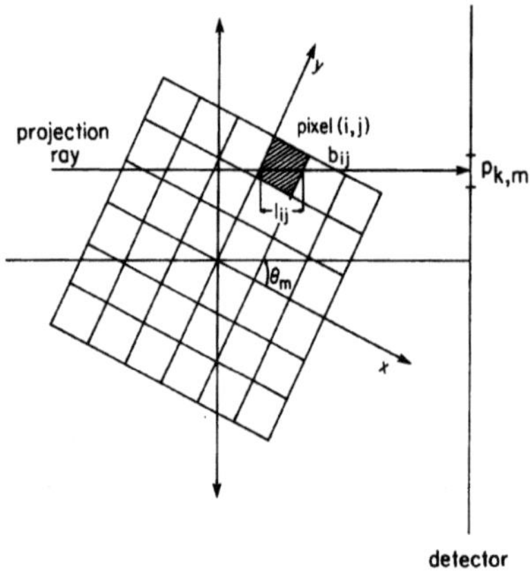

Fig. 6. Projection ray at bin k and angle θ_m through the reconstructed image matrix.

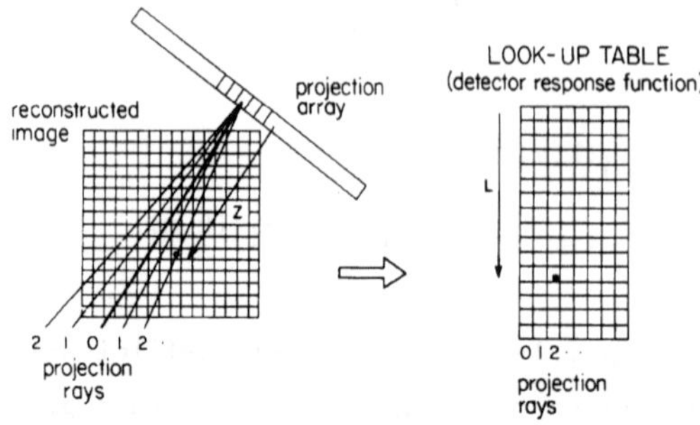

Fig. 7. A fan of projection rays originates from each projection bin to simulate the distance dependent detector response function of a rotating camera based SPECT system.

section. By including the attenuation effect, we are able to simultaneously compensate for both the attenuation and the detector response in SPECT image reconstruction.

5. COMPUTER SIMULATION

To evaluate the projector-backprojectors which incorporate the effects of attenuation and detector response, we used a computer-generated phantom to simulate the anatomy and the Tl-201 uptake in the thorax. As shown in Fig. 8, the phantom consists of the attenuation coefficient distribution of the different tissues such as muscle, lung, ribs and spine (Fig. 8 (a)), and the distribution of Tl-201 uptake in the myocardium and other tissues in the thoracic region (Fig. 8 (b)). Figs. 8 (c) and (d) show a 45° and a vertical profile through the distribution of Tl-201 uptake shown in Fig. 8 (b). These profiles serve as the reference for comparing the different reconstruction algorithms.

Projection data from 64 views over 180° were generated from the cardiac-chest phantom. The projection data included the effects of attenuation using the attenuation coefficient distribution in Fig. 8 (a) and the geometric response of a low energy general purpose collimator. The radius-

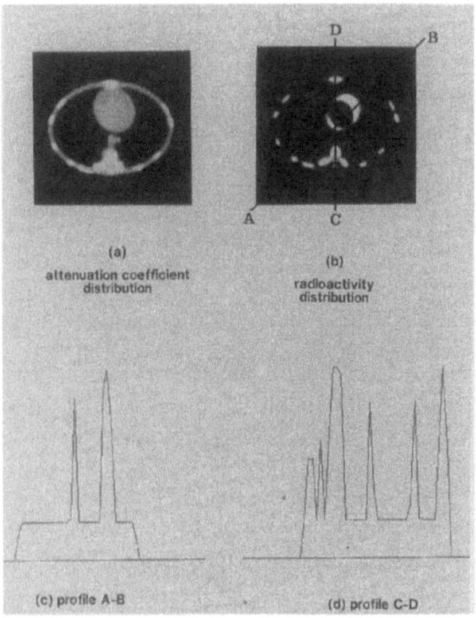

Fig. 8. A computer-generated phantom which simulate the anatomy and the Tl-201 uptake in the thorax. (a) Attenuation coefficient distribution. (b) Distribution of Tl-201 uptake. The (c) 45° and (d) vertical profiles shown in (b).

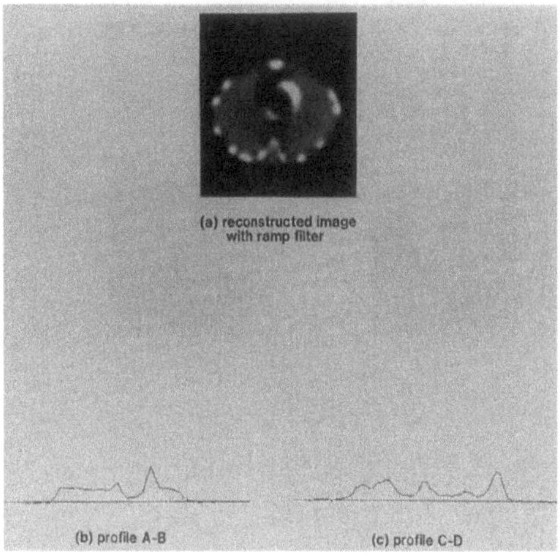

Fig. 9. (a) The reconstructed image obtained with the filtered backprojection algorithm without attenuation compensation. The (b) 45° and (c) vertical profiles are obtained from (a) corresponding to that in Fig. 8(b).

of-rotation was assumed to be 22.5 cm. Reconstructed images were obtained from the projection data using the different reconstruction algorithms. The matrix size used in the reconstruction was 64x64.

Figure 9 (a) shows the reconstructed image obtained with the conventional filtered backprojection method without attenuation compensation and Figs. 9 (b) and (c) show the corresponding profiles in Fig. 8 (b). Figures 9 (b) and (c) was plotted in the same scale as Figs. 8 (c) and (d). Besides yielding inaccurate quantitative information (Figs. 9 (b) and (c)), the reconstructed image shows distortions and artifacts especially in the central region of the thorax (Fig. 9(a)).

To demonstrate the effects of the iterative reconstruction algorithm alone, Figs. 10 (a) to (d) show the reconstructed images at the 5th, 20th, 35th and 50th iteration obtained with the iterative ML-EM algorithm without any correction, and Figs. 10 (e) and (f) show the corresponding profiles shown in Fig. 8 (b). From Fig. 10, we see that the reconstructed image

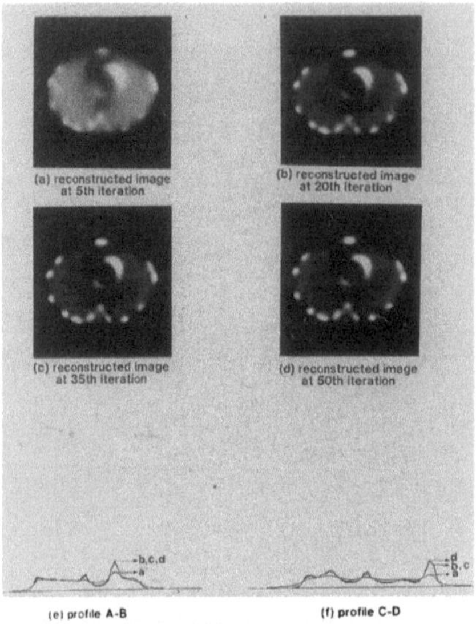

Fig. 10. (a) - (d) The reconstructed images obtained with the iterative ML-EM algorithm without correction for any physical factors. The (e) 45° and (f) vertical profiles from (a) to (d) corresponding to that in Fig. 8 (b).

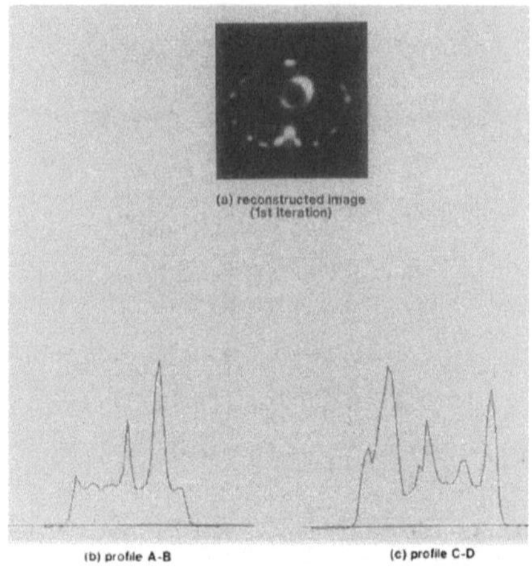

Fig. 11. (a) The reconstructed image obtained with the modified Chang attenuation compensation method after after the first iteration. The (b) 45° and (c) vertical profiles from (a) corresponding to that in Fig. 8 (b).

approaches the best image quality after about 20 iterations. By comparing Fig. 10 with Fig. 9, we find that the best reconstructed image is similar to that obtained with the conventional filtered backprojection without any correction. Also, identical distortions and artifacts are found in images obtained with both methods and are results from the lack of attenuation compensation.

Figure 11 shows the reconstructed image obtained with a modified Chang attenuation compensation method and the corresponding profiles after the 1st iteration. The algorithm was similar to that proposed by Chang (Chang 1978) except that, instead of assuming an uniform attenuator, the attenuation coefficient distribution shown in Fig. 8 (a) was used to generate the correction factors. Fig. 11 demonstrate that the modified Chang method provide good reconstructed image quality and quantitative accuracy in the noise-free case.

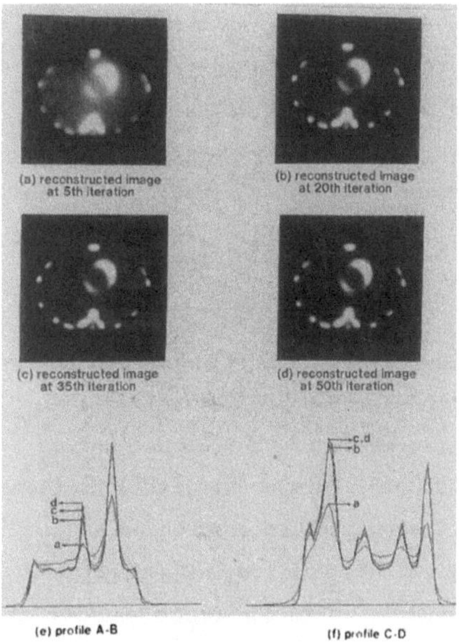

Fig. 12. (a) - (d) The reconstructed images obtained with the iterative ML-EM algorithm with attenuation correction. The (e) 45° and (f) vertical profiles from (a) - (d) corresponding to that in Fig. 8 (b).

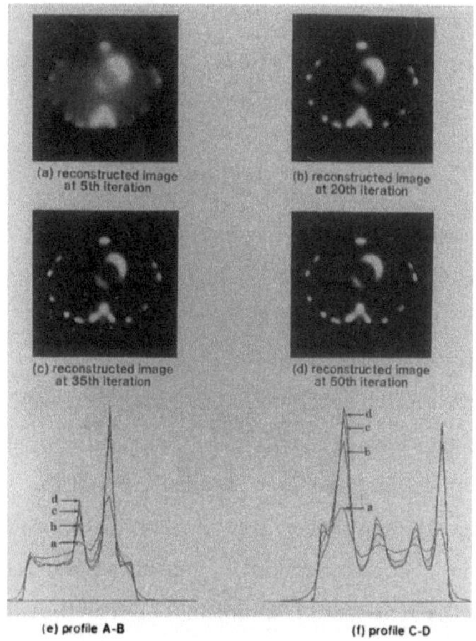

Fig. 13. (a) - (d) The reconstructed images obtained with the iterative ML-EM algorithm with simultaneous correction for both attenuation and detector response. The (e) 45° and (f) vertical profiles form (a) - (d) corresponding to that in Fig. 8 (b).

Figures 12 (a) to (d) shows the reconstructed images at the 5th, 20th, 35th and 50th iterations obtained with the iterative ML-EM algorithm with attenuation correction. The attenuation coefficient distribution shown in Fig. 8 (a) was used together with the attenuated projector-backprojector in the iterative algorithm to compensate for the attenuation effect. Figures 12 (e) and (f) show the corresponding profiles. Figure 12 demonstrate that the ML-EM algorithm gives the best image quality and quantitative accuracy between the 20th and the 35th iterations. By comparing Fig. 12 with Figs. 9 and 10, we see that proper attenuation correction results in the disappearance of distortions and artifacts, and improved quantitative accuracy in the reconstructed image.

Figure 13 shows the reconstructed images and the corresponding profiles obtained with the iterative ML-EM algorithm with simultaneous correction for both attenuation and detector response. The projector-backprojector which models exactly both the non-uniform attenuation distribution and the distance-dependent detector response was used in the iterative algorithm. Figure 13 shows that the algorithm gives reconstructed images with good image quality and no apparent distortions and artifacts after about 25 iterations. The simultaneous compensation for both attenuation and detector response provides further improvement in spatial resolution when comparing Figs. 13 (e) and (f) with Figs. 12 (e) and (f). The improvement appears to continue beyond the 50th iteration.

By comparing Figs. 11 (b) and (c) with Figs. 12 (e) and (f), we find that the modified Chang attenuation compensation method performs well in the noise-free situation. To evaluate the performance of the reconstruction algorithm in the presence of noise, we added 10% Poisson noise fluctuations to the simulated projection data. Fig. 14 shows the reconstructed image and the corresponding profiles obtained with the modified Chang algorithm after the 1st iteration. Large noise fluctuations in the reconstructed image are apparent. Further increase in the image noise is found as the iteration number increases.

On the other hand, Fig. 15 shows the reconstructed images and the corresponding profiles obtained with the iterative ML-EM algorithm with attenuation correction. The reconstructed images and the profiles demonstrate the good image quality with lower noise fluctuations.

6. PATIENT STUDY

To further evaluate the different reconstruction algorithms, we used clinical data from a Tl-201 stress study. The patient was first subjected to cycling exercise. At the peak of stress, a 2 mCi dose of ^{201}Tl chloride was injected intravenously and the exercise continued at the same level for 90 seconds. Following this the patient was transferred to a GE 400 AC/T SPECT system for the emission scan. The time lapse between injection and the onset of the emission scan was less than 5 minutes. The emission projection data were acquired in 64x64 matrices. Data were summed over three energy windows, i.e. 67.5 to 82.5 kev, 121.5 to 148.5 kev and 150.3 to 183.7 kev. Projection data were acquired at 30 sec/view with a total scan time of 32 minutes. The starting angle was 45° left posterior oblique with the camera rotating 180° right anterior oblique.

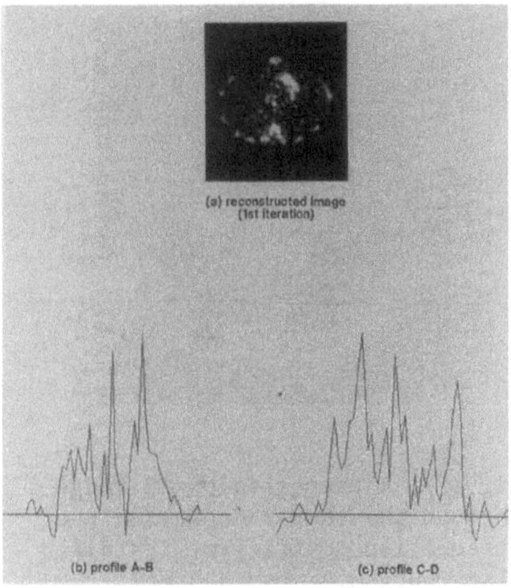

Fig. 14. (a) The reconstructed image obtained with the modified Chang attenuation correction method after the first iteration. Poission noise fluctuations were added to the simulated projection data. The (b) 45° and (c) vertical profiles from (a) corresponding to that in Fig. 8(b).

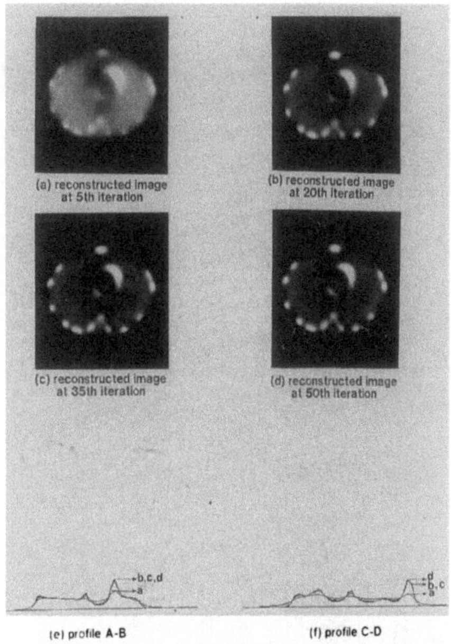

Fig. 15. (a) - (d) The reconstructed images obtained with the iterative ML-EM algorithm with attenuation correction. Poission noise fluctuations were added to the simulated projection data. The (e) 45° and (f) vertical profiles from (a) to (d) corresponding to that in Fig. 8(b).

To acquire the attenuation coefficient distribution of the patient, a transmission CT study was performed after the emission scan. As shown in Fig. 16, a collimated sheet source was installed on the gantry of the SPECT system at the side of the patient which was opposite to the camera, and rotated around the patient with the camera. The sheet source was filled with 20 mCi of Tc-99m. A 20% energy window was used. Without moving the patient, the transmission projection data were acquired in 64x64 matrices and in 64 views over 180°. At 30 sec/view, the total transmission scan time was 32 minutes.

An example of the transmission CT images depicting the attenuation coefficient distribution is shown in Fig. 17. Regions with different attenuation coefficients such as the muscle, lungs and myocardium are demonstrated. The total counts per slice was about 288,000.

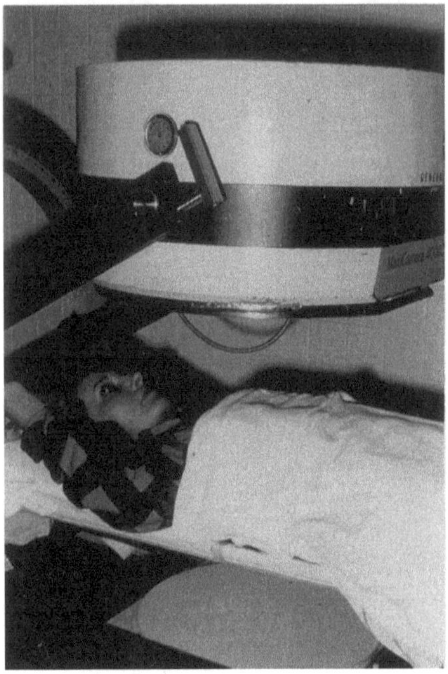

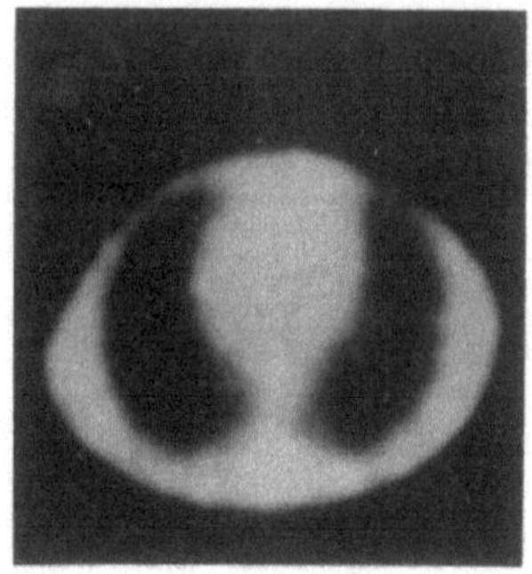

Fig. 17. A transmission CT image through the thorax depicting the attenuation coefficient installed distribution.

Fig. 16. Experimental set-up used in the transmission study.

Figures 18 to 22 show examples of the transaxial emission CT images obtained with the conventional filtered backprojection and the iterative ML-EM algorithm without attenuation correction, the modified Chang and the iterative ML-EM algorithm with attenuation correction, and the iterative ML-EM algorithm with simultaneous correction for both attenuation and detector response, respectively. The reconstructed images are obtained from the same image slice as the transmission CT image shown in Fig. 17. The attenuation coefficient distribution was used in the attenuation correction schemes. The total counts per emission image slice was about 100,000.

The results from the Tl-201 study demonstrate the importance of proper attenuation correction in clinical data. The conventional filtered backprojection method without attenuation correction produces more noise fluctuations and noticeably fewer streak artifacts in the background near the ventricular apex. The reconstructed images from the iterative ML-EM algorithm without attenuation correction give lower image noise

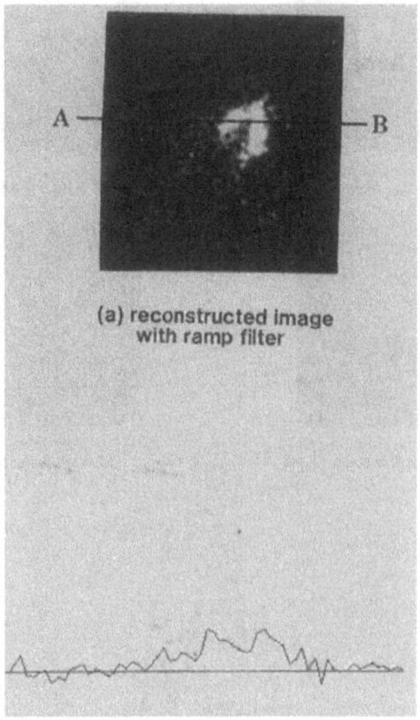

(b) profile A-B

Fig. 18. (a) The Tl-201 emission reconstructed image obtained with the filtered backprojection algorithm without attenuation correction. (b) Plot of the horizontal profile shown in (a).

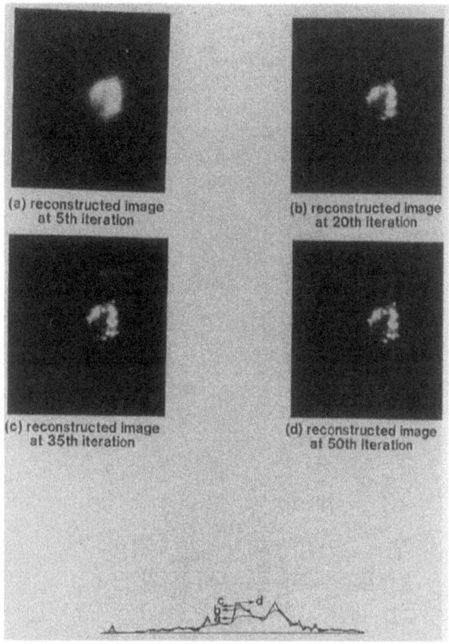

(e) profile A-B

Fig. 19. (a) - (d) The Tl-201 emission reconstructed images obtained with the iterative ML-EM algorithm without correction for any physical factors. (e) Plot of the horizontal profiles from (a) to (d) corresponding to that in Fig. 18 (a).

fluctuations but with the same image artifacts. The attenuation methods give better quantitative information. Among the two correction methods, the iterative ML-EM algorithm provides lower image noise fluctuations and overall superior image quality compared with the modified Chang algorithm. Simultaneous attenuation and detector response correction using the iterative ML-EM algorithm yields improved spatial resolution in the reconstructed image.

7. CONCLUSION AND DISCUSSION

In summary, our studies show that correction for non-uniform attenuation is important in cardiac SPECT studies. The conventional filtered backprojection method produces reconstructed images with distortions and

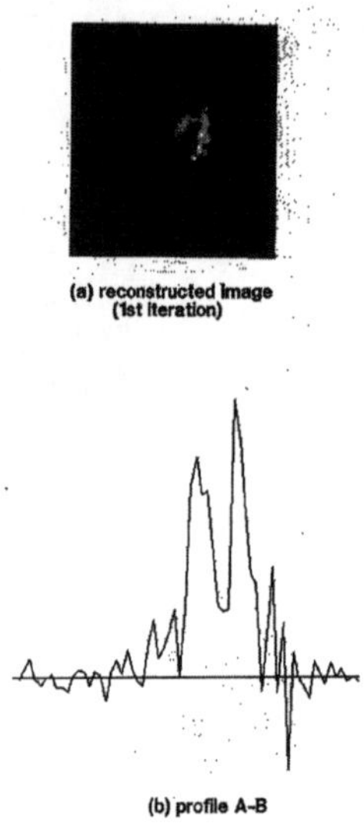

Fig. 20. (a) The Tl-201 emission reconstructed image obtained with the modified Chang attenuation correction method after the first iteration. The attenuation coefficient distribution of the same image slice shown in Fig. 17 was used in the correction scheme. (b) Plot of the horizontal profile from (a) corresponding to that in Fig. 18 (a).

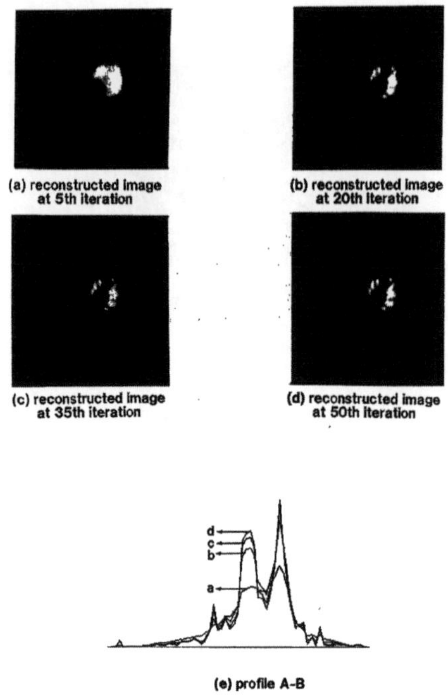

Fig. 21. (a) - (d) The Tl-201 reconstructed images obtained with the iterative ML-EM algorithm with attenuation correction. The attenuation coefficient distribution of the same image slice shown in Fig. 17 was used in the correction schems. (e) Plot of the horizontal profiles from (a) to (d) corresponding to that in Fig. 18 (a).

artifacts, and inaccurate quantitative information. Iterative reconstruction algorithms alone can not reduce these effects. Attenuation correction using the modified Chang and the iterative ML-EM algorithms in SPECT results in improved quantitative accuracy. However, the modified Chang algorithm tends to amplify statistical fluctuations especially at higher iterations more than the iterative ML-EM algorithm.

Simultaneous correction of non-uniform attenuation and detector response is possible using the iterative reconstruction algorithms. The exact detector response function can be incorporated in the projector-backprojector operator. Improved spatial resolution in the reconstructed image can be found at high iteration number.

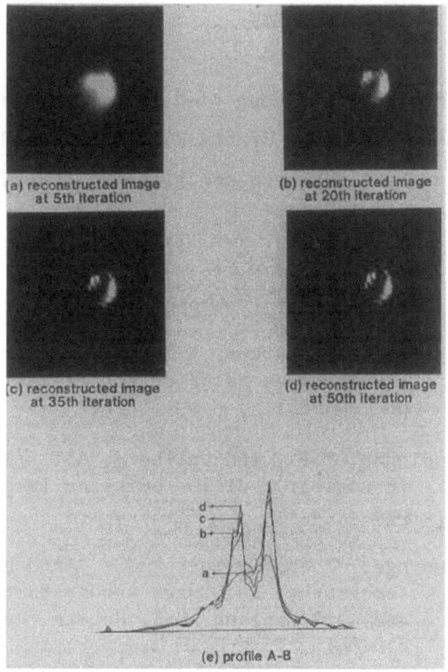

Fig. 22. (a) - (d) The Tl-201 reconstructed images obtained with the iterative ML-EM algorithm with simultaneous correction for both attenuation and detector response. The attenuation coefficient distribution of the same image slice shown in Fig. 17 was used in the correction scheme. (e) Plot of the horizontal profiles from (a) to (d) corresponding to that in Fig. 18 (a).

The use of the correction method requires knowledge of the attenuation coefficient distribution which was obtained from a transmission tomographic study. In our present study, the transmission CT study requires extra imaging time which is difficult to implement with some patients. To overcome this problem, simultaneous transmission and emission data acquisition is desirable. An example is the use of isotopes which emit photons with different energies in the transmission and emission studies, and dual energy acquisition mode which is available in most SPECT systems. For example, Gd-153 with photon emissions at 99 kev can be used in the transmission source with Tc-99m being used as the emission radionuclide (Bailey et al., 1987). The problem of the dual energy acquisition method, however, is the contribution of scatter from the higher energy emission source into the transmission data.

Recently, advances have been made in the improvement of the rate of convergence and processing time of the iterative ML-EM algorithm for use in clinical practice. Acceleration methods (Lewitt et al., 1986; Kauffman, 1987; Tanaka, 1987; Metz et al., 1988) have been proposed to improve the convergence rate of the iterative algorithm. Substantial increase in processing speed has been obtained by implementing the iterative algorithms in array processors. Further increase in computational speed can be achieved through the use of more advanced array processors and special "hardwired" processors.

8. ACKNOWLEDGMENT

This work was supported in part by the Department of Health and Human Services under Grants R01-CA39463 and R01-HL39792, The Whitaker Foundation, and by a research contract from the General Electric Medical Systems Group.

9. REFERENCES

Bailey, D.L, Hutton, B.R., and Walker, P.J. (1987). Improved SPECT using simultaneous emission and transmission tomography, J. Nucl. Med. 28, pp. 844-851.

Budinger, T.F., and Gullberg, G.T. (1974). Three-dimensional reconstruction in nuclear medicine emission imaging, IEEE. Trans. Nucl. Sci. NS-21, pp. 2-10.

Budinger, T.F., Gullberg, G.T., and Huesman, R.H. (1979). Emission computer tomography. In: Image Reconstruction from Projections: Implementation and Applications, G.T. Herman (ed.), Springer-Verlag, New York, pp. 540-541.

Chang, L.T. (1978). A method for attenuation correction in radionuclide computed tomography, IEEE Trans. Nucl. Sci. NS-25, pp. 638-642.

Gordon, R. (1974). A Tutorial on ART, IEEE Trans. Nucl. Sci. NS-21, pp. 78-93.

Gordon, R., and Herman, G.T. (1974). Three dimensional reconstruction from projections: A review of algorithms, Int. Rev. Cyto. 38, pp. 111-151.

Gullberg, G.T. (1979). The Attenuated Radon Transform: Theory and Application in Medicine and Biology. Ph.D. Dissertation, University of California at Berkeley.

Gullberg, G.T., and Budinger, T.F. (1981). The use of filtering methods to compensate for constant attenuation in single-photon emission computed tomography. IEEE Trans. Biomed. Eng. BME-28, PP. 142-157.

Gullberg, G.T., Huesman, R.H., Malko, J.A. Pelc, J.J., and Budinger, T.F. (1985). An attenuated projector-backprojector for iterative SPECT reconstruction, Phys. Med. Biol. 30, pp. 799-816.

Herman, G.T., Lent, A., and Rowland, S. (1973). ART: Mathematics and applications. A report on the mathematical foundations and on the applicability to real data of the Algebraic Reconstruction Techniques, J. Theor. Biol. 42, pp. 1-32.

Hsieh, R.C., and Wee, W.G. (1976). On methods of three-dimensional reconstruction from a set of radioisotope scintigrams, IEEE Trans. Syst. Man Cybern. SMC-6, pp. 854-862.

Huesman, R.H., Gullberg, G.T., Greenberg, W.L., and Budinger, T.F. (1977). RECLBL library users manual. Lawrence Berkeley Laboratory, University of California.

Kaufman, L. (1987). Implementing and accelerating the EM algorithm for positron emission tomography. IEEE Trans. Med. Imaging MI-6, pp. 37-51.

Kay, D.B., and Keyes, J.W. (1975). First order corrections for absorption and resolution compensation in radionuclide Fourier tomography. J. Nucl. Med. 16, pp. 540-541.

King, M.A., Schwinger, R.B., Doherty, P.W. (1984). Two-dimensional filtering of SPECT images using the Metz and Wiener filters, J. Nucl. Med. 25, pp. 1234-1240.

Lewitt, R.M., and Muehllehner, G. (1986). Accelerated iterative reconstruction for positron emission tomography based on the EM algorithm for maximum likelihood estimation. IEEE Trans. Med. Imaging MI-5, pp. 16-22.

Madsen, M.T., and Park, C.H. (1985). Enhancement of SPECT images by Fourier filtering the projection image set. J. Nucl. Med. 26, pp. 395-402.

Metz, C.E., and Chen, C.T. (1988). On the acceleration of maximum likelihood algorithms. SPIE, in press.

Ortega, J.M. and Rheinboldt, W.C. (1970). Iterative solution of nonlinear equations in several variables, Academic Press, New York.

Shepp, L.A., and Vardi, Y. (1982). Maximum likelihood reconstruction for emission tomography. IEEE Trans. Med. Imaging MI-1, pp. 113-122.

Tanaka, E. (1987). A fast reconstruction algorithm for stationary positron emission tomography based on a modified EM algorithm. IEEE Trans. Med. Imaging MI-6, pp. 98-105.

Tretiak, O.J., and Delaney, P. (1978). The exponential convolution algorithm for emission computed axial tomography. In: Rev. Inform. Process. Med. Imaging. Proc. 5th Int. Conf., Vanderbilt Univ., Nashville, TN, A.B. Brill and R.R. Price (ed.), Oak Ridge National Laboratory Rep. ORNL/BNCTIC-2, pp266-278.

Tretiak, O.J., and Metz, C.E. (1980). The exponential Radon transform. SIAM J. Appl. Math., 39, pp. 341-354.

Walters, T.E., Simon, W., Chesler, D.A., Correia, J.A. and Riederer, S.J. (1976). Radionuclide axial tomography with correction for internal absorption. In: Information Processing in Scintigraphy. C. Raynaud and A. Todd-Pokropek, (eds.), Orsay, France: Comissariat a L'energie Atomique, pp. 333-342.

Webb, S., Long, A.P., Ott, R.J. (1985). Constrained deconvolution of SPECT liver tomograms by direct digital image reconstruction, Med. Phys. 12, pp. 53-58.

COMPUTER SIMULATED CARDIAC SPECT DATA FOR USE IN EVALUATING RECONSTRUCTION ALGORITHMS

D.R. Gilland, B.M.W. Tsui, H-B Hu, J.R. Perry

The University of North Carolina at Chapel Hill
USA

ABSTRACT

The purpose of this work was to develop a technique for simulating SPECT projection data which realistically model a cardiac Tl-201 study. The data are intended for use in evaluating image reconstruction and processing algorithms using ROC analysis. The significant physical effects of Tl-201 patient acquisitions have been incorporated into the simulation, including patient anatomy, attenuation, scatter, detector response, and statistical noise fluctuations. The simulation technique involves first generating a 3D source distribution which models the radioisotope uptake of a clinical Tl-201 study. A patient CT image provides an anatomical model for generating the distribution. Next, projections are calculated as ray-sums in a way which incorporates photon attenuation and detector response. Detailed attenuation information is provided by the CT image. The effects of scatter are then incorporated by filtering the projection data at each angle with an estimate of the scatter response function specific for that angle. Finally, Poisson noise is added to the projections. The results have shown that the simulation technique generates data closely modeling clinical Tl-201 data.

1. INTRODUCTION

The purpose of this work was to develop a technique for computer simulating SPECT projection data which could be used in studies to evaluate reconstruction methods. Data are simulated with and without lesion so that reconstructed images can be used in ROC (receiver operating characteristic) studies testing lesion detectability. The ROC experiments using the simulated data will determine the relative efficacy, or rank the quality, of a set of reconstruction methods in terms of observer performance in lesion detection.

In order to provide ROC results of clinical significance, the simulated data are intended to closely model actual patient data. The measure of quality for this simulation is the extent to which the rankings of the reconstruction methods based on the ROC results using the simulated data match the rankings which would be obtained if actual patient data were used. This measure should depend on the degree to which the simulated data are realistic projection measurements of objects which are representative of source distributions in patients for both normal and diseased cases. Realistic projection measurements include the effects of physical factors such as attenuation, detector response, scatter, and noise. The objective in this simulation was to accurately model patient data with regard to both the source distribution and the projection measurements.

Simulated data has a number of important advantages over patient data in ROC studies. Simulation offers a practical means of generating the many images necessary for an ROC study. It is often difficult to locate enough clinical cases for ROC studies, and, even if the clinical cases are available, verifying the actual state of the patient, whether diseased or normal, is not always possible. By computer simulation, the true state of the image is always known.

Since the efficacy of a reconstruction technique can depend to a large degree on the projection data to be reconstructed, it is important to focus an evaluation study on a particular clinical study or class of objects. The clinical study that we have chosen to model is cardiac Tl-201 data because of the importance of this technique in SPECT imaging. The simulated data will closely model the distribution of the Tl-201 radioisotope in patients and realistically incorporate the significant physical factors of a SPECT acquisition including attenuation, scatter, detector response, and noise. Simulating cardiac projection data in SPECT is particularly challenging due to the relatively severe nonuniformity of tissue density in this region.

The next section of this paper describes in greater detail the simulation technique. Following this, the results of the simulation are presented in the form of reconstructed images, and these images are compared with reconstructions of actual clinical data. Finally, this work is summarized, and a discussion of the application of the simulation technique is presented.

2. SIMULATION METHOD

The data simulation involves first defining a 3D source distribution, then calculating projection data in a way which includes the effects of attenuation, detector response, scatter, and noise. The novel idea in this simulation is to use a patient CT study to generate the source distribution in order to realistically model actual patient anatomy, and then use the attenuation information provided by the CT image to calculate the attenuated projections of the source distribution.

The projection operator used is an extension of a ray-tracing projector/backprojector algorithm which was originally designed to model attenuation (Gullberg et al, 1985) and was subsequently modified to also model the distance dependent detector response (Tsui et al, 1988). This algorithm computes one dimensional projection arrays from 2D source and attenuation distributions over a number of angles. This algorithm has been extended in this simulation to compute the projection arrays from a 3D source distribution in order to incorporate resolution effects in the axial direction.

The effects of scatter are incorporated into the projection data by estimating for each projection angle a scatter response function and filtering the data with the scatter response function specific for each angle. Finally, Poisson noise is added to the projection data. The data simulation, then, involves the following procedure:

1. Define a 3D source distribution using CT image,
2. Calculate projection data with attenuation and detector response,
3. Incorporate effects of scatter,
4. Add Poisson noise.

2.1 Simulated Source Distribution

The simulated source distribution should model the Tl-201 distribution in patient studies. In Fig. 1 are transaxial reconstructions of two patient Tl-201 studies. The Tl-201 isotope can be seen primarily in myocardial tissue around the left ventricle. The horseshoe shaped intensity defines the septal (left side of horseshoe) and lateral (right side) walls of the left ventricle. To a lesser extent, the lateral wall of the right ventricle can be seen above and to the left of the septal wall. There is also a low level source distribution in areas outside the myocardium, namely the lung and mediastinum fields. This is the result of a small amount of activity which is not cleared from the blood or which localizes in the lung parenchyma or the smooth muscle of the major blood vessels of the lung and mediastinum.

In order to realistically simulate the 3D source distribution within the myocardium, multiple-slice patient CT studies provide an anatomical model. In CT images at the level of the left ventricle, the myocardium of the left ventricle, and to a lesser extent that of the right ventricle, is discernible from surrounding tissue. Figure 2 is an example of such a CT image from a study which used intravenuous contrast medium. Contours can be assigned manually to the myocardium of the left and right ventricles in these CT images as shown in Fig. 3. Using these contours, a 2D source distribution is generated by assigning a uniform intensity to the pixels within the contours. The relative intensity assigned to the left and right ventricular myocardium is based on the measured intensities of reconstructed patient data. This technique is performed on five contiguous CT slices to produce a 3D source distribution which can be seen in Figs. 4a-e. With a slice thickness of one cm, the five slice source distribution spans in the axial direction the resolution of most general purpose camera/collimators. The simulated source distribution is defined in the transaxial plane in a 512x512 matrix.

Myocardial defects for use in lesion detection experiments are simulated by generating a decreased intensity along the lateral or septal wall of the left ventricle. The intensity decrease is isotropic and Gaussian-shaped. A source distribution similar to that in Figs. 4a-e has been simulated with a defect in the septal wall, and the center slice of that distribution is shown in Fig 4f. The position of the defect can be varied along the myocardium. Both transmural defects, in which the lesion extends across the myocardial wall, and subendocardial defects, in which the lesion is centered near the inner edge of the myocardial wall, can be simulated.

The Tl-201 distribution in the areas of the lungs and mediastinum in reconstructed images of patients is typically 10-20% of the maximum intensity in the myocardium. This background distribution varies greatly across patients, and at this intensity level, it is sometimes difficult to distinguish radioisotope concentration from reconstruction artifact. For these reasons, modeling the Tl-201 background distribution is a difficult task. However, we have developed a technique for simulating a background which reasonably resembles the background in Tl-201 images, although it does not precisely model the radioisotope uptake in patients. The intensity assigned to a given pixel in the lung fields is a function of the CT number of the corresponding pixel in the CT image. The function which relates the source intensity to the CT number is Gaussian-shaped and is

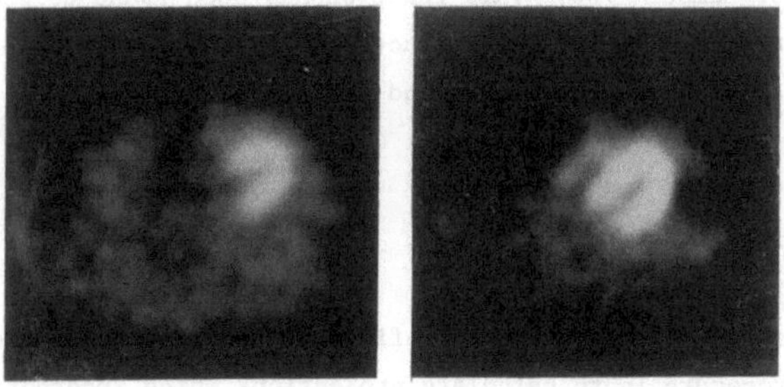

Fig. 1. Transaxial reconstructions of patient Tl-201 data. Radioisotope localizes primarily in the myocardium surrounding the left ventricle.

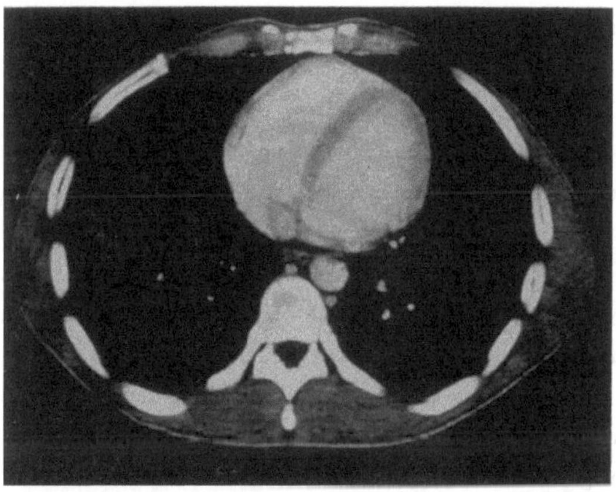

Fig. 2. Patient CT image. Myocardium is discernible from surrounding tissue.

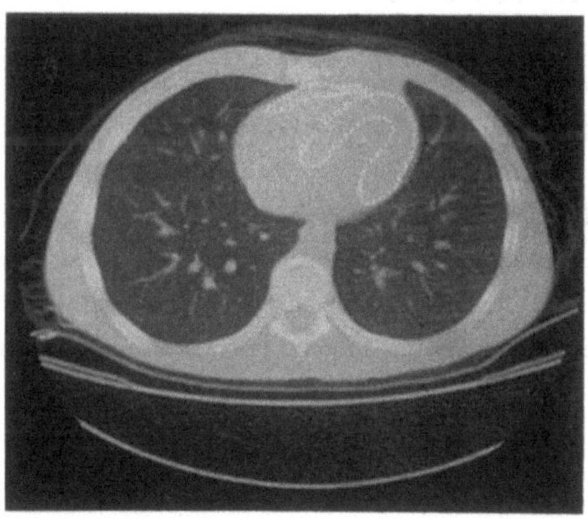

Fig. 3. Patient CT image with contour assigned to myocardium of left and right ventricles.

centered at CT number -700. This function has been found to provide a background structure in the reconstructed image similar to that in clinical images. In Figs. 4a-e, the background distribution is apparent.

2.2 Projection

Once the 3D source distribution is defined, the next step in the data simulation procedure is to calculate projections which incorporate photon attenuation and the detector response. This is accomplished by means of a ray-tracing projection algorithm which has been extended to allow projection from a 3D distribution. This section describes in detail that projection algorithm.

The algorithm calculates projection values by tracing rays across the object space and computing ray-sums. The unique feature of this algorithm is that projection values are computed as the sum of a set of rays which converges to the center of a projection bin and forms a "fan" across the transaxial plane in the object space (Tsui, 1988). This is illustrated in Fig. 5. Each ray of the fan is weighted to accurately model the blurring of the detector response. The weighting of the individual rays is based upon theoretical calculations of the geometric component of a detector response function (Metz et al, 1980) assuming a general purpose collimator. The angular spacing of the rays is such that at a distance from the projection array equal to the radius of rotation, the rays have a one pixel separation. Twenty three rays per fan are used to span the acceptance angle of the simulated collimator.

In addition to modeling the detector response by means of the fan of rays, the projection algorithm models photon attenuation by computing attenuated ray-sums based on an algorithm developed by Gullberg (1985). The contribution of a pixel in the object space to the ray-sum is weighted according to the accumulated attenuation between the pixel and the projection bin. The attenuation information is provided by the CT image after converting the CT numbers to linear attenuation coefficients for photon energy equal to Tl-201 primary peak energy, 75 kev.

The conversion of CT number to attenuation coefficient for a specific energy is complicated due to the energy dependence of the attenuation coefficient and the polychromicity of the CT x-ray beam. If one assumes for the CT beam an effective energy, the monochromatic x-ray energy which

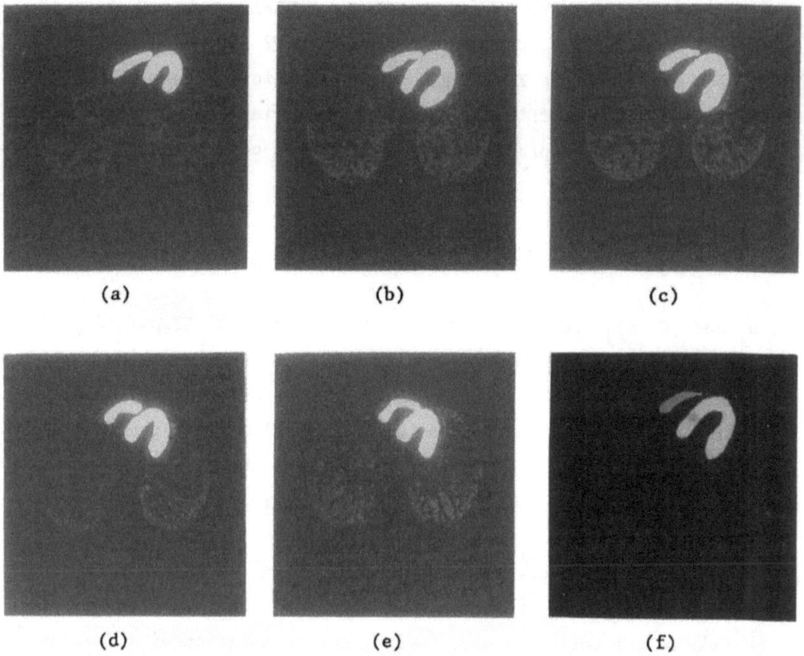

Fig. 4. Source distribution in transaxial plane. (a-e) Five contiguous slices representing the 3D source distribution. (f) Simulated lesion in septal wall is apparent.

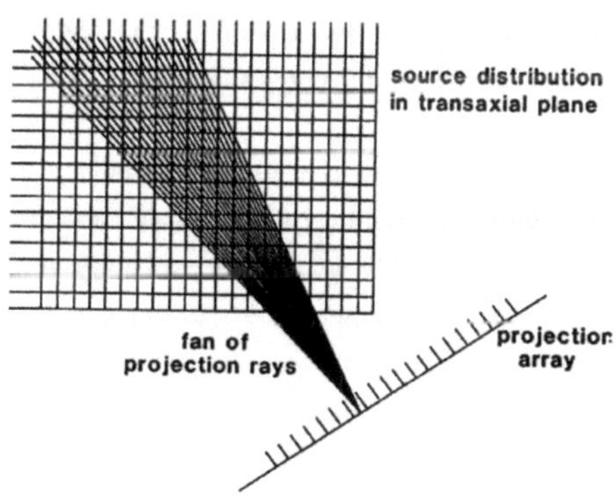

Fig. 5. Fan of projection rays. Each ray in the fan is weighted to model the detector response.

would result in equal energy absorption, the problem can be simplified. For these CT acquisitions, the effective energy, 72 kev, is close to the primary peak energy of the Tl-201 emission. Therefore, converting CT number to attenuation coefficient using the following linear relationship is expected to closely approximate the attenuation coefficient distribution for the Tl-201 energy,

$$\mu_j = (CT\ \#)_j/1000 * \mu_{water} + \mu_{water} \qquad (1)$$

where μ_j and $(CT\ \#)_j$ are the attenuation coefficient and CT number, respectively, of pixel j, and μ_{water} is the effective attenuation coefficient of water at 75 kev. Due to the detection of scattered events, the effective attenuation coefficient of water will be less than the narrow beam attenuation coefficient. In this simulation, the effective attenuation coefficient of water at 75 kev was estimated to be 0.146/cm.

The projection algorithm was modified to incorporate resolution effects in the axial direction using the off-center slices of the source distribution. This is done in the following way. During the calculation of a ray-sum, the value of a pixel which is to be included in the ray-sum is replaced by a weighted average of that pixel and the corresponding pixel in each of the four off-center slices of the source distribution. This weighting function, like the weighting function for the fan of rays, is based upon theoretical calculations of the geometric detector response function assuming a general purpose collimator and accounts for the resolution degradation with distance from the detector face.

Before performing the projection operation outlined above, the source distribution in each transverse plane is interpolated down to 256x256 matrices to reduce the computational cost of the projection operation. The projections are digitized into 256 element arrays and can be generated over any number of angles. Before reconstructing the data, the projection arrays can be averaged into 128 or 64 elements, typical sampling in SPECT imaging. The projection arrays are calculated initially with relatively fine sampling in order to approximate as closely as possible the projection data of a continuous object.

2.3 Scatter

Because the scatter response is a function of source depth as well as the attenuating properties and geometry of the scattering medium, it is extremely complex and difficult to incorporate precisely the effects of scatter into the projection data. Nevertheless, in order to incorporate in this simulation an estimate of the effects of scatter, the projection data are filtered at each angle with an "average" scatter response function calculated specifically for that angle.

The average scatter response functions are based upon Monte Carlo studies (Atkins, 1978) which estimated the series-equivalent scatter response as a function of depth of a point source in a uniform scattering medium of density equivalent to water. In this simulation, a source depth is computed for each projection angle, and this depth is used to generate the scatter response function specific for that angle. However, because in this simulation the source is distributed in a nonuniform scattering medium, an effective source depth is estimated for each angle by the following procedure. First, the source center of mass is calculated. Then, from the CT image, the accumulated attenuation is computed along a ray between the source center of mass and the projection array. The accumulated attenuation is then divided by the linear attenuation coefficient of water at 75 kev to give the effective source depth. Therefore,

$$\text{effective source depth} = \frac{\sum_{j \in J} \mu_j \ell_j}{\mu_{water}} , \qquad (2)$$

where J is the set of source elements j intersected by the projection ray, μ_j is the linear attenuation coefficient of source element j, ℓ_j is the length of intersection of the ray and source element j. A typical series-equivalent scatter response function for this simulation is illustrated in Fig. 6.

The extent to which the estimated scatter response function describes the blurring in the projection data due to scatter will depend in part upon the degree to which the source is distributed in the scattering medium. In this simulation, where the majority of the source is contained within the relatively small volume of the left ventricle, the estimate of the scatter response is expected to reasonably represent the blurring in the projection data due to scatter.

The final step in the simulation process is to add noise to the projection data. This is done by adjusting the count per pixel in the projection array to a level representative of count densities in patient Tl-201 studies, then adding to the projection array samples of a Poisson random number generator.

3. RESULTS

Reconstructed transaxial images illustrate the results of the data simulation. All images shown in this section were reconstructed using the filtered backprojection algorithm BKFIL in the RECLBL library (Huesman et al, 1977) without attenuation compensation and, unless otherwise stated, a ramp filter. Figure 7 provides a qualitative comparison between reconstructed simulated data (a) and patient data (b). Both images are 64x64 and were reconstructed from 64 projections equally spaced over 180 degrees (this is normal acquisition protocol for Tl-201 studies at most nuclear medicine clinics). The images have approximately the same total counts. The projection data of both studies were filtered equally to reduce noise prior to reconstruction.

Another example of simulated data illustrates how the individual effects on the reconstructed image of attenuation, scatter, detector response, and noise can be isolated in the simulation scheme. The images in Figure 8 are digitized in 128x128 matrices and were reconstructed from 180 projections over 360 degrees. Figure 8a is a reconstruction from data in which the effects of attenuation, detector response, scatter, and noise were absent. This image separates the effects of those physical factors from the effects of the reconstruction algorithm and sampling. The image in Figure 8b was reconstructed from data which included the effects of attenuation and detector response, but not scatter or noise. Note the blurring of the edges of the myocardium due to the detector response, and the decreased intensity in the septal wall relative to the lateral wall due to attenuation. Data equivalent to those from Fig. 8b, except with scatter, were reconstructed and are shown in 8c. The effects of scatter are evidenced by a decrease in contrast of the myocardium and a spread of intensity into the center of the chamber of the left ventricle. Finally, noise was added to the projection data followed by smoothing with a Hanning filter, and the image reconstructed from this data is shown in Figure 8d.

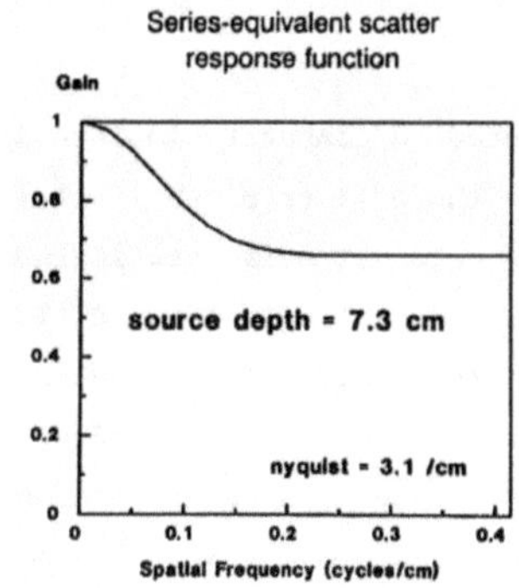

Fig. 6. A series-equivalent scatter response function is calculated for each angle and used to filter the projection data.

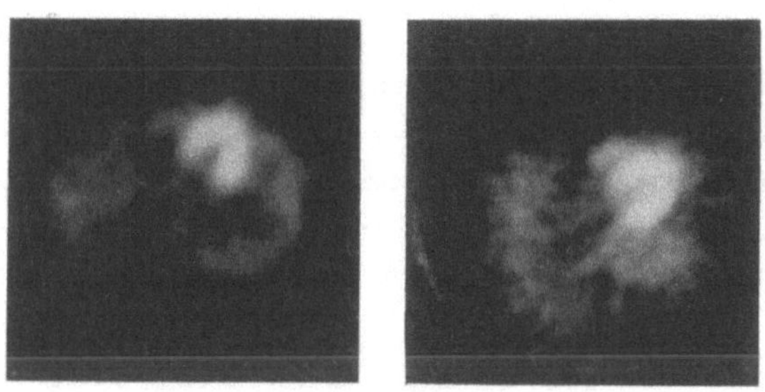

(a) (b)

Fig. 7. (a) Example reconstruction of simulated data, (b) Reconstruction of patient data.

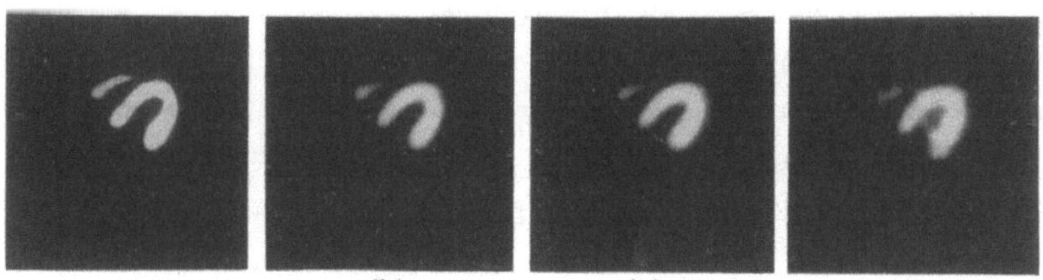

(a) (b) (c) (d)

Fig. 8. Example reconstructions of simulated data illustrating the effects of attenuation, scatter, detector response, and noise. (a) no physical effects present, (b) attenuation and detector response present, (c) same as b, plus scatter, (d) same as c, plus noise.

4. SUMMARY AND DISCUSSION

A technique has been developed to computer simulate SPECT Tl-201 projection data. The data incorporate the effects of physical factors which are inherent in actual clinical data. The factors include attenuation, scatter, detector response, and statistical noise fluctuations. In addition, the object, or source distribution, models the Tl-201 radioisotope uptake in patients including myocardial infarcts and perfusion abnormalities. Comparing simulated and patient reconstructed images, like those shown in Fig. 7, suggests a close similarity between the simulated data and actual patient data. It is reasonable to expect that the ROC results with the simulated data would be the same, in a relative sense, as the results which would be obtained if clinical data were used. Therefore, the ROC studies using the simulated data should have clinical relevance in the evaluation of reconstruction methods.

Data simulated with and without defects can be used in observer performance experiments testing lesion detectability as illustrated in Fig. 9. In this paradigm, the observer is presented with two images and asked to select which of the two images contains a defect (the actual defect in Fig. 9 is in the septal wall of the right image).

The simulated data reveal clear differences in the effects different reconstruction methods have on image quality. Figure 10 shows the results of four reconstruction algorithms applied to the same simulated data. The image in Figure 10a is reconstructed using the filtered backprojection algorithm without attenuation correction. Figure 10b is reconstructed using a method of attenuation correction introduced by Chang (1978). Figures 10c and 10d were both reconstructed with an iterative maximum likelihood algorithm (Shepp et al, 1982). No attenuation correction was incorporated in Figure 10c. The projector/backprojector of the algorithm used to reconstruct the image in Figure 10d incorporates a model for nonuniform attenuation. All images in Figure 10 are 128x128 and reconstructed from 180 angles equally spaced over 360 degrees.

The simulated data should serve as general tool to evaluated many aspects of SPECT imaging. The flexibility of the simulation technique allows investigation into the effects of factors such as center-of-rotation errors, 180 vs. 360 degree acquisitions, and changes in detector response, noise level, and lesion size.

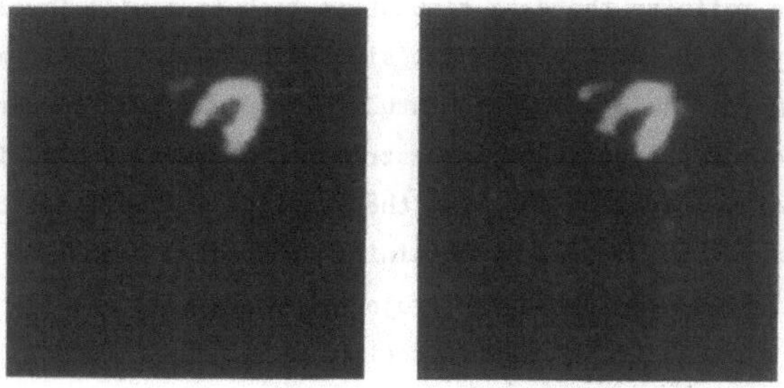

Fig. 9. Lesion detection experiment. Observer must select which image has the lesion.

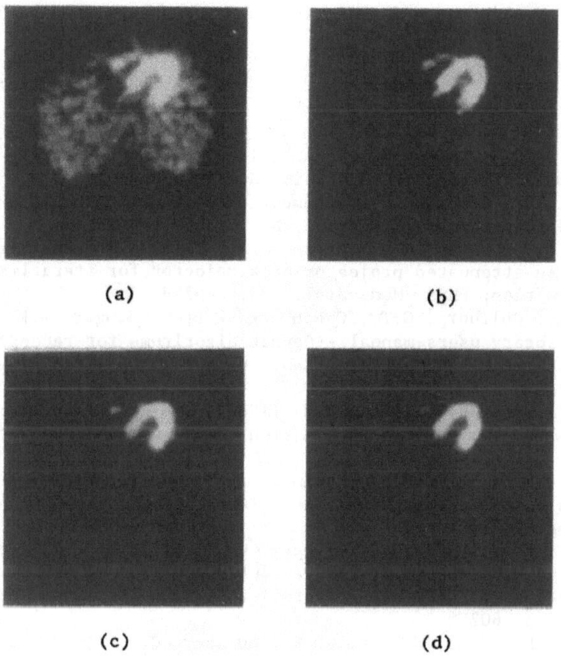

Fig. 10. Four reconstruction algorithms applied to simulated data. A. Filtered backprojection, no attenuation correction. B. Chang attenuation correction. C. Iterative maximum likelihood. D. Iterative maximum likelihood with nonuniform attenuation compensation.

Future work in refining the data simulation technique will involve, first of all, verifying the accuracy of the simulation method. This might be done through Monte Carlo simulation studies, however, due to the complexity of both the source and attenuation distributions, this is a difficult task. Another area of improvement concerns the sampling in the axial direction. This sampling can be improved by obtaining CT studies with 5 mm slice thickness. Finally, generating 2D projection data will enable investigation into 3D image reconstruction techniques in SPECT.

ACKNOWLEDGEMENT

The authors are grateful for the informative discussions and helpful input of Drs. Charles E. Metz and Ronald J. Jaszczak. This work was supported by the Dept. of Health and Human Services under grant R01-CA3 9463

REFERENCES

Atkins, F.B. (1978). Monte Carlo analysis of photon scattering in radionuclide imaging. Ph.D Thesis. U. of Chicago.

Chang, L. (1978). A method for attenuation correction in radionuclide computed tomography. IEEE Trans. Nucl. Sci., NS-25, No. 1.

Gullberg, G.T., Huesman, R.H., Malko, J.A., Pelc, N.J., Budinger, T.F. (1985). An attenuated projector-backprojector for iterative SPECT reconstruction. Phys. Med. Biol., 30:8, p799.

Huesman, R.H., Gullberg, G.T., Greenberg, W.L., Budinger, T.F. (1977). RECLBL library users manual - Donner algorithms for reconstruction tomography. Lawrence Berkeley Lab., Berkeley, CA, Tech. Rep. PUB-214, pp 226-227.

Metz, C.E., Atkins, F.B., Beck, R.N. (1980). The geometric transfer function component for scintillation camera collimators with straight parallel holes. Phys. Med. Biol., 25:6, p1059.

Shepp, L.A., Vardi, L. (1982). Maximum likelihood reconstruction for emission computed tomography. IEEE Trans. on Med. Imaging. Vol. MI-1, No. 2, pp 113-122.

Tsui, B.M.W., Edgerton, E.R., Gullberg, G.T., Gilland, D.R., Perry, J.R., McCartney, W.H., Berg, J. (1986). Correction for nonuniform attenuation distribution in SPECT using iterative maximum likelihood method. Med. Phys. 13, p. 602.

Tsui, B.M.W., Hu, H.B., Gilland, D.R., Gullberg,G.T. (1988). Implementation of simultaneous attenuation and detector response correction in SPECT. IEEE Trans. Nucl. Sci., 35:1, p778.

COLLIMATOR ANGULATION ERROR AND ITS EFFECT ON SPECT

R. E. Malmin, P.C. Stanley, and W. R. Guth

Siemens Gammasonics, Inc.
U.S.A.

ABSTRACT

Angulation errors in collimator septa degrade reconstructed image resolution in single photon emission computed tomography (SPECT) due to the mispositioning of events in the projection data. These errors should be small compared to other determinants of SPECT resolution: camera and collimator resolution, gantry alignment, and algorithmic smoothing. We estimate a tolerance level for collimator septal deviations by means of a simple computer simulation, and describe a radioisotope technique to quantitatively determine angulation errors in collimator septa. We also describe a method whereby septal orientations of constant magnitude and direction can be corrected during SPECT reconstruction.

1. INTRODUCTION

At clinical imaging distances (15-25 cm), angulation errors cause small shifts in the apparent position of a point source. At extremely large source-camera distances (~4-9 meters), however, the image of a point source fills the field of view, and septal deviations manifest themselves as changes in count density. Small deviations, well below the level which produce measurable effects in SPECT images, can easily be visualized and measured. If the septal errors are directionally correlated (as in corrugated foil collimators), they appear as faint linear "streaks". Randomly directed septal errors (as in cast collimators), though equally measurable, are less easily visualized because they produce a "mottled" appearance which depends on the relative position of the source and septum.

1.1 Theory

Our technique measures septal errors by comparing the experimental point source response function (PSRF) with a least squares fit to a model of the theoretical PSRF (Barrett and Swendell, 1986). The PSRF of a point source sufficiently distant to illuminate the entire field of view of a gamma camera (~3-5 m) is well approximated by a conical shape with a small quadratic term (Fig. 1).

$$\mathrm{PSRF(r)} = \mathrm{N(r)/No} = 1 - \mathrm{r/Rm} + \mathrm{a(r/Rm)}^2 \qquad (1)$$

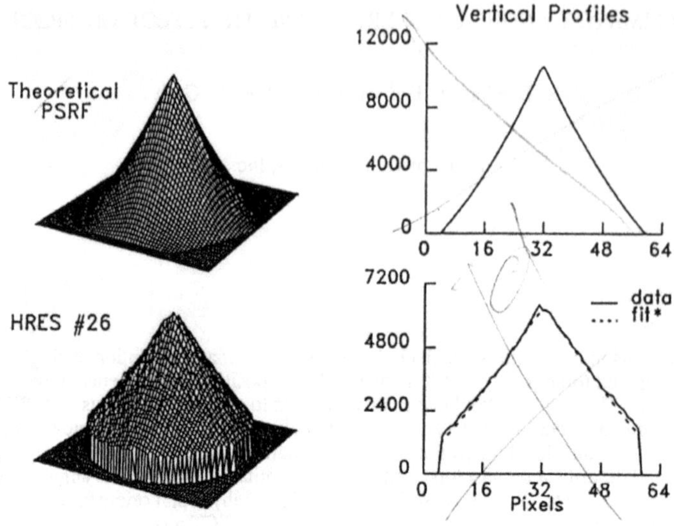

Fig. 1. Left: Theoretical and measured PSRF at 4.5 meters.
Right: Profiles and fit using Eq. (1).

In this model, deviations in the measured PSRF can be related to changes in the angle, Υ, defined by the septal direction vector, H, and the vector from the septum location to the point source, P (Fig. 2).

$$\cos(\Upsilon) = \cos(\Upsilon_0)\cos(\beta) + \sin(\Upsilon_0)\sin(\beta)\cos(\phi) \qquad (2)$$

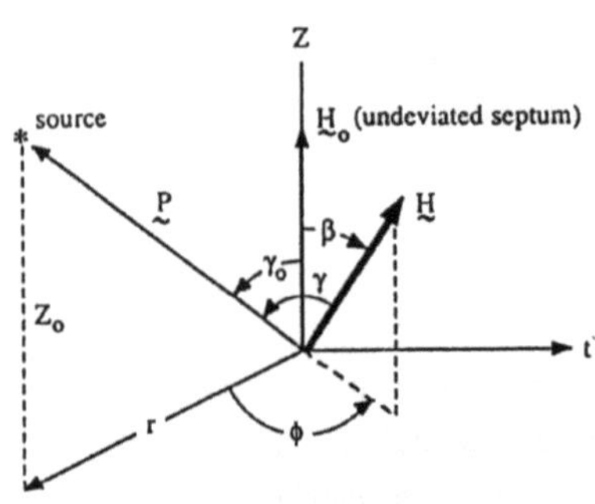

Collimator Surface

Fig. 2. Coordinate system describing septal tilt (see text)

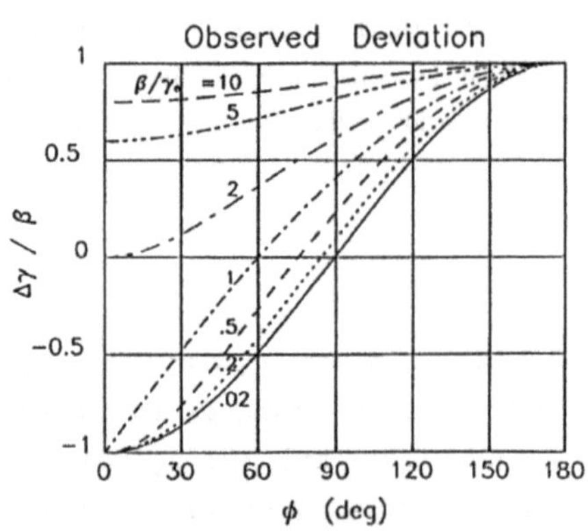

Fig. 3. Change in source-septum angle (γ) versus tilt direction (ϕ).

Observed deviations are largest when a septum is tipped directly towards or away from the point source ($\phi = 0,\pi$), in which case $\Upsilon = \Upsilon_0 \pm \beta$. The change in this angle, $\Delta\Upsilon = \Upsilon - \Upsilon_0$, for a septum deviating from the normal by polar angle β is given by

$$\frac{\Delta\Upsilon}{\beta} = \frac{-\Upsilon_0}{\beta} + \left[\left(\frac{\Upsilon_0}{\beta}\right)^2 - 2\left(\frac{\Upsilon_0}{\beta}\right)\cos\phi + 1 \right]^{1/2} \tag{3}$$

Equation (3) is plotted in Fig. (3). Observable deviations in the PSRF depend on the angular distance from the source to the septum (Υ_0), and on both the magnitude (β) and relative direction (ϕ) of the septal error.

1.2 SPECT simulation

A normal distribution of errors was randomly added to the image location of a point source in each of 180 projections. The contribution to the reconstructed resolution due only to event mispositioning (collimator, intrinsic camera resolution, and algorithmic effects removed) is shown in the bottom curves of Fig.4. The upper curves show the results for a LEAP collimator when all effects (except attenuation) are included. The radii of rotation are appropriate for scans of the head and thorax. In both cases, the reconstructed resolution increases by ~ 1% when the septal error reaches 0.6° FWHM (0.25° RMS). We take 0.55° as the upper limit or maximum allowable septal deviation, and 0.14° as the maximum allowable RMS septal error.

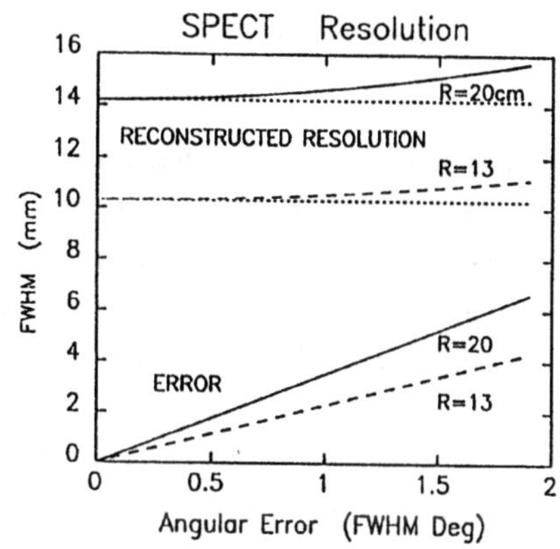

Fig. 4. Results of SPECT simulation.

2. RESULTS

The technique has been verified through measurements on corrugated foil collimators*. On the right hand side of Figure 5 is the image of a point source centered in the field-of-view and 4 m distant from the LEAP collimator. The point in the left hand image is 9 meters distant and ~ 5 cm beyond the FOV (toward the top of the figure). Figure 6 shows maps of the corresponding septal errors computed from non-linear least squares fits to these data using Eq. (1-3). The size of the septal error is indicated by the grey scale at the top of the figure; it ranges from $\beta = .1°$ to $.6°$ in $.1°$ steps. Figure 7 presents a scatter plot comparing the results in Figure 6a and 6b. The excellent agreement between the two independent

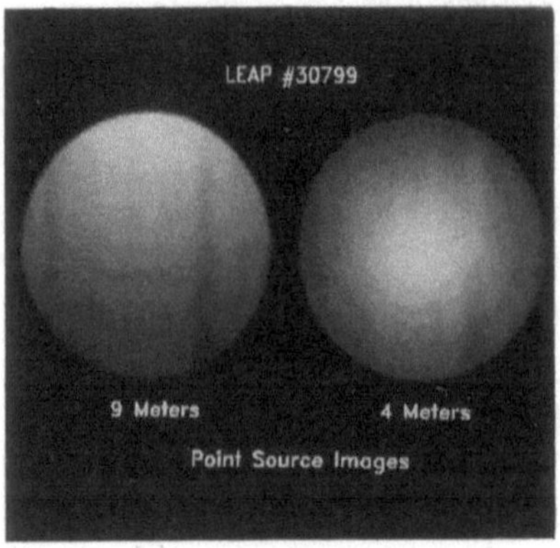

Fig. 5. Raw point source images.

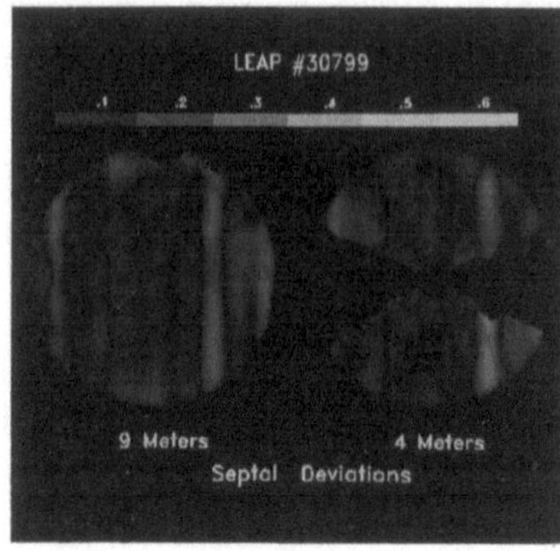

Fig. 6. Functional (error map) images.

techniques ($\sigma = 0.04°$) is remarkable, especially considering the wide variation in the parameters (ϕ, Υ_0, and $\Delta\Upsilon$) appropriate to the two data sets. We interpret this as a strong confirmation of the theory and as a measure of the absolute accuracy of the technique ($\sigma = \pm 0.04°$). Results were found to be count-independent above 30 million total counts, and were not significantly altered by 100 million count uniformity flood correction.

This technique has been implemented as a standard quality control procedure in the production of Siemens Autoform collimators. Figure 8 is a composite histogram of the septal errors measured on 54 high resolution collimators. The integral of this distribution shows that 99%, 99.9%, and 99.99% of the approximately 6900 pixels measured for each collimator fell below .25, .35, and .45 degrees, respectively. Figure 9 gives the distribution of the average, RMS, and maximum error for the 54 collimators.

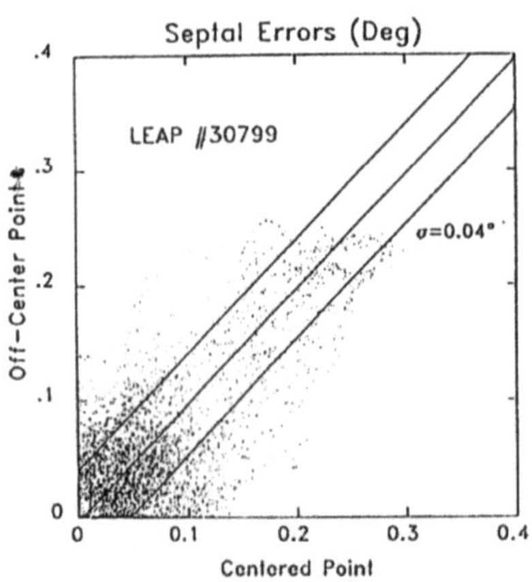

Fig. 7. Scatter plot of data in Fig. 6.

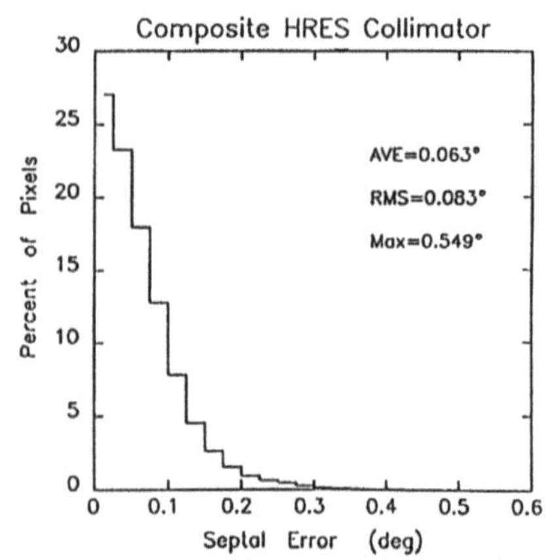

Fig. 8. Error distribution of 54 collimators.

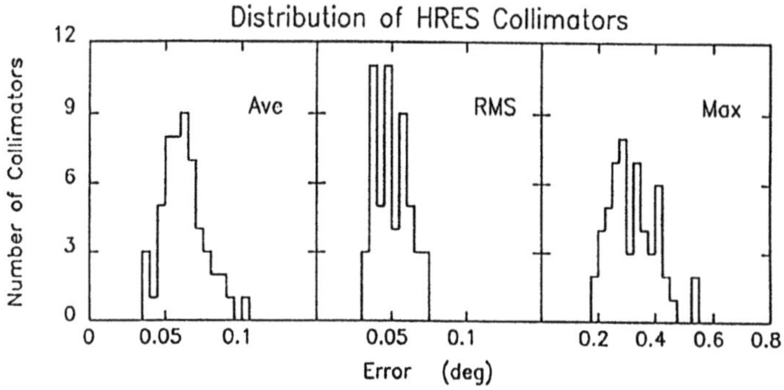

Fig. 9. Distribution of ⟨β⟩, β(RMS), and β(max) for 54 collimators.

3. DISCUSSION

Figures 5 and 6 demonstrate three attributes characteristic of these corrugated foil collimators. First, the faint vertical streaks visible in the raw images, and septal deviations associated with them, are preferentially oriented in a direction perpendicular to the planes of the individual lead foil strips (which are in the horizontal plane of the Fig. 5). They are <u>not</u> associated with an individual strip, as might intuitively be expected. Secondly, the deviations are quite uniform and propagate across the entire collimator with approximately constant magnitude. And finally, the magnitude of the deviations is small: ⟨β⟩ = .061°, β(RMS) = .082°, β(max) = .33°.

The high degree of uniformity (in both magnitude and direction) exhibited by these collimators suggests a method whereby potential image degradation due to septal misorientation can be removed. If the corrugated strips composing the collimator are aligned parallel to the axis of rotation, then the septal angulation errors will lie in a planes transverse to the axis. The net effect is to introduce an additional center of rotation (COR) offset for those transverse slices equal to the radius of rotation times the sine of the angular error (β). This can be fully removed by any SPECT software which computes an independent COR correction for each slice. It should be emphasized that the errors in positioning caused by collimators with angular errors satisfying the limits chosen (β(max) ≤ 0.55°,⟨β⟩ ≤ 0.11°, and β(RMS) ≤ 0.14°) are already very small, well below visual means of identification in tomographic images.

* Collimators studied in this work were limited to Siemens Autoform Collimators.

REFERENCES

Barrett, H.H., Swindell, W. (1986) Radiological Imaging, Academic Press, New York

THE DESIGN AND IMPLEMENTATION OF MODULAR SPECT IMAGING SYSTEMS

R. K. Rowe[1], J. N. Aarsvold[1], H. H. Barrett[1], J. Chen[1], J. N. Hall[1], A. L. Landesman[1], L. S. Mar[1], T. D. Milster[2], B. A. Moore[1], D. D. Patton[1], T. J. Roney[1]

[1]University of Arizona, U.S.A.
[2]IBM Corporation, Tucson, U.S.A.

ABSTRACT

This paper describes a small, digital, modular gamma camera that can be used individually as a small organ, planar imaging system or can be combined to create modular single-photon emission computed tomography (SPECT) systems that contain no moving parts. We describe two systems currently under design: a cardiac imager capable of multi-slice, two-dimensional SPECT, and a full three-dimensional brain imager. Finally, we present a two-dimensional tomographic image generated by a prototypical modular camera SPECT system.

1. INTRODUCTION

Modular scintillation cameras are small, minimal-configuration, digital gamma cameras which can be used individually or arranged in various geometries to perform specific single-photon emission computed tomography (SPECT) imaging tasks. A prototype of our modules was introduced by Milster et al (1983), and various aspects of its design and operation have been published previously (Arendt, 1980; Selberg, 1984; Milster et al, 1984; Milster et al, 1985; Milster, 1987). Each module is both mechanically and electronically independent of any other. This feature allows us to design systems that record all views simultaneously and thereby contain no moving parts. Furthermore, the entire crystal surface of each module can be used for imaging, allowing modules to be closely packed and thus, form a large effective crystal area. The high collection efficiency of such a modular system, in conjunction with stationary data collection, allows us to design SPECT systems that can collect an entire data set much faster than rotating SPECT gamma cameras. Moreover, these same features can be exploited to design "stop-action" tomographic systems, such as one to collect SPECT data from cardiac first-pass studies. Finally, a single module can be used with a collimator to make a small, portable, planar gamma camera useful for imaging small organs.

2. MODULAR CAMERA HARDWARE AND OPERATION

A module consists of one 10 cm square by 5 mm thick NaI(Tl) crystal, a 19 mm thick exit window/light guide, and four 5 cm square photomultiplier tubes (PMTs), as shown in Fig. 1. In addition, each of the four PMTs is connected to an electronic interface circuit which conditions the signal from the PMT, digitizes it, and processes the digitized signal, as described below.

When a photo-event occurs in the crystal, each of the four PMTs responds with an analog signal which the interface electronics digitizes to eight bits, as shown schematically in Fig. 2. The digitized signal then passes through an erasable, programmable, read-only memory (EPROM) which performs a non-linear mapping to compress the eight-bit signal to five bits, as discussed by Aarsvold et al (1988). The five-bit signals from each of the four PMTs are then concatenated to form a 20-bit signal.

Prior to using the module for imaging, its particular response to gamma ray events across the crystal face must be characterized. Two methods of characterizing the module have been used by us; the first was described by Aarsvold et al (1988) and worked well over much of the crystal face. This technique was predicated on the assumption that the signals from the four PMTs were separable. We have since determined that the PMT signals are somewhat correlated and therefore, not separable. The second method of module characterization, which we will describe here, does not assume signal separabilty, and thus yields slightly better module performance over the entire crystal (Milster, 1987).

To implement this method of module characterization, we use a computer-controlled X-Y table to position a collimated technetium-99m source under each of the 4096 (64 by 64) resolution elements of the crystal. At any given position of the source, we collect a large, fixed number of counts. Each time a photo-event occurs, a position in memory that corresponds to the resulting 20-bit signal is incremented, thus forming a histogram of digitized PMT responses for a given source location.

When all the counts for a given position have been collected, the resulting histogram may be interpreted as being proportional to the probability distribution of digitized PMT responses for a given source location. Under this interpretation, we can set a threshold probability, which we refer to as the likelihood threshold. Any set of PMT responses that occur with less frequency than this threshold value are probably due to background or scatter

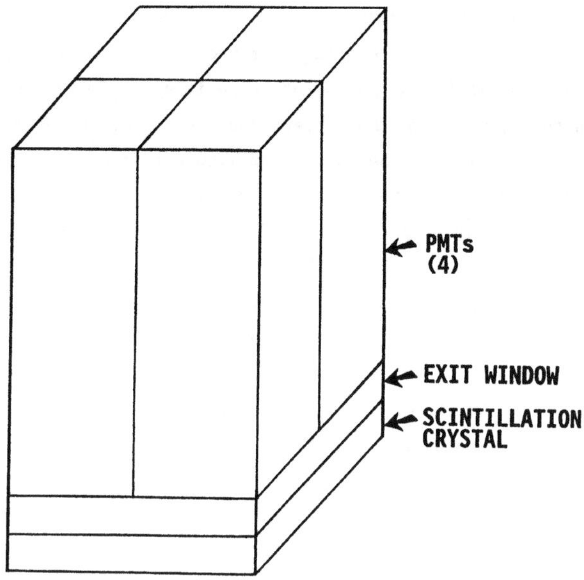

Fig. 1. Components of a module: a NaI(Tl) scintillation crystal, an exit window, and four photomultiplier tubes (PMTs).

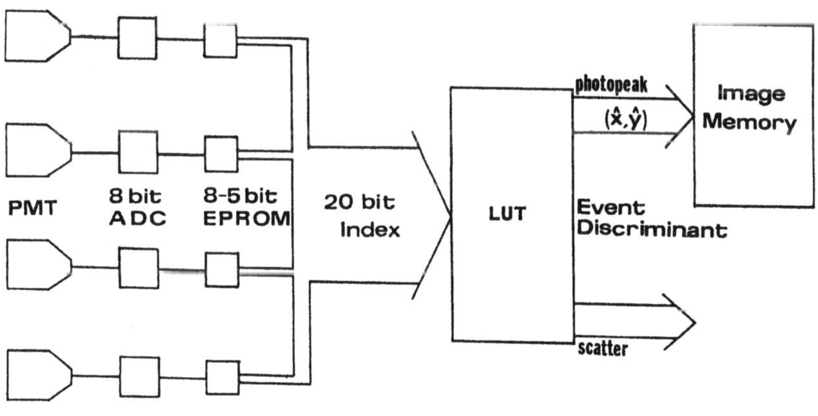

Fig. 2. Schematic showing the major elements used in processing the photomultiplier tube (PMT) signals when module is being used for imaging. Elements include four eight-bit analog-to-digital converters (ADCs), four erasable, programmable, read-only memories (EPROMs), memory allocated as a look-up table (LUT) and memory used for building the image.

events and are, therefore, ignored. We then scan the histogram and search for PMT responses that occurred with greater frequency than the threshold value. When such a response is found, and no other position estimate has already been assigned to it, we store the current coordinates of the source in a table entry corresponding to the particular module response. We also store the probability value associated with that position estimate in a separate table. If a probable PMT response is found that already has a different source location associated with it, the corresponding probabilities are compared and the most probable position estimate is then associated with the response. This process begins anew at the next source location. At the conclusion of this process we have a look-up table (LUT) which associates each of the possible PMT responses with either a maximum likelihood position estimate of the corresponding gamma ray location, or a flag that indicates that the response is probably due to a scatter event and thus, should be ignored.

When the module is being used for imaging, the 20-bit signal, which is generated whenever the module detects a photo-event, acts as an index into a previously built LUT. If the event is determined to be due to scatter or background radiation, the event is ignored and the module waits for the next photo-event. Otherwise, the electronics increments a location in memory which corresponds to the position estimate stored at the particular location in the LUT. In this manner, an image may be built at rates greater than 100,000 counts/second per module.

3. MODULAR CAMERA SYSTEMS

We are in the process of designing and building two coded-aperture SPECT imaging systems based on the modular camera. The first is a 12-module cardiac imager which will simultaneously collect two-dimensional tomographic data for ten cross-sectional slices, with a slice thickness of 7 mm. The modules will be mounted on a pedestal and arranged in a semicircle about the patient's chest (refer to Fig. 3). The coded aperture will be made of tungsten and will lie in a semicircle between the patient and the modules. The aperture assembly will consist of a series of narrow slits, which will define the code, and a slat collimator, which will define the slice thickness.

The second system to be constructed for clinical use is a brain imager that will be capable of collecting true three-dimensional SPECT data sets. As currently envisioned, the system will consist of 18 modules arranged in a hemispherical pattern about the patient's head. Between the patient and the modules will lie a hemispherical shell with a coded-pinhole array, as well as a water-filled cushion next to the patient's head, which will help equilibrate the attenuation present in different size heads.

The third clinical system under development is a single-module, stand-alone unit that will be used to obtain planar images of small organs such as the thyroid, superficial lymph nodes, and the small bones of the wrist and feet. The design consists of a module and collimator mounted on a moveable arm that can be flexed to position the module over the region of interest. The module will be connected to a small, dedicated console which will contain all the electronics necessary to collect and process the data, and display the resulting images. All the components of the stand-alone system will be portable so that they can be easily moved for use in the operating room, clinic or the patient's room.

We have built several prototypical systems that are direct forerunners of the cardiac imager. Fig. 4 shows a system of 16 modules arranged in a square and used to perform single-slice, two-dimensional SPECT. We emulated the full 16-module system using just one side of the square (i.e., four modules). To do so, we first collected a four-module, coded data set for a particular object and then rotated the object 90 degrees and collected a second data set. After repeating this process two more times, we had four data sets that were the same as those that would have been collected by a full 16-module, square system with each bank of modules having identical characteristics.

Before we could use the system for imaging, we had to determine the system response function, i.e., the response of the modules to each object-space pixel. To do so, we used an X-Y table under computer control to sequentially step a technetium-99m point source to each pixel. We then collected data for a fixed time, at the end of which, we stored the resulting coded image due to the point source at that particular location. After repeating this procedure for each pixel in object space, and applying a multiplicative correction to account for source decay, we effectively had an approximation to the system matrix. The approximate nature of these data was due to presence of statistical fluctuations in the number of counts collected during any finite time. We chose an optimal collection time that yielded a large number of counts for each source position and still allowed us to collect all the data in a reasonable amount of time.

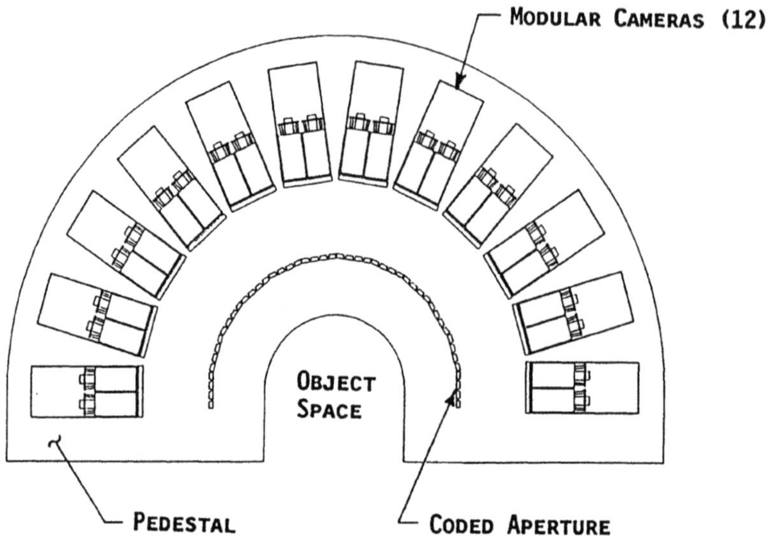

Fig. 3. Cardiac imaging system, which will be capable of two-dimensional, multi-slice dynamic tomography

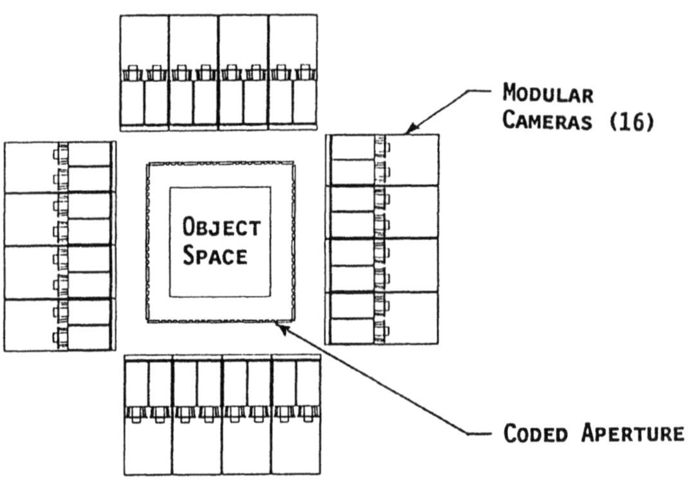

Fig. 4. Square prototypical imaging system, currently emulated using one side (i.e., four modules) of the square.

We have used the prototypical systems to perform two-dimensional, tomographic imaging of several phantoms. Fig. 5a shows the form of one such object; a cardiac left ventricle phantom that consists of an irregular groove of uniform depth and width (12 mm by 12 mm) milled into a piece of acrylic. The overall dimensions of this phantom are approximately 7.5 cm by 10.0 cm. We imaged this phantom in the 4-camera emulation of the system shown in Fig. 4, using technetium-99m. The object space for this system was 19.5 cm square and was divided into 3025 (55 by 55) pixels. The distance from the nearest edge of the object space to the aperture plane was 5.4 cm, and the distance from the aperture plane to the crystal plane was 7.6 cm. We collected approximately 100,000 counts in the coded image, and reconstructed the object using a simulated annealing algorithm (van Laarhoven and Aarts, 1987). Fig. 5b shows the resulting reconstruction.

4. CONCLUSION

We have developed a small, minimal configuration, modular gamma camera that can be used as a stand-alone planar imaging system, or can be combined to form stationary SPECT imaging systems. These modular systems promise to be very efficient due to the mechanical and electrical independence of the modules. The systems we are designing include a cardiac imager that will perform multi-slice, two-dimensional SPECT, including the ability to do dynamic tomography of cardiac first-pass studies. We are also designing a brain imager that will be able to perform full three-dimensional SPECT, as well as a stand-alone planar imager for use on small organs. Several prototypical systems have been built and used to generate two-dimensional tomographic phantom reconstructions.

(a) (b)

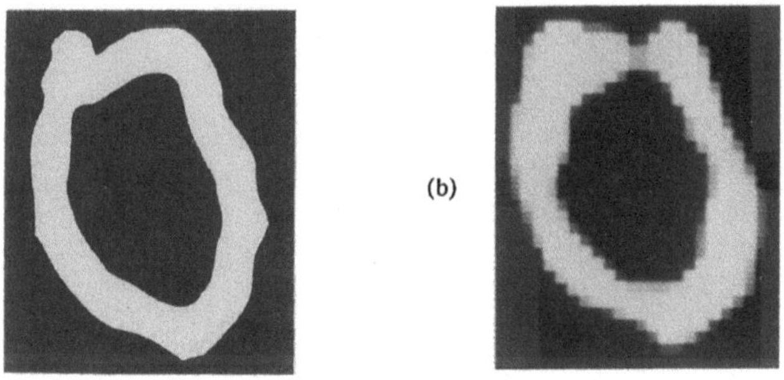

Fig. 5. (a) Shape of the left ventricle phantom, which has overall dimensions of approximately 7.5 cm by 10.0 cm. (b) Reconstruction of the phantom from a coded image containing approximately 100,000 counts. Reconstruction was performed using a simulated annealing algorithm.

ACKNOWLEDGEMENT

We wish to gratefully acknowledge the support of the National Cancer Institute, who funded this work under grant no. P01 CA 23417.

REFERENCES

Aarsvold, J.N., Barrett, H.H., Chen, J., Landesman, A.L., Milster, T.D., Patton, D.D., Roney, T.J., Rowe, R.K., Seacat, III, R.H. and Strimbu, L.M. (1988). Modular scintillation cameras: a progress report, SPIE - Medical Imaging II: Image Formation, Detection, Processing, and Interpretation 914, pp. 319-325.

Arendt, J. (1980). The Design and Analysis of a 1D Scintillation Detector, M.S. Thesis, University of Arizona.

van Laarhoven, P.J.M. and Aarts, E.H.L. (1987). Simulated Annealing: Theory and Applications, D. Reidel Publishing Co., Dordrecht.

Milster, T.D., Selberg, L.A., Barrett, H.H., Easton, R.L., Rossi, G.R. and Arendt, J. (1983). A Modular Imaging System for use in Nuclear Medicine, Scientific Exhibit, Soc. of Nucl. Med. Annual Meeting.

Milster, T.D., Selberg, L.A., Barrett, H.H., Easton, R.L., Rossi, G.R., Arendt, J. and Simpson, R.G. (1984). A Modular Scintillation Camera for Use in Nuclear Medicine, IEEE Trans. Nuc. Sci. NS-31, pp. 578-580.

Milster, T.D., Selberg, L.A., Barrett, H.H., Landesman, A.L. and Seacat, III, R.H. (1985). Digital Position Estimation for the Modular Scintillation Camera, IEEE Trans. Nuc. Sci. NS-31, pp. 748-752.

Milster, T.D. (1987). Design and Construction of a Modular Gamma Camera, Ph.D. Dissertation, University of Arizona

Selberg, L.A. (1984). Design Studies for a Modular Scintillation Camera, M.S. Thesis, University of Arizona.

COMPUTER EVALUATION OF CARDIAC PHASE IMAGES USING CIRCULAR STATISTICS AND ANALYSIS OF VARIANCE

A. S. Houston and A. Craig

Royal Naval Hospital Haslar and University of Sussex
United Kingdom

ABSTRACT

Circular statistics are used to analyse phase image data in the region of the left ventricle for a series of patients with known diagnosis. Estimates of mean, standard deviation, skewness and kurtosis are obtained from the phase histograms, together with a covariance based on spatial information. The same data are also analysed using linear statistics with an appropriate break in the circular phase histogram. Analysis of variance and signal detection theory are applied to the results and indicate that circular statistical analysis is a useful tool for analysing phase image data. In particular, it is shown that circular standard deviation, when used as a statistic for discriminating between normal and abnormal cardiac studies, is on a par with the human observer.

1. INTRODUCTION

Phase images obtained from multiple gated cardiac studies were used recently in a series of perception experiments (Houston et al, 1988). Thirty-two patient studies were involved. Based on the final diagnosis of the referring clinician, sixteen of the studies were classified as pathological - including examples of hypokinesis, aneurysm and valvular incompetence - while the remaining sixteen were classified as normal. Two views - anterior (ANT) and left anterior oblique (LAO) - were obtained for each patient and phase images generated using five methods.

These methods involve analysing time/activity curves obtained from each pixel using the following techniques:

(a) single-harmonic Fourier analysis (denoted by FOU) (e.g. Geffers et al, 1978);

(b) time-to-minimum (converted to degrees) following a two harmonic Fourier fit (denoted by 2H) (e.g. Bacharach and Green, 1982);

(c) time-to-minimum (converted to degrees) following a weighted four harmonic Fourier fit (denoted by 4HW) (Wendt et al, 1982);

(d) factorial phase imaging, where principal components are calculated

from the study under examination only (denoted by FI) (Houston et al, 1982);
(e) factorial phase imaging, where principal components are calculated using a series of studies which may or may not include the study under examination (denoted by FC) (Houston and Macleod, 1984).

The resulting 320 images were rated in pairs, ANT and LAO views of each patient being taken together, by 6 'experts' with experience in nuclear medicine, and by 12 'students' whose only previous experience in nuclear medicine was a short practice session prior to the experiment. In order to assist the 'students' an outline corresponding to the edge of the left ventricle at end diastole was superimposed on each phase image.

Although there were statistically significant differences between the methods (with a preferential ordering of FC, 2H, 4HW, FI and FOU), no significant difference was found between the results of 'experts' and 'students'. This result suggests that clinical experience, by itself, has little influence, and that judgements are dominated by a perceptual cue from within the defined region containing the left ventricle. This cue may derive from some simple statistical parameters which, if identified, would prove useful for discriminating between normal and abnormal cardiac studies.

This paper reports an attempt to identify these potential cues using circular statistics which are, a priori, more appropriate for describing cyclic phenomena, such as phase, than are the more usual linear statistics.

2. CIRCULAR STATISTICS

The left ventricular (LV) phase histogram is obtained using phase information within the defined region of the left ventricle. Statistical parameters which derive from this histogram include mean, standard deviation (SD), skewness and kurtosis. Since none of these contains spatial information, the first term of the correlation function, i.e. the covariance of adjacent pixels in units of the variance, may be used as indicative of 'structure'.

A problem which exists is that the LV phase histogran is cyclic. For normal studies, where the range of phases is small, this may not prove to be a serious problem. However, in the case of, say, an aneurysm, where phase may take values around the full 360° cycle, an alternative approach is necessary.

The theory of circular statistics or, more precisely, the statistics of directional data, has been developed over the last three decades (Mardia, 1972; Batschelet, 1981) with biological applications including animal migration and homing pigeon theory. It may be applied to time series with a fixed cycle period. Although this is not the case in equilibrium cardiac studies, the process of gating creates artificially a data set ideally suited to a circular statistical analysis.

2.1 Circular mean

The concept of preferred direction is adopted from biology to define the experimental mean. If we have N measured directions, then the ith direction may be represented by a unit vector, $\mathbf{v}_i$, emanating from the centre of a circle of unit radius. The mean vector, **r**, is

$$\mathbf{r} = \frac{1}{N} \sum_{i=1}^{N} \mathbf{v}_i \; . \tag{1}$$

The direction of vector **r** is defined as the preferred direction and the mean angle Φ is measured relative to some arbitrary axis. The mean vector is conventionally displayed as emanating from the centre of the circle of unit radius.

Mean angle is best calculated by averaging the x and y components of the individual unit vectors. If the ith unit vector has a direction given by angle Φ_i with respect to the axis, then the components of mean vector **r** are given by

$$r_x = \frac{1}{N} \sum_{i=1}^{N} \cos \Phi_i \; ; \qquad r_y = \frac{1}{N} \sum_{i=1}^{N} \sin \Phi_i \; . \tag{2}$$

The mean angle, Φ, is then given by

$$\Phi = \begin{cases} \arctan(r_y/r_x) & \text{if } r_x > 0 \\ 180^\circ + \arctan(r_y/r_x) & \text{if } r_x < 0 \\ 90^\circ & \text{if } r_x = 0 \text{ and } r_y > 0 \\ 270^\circ & \text{if } r_x = 0 \text{ and } r_y < 0 \\ \text{undefined,} & \text{if } r_x = 0 \text{ and } r_y = 0 \; . \end{cases} \tag{3}$$

The magnitude, r, of the mean vector is simply

$$r = \sqrt{(r_x^2 + r_y^2)} \quad (4)$$

2.2 Circular standard deviation

It is desirable that any definition of circular standard deviation should approximate to linear standard deviation when the range of the distribution is small. It is clear, also, that the magnitude, r, of the mean vector will be inversely related to the spread of the distribution. Using these criteria Batschelet (1981) defines circular standard deviation, measured in radians, as

$$SD_1 = \sqrt{\{2(1-r)\}} \ . \quad (5)$$

Alternatively, Mardia (1972) suggests the use of

$$SD_2 = \sqrt{\{-2 \ln(r)\}} \ . \quad (6)$$

Note that for values of r close to 1, the difference between SD_1 and SD_2 will be negligible. However, as r tends to 0, SD_2 tends to ∞, while SD_1 remains finite (tending to $\sqrt{2}$ radians).

In this paper the parameter SD_1 is adopted throughout.

2.3 Higher moments

Higher moments are defined for circular distributions using the appropriate multiple of the angle. For example, the second moment is obtained by doubling the angles and proceeding as before. Let the angle of the ith direction with respect to the same choice of axis be Φ_i. Then the vector, v_{2i}, will have components

$$\{v_{2i}\}_x = \cos 2\Phi_i \ ; \quad \{v_{2i}\}_y = \sin 2\Phi_i \ . \quad (7)$$

A mean vector, corresponding to the second moment, is given by

$$r_2 = \frac{1}{N} \sum_{i=1}^{N} v_{2i} \ . \quad (8)$$

This vector has magnitude r_2 and direction given by the angle Φ_2 and is used in the definitions of circular skewness and circular kurtosis, as follows:

$$\text{Skewness} = r_2\sin(\Phi_2-2\Phi)/(1-r)^{3/2}\ ; \qquad (9)$$

$$\text{Kurtosis} = (r_2\cos(\Phi_2-2\Phi)-r^4)/(1-r)^2\ . \qquad (10)$$

2.4 Correlation function

Unfortunately, although the theory of circular statistics is now fairly advanced (including periodic regression and correlation of samples), the authors have been unable to find any reference in the literature to circular covariance, necessary in any definition of correlation function for the circular case. An attempt is made here to define the first term of the correlation function as follows:

If Φ is the mean of N phase angles, Φ_i, and j and k are any pair of horizontally, vertically or diagonally adjacent pixels, then the first term of the correlation function is given by

$$\text{Corr.Fn.} = \frac{\sum\limits_{\substack{\text{all}\\(j,k)}} \sin(\Phi_j-\Phi)\sin(\Phi_k-\Phi)}{\sum\limits_{i=1}^{N} \sin^2(\Phi_i-\Phi)}\ . \qquad (11)$$

This definition is not rigorous and can readily be criticised. In particular, the adopted definition of standard deviation may be written as

$$SD^2 \equiv SD_1{}^2 = \frac{4}{N}\sum_{i=1}^{N}\sin^2\frac{(\Phi_i-\Phi)}{2}\ . \qquad (12)$$

It might therefore seem that angular deviations from the mean should first be halved in any definition of correlation function. However, this leads to a discontinuity at $\Phi \pm 180°$, resulting in correlated values on either side producing a highly negative contribution to the correlation function. The adopted definition ensures that adjacent pixels with similar values will contribute little if around Φ or $\Phi \pm 180°$, but will contribute a maximum positive contribution if around $\Phi \pm 90°$. Similarly, a maximum negative contribution will be obtained from adjacent pixels with values of $\Phi + 90$ and $\Phi - 90°$. Clearly problems still arise if many values lie outside the range $\Phi \pm 90°$, but the effects are rather less drastic. The assumption in geometric terms is that the unit circle may be folded across the diameter defined by this range and treated as a semi-circle.

Estimates of these parameters were obtained from each of the 320 phase image data sets. For comparison, the equivalent parameters based on linear statistics were also estimated. To make the linear estimates, an artificial break in the circular histogram was chosen at ± 180° from the phase of a cosine curve. This procedure does not validate the use of linear statistics, but is similar to techniques used previously (e.g. Houston and Macleod, 1984).

3. ANALYSIS OF VARIANCE

The aim of the investigation was to determine if any of the parameters could distinguish usefully between normal and abnormal patients, and if the sensitivity would be influenced by either the method used to generate the phase images or the view aspect, or both. Interest lay in comparing the outcome with the results of the perception experiment.

Towards these ends the data were first subjected to analysis of variance (Fisher, 1935). This technique recognises that the total variability in a set of data can be separately assigned to random and non-random sources. In the present case, the possible non-random sources are Group (normal/abnormal), Method (FOU/2H/4HW/FI/FC), and View (ANT/LAO), along with their interactions. The data were analysed as in a factorial design with three factors, Group, Method and View, and with repeated measures on the last two of these (Winer, 1962). The ratio of the variance of non-random and random sources, F, provides an appropriate index of a source's influence. Of particular relevance in the present context, a high F-ratio for the effect of Group would signify wide separation between normal and abnormal scores, and so correspond to a high level of discriminability, while the F-ratio for the Group-by-Method interaction will indicate the strength of any differential discriminability between methods.

Analysis of variance, itself, belongs to the class of linear models, and has not yet been adequately developed for the case of cyclic data. Of the parameters being analysed here, only Mean is cyclic and may present problems regarding validity. These are probably minimised in the present case, however, because the range of Mean is well contained within the cut-off limits.

4. RESULTS

The results of the analyses of variance (ANOVAs) are indicated in Tables 1 and 2. Table 1 presents the F-ratios for the main effect of the Group factor and shows clearly that SD and, to a lesser extent, Mean and Skewness distinguish between normal and abnormal patients. It is acknowledged that, for these particular data, the circular statistics are not significantly superior to their linear equivalents.

Also included in Table 1 are estimates of the signal detection theory measure, d_e' (Swets and Pickett, 1982), indexing the overall discriminability of each of the statistical parameters. This measure was used in the observer trials (Houston et al, 1988) where the values averaged over the five methods were 2.09 for 'experts' and 1.99 for 'students'.

Here, signal detection theory has been applied to determine the discriminating potential of the various circular and linear parameters. For each parameter a receiver operating characteristic curve was generated by first placing all 320 values in increasing order of magnitude and then applying decision criteria that cut off 25%, 37½%, 50%, 62½% and 75% of the total sample. True and false positive rates were calculated at each decision level and values of d_e' calculated using Dorfman's algorithm as given in Swets and Pickett (1982). These are the values listed in Table 1. In considering them it should be noted that with this procedure, cases where low values are likely to indicate an abnormality will result in a negative value of d_e'; for these cases the absolute value of d_e' should be used as the measure of discrimination.

The main points to note about the Table 1 estimates are that they mirror the magnitude of the F-ratios and also that the absolute value for SD is almost identical to that obtained from the human observers.

The strength of the Group effect on the SD statistic in the ANOVA indicates that SD would indeed provide a useful means of distinguishing between normal and abnormal cases. The view that it provides such a cue for the judgements by the human observers is prompted by the observation that the human and SD levels of discriminability as indexed by d_e' are so similar.

Judgements by the human observers, however, had been affected by the method used to generate the images. Such an influence would correspond to a Group-by-Method interaction in the present analysis. As can be seen in Table 2, this interaction is evident ($p<.001$) for both Mean and SD.

However, when separately assessed for each method, the d_e' estimates did not reflect the ordering given by the human observers. This failure is not too surprising since the present estimates for each method are based on only 64 images (or 32 when assessed separately for each Method and View) and so are unreliable: a single reversal of the order of adjacent normal and abnormal images may cause a 12% increase or decrease in the value of d_e'. For each statistical parameter a more stable index of discriminabilty in the data subsets is provided by the difference between the means of the abnormal and normal distributions (linear arithmetic means were used, consistent with the linear ANOVA model). This index is simple and relatively unaffected by the small sample size. Values of the difference index are listed in Table 3 for highly significant interactions ($p<.001$).

Values obtained for SD give an ordering of the Methods as 2H, FC, 4HW, FI and FOU. Apart from an exchange of places involving the two leading methods, this is the same order as obtained in the observer trials and strengthens the view that SD may provide the cue for the human judgements. Values using the Mean do not show the same order as the human judgements.

The small difference obtained using Mean for Method FI is a consequence of the method. Factor analysis provides a Mean close to 180° for the set used to generate the principal components; for FI this is also the data set, so that the values of Mean for both normal and abnormal studies lie close to 180°. Nevertheless, it is interesting that Method FC, which is also based on factor analysis, yields the highest value for the difference between Means. Method FC was favoured by the human observers. Perhaps, then, they place some reliance on the information provided by the Mean, as well as on SD.

5. DISCUSSION AND CONCLUSIONS

The methods of circular statistics are shown to be applicable to cardiac phase image data. However, while the validity of applying linear statistics is at best questionable, an appropriate artificial break in the circular phase histogram gives results which are almost as good as those obtained using circular statistics. For normals, the sample of phase values lay well away from the break point, whereas this was not always

	ANALYSIS OF VARIANCE F-ratio : d.f. 1 & 30		SIGNAL DETECTION THEORY d'_e using all values	
	CIRCULAR	LINEAR	CIRCULAR	LINEAR
MEAN	8.80	12.03	0.77	0.91
SD	34.78	28.93	1.98	1.98
SKEWNESS	10.07	7.01	-0.78	0.63
KURTOSIS	3.87	2.36	-0.13	-0.16
CORR.FN.	6.65	1.79	-0.72	0.22

Table 1. Values of F-ratio for effect of Group obtained from analysis of variance and d'_e obtained from signal detection theory for five statistics derived using both circular and linear statistics.

ANALYSIS OF VARIANCE (CIRC. STATS. ONLY)

	p<.001	p<.01	p<.05
MEAN	METHOD METHOD×GROUP	GROUP	-
SD	GROUP METHOD VIEW METHOD×GROUP	-	-
SKEWNESS	METHOD×VIEW×GROUP	GROUP	METHOD
KURTOSIS	-	-	-
CORR.FN.	VIEW	-	GROUP METHOD×VIEW METHOD×VIEW×GROUP

Table 2. Significance levels found for factors and their interactions using analysis of variance.

	METHOD×GROUP (BOTH VIEWS)		METHOD×VIEW×GROUP SKEWNESS	
	MEAN	SD	ANT	LAO
FOU	33.2	22.1	-0.57	-1.68
2H	35.1	28.1	-1.20	-0.09
4HW	32.6	25.9	-1.43	-0.28
FI	4.9	24.3	-0.77	-1.49
FC	41.2	26.4	-0.67	-1.61

Table 3. Differences between the mean values of abnormal and normal distributions for interactions significant at the 0.1% level.

true for abnormals. As a consequence, sometimes the very invalidity of the linear statistics exaggerated the differences between normal and abnormal groups, so that good discriminability was achieved. This applies particularly to SD and Skewness. For Skewness, the effect is evidenced by the fact that the circular and linear d_e' values, while of similar magnitude, are opposite in sign (see Table 1).

Circular standard deviation is shown to be an excellent discriminator of normal and abnormal cardiac studies. It not only provides discrimination on a par with non-experienced observers and marginally inferior to experienced observers, but gives an ordering of methods of obtaining phase images similar to that of human observers. It is felt that the use of circular standard deviation, perhaps in conjunction with another statistic, might provide the basis of a knowledge based system for analysing cardiac phase images. This will also depend on whether the role of circular statistics can be extended to include differential diagnosis. This possibility is currently under investigation.

REFERENCES

Bacharach, S.L., and Green, M.V. (1982). Data processing in nuclear cardiology: measurement of ventricular function, IEEE Trans. on Nucl. Sci. NS-29, pp. 1343-1354.

Batschelet, E. (1981). Circular statistics in biology, Academic Press, London and New York.

Fisher, R.A. (1935). The design of experiments, Oliver and Boyd, Edinburgh.

Geffers, H., Adam, W.E., Bitter, F., Sigel, H., and Kampmann, H. (1978). Data processing and functional imaging in radionuclide ventriculography. In: Information proceesing in medical imaging, A. B. Brill (ed.) ORNL/BCTIC-2, Springfield, pp. 322-332.

Houston, A.S., Craig, A., and Macleod, M.A. (1988). A perception experiment to assess the clinical accuracy of various methods of obtaining cardiac phase images, submitted to Phys. Med. Biol.

Houston, A.S., Elliott, A.T., and Stone, D.L. (1982). Factorial phase imaging: a new concept in the analysis of first-pass cardiac studies, Phys. Med. Biol. 27, pp. 1269-1277.

Houston, A.S., and Macleod, M.A. (1984). A comparative review of methods of obtaining amplitude/phase and related images for gated cardiac studies. In: Information processing in medical imaging, F. Deconinck (ed.) Martinus Nijhoff, The Hague, pp. 223-237.

Mardia, K.V. (1972). Statistics of directional data, Academic Press, London and New York.

Swets, J.A., and Pickett, R.M. (1982). Evaluation of diagnostic systems: methods from signal detection theory, Academic Press, London and New York.

Winer, B.J. (1962). Statistical principles in experimental design, McGraw-Hill, New York.

Part 3 Applications

Section 3.2 Magnetic Resonance

A METHOD FOR CORRECTING ANISOTROPIC BLURS IN MAGNETIC RESONANCE IMAGES

D. W. Ro, P. M. Joseph, R. M. Lewitt, and G. T. Herman

University of Pennsylvania
U. S. A.

ABSTRACT

All magnetic resonance images are blurred as a result of an inherent decay of spin-echo signals during data acquisition due to spatially-varying and object-dependent transverse relaxation. The extent of the blur depends on the transverse relaxation time (T2) of the object and the data acquisition time used. An iterative method that corrects for such object-dependent anisotropic blurs in hydrogen magnetic resonance images is described. The method involves mathematical generation of the components of the spin-echo signals that gives rise to blurs in the images -- using initial estimates of the distribution of spin-density and transverse relaxation time -- and subtracting those component from the distorted spin-echo signals. Images reconstructed from such corrected signals yield improved estimates of the T2-weighted spin-density distributions. The method was tested on mathematical phantoms.

1. INTRODUCTION

One of the major reasons for magnetic resonance imaging (MRI) is that it offers a wealth of information about the physiological and/or pathological state of a living tissue via three intrinsic parameters: spin-density (*SD*), longitudinal relaxation time (*T*1), and transverse relaxation time (*T*2). Accordingly, an optimum imaging system should produce images that accurately reflect *SD*, *T*1, and *T*2 individually, so that the importance of each parameter can be separately evaluated. In practice, magnetic resonance images do not represent these parameters separately, but represent distributions of *SD* weighted by functions of *T*2 and/or *T*1.

One of the principal spatially-varying and object-dependent blurrings present in magnetic resonance images of the body obtained with current MRI systems is due to *T*2 effects. Magnetic resonance images are blurred by, among other factors, an inherent decaying of magnetic resonance signals during data acquisition due to transverse relaxation. These $T2$ distributions are difficult to estimate in MR imaging and, consequently, their effects are not easily removed. We present an algorithm that corrects for object-dependent blurrings of $T2$-weighted SD images. The algorithm generates not only accurate $T2$ distributions but also accurate SD distributions of hydrogen. The algorithm involves reconstruction of corrected $T2$-weighted SD images through the synthesis and subtraction of distortion-causing components of the MR signal from the original MR signal. The corrected images are then used to compute accurate SD and $T2$ images.

In this paper we consider Fourier transform reconstruction (Kumar et al. 1975) of hydrogen magnetic resonance images. Some theoretical aspects of the problem, simulations, and experimental studies using clinically relevant imaging protocols are presented to define the nature of the object-dependent blur in the image due to the decay of MR signals during data acquisition. To consider a specific example, the spin-echo signals arising from spin-warp pulse sequences (Edelstein et al. 1980) are distorted by exponential decay due to the presence of spatially-variant $T2$ values in the plane of the object. If the decay is rapid compared to the data acquisition time, then high frequency components of the image, which are contained in the tails of the spin-echo signals, will be distorted leading to anisotropic (or direction-dependent) blurring (Ro et al. 1986) of sharp boundaries and small details in the Fourier transform reconstructed image of the $T2$-weighted SD distribution.

In the next sections, we describe the spin-warp imaging method and examine the effect of spatially varying $T2$ distribution on the reconstructed image using this method. We then present some results obtained from both simulations and experimental studies, carried out to characterize object-dependent blurs arising from $T2$ relaxation. Our results are that, using physiologically significant $T2$ values and practical pulse sequences for the spin-warp imaging method, anisotropic blurring is observed in the images but at an insignificant level. However, using a modified spin-warp imaging method with a very short echo time, significant blurring is observed. An algorithm that corrects for such blurs in images derived from the modified spin-warp imaging method is presented.

2. SPIN-WARP IMAGING METHOD

In this section, we summarize the basic elements of magnetic resonance imaging (MRI) in order to establish the notation and equations needed for the sections that follow. The mathematical models developed in this work depend on well-established fundamental assumptions (Hinshaw and Lent 1983). First, we assume that the behavior of the macroscopic magnetization associated with an elementary volume is described by the Bloch equation (Bloch 1946). Second, we assume that the transverse relaxation is mono-exponential for hydrogen nuclei in tissue. In addition to these fundamental assumptions, we assume that the longitudinal relaxation time is much greater than the transverse relaxation time for hydrogen in tissue (Bottomley et al 1984), i.e., $T1 \gg T2$.

In all magnetic resonance imaging methods, the spatial information of the structure of the scanned object is encoded in the detected magnetic resonance signal by means of a series of gradient pulses (Joseph et al. 1987). The method mostly widely used in clinical scanners is known as the two-dimensional Fourier transform method, or the spin-warp method. In the method, a selective slice excitation of the object is carried out first by applying a 90° radio-frequency (RF) pulse in the presence of a gradient (usually along the z-axis), so that the (resonant) signal originates only from a predetermined slab in the object.

The phase information in the signal which provides spatial resolution in one direction is encoded by applying a phase-encoding gradient in that direction (y-axis). Once phase-encoding is done, a 180° RF pulse is applied to refocus the spins to form an echo signal. The signal is detected in the presence of a constant frequency-encoding (or read-out) gradient used in the other direction (x-axis). For all measurements of the signal, a constant read-out gradient is used, while the magnitude of phase-encoding gradient is varied in uniform steps from measurement to measurement.

The detected signal, known as the spin-echo signal (Hahn 1950), is expressed in complex phasor notation

$$S(p,m) = K \iint I(x,y)\, e^{-\frac{p}{T2(x,y)}}\, e^{-i\gamma G_x x p}\, e^{-i\gamma G_y y m}\, dx\, dy \tag{1}$$

where $-\tau/2 \leq p \leq \tau/2$ (τ is data acquisition time) and $-n \leq m \leq n$ (m is an integer). G_x is the read-out gradient and mG_y is the phase-encoding gradient. γ is the gyromagnetic ratio, which is a constant for a given nucleus. In the equation, where $M_o(x,y)$ and $T2(x,y)$ are the spatial distributions of spin-density and transverse relaxation, respectively. $I(x,y)$ is called the $T2$-weighted spin-density distribution. TE is the echo time, i.e., the time when the center of the spin-echo signal is observed.

For the moment, let us ignore the term involving $T2(x,y)$ in (1). Then the right side is simply the Fourier transform of $I(x,y)$, evaluated at the point $(\gamma G_x p, \gamma G_y m)$, i.e., $S(p,m) = \widehat{I}(\gamma G_x p, \gamma G_y m)$. We can see that if m is fixed, say $m = 0$, then the signal $S(p,0)$ determines the transform along a line of a Cartesian grid. Now, if the value of m (so-called the pseudo time variable (Edelstein et al. 1980)) is varied in uniform steps from measurement to measurement, then samples of transform of $I(x,y)$ are obtained in a Cartesian grid of the spatial frequency space (k_x, k_y). So, for each m, $S(p,m)$ is measured for uniform samples of p and is put into the m'th row of a two-dimensional array. The completed array is inverse Fourier transformed to yield an estimate of $I(x,y)$.

3. EFFECT OF TRANSVERSE RELAXATION

We now examine the effect of a spatially varying $T2(x,y)$ on this imaging technique. Note that the high spatial frequency components of $I(x,y)$ (i.e., the parts of $\widehat{I}(k_x,k_y)$ that are far from the transform origin $k_x = 0$) are determined by $S(p,m)$ for large $|p|$. It follows that these components are the ones most affected by the factor $e^{-p/T2}$. To see the effect, assume that we can collect data (i.e., samples of $S(p,m)$) for $-\tau/2 \le p \le \tau/2$, where τ is the data acquisition time for a signal beginning at $p = -\tau/2$. If $p/T2(x,y)$ is small for all x and y, then the signal accurately represents the Fourier components of $I(x,y)$ being imaged.

Now consider an object composed of 2 regions, denoted by A and B, each having a uniform transverse relaxation time $T2(A)$ and $T2(B)$, respectively. Let $T2(A)$ be long and $T2(B)$ be short (compared to τ), so that $e^{-\frac{\tau/2}{T2(A)}} \approx 1$ and $e^{-\frac{\tau/2}{T2(B)}} \approx 0$. The integral expression for $S(p,m)$ in (1) shows that $S(\frac{\tau}{2},m)$ depends almost entirely on the $T2$-weighted spin density distribution in region A and has virtually no information on the $T2$-weighted spin density distribution within region B. Since $S(\frac{\tau}{2},m)$ represents high-frequency components (those due to small structures and sharp edges in the x direction), these components from region B will be suppressed in the imaging process. Thus, in the reconstructed image, region B will appear more blurred than region A. In this kind of example, one would expect to find a position-variant blurring in the image, due to the position-variant nature of $T2(x,y)$.

An interesting consequence of this argument is that images obtained from the spin-warp method exhibit a $T2$-dependent blurring which is orientation-dependent, as well as position-variant. An edge (in the image) parallel to the x-axis should be resolved more sharply than an edge that is parallel to the y-axis, in the presence of a relatively short $T2$. In order to investigate these effects, both simulations and experimental studies were carried out. The following sections describe the methods used and present results from such studies.

4. ANISOTROPIC BLURRING IN HYDROGEN IMAGES - SIMULATION STUDIES

The detected spin-echo signals for spin-warp data were analytically calculated for two-dimensional phantoms consisting of regions of uniform spin-density and $T2$. Consider a phantom made up of non-overlapping subregions. Let $R_j(x,y)$ be the characteristic function of the j'th subregion, that is, the function having value 1 for (x,y) in the j'th subregion and value 0 for all other (x,y). For the j'th subregion, we will take the spin-density $M_o(x,y)$ to have uniform value $M_o(j)$ inside the subregion and zero value outside. We will also take $T2$ to have uniform value inside the j'th subregion, so that $T2(x,y) = T2(j)$. Due to the j'th subregion alone, the detected signal $S_j(p,m)$ arising from a spin warp pulse sequence is

$$S_j(p,m) = K \iint M_o(j)\, e^{-\frac{TE}{T2(j)}}\, e^{-\frac{p}{T2(j)}}\, R_j(x,y)\, e^{-i\gamma G_x x p}\, e^{-i\gamma G_y y m} dx\, dy \tag{3}$$

and the total signal $S(p,m)$ from the phantom is the sum of the signals due to the subregions:

$$S(p,m) = K \sum_j I(j)\, e^{-\frac{p}{T2(j)}}\, \hat{R}_j(\gamma G_x p, \gamma G_y m)\,, \tag{4}$$

where $\hat{R}_j(.,.)$ is the analytically calculated Fourier transform of the characteristic function of the j'th subregion. For each m, $S(p,m)$ is explicitly calculated for uniform samples of p and is put into the m'th row of a two-dimensional array. The completed array is inverse Fourier transformed to yield an image of the $T2$-weighted SD distribution. Simulated spin-echo signals from a square object (20 by 20 millimeters) were used to reconstruct the 33 by 33 millimeters $T2$-weighted SD image (sampled at one millimeter intervals). Fig. 1 show examples of the reconstructed phantom for various values of $T2$. TE=20 milliseconds and τ=20 milliseconds were used for this example. With the display white level at 0.1 of the maximum image level, we find horizontal blurring of the $T2$-weighted SD distribution

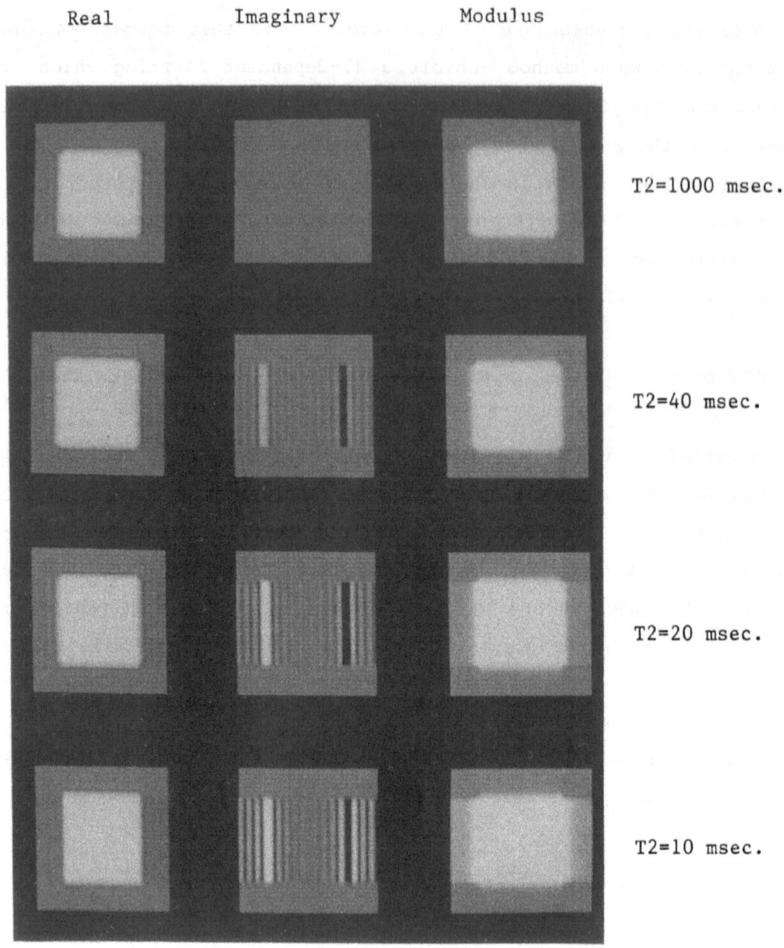

Fig. 1. Effects of different values of transverse relaxation time on reconstructed mathematical phantoms

in the read-out gradient direction (x-axis). Although not shown in Fig. 1, the application of a Hanning filter to the spin-echo signals prior to reconstruction diminished the horizontal blurring but at the expense of significant decrease in the resolution of the image.

5. EXPERIMENTAL STUDIES TO DEMONSTRATE ANISOTROPIC BLURRING

To see if such anisotropic blurring of hydrogen images is actually present in MR images of $T2$-weighted SD distributions, experiments using short $T2$ hydrogen phantoms were carried out on an experimental MR system. The system, a 1.9 Tesla superconducting magnet which has a 31 centimeter bore, operates at 80.92 MHz for hydrogen imaging.

Three vials (1 cm in diameter and 7 cm in length) containing 1, 2 and 4 millimolars of aqueous solution of manganese chloride ($MnCl_2$) with agarose concentration of 1% (by weight) were first made to produce short $T2$ hydrogen phantoms. Manganese ions are paramagnetic and they are hydrogen relaxation reagents. They influence the magnetic resonance properties of hydrogen by shortening their relaxation times (Barnhart and Berk 1986). The transverse relaxation time of protons is shortened further by the addition of agarose gels (which make the solution of manganese chloride into a gelatinous mass).

The true $T2$ of these phantoms were then measured with a simple projection technique, similar to the method used in (Joseph et al. 1987). In this technique the projection of transverse magnetization along the direction of read-out gradient was measured for different settings of TE. The $T2$ values were determined by least-squares fit of the measured signals to the equation

$$\text{Measured signal} = M_o\, e^{-\frac{TE}{T2}}\,. \tag{5}$$

The measured relaxation times using this technique yielded $T2$ of 15 milliseconds, 11 milliseconds and 7 milliseconds for the 1, 2, and 4 millimolars phantoms, respectively. A spin-warp sequence was used to image the phantoms at echo time TE=20 milliseconds with various data acquisition time (τ=3, 8, 13, 18, and 23 milliseconds). A non-slice selective 180° pulse was used to avoid flip angle inaccuracies due to the selective pulse (Joseph et al. 1984). The resolution in the images, which was 1.5 millimeter, was kept constant from experiment to experiment.

Comparison of 128 by 128 modulus images of the phantoms, reconstructed by Fourier transform, showed clearly that blurring does arise along the

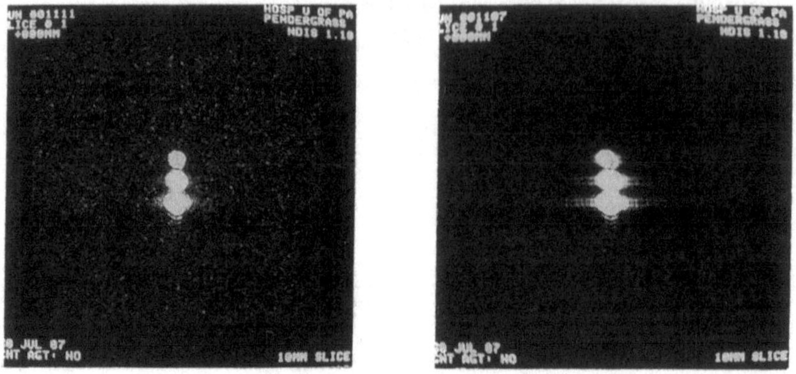

(a) (b)

Fig. 2. Experimental studies with $MnCl_2$ phantoms to demonstrate anisotropic blurring: (a) data acquisition time = 3 msec. (b) data acquisition time = 23 msec.

direction of the read-out gradient (in this case, horizontal blurring) and the degree of blurring increased as the duration of the data acquisition time was increased. Fig. 2a and 2b are modulus images of the phantoms for τ=3 and 23 milliseconds, respectively. As the data acquisition time becomes longer, an improvement of signal-to-noise ratio is observed in the images due to the smaller bandwidth.

6. ALGORITHM

In this section we present an iterative method for accurate hydrogen imaging that takes two or more spin-echo signals as input and produces as output an image of the spin-density distribution and an image of the distribution of transverse relaxation time. A key requirement of the method is the existence of an accurate mathematical model of the data acquisition process, which enables simulation of the signal that would be obtained from specified distributions of SD and $T2$. Apart from this requirement, the method itself is quite general - it is not restricted to any particular data acquisition strategy, e.g. the spin-warp, line-integral, or more general Fourier space trajectories (Twieg 1983). An MR image may be reconstructed either by a standard Fourier transform reconstruction method or by a conventional projection reconstruction method (Shepp 1980), depending on the data acquisition strategy used.

In our description of the following algorithm, we assume that images are derived from a modified spin-warp imaging method which uses the Fourier transform reconstruction technique. In this imaging method, gradients are applied such that the center of the spin-echo signal appears close to the 90° pulse (implying short echo time). Only one half side of the signal is acquired during each measurement (signals at $p < 0$ are not sampled due to transient effects). Similar methods have been used for sodium imaging by (Ra et al. 1986). The other half of the signal is obtained from a separate measurement with the read-out gradients reversed. The two signals are combined together to form one complete spin-echo signal. To increase the signal-to-noise ratio in the images, the signals are sampled over a long data acquisition time. However, the images suffer, as a consequence of long acquisition time, from significant anisotropic blurring.

We now describe the algorithm designed to correct for such $T2$-dependent blurring in hydrogen images. In the algorithm, we have not taken account which cause other types of distortions in the images, such as those due to magnetic field inhomogeneity. The general idea behind the algorithm is to iteratively refine our estimate of the error signal which gives rise to spatially varying $T2$-dependent blurring in the images. If a perfect error signal Err_i is subtracted from a corresponding spin-echo signal S_i, the result should yield an ideal signal C_i, e.g., for TE_1

$$C_1(p,m) = S_1(p,m) - Err_1(p,m)$$

$$= K \iint I_1(x,y)\, e^{-i\gamma G_x x p}\, e^{-i\gamma G_y y m}\, dx\, dy \tag{6}$$

where

$$S_1(p,m) = K \iint I_1(x,y)\, e^{-\frac{|p|}{T2(x,y)}}\, e^{-i\gamma G_x x p}\, e^{-i\gamma G_y y m}\, dx\, dy \tag{7}$$

and

$$Err_1(p,m) = K \iint I_1(x,y)\, [e^{-\frac{|p|}{T2(x,y)}} - 1]\, e^{-i\gamma G_x x p}\, e^{-i\gamma G_y y m}\, dx\, dy\,. \tag{8}$$

The ideal signal when inverse Fourier transformed should yield a $T2$-weighted SD image, free of $T2$-dependent artifacts.

To illustrate the iterative process of refining estimates of the error signal, Fig. 3 presents the algorithm in flow-chart form when only two spin-echo signals (i=1,2) are used. We now describe the k'th iteration

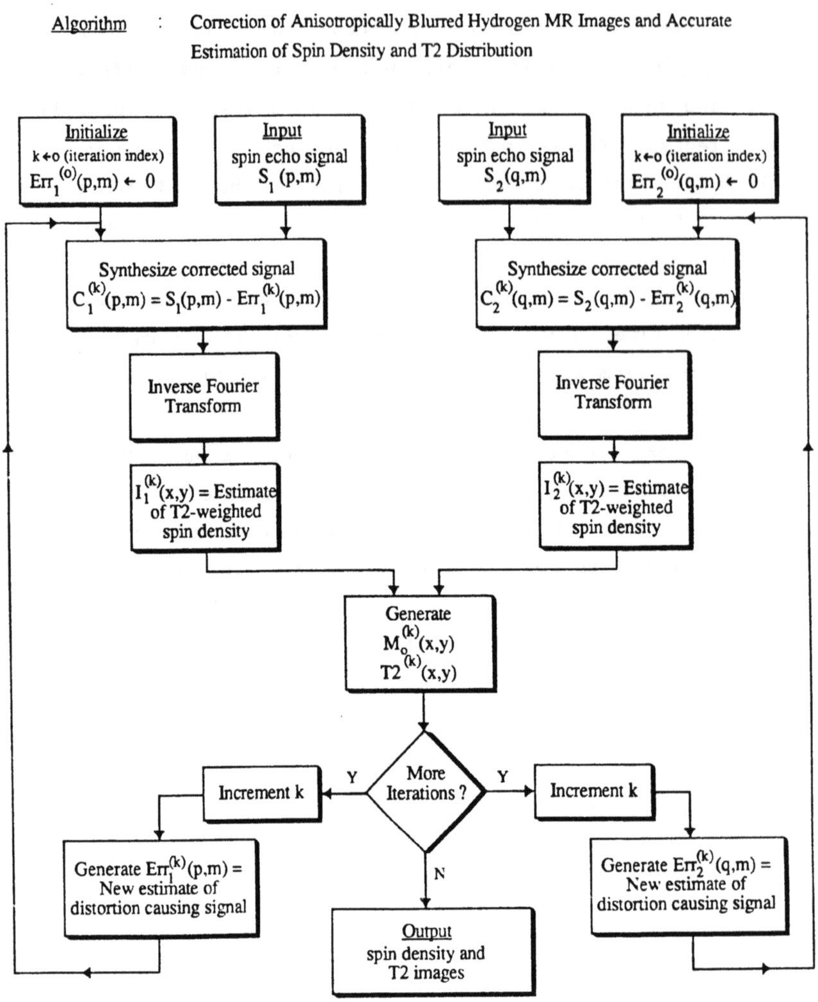

Fig. 3. Flowchart illustrating the iterative process of the algorithm.

of the algorithm, following the flow-chart. Each iteration makes use of the input data S_i. At the beginning of the loop, estimates $Err_i^{(k)}$ of the error components are available, either from the previous step of the algorithm, or, in the case of the initial (k=0) iteration, from an initial assignment to zero. $Err_i^{(k)}$ are subtracted from the corresponding spin-echo signals to yield $C_i^{(k)}$, which are the present estimates of C_i.

We apply the conventional reconstruction technique (inverse Fourier transformation F^{-1}) to $C_i^{(k)}$, obtaining images which we denote by $I_i^{(k)}$.

By definition,

$$I_i^{(k)} = F^{-1}[S_i - Err_i] \, . \tag{9}$$

We now define the k'th estimate $M_o^{(k)}(x,y)$ of the spin-density $M_o(x,y)$ and the k'th estimate $T2^{(k)}(x,y)$ of the transverse relaxation time $T2(x,y)$ as the solutions to the following equation

$$I_i^{(k)}(x,y) = M_o^{(k)}(x,y)\, e^{-\frac{TE_i}{T2^{(k)}(x,y)}} \, . \tag{10}$$

Solving the equation for $T2^{(k)}(x,y)$, we obtain

$$T2^{(k)}(x,y) = \frac{TE_2 - TE_1}{\ln(|I_1^{(k)}(x,y)|/|I_2^{(k)}(x,y)|)} \tag{11}$$

where $TE_2 > TE_1$ and $|I_1^{(k)}(x,y)| > |I_2^{(k)}(x,y)|$. We then calculate the k'th estimate of $M_o(x,y)$ by using (10).

If this is the last iteration, then $M_o^{(k)}(x,y)$ and $T2^{(k)}(x,y)$ are output, and the algorithm terminates. Otherwise, we use these estimates to compute the next ($k+1$'th) estimates of Err_i, i.e., for TE_1

$$Err_1^{(k+1)}(p,m) = K \iint M_o^{(k)}(x,y)\, e^{-\frac{TE_1}{T2^{(k)}(x,y)}} \,[e^{-\frac{|p|}{T2^{(k)}(x,y)}} - 1]\, e^{-i\gamma G_x x p}\, e^{-i\gamma G_y y m}\, dx\, dy \, . \tag{12}$$

We now increment k, so that the new estimates of Err_i will be denoted by $Err_i^{(k)}$, and the algorithm returns to the top of the loop.

The algorithm that we have just described has much in common with iterative algorithms for beam-hardening correction in x-ray CT (Herman and Trivedi 1983, and Joseph and Spital 1978). The theoretical foundations of algorithms of this type have been studied recently by (Censor et al. 1985). Convergence properties and the factors that determine the behavior of the initial iterates are also discussed in (Censor et al. 1985).

7. RESULTS

Fig. 4 shows examples of the application of the algorithm to a blurred reconstruction of 129 by 129 $T2$-weighted spin-density image. The detected spin-echo signals for modified spin-warp data were analytically calculated for 15 millimeter square hydrogen objects: square object (top) with SD=100

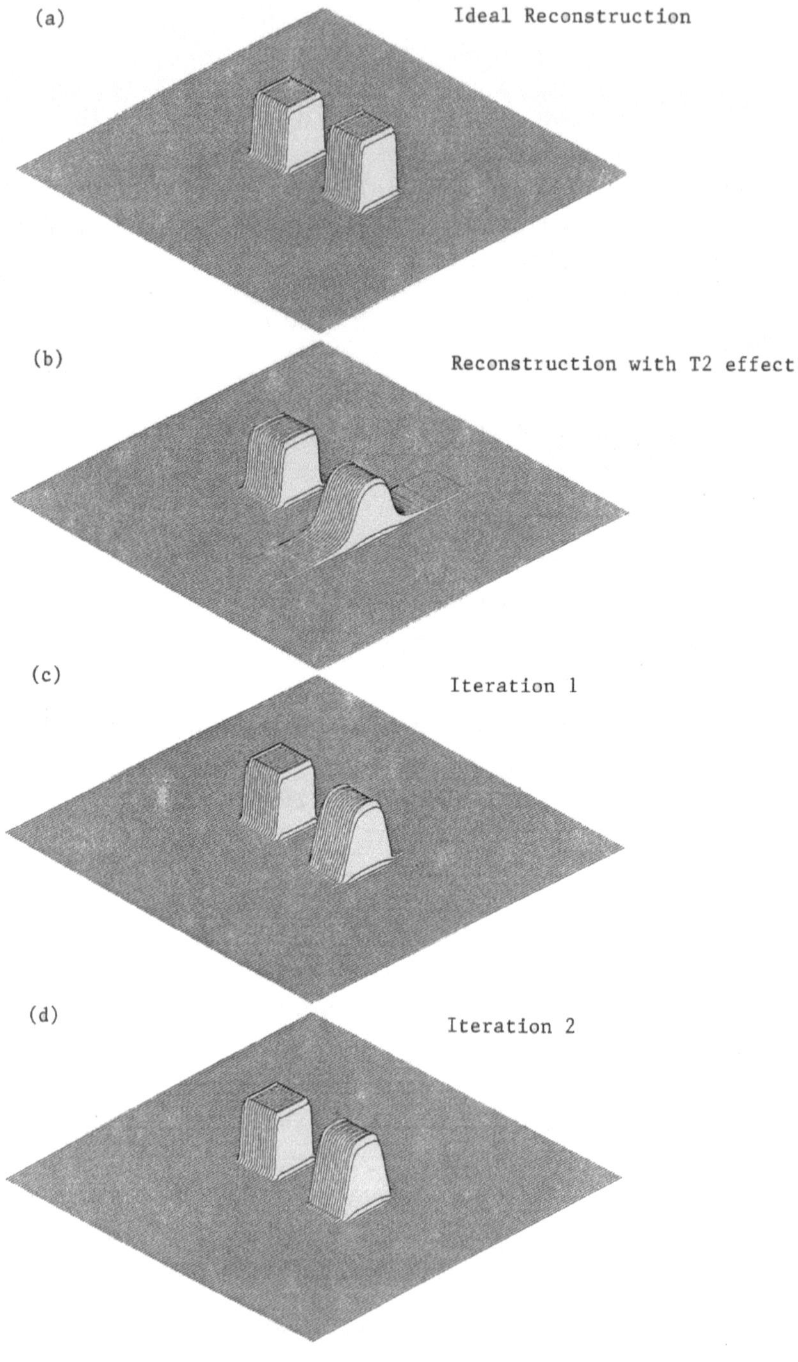

Fig. 4. Application of the algorthim to blurred reconstruction of two-dimensional hydrogen phantoms.

units and $T2$=30 milliseconds, and square object (bottom) with SD=300 units and $T2$=3 milliseconds. The modified spin-warp pulse sequence allowed us to use TE_1=2 milliseconds and TE_2=4 milliseconds with effective data acquisition time of τ=40 milliseconds.

Fig. 4a and 4b show the reconstructed phantoms using the ideal signal C_1 and the detected spin-echo signal S_1, respectively. Comparison of the objects in Fig. 4b clearly shows that the object with short $T2$ had significant anisotropic blurring while the object with relatively long $T2$ had a minimal amount of blurring. Fig. 4c and 4d show improvements of the reconstructed images after one and two iterations of the algorithm, respectively.

8. DISCUSSION AND CONCLUSION

Both simulations and experimental studies using practical pulse sequences for the spin-warp imaging method show that reconstructed hydrogen phantoms with physiologically relevant $T2$ values exhibit insignificant anisotropic blurring. Using a modified spin-warp method, reconstructed phantoms with short $T2$ values (less than 10 milliseconds) had significant blurring. The proposed algorithm succeed in removing these spatially-varying $T2$-dependent blurs in the images.

Since the modified spin-warp method is not usually used for hydrogen imaging, and since the $T2$ values of living tissues are greater than 20 milliseconds, $T2$-dependent blurring will not be a factor in image quality. However, such a modified spin-warp method is useful for imaging sodium in tissue with short $T2$. The short echo time (2-3 milliseconds) allows us to obtain samples of the rapidly decaying portion of the sodium signal - due to the short $T2$ component of sodium in tissue (0.7-3.0 milliseconds). The algorithm described above may then be modified to take into account the bi-exponential decay of the sodium signal and to correct for blurring in sodium images. We might add that the algorithm may also help to alleviate a major problem in fast imaging methods, such as the echo-planar method, which are currently limited by $T2$-dependent blurring due to the use of long data acquisition times.

ACKNOWLEDGMENT

This work was supported by the National Institutes of Health under Grant HL28438.

REFERENCES

Barnhart, J.L. and Berk, R.N. (1986). Influence of paramagnetic ions and pH on proton NMR relaxation of biologic fluids, Investigative Radiology 21, pp. 132-136.

Bloch, F. (1946). Nuclear Induction, Physical Reviews 70, pp. 460-474.

Bottomley, P.A., Foster, T.H., Argersinger, R.E. and Pfeifer, L .M. (1984), A review of normal tissue hydrogen NMR relaxation mechanisms from 1-100 MHz: dependence on tissue type, NMR frequency, temperature, species, excision and age, Medical Physics 11, pp. 425-448.

Censor, Y., Elfving, T. and Herman, G.T. (1985), A method of iterative data refinement and its applications, Mathematical Methods in the Applied Sciences 7, pp. 108-123.

Edelstein, W.A., Hutchison, J.M.S., Johnson, G. and Redpath, T. (1980), Spin warp NMR imaging and applications to human whole-body imaging, Physics in Medicine and Biology 25, pp. 751-756.

Hahn, E.L. (1950), Spin echos, Physical Reviews 80, pp. 580-594.

Herman, G.T. and Trivedi, S.S. (1983), A comparative study of two post-reconstruction beam hardening correction methods, IEEE Transactions on Medical Imaging MI-2, pp. 128-135.

Hinshaw, W.S. and Lent, A.H. (1983), An introduction to NMR imaging: from the Bloch equation to the imaging equation, Proceedings of IEEE 71, pp. 338-350.

Joseph, P.M. and Spital, R. (1978), A method for correcting bone induced artifacts in CT scanners, Journal of Computer Assisted Tomography 2, pp. 100-108.

Joseph, P.M., Axel, L. and O'Donnell, M. (1984), Potential problems with selective pulses in NMR imaging systems, Medical Physics 11, pp. 772-777.

Joseph, P.M., Shetty, A. and Bonaroti, E.A. (1987), Methods of reducing motion induced errors in T2-weighted magnetic resonance imaging, Medical Physics 14, pp. 608-615.

Kumar, A., Welti, D. and Ernst, R. (1975), NMR Fourier zeugmatography, Journal of Magnetic Resonance 18, pp. 69-83.

Ra, J.B., Hilal, S.K. and Cho, Z.H. (1986), A method for in vivo MR imaging of the short T2 component of sodium-23, Magnetic Resonance in Medicine 7, pp. 11-22.

Ro, D.W., Joseph, P.M., Lewitt, R.M. and Herman, G.T. (1986), Anisotropic blurring in sodium-23 images. In Works in Progress: Fifth Annual Meeting of SMRM, Montreal, pp. 241-242.

Shepp, L. (1980), Computerized tomography and nuclear magnetic resonance, J. of Computer Assisted Tomography 4, pp. 94-107.

Twieg, D.S. (1983), The k-trajectory formulation of the NMR imaging process with applications in analysis and synthesis of imaging methods, Medical Physics 10, pp. 610-621.

ICONIC FUZZY SETS FOR MR IMAGE SEGMENTATION

Wido Menhardt

Philips Research Labs Hamburg
West Germany

ABSTRACT

In this paper we present an approach to the top-down analysis of magnetic resonance images using the concept of iconic fuzzy sets. A symbolic model is used to describe the properties and relationships of scene objects. Using iconic fuzzy sets as pixel level representations of image objects, and fuzzy operationalizations of relations of these objects, a new framework for image analysis can be defined. We will demonstrate the feasibility of the approach with a detailed example of a supraorbital transaxial MRI brain slice.

1. INTRODUCTION

The goal of image analysis is the labeling of the pixels in an image with the names of objects in the depicted scene. The standard approach to this task is the segmentation of the input image into a set of image primitives, which are then matched to a symbolic image description (e.g. Ballard and Brown, 1982).

But the problem here is the extraction of primitives, i.e. homogenous regions or significant edges. The definition of these items involves very hard decisions in an early stage of the analysis process.

An alternative are fuzzy image subsets (Rosenfeld, 1979 and 1984), *iconic fuzzy sets*, as a basis for an image analysis system. Instead of defining crisp regions (pixel sets), fuzzy sets (Zadeh, 1965), which contain pixels to a certain degree of membership, can be used to model inherent uncertainty and the hypothetical character of intermediary results on the pixel level.

In the following sections we will present our image analysis system, which is based on iconic fuzzy sets. We will address knowledge representation (2), the principles of iconic fuzzy sets and some definitions of functions over these sets (3) and, finally, control (4).

2. KNOWLEDGE REPRESENTATION

Knowledge about the structure and the intensities of the images under inspection has to be explicitly respresented in an image anlysis system in order to make it easily adaptable to other situations. Several representation schemes are commonly used in image interpretation. We have used semantic networks, graphs with image objects as nodes and relationships between these as arcs (Menhardt 1988). In this paper, we use a functional description because this matches the concept of iconic fuzzy sets, and functions over these sets, very closely.

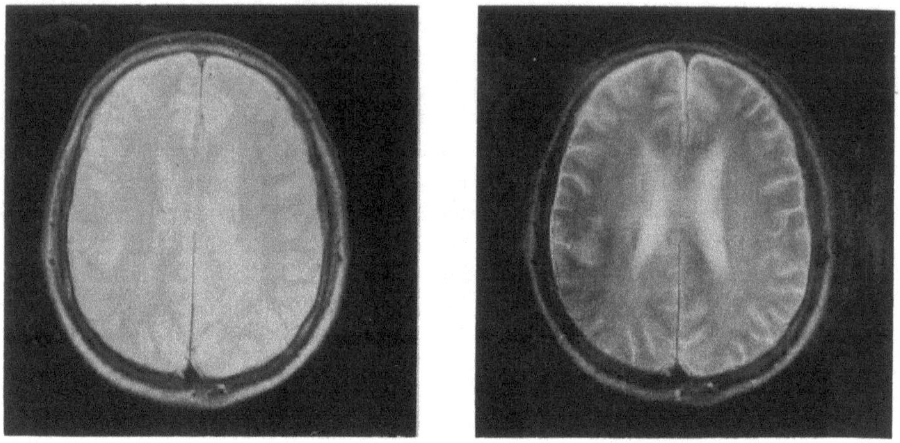

Figure 1: First and second Spin Echo of example case

We start with a natural language description of the MR images in figure 1:

> "In the *1st Spin Echo*, *soft tissues* have a *high* intensitiy with respect to the entire *image*. *Non soft tissues*, on the other hand, have a *low* intensity in the *1st Spin Echo*. The *skull* is part of *non soft tissues and inside* of *tissues*. The *intracranial* region is *inside* of the skull. *skin* is a *soft tissue, but not intracranial. Background* is *outside* of the *skin*. *CSF* is part of the *intracranial* region, *and* it has a *very high* intensity - with respect to the *intracranial* region - in the *2nd Spin Echo*. *Brain* is in the *intracranial* region, *but not CSF*. The *ventricles* are part of *CSF*, *and* a *large component* of the cavities *inside* the *brain*. *External CSF* is part of *CSF*, *but not* part of the *ventricles*."

Such a text can easily be rewritten in a functional description (figure 2): this is the form of the knowledge base in our system. The first item in every expression in brackets is a *function* symbol, and the other items are the *arguments* of the function. In nested expressions, functions return their results to the surrounding expression. "IS" is just an assignment operation for sake of readability and accessibility of the iconic fuzzy sets for the desired image objects.

Figures 3 and 4 show an iconic version of the network after application of the knowledge base to our example case.

```
( IS SOFT.TISSUES ( HAS.A.HIGH.R SPIN-ECHO-1 IMAGE ) )

( IS NON.SOFT.TISSUES ( HAS.A.LOW.R SPIN-ECHO-1 IMAGE ) )

( IS SKULL ( AND NON.SOFT.TISSUES ( INSIDE.G SOFT.TISSUES ) ) )

( IS SKIN ( BUT.NOT SOFT.TISSUES INTRACRANIAL ) )

( IS BACKGROUND ( OUTSIDE.G SKIN ) )

( IS INTRACRANIAL ( INSIDE.G SKULL ) )

( IS CSF ( AND INTRACRANIAL
                  ( HAS.A.HIGH.R SPIN-ECHO-2 INTRACRANIAL ) ) )

( IS BRAIN ( BUT.NOT INTRACRANIAL CSF ) )

( IS VENTRICLES ( AND CSF
            LARGE.COMPONENT ( INSIDE BRAIN ) ) )

( IS EXTERNAL.CSF ( BUT.NOT CSF VENTRICLES ) )
```

Figure 2: The knowledge base

Note, that the functional representation in figure 2 has no cycles. Each description depends only on image structures or input images which are either known in advance (*image* and *spin-echo*) or have been defined elsewhere. In section 4, the handling of cycles and the combination of evidence from different sources will be discussed.

3. ICONIC FUZZY SETS

As mentioned in the introduction, we define our image analysis task as the identification of those pixel sets, which correspond to image structures in a symbolic scene model. Instead of crisp sets, we use iconic fuzzy sets, i.e. mappings of the set of all pixels, U, to $[0,1]^n$, where n is the cardinality of U:

$$f : U \mapsto [0,1]^n$$
$$0 \leq f(\vec{x}) \leq 1. \qquad (1)$$

We associate such a mapping with each image structure in our symbolic representation. A mapping $skull : U \mapsto [0,1]^n$ expresses the "plausibility" of being *skull* for each individual pixel. Moreover, we can represent the plausibility of pixels being *inside skull* in a similar way:

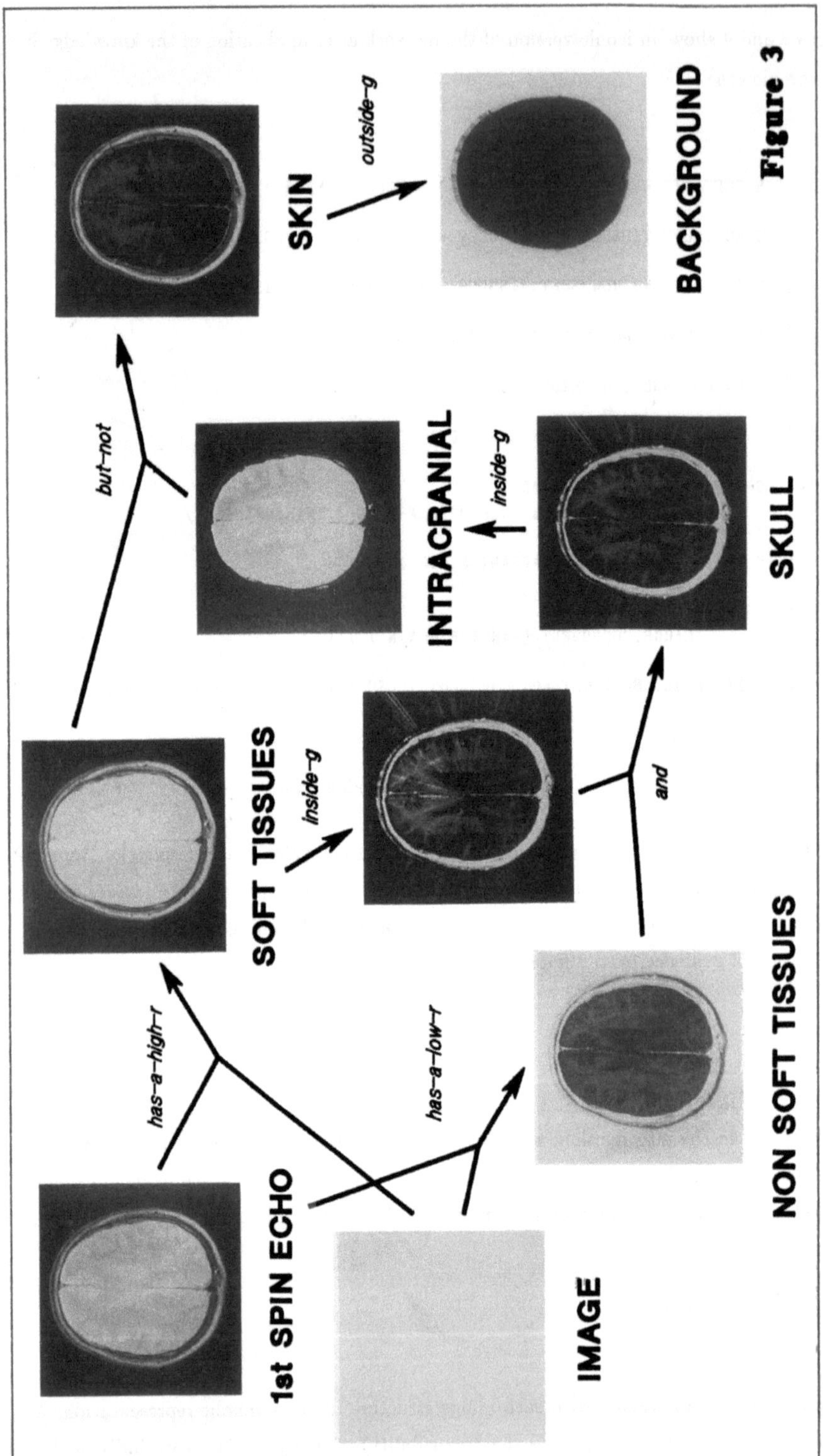
1st SPIN ECHO
IMAGE
has-a-high-r
has-a-low-r
SOFT TISSUES
NON SOFT TISSUES
inside-g
and
but-not
INTRACRANIAL
SKULL
inside-g
SKIN
outside-g
BACKGROUND

Figure 3

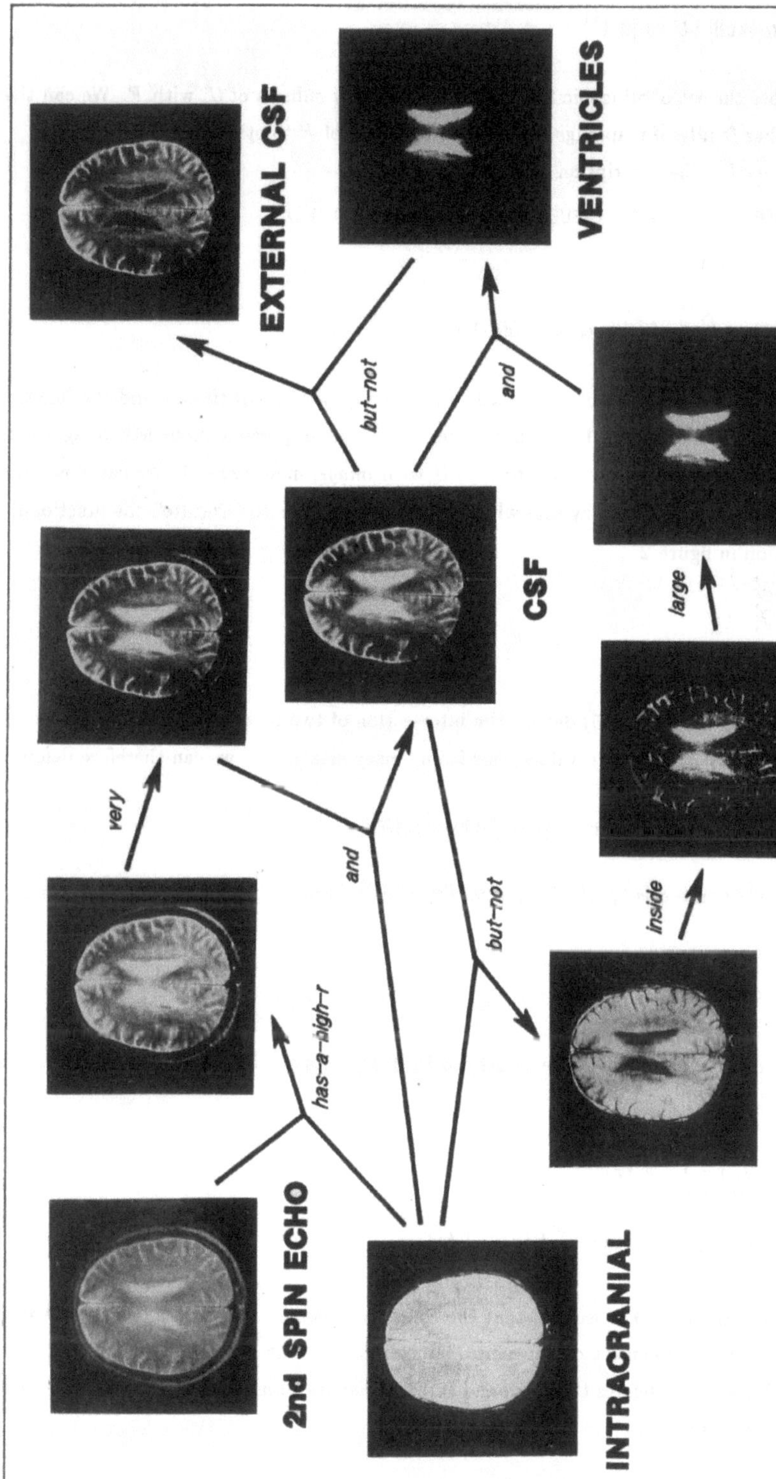
2nd SPIN ECHO
INTRACRANIAL
has-a-high-r
very
and
but-not
BRAIN
inside
large
CSF
but-not
and
EXTERNAL CSF
VENTRICLES

Figure 4

$$Inside[\text{skull}] : U \mapsto [0,1]^n \tag{2}$$

Let us denote the set of all iconic fuzzy sets, i.e. all fuzzy subsets of U, with F. We can then define another family of mappings, which map elements of F to other elements of F, e.g. $Outside : F \mapsto F$. The description

```
( IS SKIN ( AND SOFT.TISSUES ( OUTSIDE SKULL ) ) )
```

then takes the form

$$\text{skin}(\vec{x}) = And[\text{soft.tissues}, Outside[\text{skull}]](\vec{x}). \tag{3}$$

The iconic fuzzy set *skin* can be computed, if the sets *skull* and *soft.tissues*, and the functions *And* and *Outside*, are known. Let us finally denote the set of possible input MR images with I, and the iconic fuzzy set which is identical to U with *image*, $\text{image}(\vec{x}) \equiv 1$. We can now define a set of functions over iconic fuzzy sets which will then enable us to "execute" the functional representation in figure 2.

3.1 AND and PART-OF

Fuzzy set theory (Zadeh, 1965) defines the intersection of two fuzzy sets as the pointwise minimum of fuzzy membership values. For iconic fuzzy sets $f_i \in F$ we can therefore define

$$And[f_1, f_2, .., f_n](\vec{x}) = \min(f_1(\vec{x}), f_2(\vec{x}), ... f_n(\vec{x})). \tag{4}$$

If an image structure a is "part-of" another image structure b, this means that $a(\vec{x}) \leq b(\vec{x})$.

3.2 NOT and BUT.NOT

The complement and the difference as defined for fuzzy sets can be readily extended to iconic fuzzy sets:

$$Not[f_1](\vec{x}) = 1 - f_1(\vec{x}) \tag{5}$$

$$But.Not[f_1, f_2](\vec{x}) = \max(0, f_1(\vec{x}) - f_2(\vec{x})) \tag{6}$$

Whereas *Not* can be used to express that the contrary of some constraint is true, *But.Not* is used to express a complement relationship. An example is "skin is a soft tissue but not intracranial", which indicates that *skin* and *intracranial* are complementary structures with respect to *soft.tissues*.

3.3 HAS.A.HIGH.R and HAS.A.LOW.R

Functions which relate input image intensities to image structures can be viewed as classical fuzzy sets associated with the linguistic variables high and low: they can be operationalized using pixel-by-pixel mappings of signal values to the interval [0,1]. *Has.A.High.R* denotes a special mapping, which takes the distribution of signal values in a certain image structure into account. A relation `( HAS.A.HIGH.R I F )` maps low values of I in F to low membership values and vice versa. A natural operationalization of this is the following procedure: first we calculate the mean and variance of I in F by giving high weights to pixels which have a high membership value for F:

$$\mu_{i|f} = \frac{\sum_{\vec{x}} i(\vec{x}) f(\vec{x})}{\sum_{\vec{x}} f(\vec{x})}, \qquad \sigma^2_{i|f} = \frac{\sum_{\vec{x}} (i(\vec{x}) - \mu_{i|f})^2 f(\vec{x})}{\sum_{\vec{x}} f(\vec{x})}, \qquad i \in I, \quad f \in F. \tag{7}$$

With these, we define

$$Has.A.High.R[i,f](\vec{x}) = \int_0^{i(\vec{x})} e^{\frac{-(p-\mu_{i|f})^2}{2\sigma^2_{i|f}}} dp \tag{8}$$

and similarly

$$Has.A.Low.R[i,f](\vec{x}) = 1 - Has.A.High.R[i,f](\vec{x}). \tag{9}$$

3.4 INSIDE and OUTSIDE

For crisp pixel sets, one can define that a pixel $\vec{x}$ is *inside* of a structure f, if there is no path of 4-connected pixels from $\vec{x}$ to the edge of the image, which *does not* cross f. Following the definition of *connectedness* and *surroundedness* in (Rosenfeld, 1979), we can extend this definition to fuzzy structures. Let us denote the set of pixels on a path from $\vec{x}$ to the edge of the image with $P_{\vec{x}}$. The form

$$\max_{\vec{y} \in P_{\vec{x}}} f(\vec{y}) \tag{10}$$

yields the maximal value of f along P, which we will call *cost* of the path. The minimal possible cost for a point,

$$\min_{\forall P_{\vec{x}}} \max_{\vec{y} \in P_{\vec{x}}} f(\vec{y}), \tag{11}$$

tells us, to which degree point $\vec{x}$ is in or inside f. This value is high for points of f itself. As we are interested in points which are *inside* of f (the holes of f), we define

$$Inside[f](\vec{x}) = (\min_{\forall P_{\vec{x}}} \max_{\vec{y} \in P_{\vec{x}}} f(\vec{y})) - f(\vec{x}), \tag{12}$$

and, in a similiar way,

$$Outside[f](\vec{x}) = 1 - (\min_{\forall P_{\vec{x}}} \max_{\vec{y} \in P_{\vec{x}}} f(\vec{y})). \tag{13}$$

Although these definitions are not directly implementable, they can be rewritten as simple fill operations which lend themselves quite well for efficient implementation.

An illustrative description of the definition of *inside* is as follows: view the iconic fuzzy set f as a terrain map. The edge of the image is the shore to an ocean (an infinite reservoir). When rain falls on the image, lakes will be formed at some places: the depth of these lakes are measures for insidedness.

3.5 INSIDE.G and OUTSIDE.G

In some cases, the above definition does not operationalize what humans really mean with *inside*. If, for example, the surrounding object has small gaps (which allow water to penetrate), humans may still say that there is something inside. To account for this, we will give another definition, which is valid for a certain class of objects, and, for the ideal case, reduces to the definition given above.

We first compute the *center of gravity* for fuzzy objects:

$$\vec{c}(f) = \frac{\sum_{\vec{x}} \vec{x} f(\vec{x})}{\sum_{\vec{x}} f(\vec{x})}. \tag{14}$$

We then define a function *Inside.G* similar to *Inside*, with the difference, that a path from a point to the edge of the image may only be on the line defined by the point and the center of gravity:

$$Inside.G[f](\vec{x}) = \max_{\lambda>0}(f(\vec{x} + \lambda(\vec{x} - \vec{c}))) - f(\vec{x}). \tag{15}$$

The values for *Inside.G* can be normalized per line using $\max_{\lambda>0} f(\vec{c} + \lambda(\vec{x} - \vec{c}))$. Similarly, *Outside.G* can be defined by

$$Outside.G[f](\vec{x}) = 1 - \max_{\lambda>0}(f(\vec{x} + \lambda(\vec{x} - \vec{c}))). \tag{16}$$

These definitions can be used in those cases, where the question of insidedness can be solved on lines between the center of gravity and points on the edge of the image. In our example case, this is true for e.g. the skull and background, but not for the ventricles.

Still another alternative definition, ***inside.ch***, **assigns** a high insidedness to all points, which are a member of the convex hull of an object, but not a member of the object itself. This enables us to determine the inside of e.g. a cup, which does not entirely *surround* its inside.

3.6 LARGE.COMPONENT

In crisp pixel sets, connected components can easily be defined: two points of a set belong to the same component, if there exists a path between them, whose elements are all element of the pixel set. As all pixels are an element of each iconic fuzzy set with a certain degree of membership, all components of the set are connected in this sense. It is therefore not straightforward to determine connected components in iconic fuzzy sets.

Deviating from (Rosenfeld, 1979), we associate a component $C_{\vec{x}}$ with each pixel $\vec{x}$ of a set f. Each component is itself a fuzzy subset of f:

$$C_{\vec{x}}(\vec{x}') = \max_{\forall P_{\vec{x}\vec{x}'}} \min_{\vec{y} \in P_{\vec{x}\vec{x}'}} f(\vec{y}), \tag{17}$$

with $P_{\vec{x}\vec{x}'}$ as a path from $\vec{x}$ to $\vec{x}'$. The area of a fuzzy set is defined in (Rosenfeld, 1984) by

$$A(f) = \sum_{\vec{x}} f(\vec{x}). \tag{18}$$

We can now give a definition for a function *Large.Component*, which tells us for each pixel, to which degree it is a member of a large component.

$$\mathit{Large.Component}[f](\vec{x}) = \frac{A(C_{\vec{x}})}{A(f)} \tag{19}$$

3.7 Other Functions

We only described the functions which have been used for the example in this paper. Many relations known for crisp pixel sets can be extended to iconic fuzzy sets. A variety of other definitions for the geometry and topology of iconic fuzzy sets can be found in (Rosenfeld, 1979 and 1984) or in the references in (Pal and Rosenfeld 1988).

4. CONTROL

In this section we will discuss some issues which are essential for our image analysis system based on iconic fuzzy sets. *Normalization* is necessary to avoid problems with the minimum operation used to constrain fuzzy sets, which only allows a decrease of membership values (Rosenfeld et al, 1976). The *evaluation* of results is necessary for control, especially for the case of cyclic descriptions in the knowledge base, and as a termination condition. *Redundancy* in image descriptions can be used to combine evidence from different sources. Finally, we will present a *control algorithm*, which uses a heuristic to estimate the amount of information which can be gained by the application of a function to iconic fuzzy sets.

4.1 Normalization

As in other approaches (Dubois and Prade, 1988) we decide to use *normalized* fuzzy sets, which means that for each iconic fuzzy set, at least one pixel has the maximal membership value, 1.0 . This is justified, if each iconic fuzzy set represents an image structure which is known to exist in the image under consideration.

Each iconic fuzzy set which is generated by one of the functions which we defined in the last section, is subjected to normalization. This corresponds to "a non-monotonic change in the representation of our belief" (Dubois and Prade, 1988).

4.2 Evaluation

Although the representation in figure 2 does not cause difficult control problems, they will occur with knowledge bases which contain redundant information or even cycles. For that case, we need a measure to assess the quality of a result. Even for image analysis with fuzzy sets, the ultimate goal is to generate fuzzy sets with a low fuzziness, or, a high crispness. A measure for fuzziness has been defined by Kosko (1986):

$$\mathcal{F}(f) = \frac{\sum_{\vec{x}} f(\vec{x}) \cap \bar{f}(\vec{x})}{\sum_{\vec{x}} f(\vec{x}) \cup \bar{f}(\vec{x})} \tag{20}$$

where f is an iconic fuzzy set and $\bar{f}$ is its complement.

Another interesting question is, how the action of a particular operation on an iconic fuzzy set can be measured. If the action decreases the fuzziness of the set, we can say that we gained information. We can regard fuzziness as a measure of uncertainty. Following Shannon (1948), we can than say that the information we gained is equal to the decrease in fuzziness:

$$\mathcal{I} = -\Delta \mathcal{F} . \tag{21}$$

4.3 Redundancy

Very often, image objects can be described in several ways, which means that several sets of relations form unambiguous image object descriptions. An example is

```
( IS SKIN ( AND SOFT.TISSUES ( OUTSIDE SKULL ) ) )
( IS SKIN ( AND ( OUTSIDE SKULL ) ( INSIDE BACKGROUND ) ) ).
```

We represent redundant information with

```
( IS SKIN ( ANYOF ( AND SOFT.TISSUES ( OUTSIDE SKULL ) )
                  ( AND ( OUTSIDE SKULL ) ( INSIDE BACKGROUND ) ) ) ).
```

This means that *skin* can be computed, if either the iconic fuzzy sets for *soft tissues* and *skull* or those for *skull* and *background* have been generated before. If all fuzzy sets are known, and both skin_1 and skin_2,

$$\text{skin}_1 = \textit{And}(\text{soft.tissues}, \textit{Outside}(\text{skull}))$$

$$\text{skin}_2 = \textit{And}(\textit{Inside}(\text{background}), \textit{Outside}(\text{skull})) \tag{22}$$

are computed, a decision has to made, whether skin_1 or skin_2 is assigned to *skin*. We examined two possibilities: (1) choose either skin_1 or skin_2 based on their fuzziness values and (2) combine skin_1 and skin_2 by pointwise selection of that value, which minimizes the fuzziness in *skin*. Although the second possibility yields crisper results, the first alternative results in a robuster behaviour. Because the redundancy does not yield significantly better results for simple cases, we do not display an example.

4.4 The control algorithm

If redundancy is introduced in the knowledge representation, a control framework for the control of cycles and termination becomes necessary. The simple question is: which function should be executed next ? The control framework is based on the following data structure (see fig. 5 for the example of `( AND B C D )`):

Let us denote an iconic fuzzy set after the n-th iteration with f^n. Each iconic fuzzy set has a fuzziness value $\mathcal{F}$, which is initialized with 1 (maximally fuzzy) for f^0.

Each relation has an information bin I for each iconic fuzzy set which serves as an input to the related function. I is initialized with 0. Also, each relation has an associated number, I^*, which gives an estimate for the information which could be gained by execution of the related function. Currently, we set I^* to the maximum value of all information bins. This is of course a heuristic estimate, which could be replaced by e.g. the sum over all information bins.

If an iconic fuzzy set f^{n+1} is generated by a function, the decrease in fuzziness, $-\Delta\mathcal{F}$, is propagated to all relations, for which the set could serve as input. This value can be interpreted as the information of f^{n+1} with respect to f^n. The values of $-\Delta\mathcal{F}$ are *accumulated* in the information bins, and the I^*'s are updated.

The relation with the highest positive value of I^* is selected and the corresponding function is executed. A precondition (except for *Anyof*) is, that the fuzziness of all input fuzzy sets is < 1.

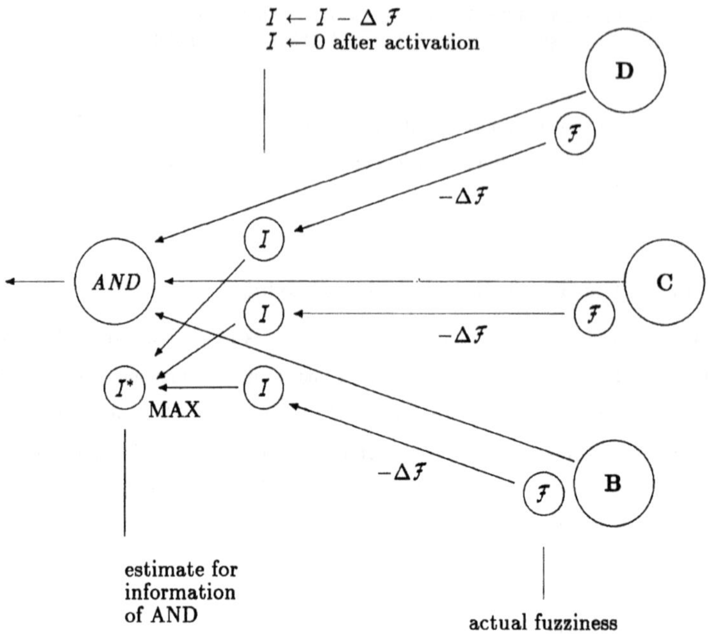

Figure 5: Control algorithm I^* for `( AND A B C )`

After execution, the information bins are cleared, and I^* is set to 0 as well. The execution of functions is terminated, when no I^* is positive, which means that no more information can be gained.

The algorithm can be changed to include the objective of efficiency (Cohen 1987) by e.g. multiplying the value for I^* with the estimated execution time of the function.

5. CONCLUSION

Figures 3 and 4 illustrate the results of the segmentation of our example image in figure 1. They also demonstrate the operation of the functions over iconic fuzzy sets which where defined in section 3.

Our present system is very well adaptable to other images, as all knowledge is explicitly represented. Of course, the system is not able to adapt to the ambiguous semantics of many natural language predicates; we discussed insidedness as an example. Therefore, the exact function definitions have to be kept in mind when designing a knowledge base.

Many further developments are possible: our system is strictly top-down, making explicit inferences from the image contents impossible. However, our functions can also be used to *test* the validity of relations between fuzzy regions, which is the basis for bottom-up approaches. Some of the functions are sensitive to noise; although this is compensated for by fuzzy values, multiresolution techniques (Pizer, 1988) may help in well-controlled smoothing of the input images. Fuzzy contours, the dual of fuzzy regions, can be based on gradient values (Menhardt and Imme, 1988).

The concept of iconic fuzzy sets has been proven to be a valuable tool for image segmentation. It becomes possible to avoid the critical region finding or edge detection preprocessing operations. In the entire segmentation process, all information is used on the full spatial resolution. Moreover, binary decisions are avoided, and all results are maintained on the basis of fuzzy membership values. This is essential in many tomographics medical imaging applications, where e.g. partial volume effects make the exact identification of edges strictly impossible.

REFERENCES

Ballard, D., and Brown, C. (1982). Computer Vision, Prentice-Hall, Englewoods Cliffs.

Cohen, P. (1987). The control of reasoning under uncertainty: A discussion of some programs, The Knowledge Engineering Review 2, pp. 6-25.

Dubois, D., and Prade, H. (1988). On the combination of uncertain or imprecise pieces of information in rule-based systems, International Journal of Approximate Reasoning 2.

Menhardt, W. (1988). Image Analysis using Iconic Fuzzy Sets. In: Proceedings of European Conference on Artificial Intelligence, Kodratoff, Y. (ed.), Pitman Publishing, London, pp. 672-674.

Menhardt, W. and Imme, M. (1988). Contour Detection using Fuzzy Sets. In: Signal Processing IV: Theories and Applications, Lacoume, J.L., Chehikian, A., Martin, N. and Malbos, J. (eds.), Elsevier Science Publishers, Amsterdam, pp. 1633-1636.

Kosko, B. (1986). Fuzzy Entropy and Conditioning, Information Sciences 40, pp. 165-174.

Pal, S.K. and Rosenfeld, A. (1988). Image Enhancement and Thresholding by Optimization of Fuzzy Compactness, Pattern Recognition Letters 7, pp. 77-86.

Pizer, S. (1988). Multi-resolution methods and the segmentation of medical images, in this volume. Rosenfeld, A., Hummel, R.A. and Zucker, S.W. (1976). Scene Labeling by Relaxation

Operations, IEEE Transactions on Systems, Man and Cybernetics 6, pp. 420-433.

Rosenfeld, A. (1979). Fuzzy Digital Topology, Information and Control 40, pp. 76-87.

Rosenfeld, A. (1984). The Fuzzy Geometry of Image Subsets, Pattern Recognition Letters 2, pp. 311-317.

Shannon, C.E. (1948). A Mathematical Theory of Communication, Bell System Technical Journal XXVII, pp. 379-423, 623-656.

Zadeh, L.A. (1965). Fuzzy Sets, Information and Control 8, pp. 338-353

Part 3 Applications

Section 3.3 Radiology

REVERSIBLE DATA COMPRESSION OF ANGIOGRAPHIC IMAGE SEQUENCES

Paul Roos

Fac. of Technical Mathematics and Informatics,
Delft University of Technology,
Delft, The Netherlands.

and

Max A. Viergever

Fac. of Medicine,
University of Utrecht,
Utrecht, The Netherlands.

ABSTRACT

We have investigated whether displacement estimation methods can be used to efficiently register time series of angiographic images for the sake of reversible data compression. Displacement estimation methods can be divided into two classes: pixel-recursive methods and block matching methods. The performance of interframe decorrelation techniques based on both classes of methods is compared to interframe compression without motion compensation, and to intraframe compression. The tentative conclusion is that registration of time series of 2D images is not advantageous for data compression purposes.

1. INTRODUCTION

In a previous study of reversible intraframe data compression methods for two-dimensional (2D) images, HINT (Hierarchical INTerpolation) emerged as the most efficient method of decorrelation (Roos et al, 1988). For time series of 2D images it is likely that more efficient decorrelation schemes can be obtained by utilizing the temporal correlation, for instance by decorrelating difference images or interpolated images rather than the original images of the series. Furthermore, image sequences may contain motion artifacts. By estimating the displacement between consecutive images these artifacts may be partially removed. This registration procedure will increase the correlation between the images of the sequences and hence may also be expected to increase the efficiency of the decorrelation schemes.

In this paper we study reversible interframe compression of image sequences, both without (Section 2) and with registration. The displacement estimation methods upon which the registration procedure is based can be divided into two classes, viz. pixel-recursive methods (Section 3) and block matching methods (Section 4). The two classes of estimation methods are compared with respect to decorrelation efficiency and complexity in Section 5. In Section 6 the results of applying various interframe decorrelation methods to two series of angiograms are presented and discussed.

2. INTERFRAME DECORRELATION WITHOUT MOTION COMPENSATION

A first and simple approach in coding a time series of images is to code the difference images. Thus instead of coding $I(\underline{x},t)$ $(t=0..T)$, the image $I(\underline{x},0)$ and the images

$$I(\underline{x},t+1)-I(\underline{x},t) \qquad (t=0..T-1) \tag{1}$$

are coded. This approach will be referred to as difference coding.

A second approach is to apply temporal interpolation. To obtain a fully reversible decorrelation technique a causal scheme must be applied. For instance, all odd images can be estimated by interpolation of the even images; thus instead of coding $I(\underline{x},t)$ $(t\ odd)$ the image

$$I(\underline{x},t)-\frac{1}{2}\cdot\left[I(\underline{x},t-1)+I(\underline{x},t+1)\right] \tag{2}$$

is coded. This technique can be refined by first estimating some even images from neighbouring even images. As an example, consider $I(\underline{x},t)$ where t=0..8. First $I(\underline{x},0)$ and $I(\underline{x},8)$ are coded. These images are used to estimate $I(\underline{x},4)$. Instead of coding $I(\underline{x},4)$ the image

$$I(\underline{x},4)-\frac{1}{2}\cdot\left[I(\underline{x},0)+I(\underline{x},8)\right] \tag{3}$$

is coded. $I(\underline{x},2)$ can be estimated using $I(\underline{x},0)$ and $I(\underline{x},4)$, and so on.

The efficiency of these two interframe decorrelation methods will be discussed in Section 6. The next three sections are devoted to displacement estimation methods underlying motion-estimated interframe decorrelation.

3. PIXEL-RECURSIVE MOTION ESTIMATION

Pixel-recursive motion estimation has originally been proposed by Netravali and Robbins (1983). In this method, for each pixel a displacement vector is calculated by a gradient search in the mask or preceding image, see Fig.1. The method assumes that there is a $\underline{D}$ for each pixel $\underline{x}$ such that

$$I(\underline{x},t) = I(\underline{x}+\underline{D},t-\tau). \qquad (4)$$

The Displaced Frame Difference (DFD) is defined by

$$DFD(\underline{x},\hat{\underline{D}}) = I(\underline{x},t) - I(\underline{x}+\hat{\underline{D}},t-\tau), \qquad (5)$$

where $\hat{\underline{D}}$ is an estimate of $\underline{D}$. There are various criteria by which $\hat{\underline{D}}$ can be estimated. The customary procedure is to minimize the squared value of the displaced frame difference iteratively:

$$\hat{\underline{D}}_{i+1} = \hat{\underline{D}}_i + \frac{1}{2}.\epsilon.\nabla_{\hat{\underline{D}}_i} \left[DFD(\underline{x},\hat{\underline{D}}_i)\right]^2, \qquad (6)$$

where ϵ is some constant. Many authors have paid attention to substituting the constant ϵ by a variable dependent on the local image statistics (see Musmann (1985) for a survey). This may influence the rate of convergence but not necessarily the quality of the estimation. Since we are interested in the efficiency of the decorrelation rather than computational speed, we used a small constant ϵ such that the obtained displacement vectors are accurate.

At this point the gradient search method is not yet reversible. In calculating the displacement vector, the value of pixel $\underline{x}$ at time t is used; this value has not yet been coded. To obtain a fully reversible decorrelation method a causal prediction window W is introduced, by which the vector $\underline{D}$ is estimated not from eq.(6), but from neighbouring displacement vectors:

$$\underline{\hat{D}}_{i+1} = \underline{\hat{D}}_i + \frac{1}{2}.\epsilon.\sum_{j\in M} W_j \left[\nabla_{\underline{\hat{D}}_i}\left[DFD(\underline{x}_j,\underline{\hat{D}}_i)\right]^2\right] \tag{7}$$

$$\sum_{j\in M} W_j = 1 \quad with \quad W_j \geqslant 0, \tag{8}$$

where M defines which pixels are included in the window, $\underline{x}_j$ are the neighbouring pixels of pixel $\underline{x}$. An example of a causal window is:

$$W = \begin{pmatrix} a & b \\ c & . \end{pmatrix} \tag{9}$$

where the dot represents the position of the pixel to be coded, and where the coefficients a, b and c may take any values satisfying eq.(8). The displacement vector $\underline{\hat{D}}$ which follows from eq.(7) is solely based on previously coded information. Note the analogy with (3D) differential pulse code modulation.

4. DISPLACEMENT ESTIMATION BY BLOCK MATCHING

In block matching, the image $I(\underline{x},t-\tau)$ is subdivided into (overlapping or non-overlapping) blocks. Each sub-image, denoted by $I_b(\underline{x},t-\tau)$, is matched with an equal-sized subimage $I_b(\underline{x},t)$ of the image $I(\underline{x},t)$ to be coded; a displacement vector $\underline{\hat{D}}$ is obtained by minimizing

$$d\left[I_b(\underline{x}+\underline{\hat{D}},t-\tau), I_b(\underline{x},t)\right] \tag{10}$$

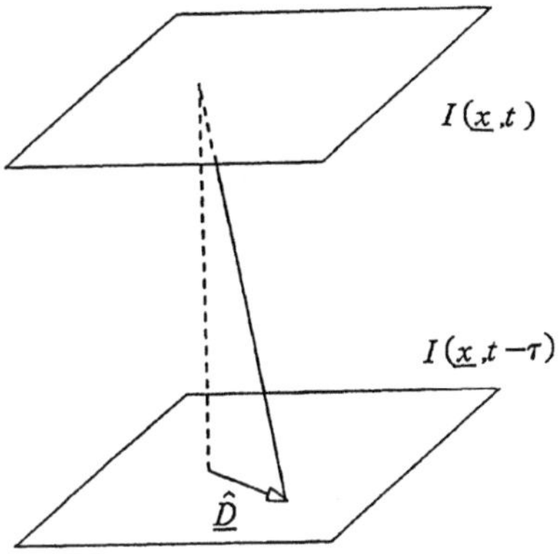

Fig.1. In pixel-recursive methods for each pixel $\underline{x}$ a displacement vector $\hat{\underline{D}}$ is obtained by minimizing the DFD squared.

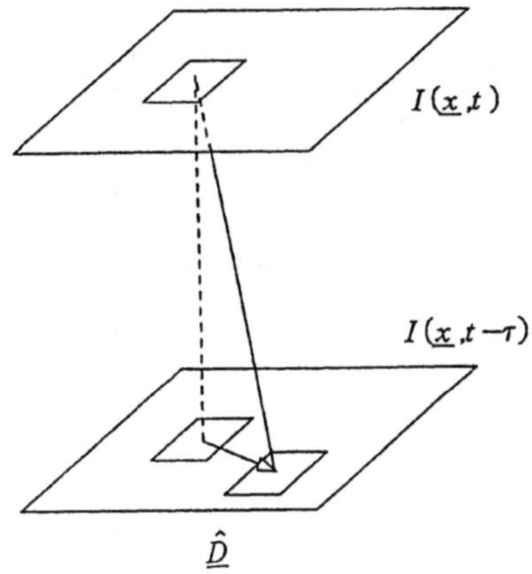

Fig.2. In block-matching methods for each image block I_b of the image at $t-\tau$, a displacement vector $\hat{\underline{D}}$ is obtained, such that the distance $d\left[I_b(\underline{x}+\underline{D},t-\tau), I_b(\underline{x},t)\right]$ is minimized.

where d is some distance measure (see Fig.2). Examples of image distance measures are Cross Correlation function (CCf), Sum of Absolute Values of the Differences (SAVD) and Discrete Sign Change (DSC) (Venot et al. 1984). We use the DSC criterion since Venot and Leclerc have shown that this criterion is more robust than the other distance measures mentioned.

A local optimum can be obtained by direct search techniques such as Hooke & Jeeves (Hooke and Jeeves, 1961), 2D-logarithmic search (Jain and Jain, 1981) and conjugate direction search (Srinivasan and Rao, 1984). We found that for all three search methods at least 30 percent of the obtained displacement vectors did not globally optimize the DSC distance measure. We therefore simply scanned all possibilities (within a sufficiently large window) to ensure the globality of the obtained optimum.

Having obtained a displacement vector for each block, a complete vector field (one displacement vector per pixel) is obtained by bilinear interpolation of the displacement vectors (Fitzpatrick and Leuze, 1987). The displacements of the four corners are specified as shown in Fig.3 by four vectors $d\,0$, $d\,1$, $d\,2$ and $d\,3$. The components of these vectors are denoted by the subindices x and y. The displacement vector $d(x,y)$ of pixel (x,y) within a block with size b is

$$d(x,y) = \frac{1}{b^2}\begin{pmatrix} d\,0_x & d\,1_x & d\,2_x & d\,3_x \\ d\,0_y & d\,1_y & d\,2_y & d\,3_y \end{pmatrix}\begin{pmatrix} (b-x)(b-y) \\ x(b-y) \\ (b-x)y \\ xy \end{pmatrix} \tag{11}$$

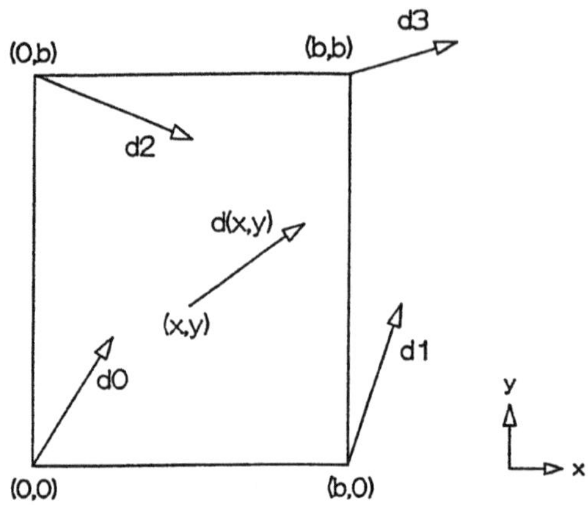

Fig.3 Bilinear interpolation scheme to calculate the displacement vector field in block matching.

Now for every pixel $\underline{x}$, a $\hat{\underline{D}}$ is obtained such that $I(\underline{x}+\hat{\underline{D}},t-\tau)$ approximates $I(\underline{x},t)$, analogously to the pixel-recursive method. However, apart from the difference image $I(\underline{x},t)-I(\underline{x}+\hat{\underline{D}},t-\tau)$ the block displacement vectors must be coded as well in order to obtain a fully reversible decorrelation technique.[1]

The segmentation into fixed blocks involves two parameters; firstly the size of the blocks and secondly the number of blocks. The choice of the size requires a compromise. A large block size yields a robust but global displacement vector, whereas a small block size leads to a local but noise-sensitive displacement vector. The number of blocks influences the reliability of the estimation of the bilinear interpolation. Consequently, application of this method to a particular class of time series of images requires optimization of both size and number of the blocks.

5. PIXEL-RECURSIVE VERSUS BLOCK MATCHING MOTION ESTIMATION

We simulated the two classes of motion estimation on a synthetic scene (computer generated), originally proposed by Netravali and Robbins (1980). The mask image $I(\underline{x},t-\tau)$ is described by

1. This additional storage may be acceptable here thanks to the relatively small number of blocks, as opposed to the corresponding storage in pixel-recursive estimation. A causal scheme similar to eq.(7)-(9) would be inappropriate here owing to the large distance between the blocks.

$$I(R)=100.e^{(-0.01R)}.\cos\left(\frac{2\pi R}{P}\right)+128, \qquad 0 \leqslant R \leqslant 60$$

$$P=\left(1-\frac{R}{60}\right)10+10 \tag{12}$$

where R is the radial distance from the centre of the circular symmetric pattern, see Fig.4. We considered two basic displacements; a uniform horizontal translation of 5 pixels (Fig.5) and a nonuniform dilation (expansion horizontally and compression vertically), see Fig.6.

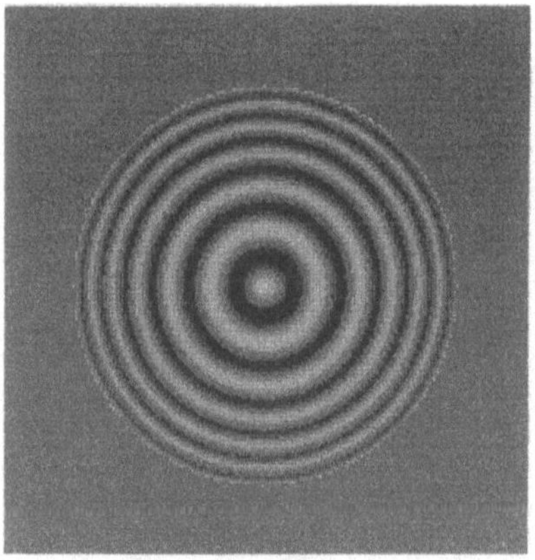

Fig.4 Undisturbed circular symmetric pattern.

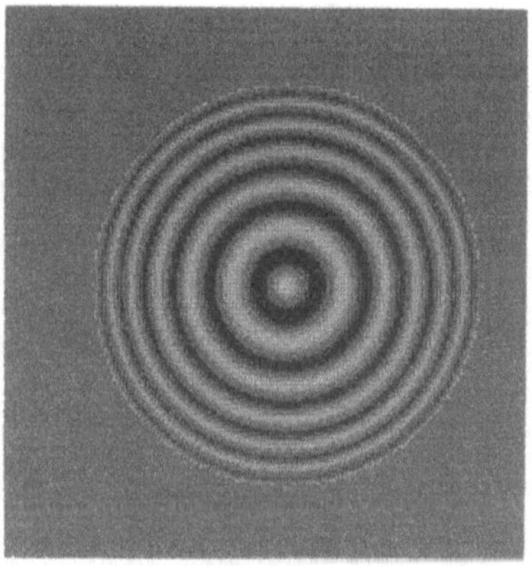

Fig.5 Uniformly translated pattern of Fig.4.

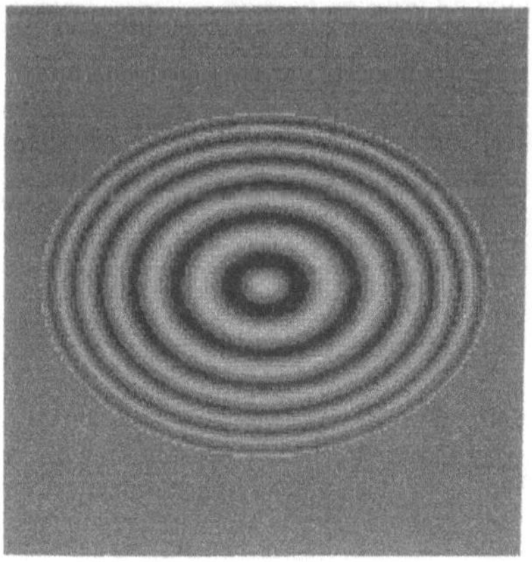

Fig.6 Nonuniformly dilated pattern of Fig.4.

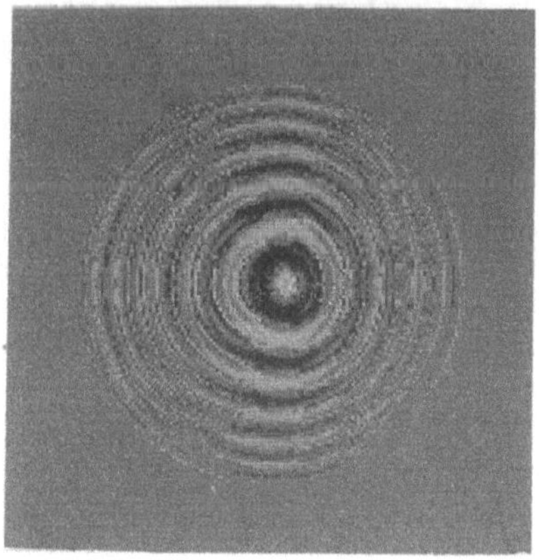

Fig.7 Pixel-recursive registered image of translated pattern.

Fig.7 and Fig.8 show the results of applying the pixel-recursive method, with $a=0, b=c=\frac{1}{2}$ in the window W of eq.(9). The images shown are the registered images $I(\underline{x}+\hat{\underline{D}},t-\tau)$, where $I(\underline{x},t-\tau)$ is the undisturbed pattern of Fig.4. In Fig.7 the registered translated pattern is presented, in Fig.8 the registered dilated pattern. From Fig.7 and 8 and from eq.(7) we conclude that the pixel-recursive method does not detect any motion in the homogeneous areas of the original image; the (spatial) gradient is zero. We furthermore observed that the gradient method has too weak a nature to accurately detect large displacements.

The uniform translation is detected perfectly by the block matching technique. Fig.9 shows the estimate of the dilated pattern by block matching registration. The block structure is visible in the contours of the ellipse.

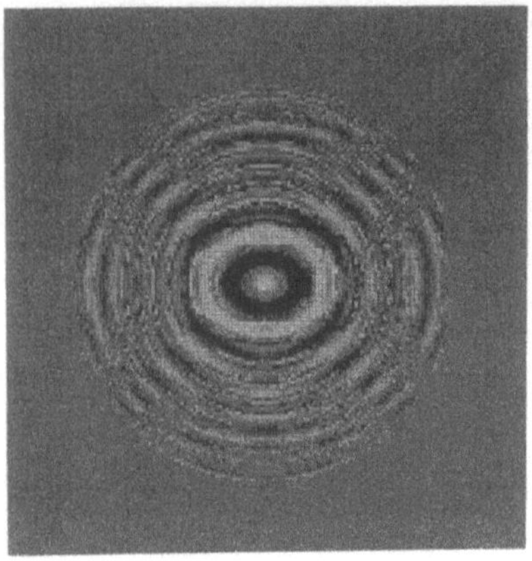

Fig.8 Pixel-recursive registered image of dilated pattern

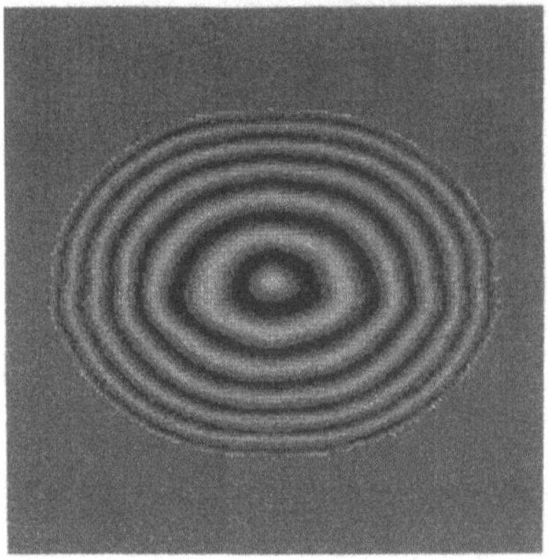

Fig.9 Block matching registered image of dilated pattern

6. INTERFRAME COMPRESSION OF ANGIOGRAPHIC IMAGES

We have applied the above mentioned techniques to two series of 9 angiograms. The first series (which we refer to as series A) is 512x512, 9 bits. The second series (series B) is 512x512, 8 bits.

In Table I the average entropies are shown for the interframe decorrelation methods without motion compensation as described in Section 2. Since the resulting images might still contain spatial correlation we have decorrelated the images further using HINT. The methods are compared to direct intraframe HINT decorrelation. The entropy of the original images is also included in table I.

TABLE I

Entropy (bits/pixel) of the original images, difference images and interpolated images, with and without post HINT decorrelation.

	Series A		Series B	
		Post HINT		Post HINT
Intraframe	7.57	2.69	6.24	2.50
Difference	3.75	2.59	3.55	2.67
Interpolation	4.14	2.50	3.79	2.53

From Table I we conclude that neither the difference nor the interpolation method decorrelates the images substantially better than intraframe HINT.

In pixel-recursive motion estimation the causal prediction window W was introduced to obtain a fully reversible decorrelation scheme. Table II gives the entropy of the motion compensated difference image for three sets of weight coefficients (eq.9).

TABLE II

Entropy (bits/pixel) of the pixel-recursive registered angiograpic images for various sets of coefficients in the causal prediction window.

	Series A		Series B	
		Post HINT		Post HINT
$a=0, b=1, c=0$	3.24	2.66	3.10	2.70
$a=0, b=½, c=½$	3.11	2.66	3.07	2.69
$a=0.26, b=0.37, c=0.37$	3.14	2.67	3.08	2.69

It is obvious from table II that image registration based on pixel-recursive motion estimation does not significantly, if at all, increase the decorrelation of the series of angiograms.

In block matching motion estimation two parameters have to be chosen: the number and size of the blocks. For the 512x512 angiographic images we have calculated the displacement vectors with an in between distance of 16 pixels. The DSC criterion is calculated over an area of 31x31 pixels. The results are given in Table III.

From Table III we conclude that the amount of decorrelation of the block matching technique is rather low in comparison to intraframe HINT and the simple difference and interpolation methods (see Table I). Thereby an additional storage of the displacement vectors (one per block)

TABLE III

Entropy (bits/pixel) for the block matching registered series of angiograms.

Series A		Series B	
	Post HINT		Post HINT
3.35	2.74	3.16	2.74

is required to obtain a fully reversible decorrelation method. We therefore must conclude that block matching is inefficient for reversible coding purposes of medical images.

In summary, none of the reversible interframe methods described in this paper decorrelates the sequences of angiographic images better than intraframe HINT. This is true even when the registration step is not stored, so that the reversibility only applies to the registered image sequences. The preliminary conclusion is that registration of time series of angiographic images is not advantageous for data compression purposes. Up to now motion estimation has only been applied in its differential form. The interpolation form of both classes of motion estimation still has to be examined but the results presented in this paper make it unlikely that this will be successful.

It appears that the temporal decorrelation destroys the spatial correlation of the images in the sequence to such an extent that further spatial decorrelation is hardly beneficial. An alternative procedure is to reverse the order of decorrelation, which was found to give better results for irreversible difference image compression (Moon et al, 1987). We are currently investigating the merits of this idea for reversible compression of image sequences.

ACKNOWLEDGEMENTS

Image Sequence A was provided by dr. H. Haas of Philips Medical Systems, Best, The Netherlands. Sequence B was provided by dr. H. Reiber, director of the laboratory of clinical and experimental image processing, Thoraxcenter of the Erasmus University and Academic Hospital Dijkzigt, Rotterdam, The Netherlands.

REFERENCES

Fitzpatrick, J.M. and Leuze, M.R. (1987). A Class of One-to-one Two Dimensional Transformations, Comp. Vision, Graphics, Im. Proc. 39, 369-382.

Hooke, R. and Jeeves, T.A. (1961). Direct Search Solution of Numerical and Statistical Problems, J.A.C.M. 8, pp. 535-552.

Jain, J.R. and Jain, A.K. (1981). Displacement Measurement and its Application in Interframe Image Coding, IEEE Trans. Commun. 29, 1799-1806.

Moon, J.H., Koh, J.S., Kim, S.D. and Kim, J. (1987). Spatial-Temporal Prediction Algorithm with Pel Recursive Motion Compensation, Globecom Tokyo, Conference Papers, 70-72.

Musmann, H.G., Pirsch, P. and Grallert, H.J. (1985). Advances in Picture Coding, Proc. IEEE. 73, 523-547.

Netravali, A.N. and Robbins, J.D. (1980). Motion-Compensated Coding: Some New Results, BSTJ 59, 1735-1745.

Robbins, J.D. and Netravali, A.N. (1983). Recursive Motion Compensation: A Review. In: Image Sequence Processing and Dynamic Scene Analysis, Huang, T.S. (ed.), Springer-Verlag, Berlin, 76-103.

Roos, P., Viergever, M.A., Dijke, M.C.A. van and Peters, J.H. (1988). Reversible Intraframe Compression of Medical Images, IEEE Trans. MI. 7, 328-336.

Srinivasan, R. and Rao, K.R. (1984). Predictive Coding Based on Efficient Motion Estimation, ICC'84 Conf. Rec., 521-526.

Venot, A., Lebruchec, J.F. and Roucayrol J.C. (1984). A New Class of Similarity Measures for Robust Image Registration, Comp. Vision, Graphics, Im. Proc. 28, 176-184.

THE MEASUREMENT OF ABSOLUTE LUMEN CROSS SECTIONAL AREA AND LUMEN GEOMETRY IN QUANTITATIVE ANGIOGRAPHY

Hawkes D.J.[1], Colchester A.C.F.[1],
de Belder M.A.[2], Norbury R.N.[2], Camm A.J.[2], Davies M.J.[2]

[1]UMDS, Guy's Hospital
[2]St. George's Hospital Medical School
United Kingdom

ABSTRACT

This paper describes our calibration technique to derive the absolute cross-sectional area of diseased blood vessel lumen from X-ray angiograms. This technique relies on the identification of a blood vessel of circular cross-section near to the diseased segment. If we assume that the concentration of contrast material and the densitometric response of the imaging system are constant over the diseased segment we may apply the calibration factor derived from adjacent circular segments to the diseased segment. Our interactive computer program, which incorporates this procedure, has been validated for segments of diseased coronary artery obtained at post mortem. Comparison with photographs of cross-sections of the arteries yielded a standard error of the estimate of 0.47mm^2 using cine fluorography with samples every 0.3mm along the artery's axis and film-screen radiography with samples every 0.1mm. The X-ray and geometric projections of the arterial cross-sections were consistent within the expected accuracy of the experiment. For the film-screen the precision of the densitometric profiles was approximately 0.1mm with systematic calibration errors of upto 0.2mm. There is considerable information in the X-ray projection data which could be used for the analysis of lumen shape, and hence the classification of vascular pathology.

1. INTRODUCTION

1.1 Pathological mechanisms in arterial disease

The development of a clinical syndrome in an individual with established atheroma occurs when blood flow past the atheromatous lesion is insufficient to meet metabolic demands or when plaque fissuring or rupture occurs. The latter leads to thrombus formation which may obstruct the vessel at that site or, together with other plaque components, may cause embolisation downstream. The development and progression of atheroma is complex and not fully understood although it is known that lesions vary in terms of the proportions of the major constituents: lipid, macrophages, smooth muscle cells and collagen (Davies, 1987).

The progression of disease is a gradual process in the majority of cases, with some acceleration occurring when endothelial damage over the plaque leads to platelet adherence, release of growth factors, smooth muscle proliferation and collagen synthesis. At some stage however, plaque fissuring and even rupture may occur with greater potential for platelet aggregation, thrombosis and haemorrhage into the plaque. Depending on the extent of these reactions in coronary arteries, unstable angina or myocardial infarction may result while stroke may result from disease in the vessels of the head and neck. In other cases the plaque may be positioned eccentrically and render that part of the vessel wall occupied by the plaque unable to respond to factors responsible for variation in coronary tone; if this segment is large enough, the normal constriction of the undiseased segment may lead to a critical narrowing of the lumen (Brown, 1978).

Non-circular and off-axis vessel lumen cross-section and the inter-reactions between the intimal surface of the lesion and the blood constituents are important features that relate to the development of a clinical syndrome. Plaque fissuring and the resulting complex lesion is probably the basis for the variation in angiographic appearances of arterial lesions. A few studies have suggested, moreover, that certain angiographic appearances predict rapid progression of the disease (Fuster et al, 1975; Alison et al, 1978; Singh, 1984; Ambrose et al, 1986).

Selection of the ideal therapeutic options for an individual patient should take account of these differing underlying processes. It has been established that post-mortem angiographic appearances of coronary artery lesions correlate with their histo-pathological features (Levin and Fallon, 1982; Thomas and Davis, 1985 a & b). The quantitative analysis of the images produced should enable an accurate prediction to be made of both the underlying disease process and the potential for further events.

1.2 Current status of quantitative arteriography

Conventionally, angiographic images are interpreted subjectively leading to considerable intra- and inter-observer variability. (Detre et al, 1975; Zir et al, 1979 and De Rouen et al, 1977). Recently there has been great interest in quantitative techniques in angiography (Brown et al, 1986). This work is mainly restricted to the estimation of relative stenosis from measurements of the width of the blood vessel lumen projected onto the imaging plane. This is often inadequate as the lumen may be non-circular at the site of the lesion; identification of an adjacent normal segment is required and this may be diseased; no account is taken of the length of arterial narrowing; and finally the haemodynamic consequences of a lesion are more closely related to the absolute not relative dimensions of the artery (Harrison et al, 1984).

Several groups have applied film or video densitometry to the measurement of blood vessel cross sectional area (Brown et al, 1986; Reiber et al, 1983; Harrison et al, 1984) and many commercial digital fluorographic systems now incorporate programs which permit the assessment of relative cross sectional area.

The aim of these techniques is to produce an image proportional to the mass of injected contrast material per unit area by careful densitometric calibration and either subtraction of a mask image, which removes detail due to superimposed anatomical structures, or by interpolating background image grey values from regions adjacent to the blood vessel image. The integral of a grey level profile across the image of the vessel is assumed to be proportional to the cross-sectional area of the lumen in a plane defined by the line of the profile in the imaging plane and the X-ray focal spot.

Unfortunately these systems all suffer from the following limitations:- the concentration of iodine is assumed to be constant along each vessel yet no check is incorporated within the program; the relationship between the image grey level and iodine mass integrated along the X-ray beam is assumed to be linear and spatially invariant and again this relationship is not checked; the whole length of the analysed blood vessel segment is assumed to have the same orientation relative to the X-ray beam; and relative not absolute lumen cross-sectional area is measured. In order to address these deficiences we undertook the work described below.

1.3 Definition of terms

We define the transverse density profile, TDP, as the plot of grey level intensity across the X-ray image in a direction perpendicular to the image of the blood vessel centre line or axis. We define the transverse chord profile or TCP, as the projection of the vessel lumen in the plane defined by the perpendicular to the axis of the image of the vessel segment and the X-ray focal spot.

The aims of this work were firstly to produce TDP's which were proportional to the TCP at a given point along the artery, secondly to derive the constant of proportionality, thirdly to integrate the TDP across the vessel image to produce a measure of absolute cross-sectional area and fourthly to produce a scheme for classifying the shape of the diseased blood vessel lumen.

2. Theory

For the conventional X-ray film-screen combination and cine fluorography, integration of X-ray intensity to produce an image occurs on the film, while image integration occurs at the TV camera pick-up tube in digital fluorography. For the large numbers of quanta which contribute to the final image (approximately 10^5 per mm^2), we may approximate this discrete process by continuous variables as follows. Neglecting the effect of X-ray scatter and the finite resolution of the imaging system we may define the differential detected energy spectrum $D_i(E)$, at pixel i and X-ray energy E by:-

$$D_i(E) = L(E)\, Q(E)\, X_i(E)\, U_i(E) \qquad (1)$$

where L(E) is the mean light energy detected at image integration per X-ray photon detected, Q(E) is the quantum detection efficiency for X-rays of energy E, $X_i(E)$ is the differential X-ray spectrum emitted from the X-ray source towards pixel i in the imaging plane (photons/keV) and $U_i(E)$ is the attenuation of the X-ray beam at energy E for the mask image along the X-ray path which maps to pixel i in the imaging plane.

The grey level in the mask image at pixel i is given by

$$G_i^m = g\{\int D_i(E)\, dE\} \tag{2}$$

where g{} is the function relating image intensitity at image formation to the calibrated grey-level. This function is determined by the transfer characteristics of the image intensifer, the sensitometric properties of the image recording device (film and/or TV camera) and the analogue to digital converter. Approximations to this function may be determined using a linear wedge of a suitable calibration material. This calibration material should have the same composition as the X-rayed tissue, but, in practice some other material such as copper or aluminium is usually used. The function, g{}, for a reasonably well filtered x-ray beam, is approximately proportional to the logarithm over the effective dynamic range of the imaging system.

The grey level in the opacified image at pixel i is given by

$$G_i^o = g\{\int D_i(E)\, \exp[(\mu_b - \mu_I)Z_i]dE\} \tag{3}$$

where μ_b is the X-ray linear attenuation coefficient of blood, μ_I is the X-ray linear attenuation coefficient of the blood and iodine mixture in the opacified blood vessel and Z_i is the projection of the blood vessel along the X-ray beam, i.e. the thickness of the vessel. To a good approximation, after calibration with a wedge, we may express the difference between G_i^o and G_i^m, the background-subtracted image grey value G_i^s as

$$G_i^s = k_i c Z_i \tag{4}$$

where c is the concentration of iodine in the blood vessel. With background subtraction it is assumed that interpolation between adjacent background grey values is a good approximation to unopacified image grey values at position i. This will only apply if no high contrast features, such as bone-soft tissue interfaces, are present. For an ideal imaging system the factor k_i will be independent of pixel position. In practice, however, the factor k_i will vary due to the effects of beam hardening, the contribution of X-ray scatter and spatial non uniformities in the imaging device. For small values of Z_i, such that $(\mu_b - \mu_I)Z_i \ll 1$, the factor k_i will be independent of Z_i, but for larger values of Z_i beam hardening effects will lead to some dependence of k_i on Z_i. In an angiogram we generally do not know the factor k_ic and therefore cannot retrieve Z_i directly.

Consider a blood vessel whose axis is orientated at an angle θ to the X-ray beam. Summation of the function G_i^S, the transverse density profile (TDP) perpendicular to the vessel axis, yields the TDP area, a. The true cross sectional area A is given by

$$A = a\ ((\cos\theta)/m)\ (1/k_ic) \tag{5}$$

where m is the geometric magnification factor relating pixel size to object size. The factor k_i is assumed to be independent of i across the image of the blood vessel. This relationship is independent of the shape of the cross-section of the blood vessel. In a vascular tree of unknown configuration the magnification factor m and the vessel orientation could be determined from the 3D reconstruction of the vascular tree (Hawkes et al, 1987).

Although there are several possible strategies to derive k_i at each point on the image, the iodine concentration, c, is generally not known and hence derivation of absolute cross-sectional area has not been possible.

We propose here a scheme to derive the product k_ic which will allow us to derive the absolute TCP from the TDP and hence calculate the absolute cross-sectional area along the vessel segment. Our technique relies on the identification of a segment of blood vessel which is assumed to have a circular cross-section. This model of a circular cross-section is then used to derive the product k_ic. If the factor k_ic does not vary along the section of the vessel then the absolute cross-sectional area along that section may be derived, regardless of whether the blood vessel is circular along its full length.

The image grey-value at each point across the TDP is given by Eq.(4). For a straight segment of blood vessel of circular cross section the maximum value of the TDP, h, is given by

$$h = k_icH/\cos\theta \tag{6}$$

where H is the diameter of the blood vessel and the area A is given by

$$A = \pi \ (H/2)^2 \qquad (7)$$

As proposed by Colchester (1985) the combination of Eqs. (5), (6) and (7) yields an expression

$$k_{\perp}c = (\pi/4)\ (h^2/a)\ m \cos\theta \qquad (8)$$

The height, h, and area, a, are calculated from the TDP and in a vascular tree of known geometry we may deduce the orientation, θ, and magnification factor, m, using our technique of 3D calibration. Hence we may deduce $k_{\perp}c$ for a vessel whose cross-section is circular. For such a section, this allows us to calculate the true cross-sectional area and the transverse chord profile.

Similar algebraic steps lead to the following interrelated expressions for the factor $k_{\perp}c$:-

$$k_{\perp}c = (4/\pi)\ (a/w^2)\ m\cos\theta \qquad (9)$$

$$k_{\perp}c = (h/w)\ m\cos\theta. \qquad (10)$$

Each of these expressions could be used to derive the absolute cross sectional area. Eq.(8) was selected in our experimental work as the height and area measurements of the TDP are insensitive to the choice of edge detection algorithm.

Although $k_{\perp}$ can vary across the image and c does not necessarily remain constant from one vessel to another, $k_{\perp}c$ is likely to remain constant along a single short segment of blood vessel containing well mixed contrast material. This leads to the important principle that vessel segments with a circular cross section which are adjacent to diseased segments can be used for calibration purposes to derive $k_{\perp}c$. This value and $(\cos\theta)/m$ are then used to derive the absolute cross sectional area using Eq.(5). The absolute TCP is derived using Eq.(4). For data acquired 'in vivo' we propose to incorporate the additional check that the value of $k_{\perp}c$ must be the same distal and proximal to a diseased segment for extrapolation to be justified. We have demonstrated this procedure with the experiments described below.

3. Method

Thirteen diseased segments of coronary artery were excised at post-mortem, perfusion fixed, filled with a gel containing 5% barium sulphate and set in wax blocks. X-ray films were taken of the wax blocks in two directions orthogonal to the vessel axis using a Du Pont Hi-Plus screen exposed at 90kVp. The factor cos was unity and the magnification factor m was derived from the image of a 2 cm copper disc placed on the radiograph. The films were digitised to give approximately 10 pixels per mm. The grey level response was linearised using a look-up table of grey values derived from an image of a copper wedge.

The TDP and its area were calculated using an interactive program that we have developed. The program operated as follows. The mask and opacified images were subtracted after re-registration, if necessary. A region was marked, using the mouse, to include the blood vessel under consideration but to exclude artefact, crossing vessels, nearby vessels and bifurcating vessels. This was to ensure that the region used to calculate the residual background contribution contained only valid data. An approximate vessel width was entered and a few points along the approximate vessel centre line were marked. The system then joined these points with a straight line, constructed perpendiculars at successive points along the straight line and computed the centre of gravity of the grey level distribution. The position of the vessel centre line was defined as the locus of this centre of gravity. The centre line locations were smoothed and the image was sampled along perpendiculars to the new centre line to yield the transverse density profile (TDP).

The edge points of the blood vessel were then calculated. A second order polynomial was fitted successively to 3, 5 or 7 points of the TDP and the first and second differential of the fitted function was computed for each point on the TDP. The edge was identified as a position in sub-pixel ($^1/_{10}$th of a pixel) coordinates where the sum of the moduli of the first and second differentials was a maximum, provided that the edge points were not both on the same side of the centre line and were not at the end of the profiles. The edge point locations were averaged by median filtering which removed individual erroneous points. The width of the vessel was stored as the real distance between the left and right edge points. Background values were subtracted by linear interpolation of the average grey value of up to 4 pixels starting 3 pixels outside the edge point along the perpendicular of the vessel centre lines. Invalid data was excluded from this calculation. If there were no background points of valid data then the previous background value was taken. The sum of the TDP was calculated to give the relative cross sectional area and stored. The area, maximum height, width, area divided by width squared, maximum height divided by width and maximum height squared divided by area were calculated for each point and stored. Graphical displays were generated of each parameter as a function of vessel length. A cursor placed on the graph and on the vessel centre line permitted visual registration of measurements with the blood vessel. Interactive interrogation of the graphs permitted the extraction of numerical data at points averaged over regions on the graph and computation of area between a hand drawn line and the graph. Areas of invalid data could be marked.

The true area of the vessel lumen was found for corresponding points by cutting the wax block containing the blood vessel segments with a microtome and photographing the exposed section with mm scales indicating the horizontal and vertical dimensions. Examples of photographs of exposed sections are shown on Figs. 4(c) to 8(c). A simple planimetry program was used to measure the area and calculate the TCP in the two orthogonal directions. Care was taken to ensure accurate registration between the TDP and the corresponding section along the long axis of the vessel.

We repeated this experiment using cine fluorography for seven of the diseased segments of coronary artery. The wax blocks were positioned on top of a skull embedded in plastic in order to provide a realistic scatter flux. Cine films were taken with a Siemens Angioskop cine x-ray system. These films were digitised and the grey level response was calibrated with a copper wedge. The magnification factor was determined using a cine image of a 1 cm grid placed on the input face of the image intensifier. The digitised films had a pixel size of approximately 3.5 pixels per mm.

Plots of the TCP registered with the corresponding TDP were obtained. X-axis scaling was derived from the magnification factor m. X-axis registration was achieved by registering to the centre of gravity of each profile. In this experiment vertical scaling was achieved by identifying by eye a circular section of each segment, finding the factor $k_{\perp}c$ using Eq.(8) and rescaling the image grey values $G^S_{\perp}$ to yield $Z_{\perp}$ using Eq.(4).

The aim of the present experiments was to validate the derivation of absolute cross sectional area and the TCP of diseased segments using values of $k_{\perp}c$ extrapolated from neighbouring segments with circular cross sections. The circular cross-sections used for calibration were identified from the cut sections by eye. We did not evaluate our methods for segments with circular cross sections from the image data alone.

4. Results

Correlation of TDP area derived from the film screen combination with planimetry area gave a correlation coefficient of 0.96 for 118 measurements from 13 segments of artery. These data are plotted on Fig. 1. From the same data we found the factor $k_{\perp}c$ from a circular section of each segment and used Eq.(5) to deduce absolute cross-sectional area. Linear regression yielded a slope of 1.126 when comparing planimetric and densitometric cross-sectional areas with a correlation coefficient of 0.97 and a standard error of the estimate (SEE) of 0.47mm² for 84 measurements from 10 segments of artery. These data are plotted on Fig. 2.

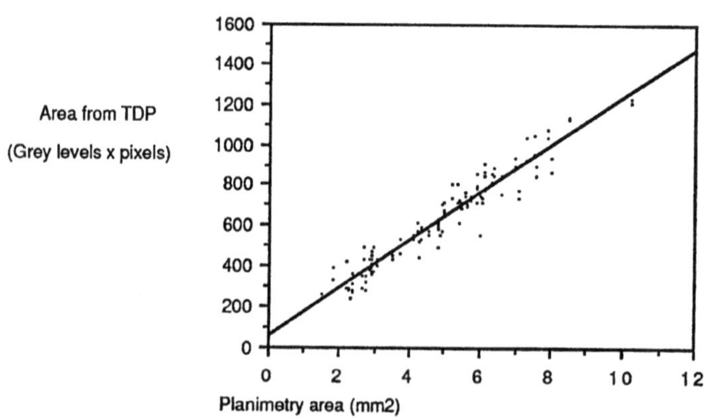

Fig.1 Plot of the area of the TDPs derived from the film-screen combination versus planimetry area for the 118 samples from 13 segments of artery.

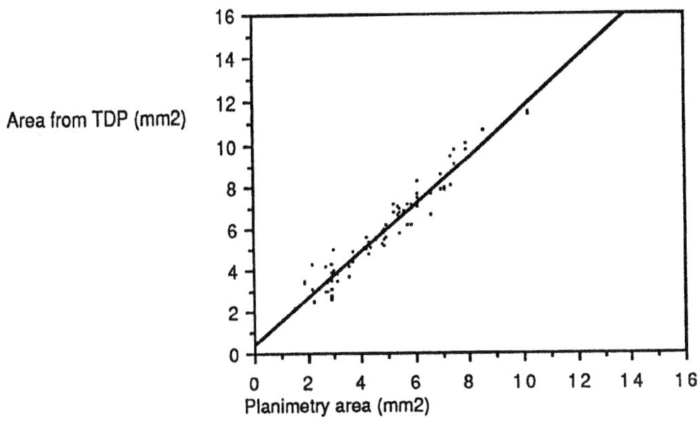

Fig.2 Plot of absolute cross-sectional area derived from the screen-film combination versus planimetry area.

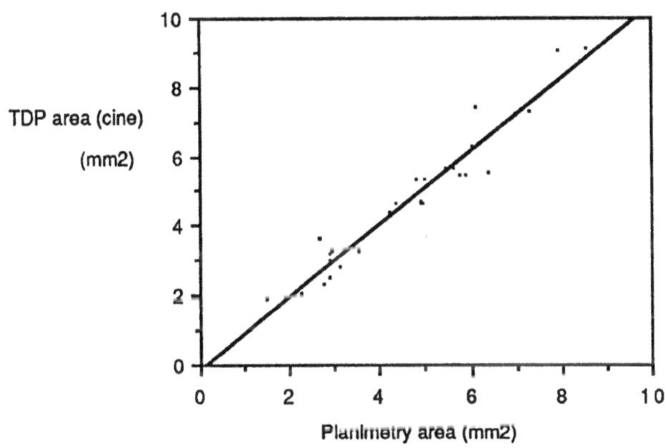

Fig.3 Plot of 29 measurements of the absolute cross-sectional area derived from cine-film versus planimetric area.

Correlation of TDP area derived from the cine films with planimetry yielded a correlation coefficient of 0.966 with an SEE of .47mm^2 for 29 measurements from 7 segments of artery. Determining $k_{\perp}c$ from one circular segment yielded a slope of 1.053 when plotting densitometric area against planimetric area. These data are plotted on Fig. 3.

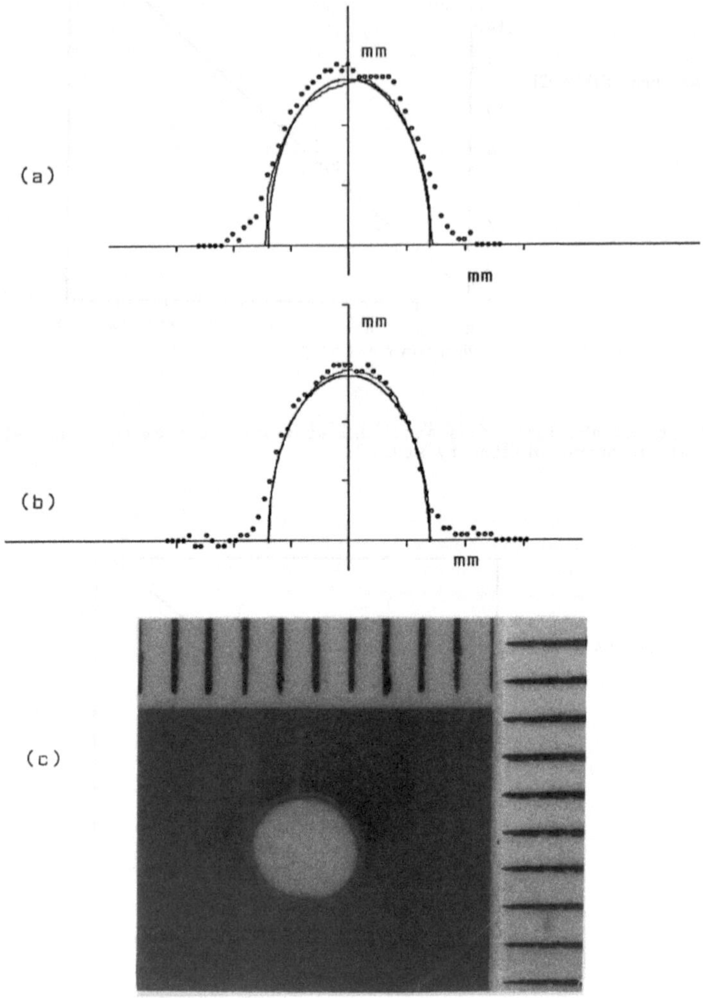

Fig.4 Plots of the TCP derived from the densitometric data (•••••) superimposed on the TCP derived from planimetric data (—) for two orthogonal views (a) and (b). Also plotted is the elliptical profile that would have been obtained for a circular section. A photograph of the corresponding cut section is shown, (c), demonstrating a near normal vessel segment.

Comparisons of the TCP derived from the screen-film data with the TCP derived from the planimetric data are shown on Figs. 4 to 8. Results for two orthogonal views are shown on (a) and (b) while photographs of the cut sections are shown for comparison on (c). There is excellent

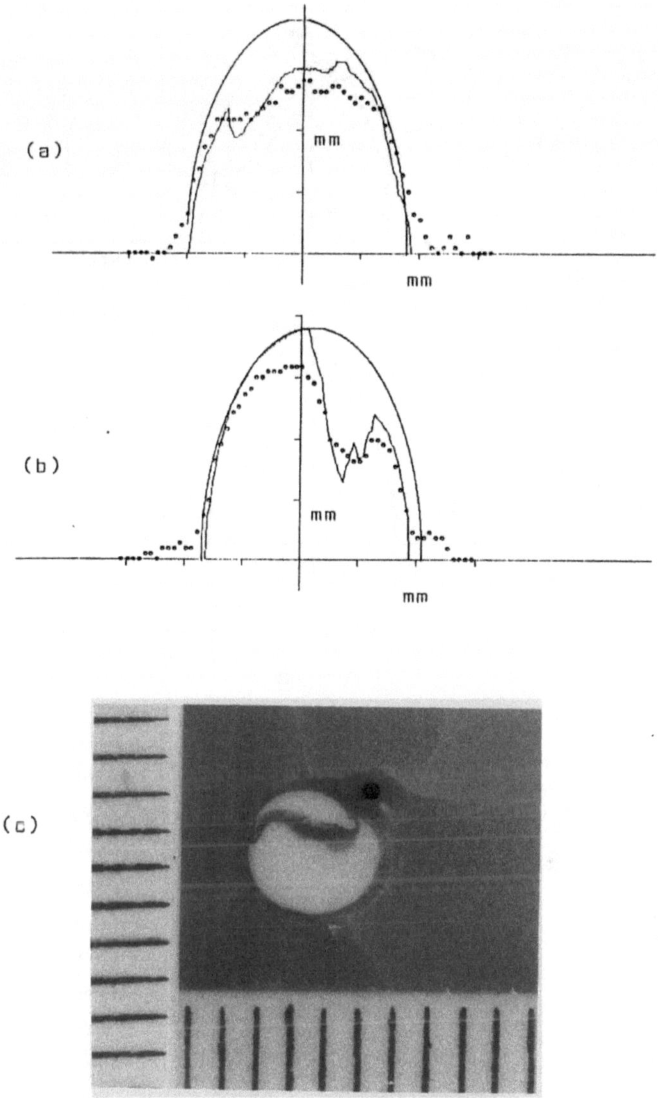

Fig.5 As in Fig. 4 but showing results for an intimal dissection.

correspondence between the two plots. The small errors are due to the addition of noise and the limited resolution of the X-ray imaging systems. Estimation of quantum noise at the exposure used in these films corresponds to an error in $Z_{\perp}$ of about 0.2mm. A full width half maximum of the line

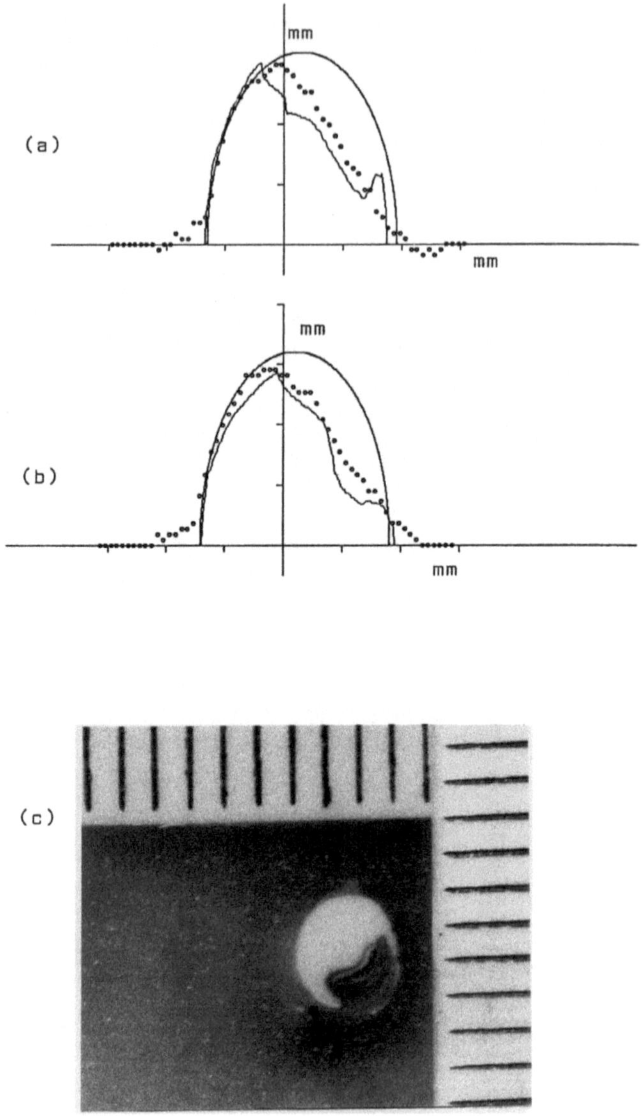

Fig.6 As in Fig. 4 but showing results for a crescentic shaped lumen 4mm further into the same lesion as that shown on Fig.5.

spread function of 0.1 mm would account for the blurring at the edges of the profile. Note that the prominent features of the planimetric profile are faithfully reproduced by the densitometric profile. Also plotted is the elliptical profile that would have been obtained by assuming that the

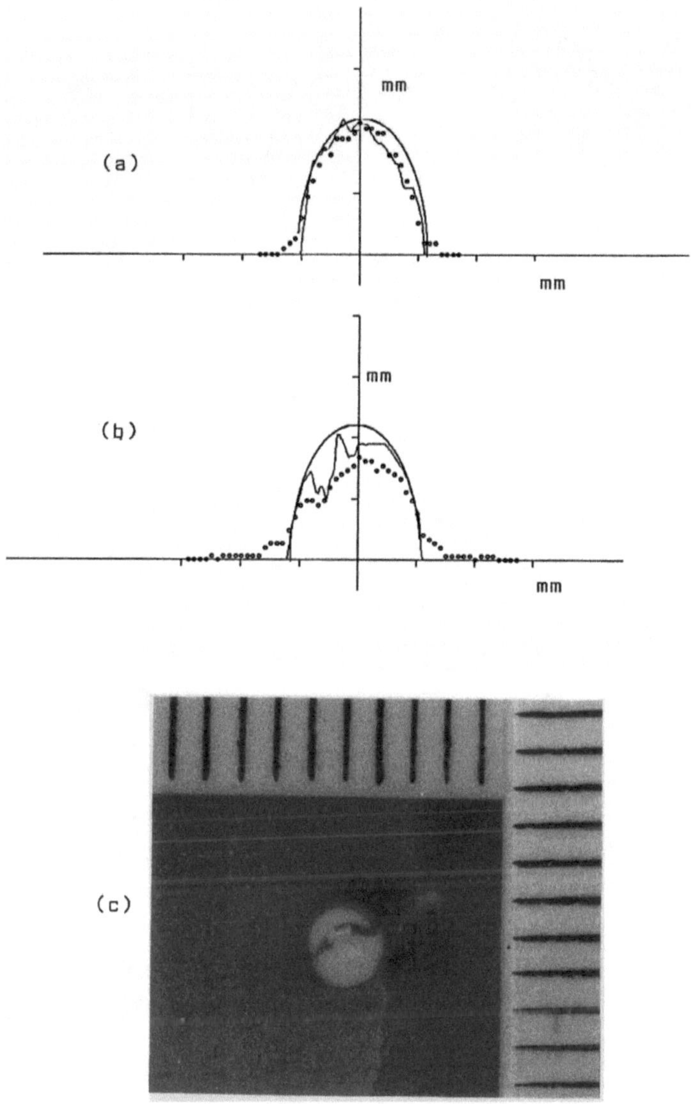

Fig.7 As in Fig.4 but showing results for a smaller diseased vessel segment.

cross-section used to derive the factor $k_{\perp}c$ was circular. The lateral position of this profile was judged by eye.

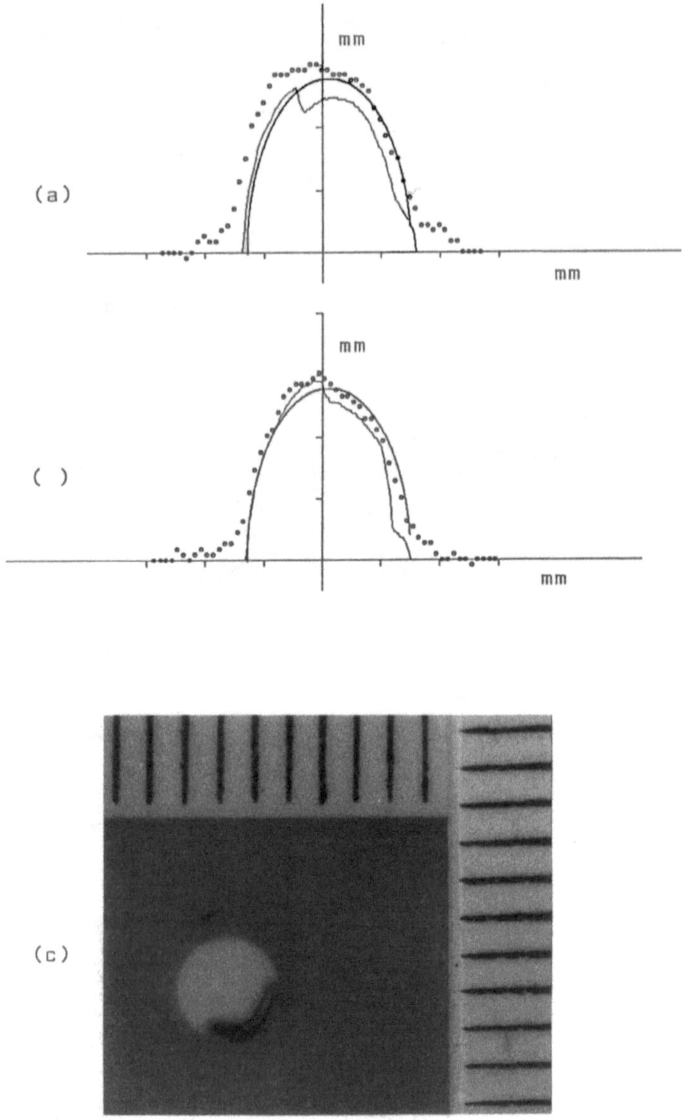

Fig.8 As in Fig. 4 but showing results for a vessel segment with only slight disease.

The section on Fig. 4 is a normal arterial segment of 2.8 mm diameter. The section on Fig. 5 shows an intimal dissection, Fig. 6 shows a section 4 mm further into the same lesion showing a crescentic shaped lumen caused by atheroma. Fig. 7 shows the results for a smaller vessel of 2.2 mm diameter.

Fig. 8 demonstrates a slight discrepancy in the densitometric calibration, probably due to slight shading across the film. Note however, that the shape of the TDP corresponds very closely to that of the TCP.

5. Discussion and Conclusions

In the experiment outlined above we were able to derive absolute cross-sectional area within the expected precision of the experiment. With both cine fluorography and screen-film we were able to measure lumen cross-sectional area with a precision of about 0.5 mm^2 at intervals of 0.1 mm for screen film and 0.3 mm for cine fluorography.

In vivo, the only data on vessel circularity is the image of the lumen. Criteria for deciding whether a blood vessel is circular include the identification of a blood vessel outline in the image as radiographically normal, i.e. having a smooth edge, uniform opacification etc.; the goodness of fit of the TDP to a semi-ellipse; and the constancy of values of h^2/a, a/w^2 or h/w versus position along the vessel axis. Assuming k_1 and c are constant, these factors will always be constant for segments with circular cross-section. Any deviations will represent deviations from a circle. As blood vessel lumens are usually circular identification of candidate circular regions close to a diseased segment should be possible remembering that our criterion is not normal vessel but a circular lumen. Even when the lumen is non-circular the values of h^2/a, a/w^2 and h/w are likely to be distributed around the values appropriate to a circle.

Variations in k_1, due to beam hardening, X-ray scatter and shading across the image, will introduce errors and we are investigating improved calibration procedures to overcome these errors. We are also developing a classification scheme to predict vascular pathology in which the lumen is modelled by a series of elliptical cones. The position and shape of the cones is derived by matching chord profiles of the cones to the TDPs. This description will be used to classify the diseased segment according to a limited range of vascular pathologies and normal vessel lumen shapes. As a first stage in this work we have shown with the results presented in this paper that we can generate the TCP from the TDP and that the TDP of normal vessel segments is very close to semi-elliptical.

ACKNOWLEDGEMENTS

We gratefully acknowledge the Leverhulme Trust for supporting DJH, the Jensen Trust for supporting RHN and the British Heart Foundation. Thanks are due to Marion Blagg for typing the manuscript, Mrs. Sue Williams of the Department of Histopatholgy at St. George's Hospital for assistance in preparing the arterial samples and the radiographic staff for assistance in obtaining the radiographs.

REFERENCES

Alison, H.W., Russell, R.O.Jr, Mantle, J.A., Kouchoukas, N.T. Moraski, R.E. and Rackley,C.E. (1978). Coronary anatomy and arteriography in patients with unstable angina pectoris, Am. J. Cardiol. 41, pp. 204-209.

Ambrose, J.A., Winters, S.L, Arora, R.R., Eng, A., Riccio, A., Gorlin, R. and Fuster, V. (1986). Angiographic evolution of coronary artery morphology in unstable angina, JACC. 7, pp. 472-478.

Brown, B.G. (1978). Coronary vasospasm. Observations linking the clinical spectrum of ischaemic heart disease to the dynamic pathology of coronary atherosclerosis. Arch. Intern. Med. 141, pp.716-722.

Brown, B.G., Bolson, E.L., and Dodge, H.T. (1986). Quantitative computer techniques for analysing coronary arteriograms. Prog. Cardiovasc. Dis. 6, pp. 403-418.

Colchester, A.C.F (1985). The effect of changing $PaCO_2$ on Cerebral artery calibre estimated by a new technique of dynamic quantitative digital angiography. Ph.D. Thesis, University of London.

Davies, M.J. (1987). Pathology of ischaemic heart disease. In: Current status of clinical cardiology. Ischaemic Heart Disease, Fox, K.M (ed.), MTP Press Ltd., pp.33-68.

De Rouen, T.A., Murray, J.A., and Owen W. (1977). Variability in the analysis of coronary arteriograms. Circulation 55, pp. 324-328.

Detre, K.M, Wright, E., Murphy, M.L., and Takaro, T. (1975). Observer agreement in evaluating coronary angiograms. Circulation 52, pp.979-983.

Fuster, V., Frye, R.L., Connolly, D.C., Davidson, M.A, Elveback, L.R., and Kurland, C.T. (1975). Arteriographic pattern early in the onset of the coronary symptoms. Br. Heart Journal 37, pp.1250-1255.

Harrison, D.G., White, C.W., Hiratzka, L.F., Doty, D.B., Barnes, D.H., Eastham, C.L. ad Marcus, M.L. (1984). The value of lesion cross-sectional area determined by quantitative coronary arteriography in assessing the physiological significance of proximal left anterior descending coronary arterial stenosis. Circulation 69, pp. 1111-1119.

Hawkes, D.J., Mol, C.B., and Colchester, A.C.F. (1987). The accurate 3D reconstruction of the geometric configuration of vascular trees from X-ray recordings. In: Physics and engineering of medical imaging. NATO ASI. Guzzardi (ed.), Martinus Nijhoff, Holland, pp. 250-256.

Levin, D.C., and Fallon, J.T. (1982). Significance of the angiographic morphology of localised coronary artery stenosis: histopathological correlations. Circulation 66, pp. 316-320.

Reiber, J.H.C., Gerbrands, J.J., and Troost, G.J. (1983). Transfer functions of the X-ray-cine-video chain applied to digital processing of coronary cine-angiograms. In: Digital imaging in cardiovascular radiology. Heintzen and Brennecke (eds.), Georg Thieme Verlag, Berlin, pp. 89-103.

Singh,R.N. (1984). Progression of coronary atherosclerosis. Clues to pathogenesis from serial coronary arteriography. Br. Heart J. 52, pp. 451-461.

Thomas, A.C., and Davies, M.J. (1985a). The demonstration of cardiac pathology using perfusion fixation. Histopathology 9, pp. 5-19.

Thomas, A.C., and Davies, M.J. (1985b). Post mortem investigation and quantification of coronary artery disease. Histopathology 9, pp. 959-976.

Zir, L.M., Miller, S.W., and Dinsmore, Gilbert, J.P. and Hartbourne, J.W. (1979). Interobserver variability in coronary angiography. Circulation 53, pp.627-632.

MULTIPLE SOURCE DATA FUSION IN BLOOD VESSEL IMAGING

G.P. Robinson & A.C.F. Colchester

Guys' Hospital London
United Kingdom

ABSTRACT

We have developed a processing module which can accept data from a number of different sources about a particular vessel segment and can reconcile and combine (fuse) this data. It can be used to test the hypothesis that two or more vessel segments from different sources are the same and combine the data if they are compatible. Each source provides a description in terms of attributes distributed along the segment. Each attribute is expressed in terms of three probability density functions, one for the attribute position, one for its longitudinal extent, and one for the type of property being measured (for example cross-sectional area or width). The reconciliation process depends on reasoning about evidence against certain property values being correct, as well as evidence in favour.

1. INTRODUCTION

When asked to make qualitative judgements about the condition of a patient, clinicians and radiologists are often presented with a great deal of (primarily) visual information from several different sources. Examples of sources include various types of angiograms, MRI scans, CT scans, and ultrasound. A computerised system which systematically combined data obtained from multiple image sources would be a powerful clinical tool which could be used in the generation of synthetic "summary" images showing salient features, and improving the accuracy and reliability of quantitative measurements.

To simplify the problems associated with data fusion, we have initially decided to concentrate on a particular level of detail in the analysis of vessel images. Nevertheless, we believe that the concepts apply at different levels of detail and to different types of image.

The combination of data from multiple sources is difficult because the data may vary greatly in the sampling locations, sampling density, and also in the confidences associated in data from different sources. For example, suppose that from two independent sources, there was data about the width of a blood vessel mid-way between two well defined branches. Source 1 states that the width is 2mm at this point, and source 2 states that the width is 1.5mm. Is it fair to say that these two measures are compatible ? If they are compatible, what is the best estimate of the true width. These questions can not be answered satisfactorily without further information. Firstly, information about the reliability is required. In addition, we require information about the exact location of the property value along the vessel segment. This location will in fact be a finite length of vessel. We have chosen to standardise our representation of the location property value measurements as two components in the input data: the central position of the measurement and the longitudinal extent of the vessel to which it applies. As with the property value itself, both the position and extent values need to be supplemented by information about their reliability. We do this by expressing the property, position and extent values as probability distributions, without any prior information about the shape of these distributions.

The organisation of the remainder of this paper is as follows. We will begin by attempting to put this work into context. This will be followed by a full description of the input to the module and the processing steps that are being carried out. A worked example will be given with graphical output showing the results of each stage. We will then briefly examine some other methods of handling uncertainty, and we will conclude with a discussion of problems and future work.

2. THE CONTEXT IN WHICH THE MODULE WILL OPERATE

We are currently developing one relatively small "module" which in turn will form part of a large artificial intelligence system for the analysis and interpretation of medical images. We envisage within this system a representation scheme for objects in the form of a connected graph as shown in Fig. 1.

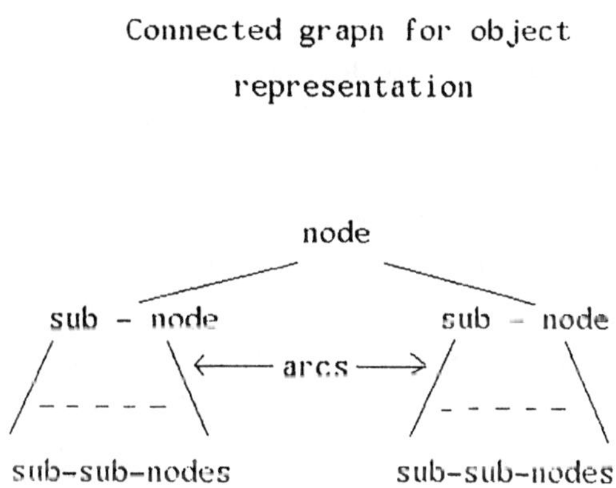

Fig. 1. Object representation as a connected graph

Within this graph representation, both the nodes and arcs can have their own associated data structure, thus allowing the system to represent a number of concepts including the physical position of objects relative to other objects, and the "pathway" between these objects.

This scheme maps onto a description of vessel trees in the following way. The nodes within the the graph would represent the branch points within the arterial tree, and these would have an associated data structure describing the number of branches originating at that point, the angles between them, their initial calibre etc. The arcs would be used to describe the segments of vessels connecting the two branch points described by the nodes at each end of the arc. For this representation there would be two types of data fusion: fusion of data about branch points (the nodes within the graph) and fusion of data about vessel segments (the arcs within the graph). In the present paper we are tackling only the arc vessel segment fusion task. There would be connection graphs generated from each source of input data. Structural models would also be stored as graphs and would be handled in the same way as current instance data.

The purpose of the module is to derive the best estimate, or instantiated model, of the longitudinal structure of a particular vessel segment. It makes explicit which parts of the segment are well specified, which are poorly specified, and which have conflicting data describing them. The output required of the module is a list of property values for each position along the vessel segment. Each value will have an associated confidence value (in the form of a probability). Low confidence in the output implies uncertain input data and/or conflict between input data. High confidence implies both high confidence in the data AND good agreement between competing input data.

We make important assumptions about prior processing before data is passed to the vessel segment fusion module. Firstly, for each source of data, it is assumed that the vessel tree has been identified and sub-divided into vessel segments, a segment being a length of vessel between two major branch points. Properties of each vessel segment have been sampled at various positions along the length. Secondly, the correspondence between vessel segments from different sources has been established (or at least an initial hypothesis of correspondence has been generated). Thirdly, the correspondence between co-ordinate systems for different sources has been established (or at least an initial hypothesis of correspondence has been generated). For example, for position values this implies that the reference point (start or end of the vessel segment) and the type of scale (fraction of segment length, mm etc.) are the same. The module operates on a particular hypothetical real world vessel segment, and is not concerned with analysing branch point configurations. Fig. 2 shows the context within which the module is being developed.

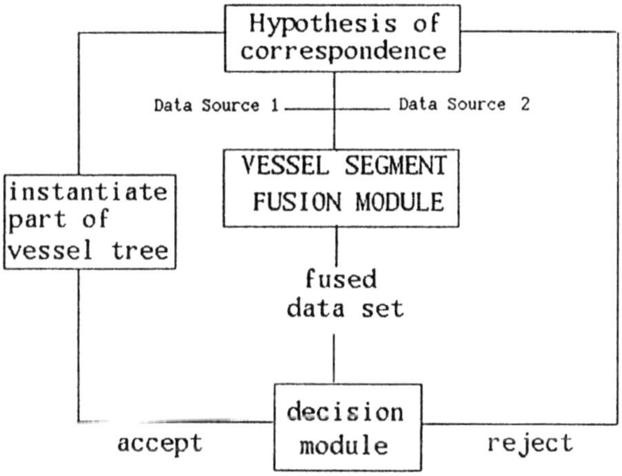

Fig. 2. The module in context

The results of the module will be used by other processing modules for high level inferencing tasks, and also for the generation of graphical displays.

3. INPUT DATA STRUCTURE.

Healthy arteries can be considered as curving, branching tubes which become narrower after branches. The lumen is usually circular in cross-section with a smooth lining. Healthy arteries thus lend themselves to a simple description in the form of an attribute list where the property (for example width) is given as a function of distance down the vessel measured from a specific point. Even in a diseased vessel, where atherosclerosis means that properties are not so regular and sharp bends are more common, it is convenient to reduce a description to an attribute list as far as possible.

Each vessel segment is described in terms of attribute lists. Each attribute is expressed as three components: firstly, the property value itself (for example, cross-sectional area value); secondly, its position along the vessel segment relative to one end and thirdly, the longitudinal extent (the size of the sub-part) of the vessel to which the property value applies. For each of these components we must provide a framework for describing the uncertainty. We have chosen to do this by means of a probability density functions of discrete values and their associated probabilities. Fig. 3 shows how a vessel segment is described in terms of property value/position/extent, and their associated probability density functions.

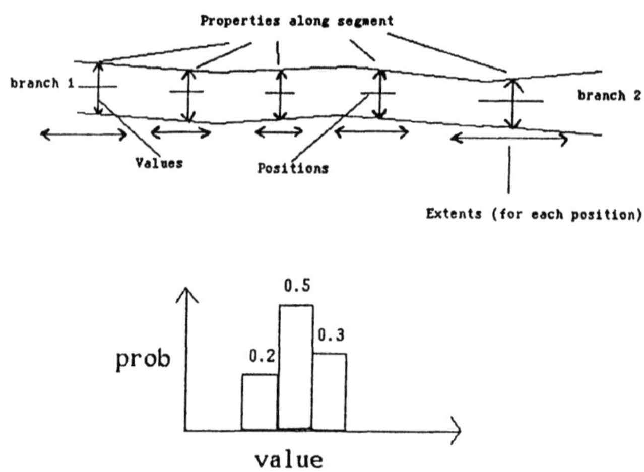

Fig. 3. Property/position/extent description

4. PROCESSING STEPS

Working under the assumptions stated, and using the data format described, the module has seven major processing steps.

1 Read input data
2 Generate spatial applicability functions (SAF)
3 Generate PRO data (evidence from each source in favour of particular property values at frequent positions along the vessel)
4 Fuse PRO data from different sources
5 Generate CON data (evidence from each source against particular property values at frequent positions along the vessel)
6 Fuse CON data from different sources
7 Combine fused PRO and CON data

These steps are described in detail below.

4.1 Read input data

The input data from each source about the vessel segment being fused is read into the module and held in an internal dynamic data structure. At this point no restriction is put on the sampling size or sampling location for any of the properties or sub-properties. It is quite possible that one source of data may provide a very well sampled data set, for example, an angiographic view where the vessel runs perpendicular to the x-ray beam, while another may provide a very sparse data set, for example an angiographic view where a vessel is considerably foreshortened.

4.2 Generate spatial applicability functions (SAF)

Every property value refers to (i.e. is applicable along) a finite length of vessel. The "position" value in the input data is a location of the centre of this zone of applicability. The "extent" value gives the size of the zone (i.e. the length of vessel over which the property value applies) but not its absolute location. It is necessary to generate a function describing the probability of a particular property value distribution applying at a series of locations along the vessel. We term this the "spatial applicability function" (SAF). If the position and extent were single values the SAF would be a square wave of amplitude 1 whose centre was fixed by the position value, and whose length was fixed by the extent value. However, the position and extent values are in fact probability density functions rather than single values and the SAF will normally have slopping sides reflecting the increasing uncertainty about whether the property value distribution applies at different distances away from the centre position. For a particular combination of position value (P(i)) and extent value (E(j)), the property value distribution would apply from positions P(i) - 1/2 E(j) to P(i) + 1/2 E(j), and the probability of the property value distribution applying within this range is given by

$$p(P(i)) * p(E(j)) \tag{1}$$

where p(P(i)) is the probability associated with position P(i) and p(E(j)) is the probability associated with extent E(j). Considering all possible combinations of positions and extents, the probability of applying at a particular location say "k" (the SAF) is given by

$$s(k) = \sum_{i=1}^{m} p(P(i) * \sum_{j=2|i-k|}^{n} p(E(j) \tag{2}$$

where m is the number of possible positions and n the number of possible extents.

Thus the combination of the extent and position takes the form of a simple convolution. At this stage a uniform sampling location is applied so as to force conformity between individual properties from a source, and also between separate sources. The maximum value for any point along the calculated SAF is one. This occurs when all possible combinations of input data of position and extent agree that the property actually applies at that point. Fig. 4 shows in graphical form the process of generating the SAF.

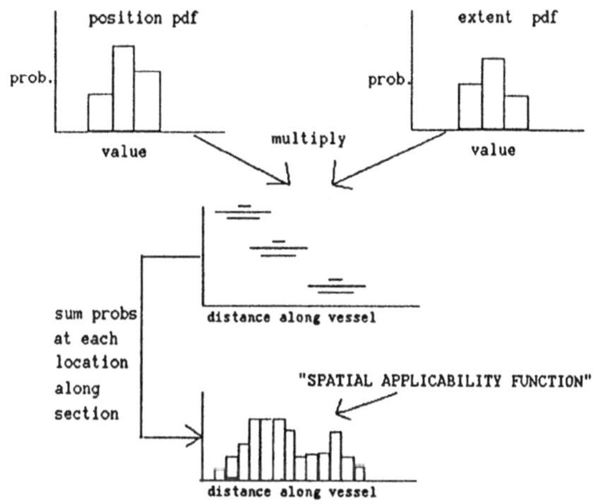

Fig. 4. Generating the SAF

4.3 Generate PRO data

For each source separately we can now calculate the probability of a particular property value applying at a particular location (event A). This is the probability that the particular property value is true (event B) AND that this attribute applies at the particular location under consideration (event C). The probability of A is then the product p[B] (obtained from the input data property value pdf) and p[C] (obtained from the SAF for this attribute). The combination is done by taking each point along the SAF and using its probability value as a scaling factor for the elements in the value-pair distribution for the property value. This gives us a 3D graph of distance along the vessel against property value against probability. If the value of the SAF at any point is unity, then the result of the combination at that point is a probability density function the same as the input data property value pdf. If the value is not unity then the result is the probability density function with the values scaled down accordingly.

This stage of the processing limits the sampling of property values to a specific size. We refer to this as the "bracket size" and this is constant over all properties and all sources.

When this combination has been done for all properties, then for each source we have a 3D graph of all the evidence in favour of values applying at each point along the vessel segment. This data is called the "PRO data".

4.4 Fuse PRO data from different sources

The next step is to combine the 3D graphs of PRO data from each source to produce a 3D graph of the same format. The required function will take two or more pieces of evidence about each value bracket and combine them to give a value greater than any one of the individual pieces of evidence. This is done in the following way.

To combine two sources of data we step along the vessel length and take each possible property value bracket in turn. We calculate the resultant probability at that bracket by the following equation

$$val(r) = val(s1) + (1-(val(s1))*val(s2)) \qquad (3)$$

where val(r) is the resultant value for that bracket, val(s1) is the value from source one and val(s2) is the value from source two. To combine a third source we do the same as above but make the previous result the new s1 and make the new source the new s2.

The result of this is to produce a function at each point along the vessel of property value against likelihood. Any individual value bracket can have a confidence value of up to one. The total area under the resultant curve is not necessarily unity, and the function is not a probability density function.

4.5 Generate CON data

Until now we have been dealing with data in favour of values applying (PRO data), and we have assumed that input data only includes PRO data. However, a future requirement would be the ability to accept CON data as input. Furthermore, a particular category of CON data can be generated from PRO input data. For each property bracket there is evidence against this property value in the form of the evidence in favour of all the other property value brackets at this location down the vessel. Thus each CON confidence is the sum of all the PRO confidences for all of the other brackets.

This is illustrated for one source at one position along the vessel in Fig. 5.

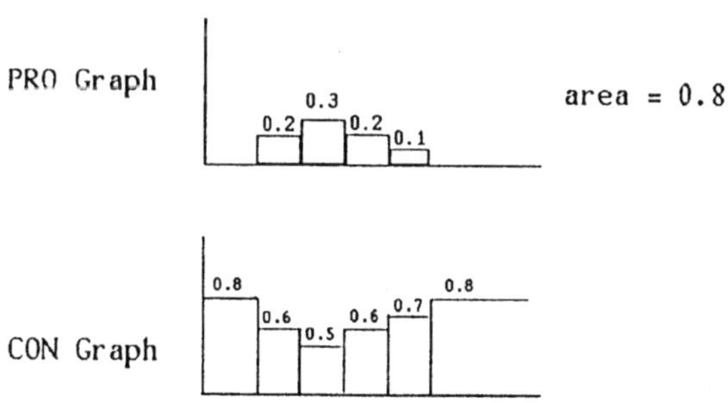

Fig. 5. Generating the CON graph

4.6 Fuse CON data from different sources

The generated CON data is next fused in exactly the same way as the PRO data was fused. We now have two 3D graphs, one of fused PRO data and one of fused CON data.

4.7 Combine fused PRO and CON data

The final stage, the combination of fused PRO and CON data about one value bracket at one location along the vessel segment is accomplished by subtracting the CON value from the corresponding PRO value.

If the resultant value is negative (minus one is the minimum) then the total evidence against outweighs the total evidence in favour of that bracket. Conversely, if the value is positive (plus one is the maximum) then the total evidence in favour of that bracket outweighs the total evidence against. If the value is close to zero then: either there was little or no evidence for that bracket, or the evidence in favour was close to the evidence against.

The result is a 3D graph of distance along vessel against property value against the net evidence in favour, where the evidence can range between minus one and plus one.

5. WORKED EXAMPLES

5.1. Fusion of complete data sets from two sources

The following series of illustrations show the results of the various stages of processing upon data which were derived from manual measurements on the two views of a bi-plane angiogram of a segment of left common carotid artery.

Fig. 6 shows the results of steps one to three. The left hand diagram is the PRO data from source one, and the right from source two. The long axis represents distance along the vessel, the short axis represents probability, and the final axis represents the property value itself (width). The areas of each graph where there are no curves indicates areas along the vessel where it was impossible to obtain measurements. The centre of the curves lie further to the right as you progress down the vessel implying a gradual widening. This was because the reference point for position measurements was at the upstream end of the vessel segment.

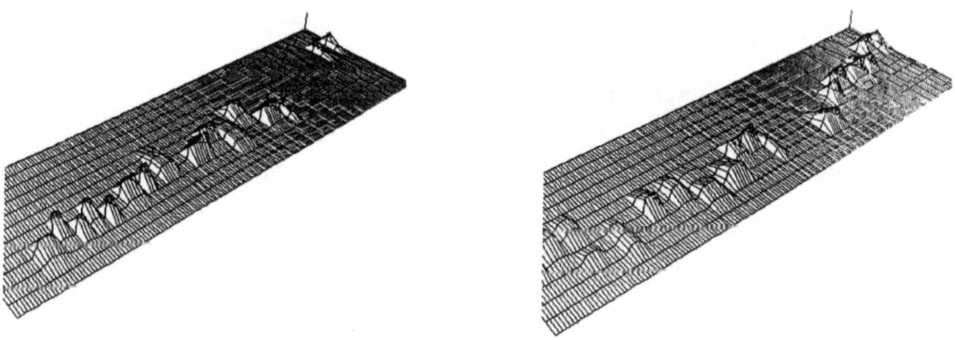

Fig. 6. PRO data from two sources

Fig. 7 shows the result of fusing the two PRO data sets. Note how the previously unsampled areas have now obtained data from the other source so as to give a description at almost all positions along the vessel.

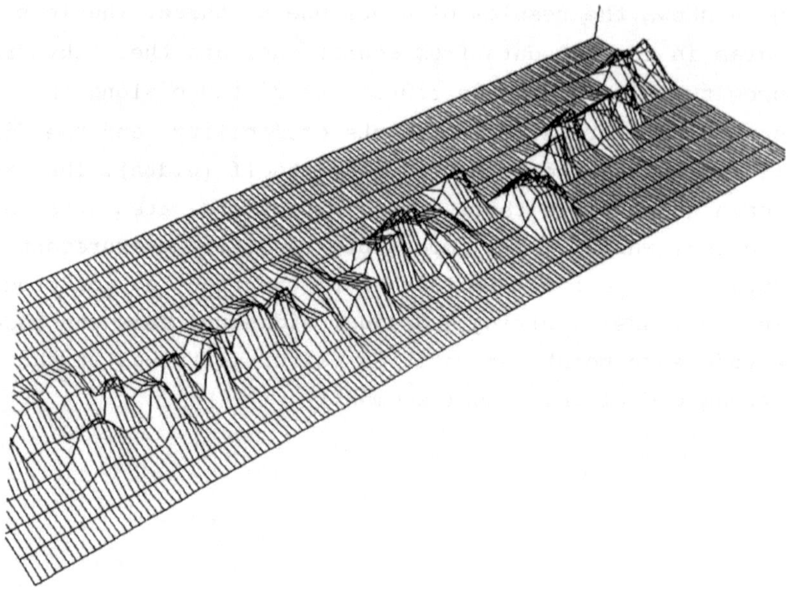

Fig. 7. Fused PRO data from two sources

Fig. 8 shows the generated CON data from each source.

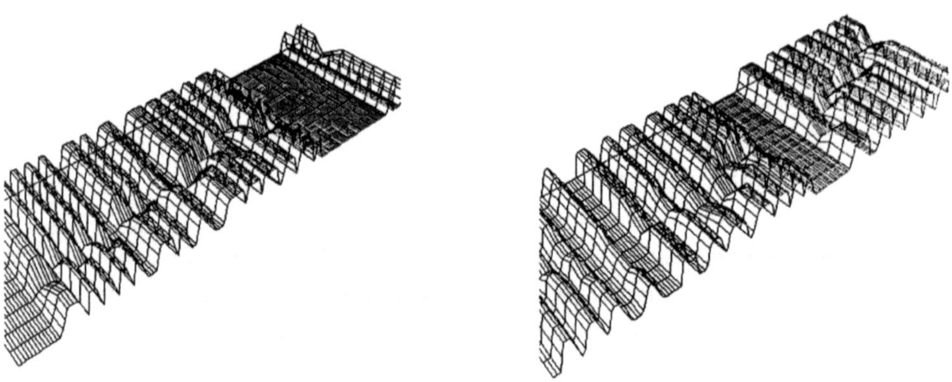

Fig. 8. CON data from two sources

Fig. 9. shows the fused CON data.

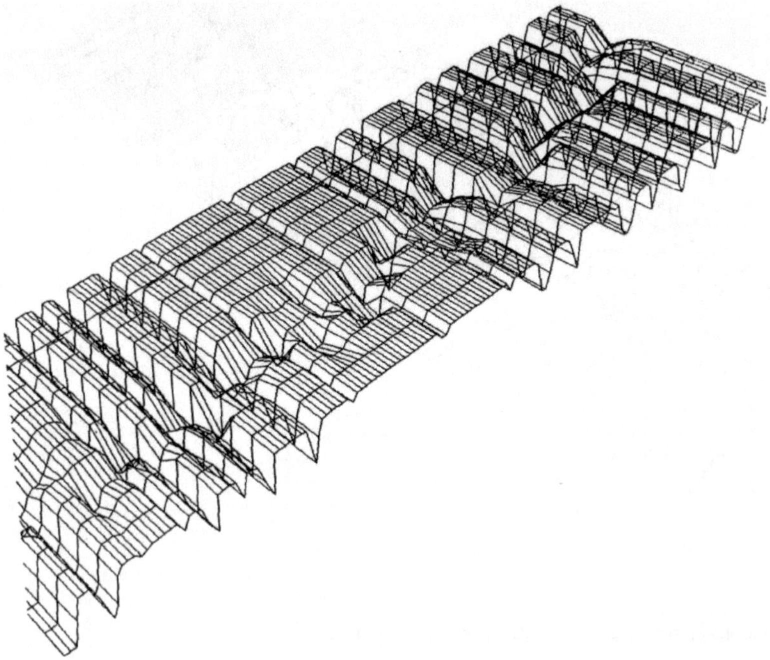

Fig. 9. Fused CON data from two sources

Fig. 10 shows the result of combining the fused PRO and CON data. Note how at certain positions along the vessel there are peaks indicating good correspondence, while at others there are large areas of low value implying a high degree of evidence against values.

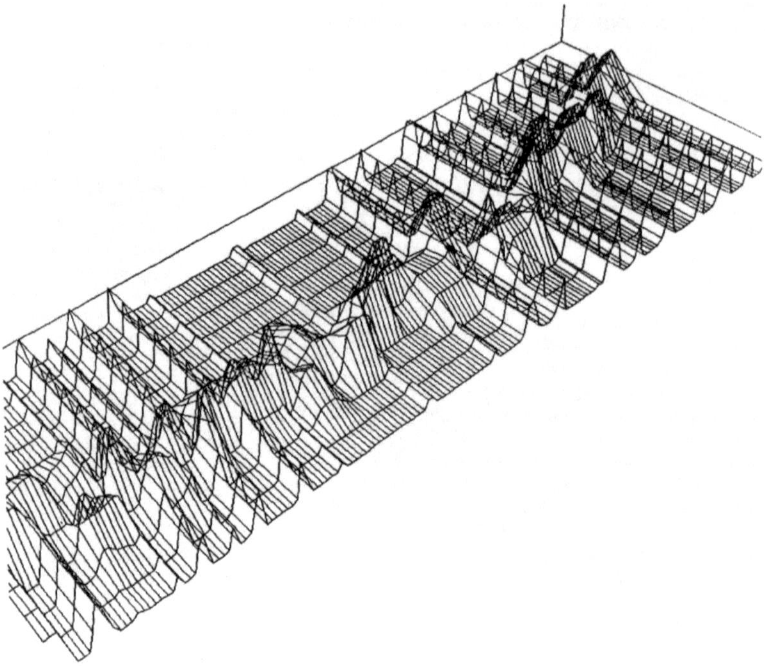

Fig. 10. Combined fused PRO and CON data

5.1 The effect of varying degrees of overlap of property values from two sources.

The function of steps 3 to 7 can be understood more clearly if we consider a particular location along the vessel, and then vary the degree of overlap of the example property value distributions from two sources. We assume here that the spatial applicability functions have been derived, and that we are looking at data from different sources at a particular position AFTER it has been scaled by the value of the SAF.

Fig. 11(a) shows two input sources which provide non-overlapping PRO data distributions. The area of source 1 PRO graph 1 is unity, and the area of the source 2 PRO graph is 0.5. The fused PRO curve which at any one point is no more that the individual input PRO data. The source 1 CON graph is unity outside the bounds of the source 1 PRO data because the area of the source 1 PRO graph is unity. The fused CON graph is thus forced to unity everywhere outside the bounds of source 1 PRO data, regardless of the PRO data from other sources. The net PRO graph does not rise above zero threshold for this example.

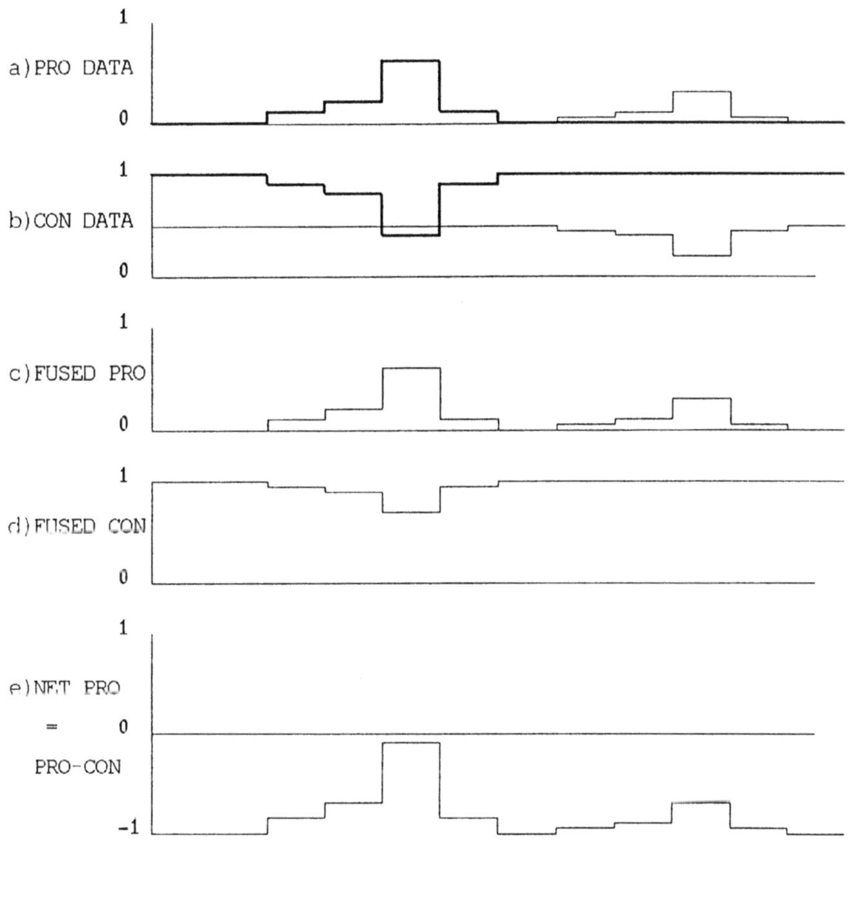

Fig. 11. Non overlapping PRO data. (In (a) and (b) the bold line represents source 1, the fine line source 2.

As we move the two distributions closer together and allow them to overlap, there is both a decrease in the fused CON values where overlap occurs, and also an increase in the fused

PRO values. This results in an increase in the final values after subtracting the fused CONS from the fused PROS. This is shown in fig. 12. Fig. 13 shows the result of giving two totally overlapping input PRO data sets; with this combination a single bracket of property values now rises above the zero threshold.

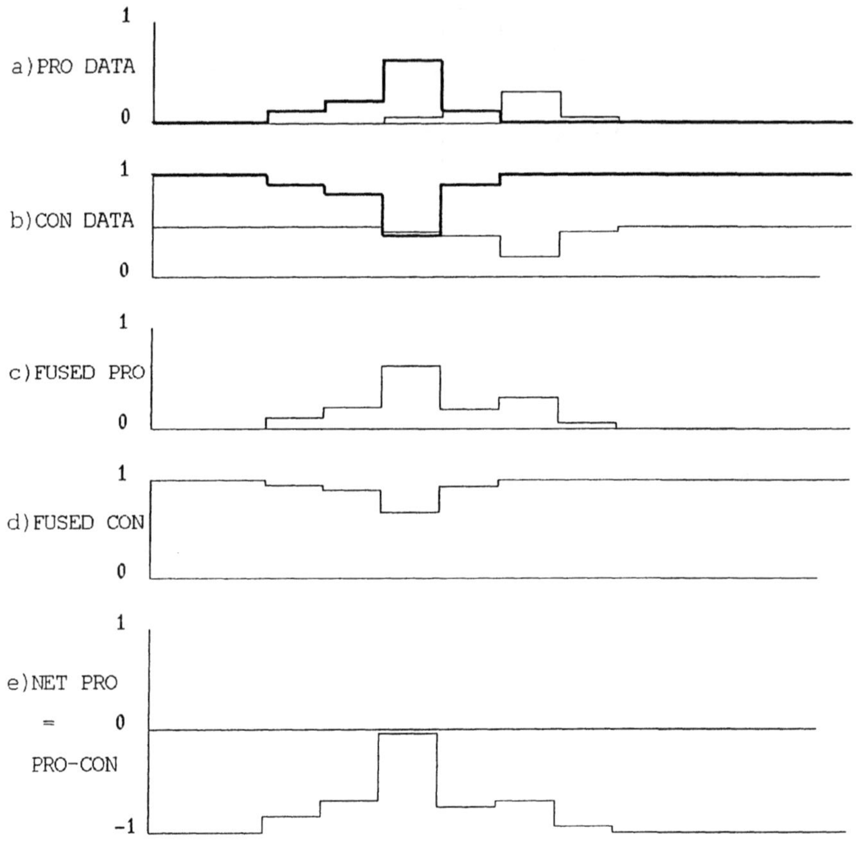

Fig. 12. Partially overlapping PRO data (of Fig. 11)

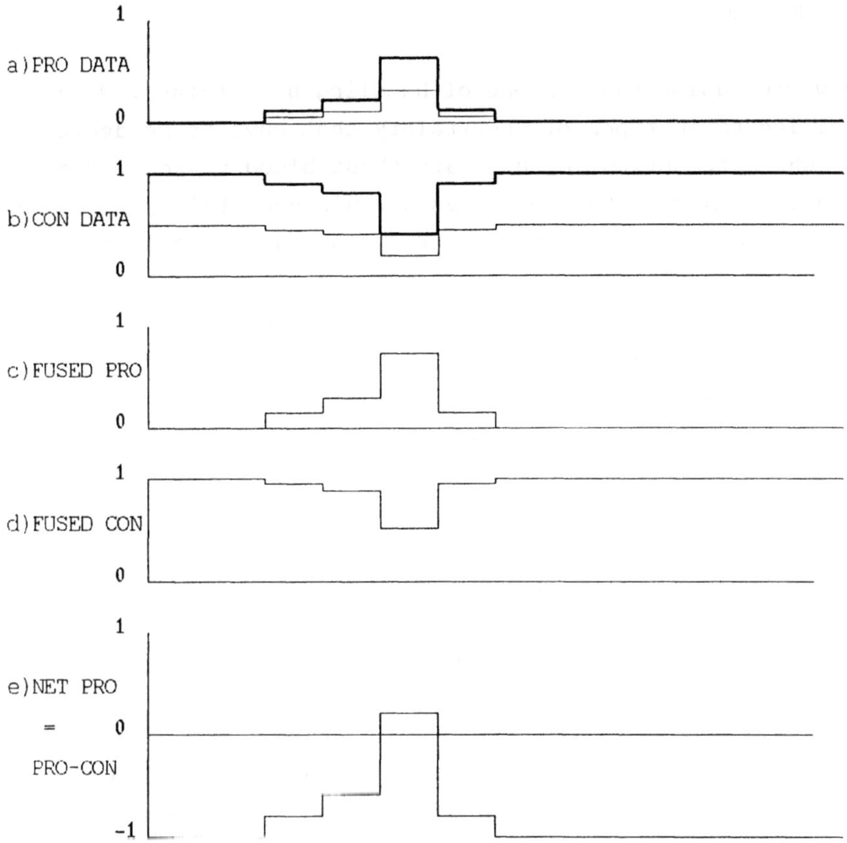

Fig. 13. Totally overlapping PRO data (of Figs. 11 & 12)

The value of the net PRO value for any one property value bracket after subtracting the fused CONS from the fused PROS will be greater than zero only if there is more total evidence in favour of that bracket than there is against it. In the case of only having one input PRO data source, then for a net PRO to be greater than zero it must have a PRO likelihood greater than half the total likelihoods from that source.

6 DISCUSSION

The whole fusion task is one of handling uncertainty. We have specified three types of uncertainty that have to be dealt with when attempting to fuse data about blood vessels. The approach we have taken to represent this uncertainty is based on standard probability theory. The combination mechanisms we employ are simple mathematical operations, the results of which have a predictable meaning at each stage. Other approaches to handling uncertainty have been developed, some based on standard probability, and others deviating or attempting to extend it in some way. The following section briefly looks at three of the more common approaches to handling uncertainty.

6.1 Other approaches to uncertainty

6.1.1 Fuzzy sets and fuzzy logic

The idea of reasoning based on "fuzzy sets" (Zadeh, 1965) was developed to overcome some of the rigidity of standard logic and probability theory. Fuzzy sets are defined such that an object is not simply a member of a set or not, rather it is given a "degree of membership" of a set. This degree of membership is given as a value in the range zero to one.

An advancement of fuzzy sets was to introduce the concept of a "linguistic variable", and from this "fuzzy logic" (Zadeh, 1975). A linguistic variable is any word occurring in natural language whose values are non-numeric. For example, "age" is a linguistic variable if its values are taken as "old"; "young"; "not very old" etc., as opposed to 10,11,12...99,100 etc. Each of the values of the linguistic variable is known as a "linguistic value" of the variable. Each linguistic value has an associated "compatibility function" which maps it onto a "base variable" in the range zero to one, as shown in Fig. 14.

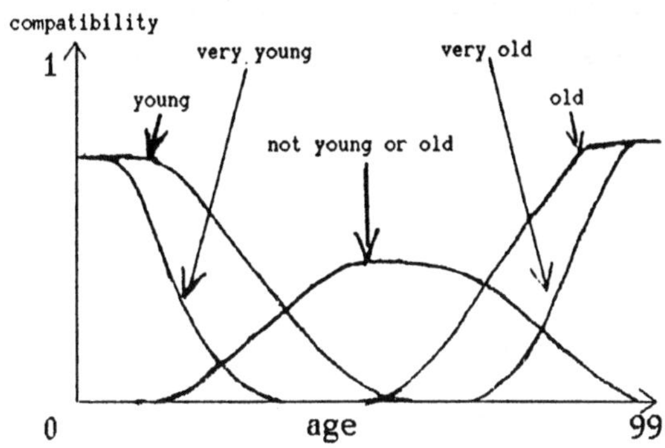

Fig. 14. Linguistic variable "age", and its linguistic values.

Treating truth as a linguistic variable ("true"; "very true" etc.) is the basis for fuzzy logic and inexact reasoning. Using probability as a linguistic variable leads to results which are as imprecise as are the underlying probabilities.

One fuzzy approach to the data fusion problem could be to define the property type being measured, width say, as a linguistic variable. Input could then be of a number of types. One example, in terms of a linguistic value, could be as "the vessel is wide" or "the vessel is narrow". Or the current style of input data could be used as a compatibility function of the variable in question. The segment of vessel covered by each property can also be viewed as a fuzzy set. This again allows input to be in a linguistic form, and provides the mapping from linguistics into a continuous numeric scale (the base variable).

The combination of input data would not be straightforward as Zadeh's combination of sets is restricted to the "max/min" of union and intersection, but he has also developed a theory of "fuzzy decision making" which allows the combination of fuzzy sets where one fuzzy set represents a goal, and others the constraints. The result of the combination can be a fuzzy set itself, or a maximised value for the combination of sets (Albin, 1975). The concepts of fuzzy decision making could be used for combining sets from different sources at the same point along the vessel.

Problems with the fuzzy approach are based round practical problems of assigning degrees of set membership, and compatibility functions. These require a great deal of a priori knowledge, and can also be highly domain specific, for example "old" when talking about trees would require a different compatibility function than when talking about cars, even though they have the same base variable.

6.1.2 MYCIN

The MYCIN approach to dealing with uncertainty (Shortliffe, 1976) was designed to be highly domain specific, and as such much be viewed as an ad hoc approach which appears to work in the domain it was designed for, namely diagnosis of blood infections. MYCIN uses what are called "certainty factors" (CF) to try and give a degree of confidence to a hypothesis, given all the evidence collected about that hypothesis. Hence CF(h,e) can be read as "the confidence in hypothesis 'h' given all the evidence 'e'".

CFs are obtained by finding the differences in the "measure of belief" of a hypothesis given all the evidence (MB(h,e)) and the "measure of disbelief" of a hypothesis given all the evidence (MD(h,e,)). Hence,

$$CF(h,e) = MB(h,e) - MD(h,e). \qquad (4)$$

MYCIN then collects both measures of evidence for and evidence against a hypothesis. These measures are constantly being updated (and hence Cfs also) in the light of new evidence. How this updating process occurs is at the heart of MYCINS reasoning.

If we assume that P(h) is the subjective belief that a hypothesis "h" is correct. Adding evidence "e" can alter the posterior probability P(h,e) in one of two ways (Jackson, 1985). Either P(h,e) will be greater than P(h), or P(h,e) will be less than P(h). If the first of these conditions is true, then the increase in belief is given by

$$\frac{P(h,e)-P(h)}{1-P(h)}. \qquad (5)$$

Otherwise, the degree of disbelief increases as

$$\frac{P(h)-P(h,e)}{P(h)}. \qquad (6)$$

The resultant Cfs are not probabilities as they can range from minus one (total disbelief) to plus one (total belief).

There are a few similarities between the MYCIN approach and our own. The major similarity is in the collection of evidence both for and against hypothesis. MYCIN assumes that all evidence collected is independent and as such can use Bayesian updating for MB or MD. One method of obtaining our CON data is not independent of the corresponding PRO data and so the methods are incompatible in this respect. On the other hand, individual PROs are assumed independent from the other (and likewise for CONs) and as such a MYCIN approach could be formalised to combine PRO data (and CON data).

One major problem with the MYCIN approach is its requirement for prior probabilities which may not be available in a particular domain. Also, although MB and MD are related to probability, the result of their combination is not. The reason MYCIN used differences between MB and MD was to rank hypothesis. Adams (1976) points out that it is possible for the ranking CFs to differ from the ones that would occur using approximation to probability theory.

6.1.3 Dempster-Shafer

The Dempster-Shafer theory (Dempster, 1968) and (Shafer, 1976) is a relatively new approach that allows the user to overcome some of the forced rigidity of Bayes by allowing "ignorance" in probabilities to be specified. Probability is treated as a mass which is distributed in some way to all values concerned. This mass is the probability supporting a value. For example, "sup(A) = 0.2" implies there is a probability of 0.2 supporting the truth of A. The support for ~A is given separately and is not necessarily 1-sup(A).

The plausibility of A is then given by

$$pls(A) = 1 - supp(\sim A) \qquad (7)$$

(Saffriotti, 1988). The ignorance in A is then defined as the gap between pls(A) and sup(A). In the case where sup(~A) = 1-sup(A) (i.e. pls(A) = sup(A)) then Dempster-Shafer is the same as Bayesian probability.

There has been some recent work on the use of Dempster-Shafer techniques to solve some of the problems of data fusion (Garvey, Lowrance et al, 1981). This work assumes that a unit mass of belief is distributed over a set of propositions for which there is direct evidence (this could easily correspond to our possible input property value). The mass is equivalent supp(a) for each proposition and the evidentual support can be calculated directly. This unit of mass is divided for each input source available. The pooled evidence one proposition from two sources is obtained by calculating the mass from source 1 with the mass from source 2, this is the same as the intersection of each proposition from the two sources. This method of combination is commutative, hence order is unimportant, and also associative.

Choosing values for supporting evidence and ignorance can be a major problem (as can choosing probabilities). Dempster-Shafer is causing a great deal of interest within the AI. community as it appears to give one approach to overcoming some of the problems of Bayes.

6.2. Future work

Future work will include attempting to derive models of vessel segments which can be treated as sources of input to the module and utilised as a priori knowledge. We are also looking at different ways of combining PRO and CON data such as log likelihood ratios. Another interesting area of development is in varying the value bracket size so as to change the amount of PRO data of a bracket size. Increasing the size of the brackets gives more PRO data to that bracket (and hence less CON data) but means that some of the precision is lost. Manipulating the bracket size could allow the module to return data to higher level processes at differing degrees of specificity depending upon the compatibility of the input data.

ACKNOWLEDGEMENTS

This work was carried out as part of Alvey MMI-134 and we would like to acknowledge the valuable comments received from members of the consortium.

REFERENCES

Adams, J.B. (1976). A probability model of medical reasoning & the MYCIN model, Mathematical Bioscience, Vol 32, pp. 177 -186.

Albin, M.A. (1975). Fuzzy sets & their application to medical diagnosis and pattern recognition, PhD Thesis Graduate Division , University of California, Berkeley.

Dempster, A.P. (1968). A generalisation of Bayesian Inference, Journal of Royal Statistical Society, Ser. B. Vol 30, pp. 205-247.

Garvey, T.D. Lowrance, J.D. Fischler, M.A. (1981). An inference technique for integrating knowledge from disparate sources, Proc. Int. Joint Conf. A.I., pp. 319-325.

Jackson, P. (1985). An introduction to expert systems, Addison -Wesely.

Saffriotti, A. (1988). An AI view of uncertainty, The Knowledge Engineering Review, Vol , pp. 75-97.

Shafer, G. (1976). A mathematical theory of evidence. Princeton University Press.

Shortliffe, E.H. (1976). Computer based medical diagnosis: MYCIN, Elsevier Science Publishing.

Zadeh, L. (1965). Fuzzy sets, Information and control, Vol 8, pp. 338-353.

Zadeh, L. (1976). The concept of a linguistic variable & its application to approximate reasoning - 1, Information Sciences, Vol 8, pp. 199-249.

A METHOD FOR MULTI-SCALE REPRESENTATION OF DATA SETS BASED ON MAXIMUM GRADIENT PROFILES: INITIAL RESULTS ON ANGIOGRAPHIC IMAGES

A C F Colchester[1]
R T Ritchings[2]
N D Kodikara[2]

[1] Dept of Neurology Guy's Hospital London
[2] Dept of Computation UMIST Manchester
United Kingdom

1. INTRODUCTION

Our previous work on knowledge based interpretation of angiographic images (Ritchings and Colchester 1986; Ritchings et al 1986; Wang et al 1987) has underlined the need to establish a hierarchical data structure which represents the image at multiple scales and which allows simplified control of the interaction between image-driven and knowledge-driven processes. Multi-resolution methods which use Gaussian blurring of different degrees in the analysis of features at multiple scales (Pizer et al 1987, 1988) have some theoretical advantages but, in practice, the selection of the most appropriate scales at which to smooth the image is difficult. We have developed a hierarchical algorithm which groups small elements into larger elements and then groups these into larger elements and so on. The size of the groups is determined by local criteria and prior decisions about scale are not required.

We consider an image as a 3-D grey level landscape with topological features (Koenderink 1988), and we treat an edge as a "hillside" or steep region on this 3-D surface. A fundamental property of the edge is the profile of grey level change along lines of maximum gradient. Adjacent lines of maximum gradient on an edge will have similar profiles and tend to be approximately parallel. The directions of these maximum gradient profiles (MGPs) are normal to the direction of the edge.

In the present paper we outline the principles of the proposed multiscale algorithms, initially for a one dimensional signal (i.e., a 1-D array of grey levels, extendable to a 2-D array of binary numbers) and then for a 2-D signal such as a grey scale image (a 2-D array of grey levels, extendable to a 3-D array of binaries). We describe the implementation of a simplified 2-D version of the algorithm and examine the results when applied to conventional angiographic images.

2. PRINCIPLES OF THE MULTI-SCALE SEGMENTATION ALGORITHMS

2.1 Algorithm for 1-D signal (Figs. 1 and 2)

2.1.1. **First generation grouping.** The gradients between adjacent signal elements (dots in Figs 1 and 2) are calculated to form a sequence of vectors. Adjacent vectors having the same sign (positive or negative gradient) are linked. These linked sets terminate where the gradient becomes zero or changes sign. Adjacent zero-gradient vectors are linked to each other to form a zero-gradient set. The centroids of each set of signal elements are calculated. These centroids form the second generation elements (small squares in Figs. 1 and 2).

2.1.2. **Second generation grouping.** The gradients between adjacent centroids are calculated to form a sequence of second generation vectors. Adjacent vectors having the same sign are linked in the same way as first generation vectors. The centroids of each set of second generation elements are calculated to form third generation elements (triangles in Figs. 1 and 2)

2.1.3. **Third and higher generation grouping.** The third and higher generation elements are grouped in the same way until a single element representing the entire signal is formed.

2.1.4. **Data structure.** The signal is represented as a network whose nodes are the elements described above. Each element carries links to the elements in the previous generation from

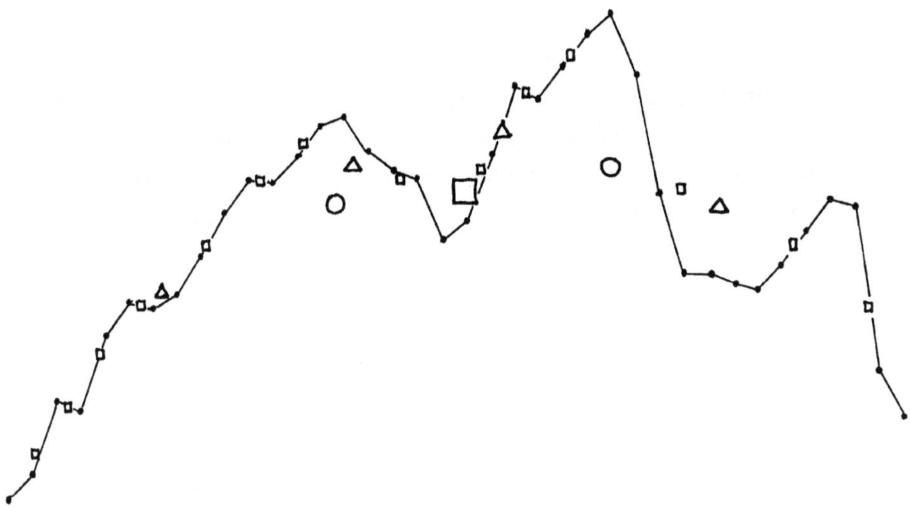

Fig. 1. Diagram showing segmentation of a one-dimensional signal. Symbols for elements : **dot** = 1st generation (input signal); **small squares** = 2nd generation; **triangles** = 3rd generation; **circles** = 4th generation; **large squares** = 5th generation.

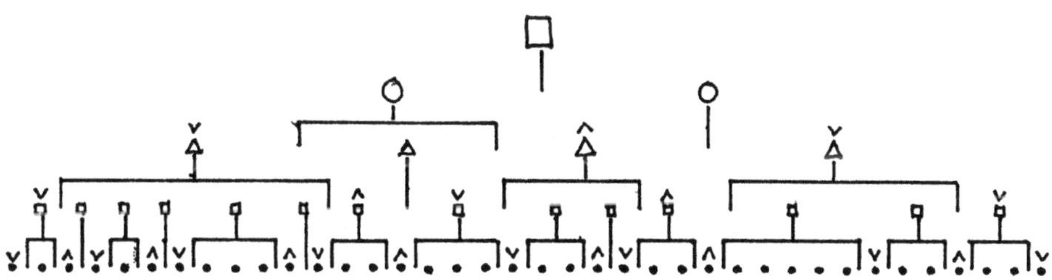

Fig. 2. Data structure showing hierarchical relationship between elements in Fig. 1. Local maxima and minima area shown as ^ and v respectively.

which it has been formed and to the element in the next generation of which it is part. These (between generation) links thus form a part hierarchy (Fig. 2). In addition each element carries links to its two neighbours within the same generation to form a linked list. Note that the sequence of construction and the numbering of the generations of the part hierarchy goes from 'children' to 'parents'.

2.2 Algorithm for 2-D signal such as an image

2.2.1. **First generation grouping.** The local direction of maximum grey-level gradient is derived for every neighbourhood from the grey values of nearest neighbour elements (pixels). Vectors are linked along the direction of maximum gradient to form maximum gradient profiles (MGP). These define local "water courses". Linking is discontinued where the vector gradient changes sign (where 'uphill' changes to 'downhill' and vice versa) or becomes zero. The centroid or other measure of central location of each MGP is calculated and defines its location. MGPs are compared with their nearest neighbours (proximity being measured by inter-centroid distances) and if adjacent (side-by-side) MGPs have similar direction they are linked to form an 'edge section' or 'hillside region'. Sets of elements with zero grey level gradients between them are linked to form a flat region or plateau. The centroid of each hillside region or plateau is calculated and defines its 3-D location. Each centroid forms a second generation element.

2.2.2. **Second generation grouping.** The vectors linking centroids of neighbouring second generation elements are calculated and the direction of local maximum gradient is established. Vectors are linked along the direction of maximum gradient to form second generation MGPs. Neighbouring MGPs with similar direction are linked side-by-side to form second generation edge sections or hillsides. Elements with zero grey level gradients between them are linked end-to-end and side-by-

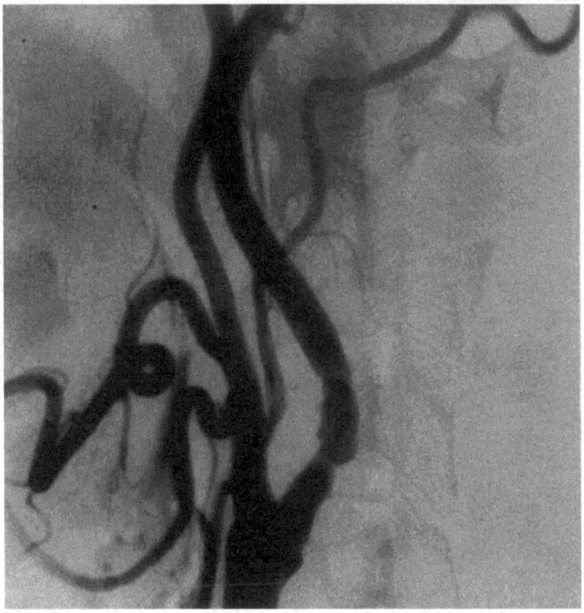

Fig. 3. Angio 1: Region of interest before digitisation

Fig. 4A. Angio 1: Perpsective view of Maximum-Gradient Profiles shown as if superimposed on the 3-D grey level surface

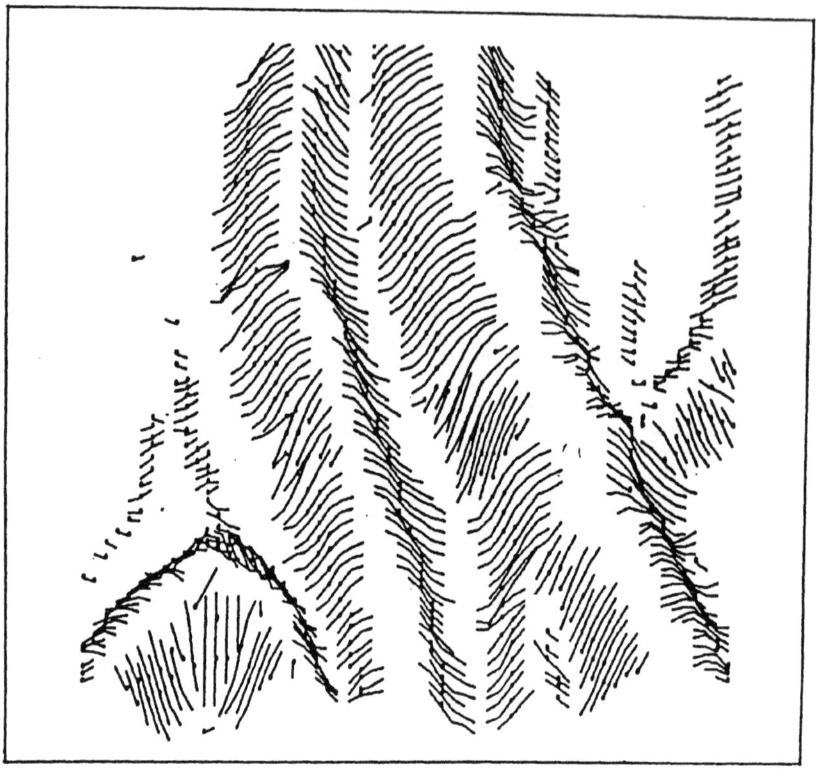

Fig. 4B. Angio 1: Enlarged view of Maximum-Gradient Profiles from region near centre of Fig. 2A.

side to form a plateau. The centroid to each of these regions is calculated and defines its 3-D location.

2.2.3. **Third and higher generation grouping** is accomplished in exactly the same way until the whole image is defined as a single region.

3. IMPLEMENTATION OF A SIMPLIFIED VERSION OF THE 3-D ALGORITHM

A simplified single-level version of the 3-D algorithm was implemented on an Apollo DN 3000 Workstation in the Department of

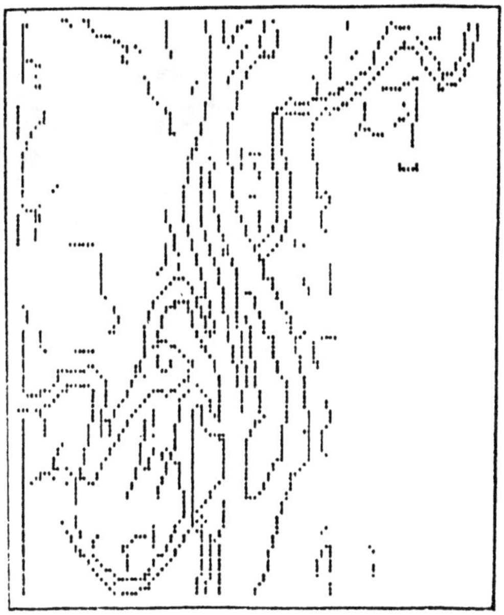

Fig. 5. Angio 1: steepest-slope points on the Maximum Gradient Profiles

Computation at UMIST. A 3x3 Sobel operator was used to establish the local direction and magnitude of maximum gradient for every pixel. For the initial implementation, which produced only a single generation of elements in the hierarchy, a threshold for these gradients was used to suppress gradients with a low value. The threshold was calculated automatically as follows. The image was divided into 10x10 pixel squares. Within each square the standard deviation (SD) of the magnitudes of all the local gradients was established. The square with the lowest SD was found and three times this SD was taken as the global threshold above which all gradients were retained. For the neighbours of each above-threshold gradient the threshold was relaxed to 2 SDs. All gradients with a magnitude less than 2 SDs were rejected.

Subsequent processing steps involved successively grouping elements to form larger elements, larger elements into still larger elements, and so on. Pixels with a local gradient above the calculated threshold were linked along the path of maximum

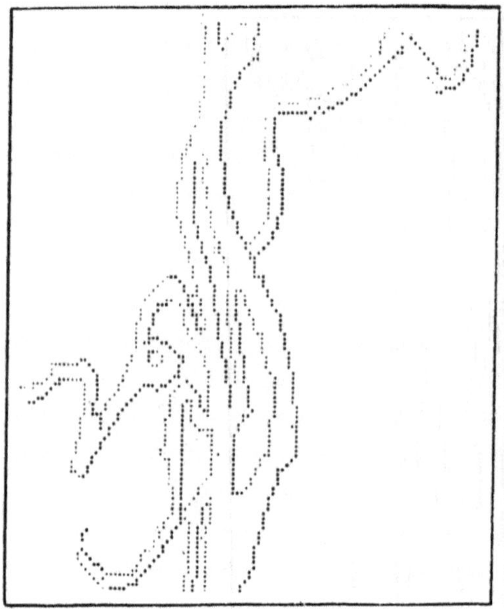

Fig. 6. Angio 1: Blood vessel boundary map derived from MGP grouping method

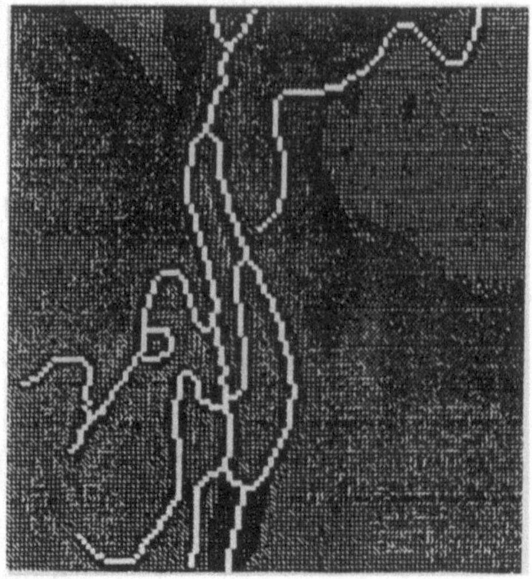

Fig. 7. Blood vessel centre lines derived from MGP grouping method

gradient to form MGPs provided that the track of an MGP did not change direction by more than 45° in total. At the ends of each MGP (i.e. where there were no further grey changes above threshold), neighbouring MGPs with the nearest end points were examined, and if the MGPs were approximately co-linear they were linked end-to-end to form a composite MGP. In this implementation the point of steepest slope of the MGP was taken as the central point of the MGP and is referred to as the edge point. MGPs whose edge points were near neighbours were compared, and those MGPs with similar direction and maximum gradient were linked to form a side by side set of MGPs. This set formed an edge section with a normal given by the mean of the MGP directions. The edge sections thus derived were compared with edge sections derived from Canny and Marr-Hildreth operators. The scale parameter of these was set at 11 pixels after some experimentation.

Processing using the edge sections derived from the MGP grouping method was continued to form larger features as follows. Edge sections were compared with near neighbour edge sections (searching over a larger scale than with single MGPs) and linked to form longer edge sections. Next, near-neighbour anti-symmetric edge sections were linked to form ridges. In the present application, where the objective was to detect blood vessels on good quality angiographic images on which blood vessels appeared as ridges on the 3-D surface, edges were not linked toform valleys, and edges without anti-symmetric partners forming a ridge were suppressed. Along the length of ridges the width of the ridge and the locus of its centre-line or symmetric axis were derived from the edge points of anti-symmetric MGP pairs.

The majority of the code was in the form of conditional rules programmed in Lisp; the Sobel operator was programmed in Pascal.

4. RESULTS

Two examples of the digitised region of interest are shown in Figures 3 and 10. Figs 4 to 7 show stages of processing of Angiogram 1 and Figs 11 to 14 stages of Angiogram 2. Perspective views of the extracted MGP's, shown as if superimposed on the 3-D grey level surface of the images, are shown in Figs 4 and 11. Fig 4B is an enlarged view of the centre of Fig 4A. When the edge points are displayed the method can be seen to function as a very effective low level edge detector (Figs 5 and 12). With further processing to extend edge sections and remove those sections without an anti-symmetric partner forming a ridge, reliable blood vessel boundary map is generated (Figs 6 and 13). The detected vessel centre lines, which corresponded very well with an expert's judgement of centre line position, are shown in Figs 7 and 14.

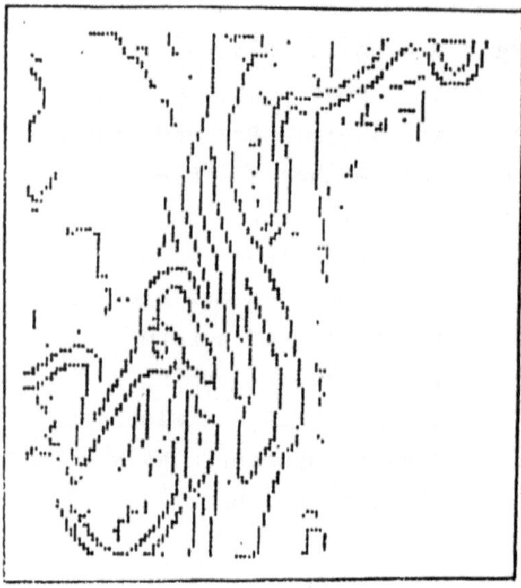

Fig. 8. Angio 1: Canny operator output with width = 11 pixels

Fig.9. Angio 1: Marr-Hildreth operator output with width = 11 pixels

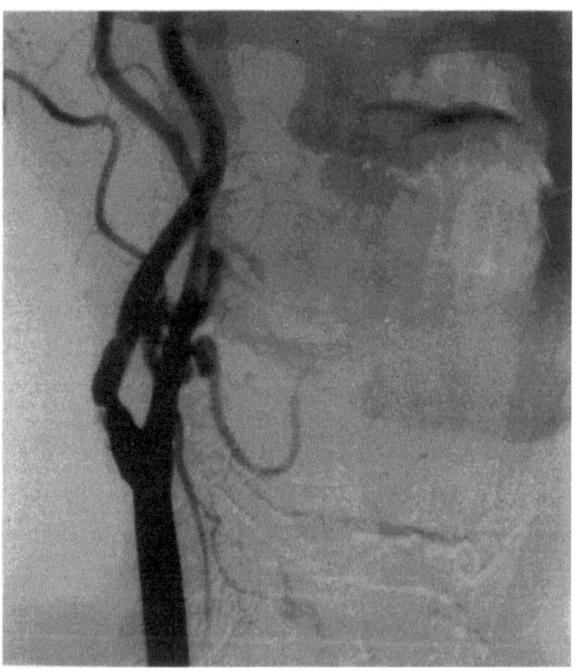

Fig.10. Angio 2: Region of interest before digitisation

Fig.11. Angio 2: Perspective view of Maximum-Gradient Profiles shown as if superimposed on the 3-D grey level surface

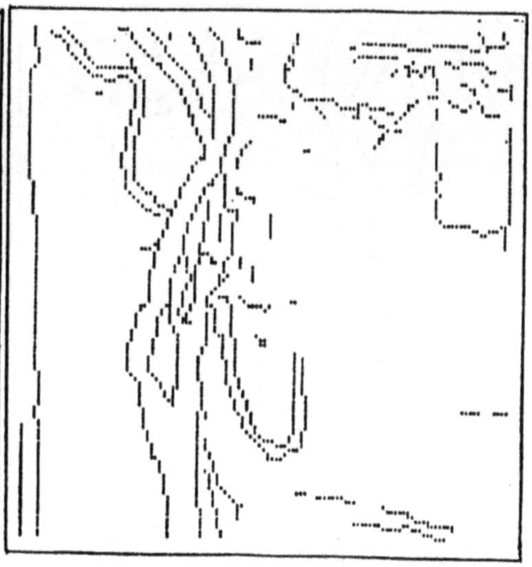

Fig.12. Angio 2: Steepest slope points of the Maximum Gradient Profiles

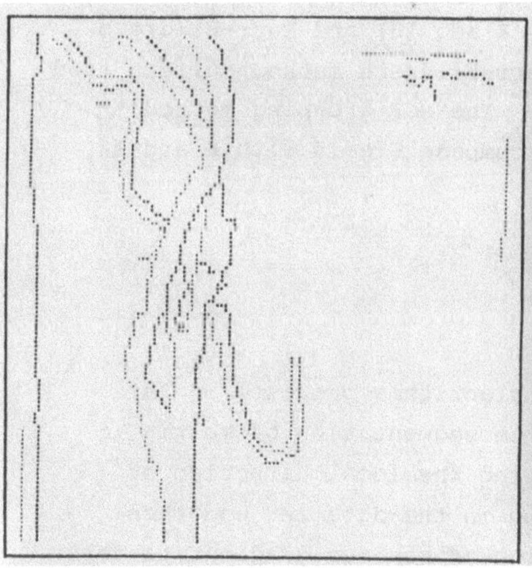

Fig.13. Angio 2: Blood vessel boundary map derived from MGP grouping method

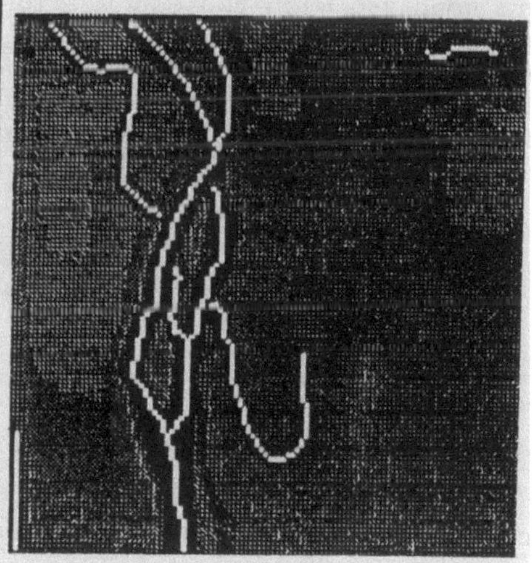

Fig.14. Angio 2: Blood vessel centre lines derived from MGP grouping method

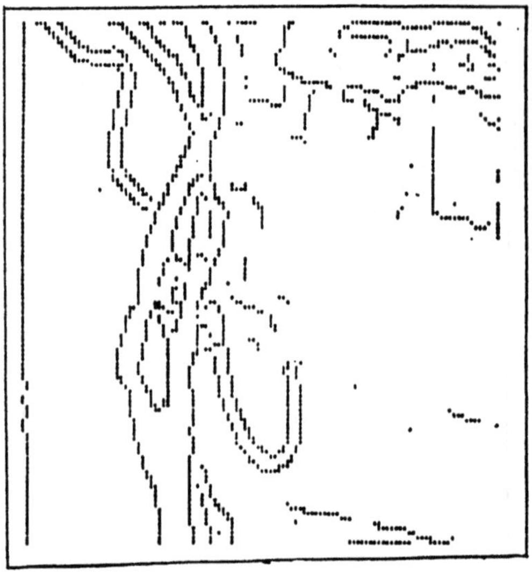

Fig.15. Angio 2: Canny operator out-put with width = 11 pixels

Fig.16. Angio 2: Marr-Hildreth operator output with width = 11 pixels

Edge sections detected by the MGP grouping method (Figs 5 and 12) were compared with the Canny (Figs 8 and 15) and Marr-Hildreth (Figs 9 and 16) operators. The Marr-Hildreth method suffers from major distortion at sharp corners. The MGP grouping method performs comparably to the Canny (compare Figs 5 with 8 and 12 with 15).

5. DISCUSSION

There are two main aspects of the algorithms proposed in this paper. The first is the approach to segmentation of multi-dimensional data sets by establishing the local direction of maximum gradient at every location in the data set and then linking elements along the direction of maximum gradient to form maximum gradient profiles. Regardless of the dimensionality of the input data, MGPs are linear lists of grey values (or other variable) which are progressively increasing or decreasing. Each value is accompanied by its own location in the input data set, and this location will be given by one or more numbers according to the dimensionality of the input data. The MGPs form string-like structuring elements. In data sets which are two or more dimensional they tend to form groups lying parallel to each other. Adjacent MGPs which are similar to each other are linked together to form a side-by-side group which describes a surface. For a two-dimensional array of grey values (as in the data presented in this paper) this will complete a single grouping cycle. For an array with three dimensions, the surfaces formed by side-by-side MGP groups will be stacked in multiple layers and the linkage of adjacent layers which are similar to each other will form a volume. The concept behind the grouping cycle is thus to take an n-dimensional array of input data, to construct one-dimensional elements (MGPs), to link adjacent elements if they are similar to form two-dimensional elements (layers or surfaces), to link these in turn to form three dimensional elements (volumes), and so on till n-dimensional elements have been constructed. All of these steps are part of a single generation grouping cycle. Within the cycle each step forms elements with increasing dimensionality (line-surface-volume). The task of the cycle as a whole is to combine n-dimensional elements which are changing in a uniform manner (positive, negative or zero change) to form a new generation of larger n-dimensional elements.

The second main aspect of the proposed algorithms is the construction of a multi-scale representation by means of hierarchical grouping rather than by filtration with different degrees of smoothing. The proposed grouping process fulfils the same function as filtration, but with our method the size of each group is set by local data properties and can vary abruptly from region-to-region. This avoids the need for prior decisions about the appropriate filter size and allows definition of boundaries at the finest scale. Another related advantage of the grouping process is that it generates explicit links between generations. This avoids the potential ambiguity in establishing the correct linkage between different levels of the hierarchy which can arise during the construction of the Gaussian pyramid as a result of only being able to apply a limited number of discrete filters. Our method is similar to other pyramidal algorithms where the same basic grouping rules are applied at each stage of the hierarchy (Rosenfeld 1984; 1987). Each level of our pyramid is one generation of the grouping cycle. Scale increases as one ascends from the lowest to the highest level. However, the level of the element in the hierarchy specifies the number of generations below it (a method of encoding relative size) rather than absolute size.

We have so far only implemented a single-generation grouping cycle and the data we have presented confirms the potential of the first aspect of the proposed algorithms (the MGP grouping approach). The performance of the MGP grouping method in the detection of edge fragments indicates that it is at least as effective as the Canny operator when applied to the angiographic images so far examined. These images have relatively strong features and we have yet to evaluate the MGP grouping method in very noisy images. In such images, a multi-generation algorithm will be necessary to provide a description of small scale features, which will be mostly noise, and larger scale features which will include the representation of the vessel structure.

A significant advantage of the MGP grouping method was expected to be in the later stages of processing and this was borne out in our implementation. The linking of co-linear and anti-symmetric edge sections and derivation of the centre-line of ridges were all natural extensions of the earlier grouping stages and very successful in operation. In many types of image, the shape of the grey-level profiles normal to edges carries important information (Sleigh 1985) which is potentially useful both during image-driven segmentation and during image interpretation using domain specific knowledge. This is particularly true in quantitative angiography, where many of the calculations about vessel width, cross-sectional area, cross-sectional shape and other parameters are derived from the profile of grey level values along a track perpendicular to the long axis of the vessel (Colchester 1985). Traditionally the construction of these transverse density profiles has required several initial steps: vessel recognition (usually with extensive user interaction); edge localisation; centre-line localisation; calculation of the normal to the centre-line; and selection of nearest pixels for grey level sampling. The MGP grouping method described here potentially provides a much more logical processing sequence.

REFERENCES

Colchester ACF (1985) The effect of changing Pa CO_2 on cerebral artery calibre estimated by a new technique of dynamic quantitative digital angiography. PhD Thesis, University of London.

Colchester ACF, Hawkes DJ, Brunt JNH, du Boulay GH and Wallis A. (1986) Pulsatile blood flow measurements with the aid of 3-D reconstructions from dynamic angiographic recordings. In Information Processing in Medical Imaging, Ed Bacharach S. Dordrecht: Martinus Nijhoff, pp 247-265.

Du Boulay GH, Brunt JNH, Colchester ACF, Hawkes DJ, Wallis A and Wicks D. (1987) Volume flow measurements of pulsatile flow by digitised cine angiography. Acta Radiologica suppl 369, 59-62.

Hawkes DJ, Colchester ACF, and Norbury R. (1988) The Measurement of Absolute Lumen Cross-Sectional Area and Lumen Shape in Quantitative Angiography. Presented at NATO ASI meeting "The Formation, Handling and Evaluation of Medical Images", Portugal.

Koenderink JJ (1988) Image structure. In Viergever MA and Todd-Pokropek A (eds) Mathematics and Computer Science in Medical Imaging, Berlin: Springer, 67-104. Pizer SM, Oliver WR, and Bloomberg SH (1987) Hierarchical shape description via the multi-resolution symmetric axis transforms. Pattern Analysis & Machine Intelligence PAMI-9 505-511.

Pizer SM, Oliver WR, Gauch JM and Bloomberg SH. (1988) Hierarchical figure-based shape description for medical imaging. In Mathematics and Computer Science in Medical Imaging, eds Viergever MA and Todd-Pokropek A. Berlin: Springer,365-387.

Ritchings RT and Colchester ACF. (1986) Detection of abnormalities on carotid angiograms using syntactic techniques. Pattern recognition Letters 4 367-374.

Ritchings RT, Colchester ACF and Wang H-Q. (1986) Knowledge based analysis of carotid arteriograms. Image and Vision Computing 3 217-222.

Rosenfeld A (1984) Multi-resolution image representation. In Digital Image Analysis ed Levialdi S. Pitman.

Rosenfeld A (1987) Pyramid algorithms for efficient vision, TR 1866, Center for Automation Research, Univ. Maryland, College Park, June 1987.

Shirai Y (1973) A context sensitive line-finder. Artificial Intelligence 4: 95-119.

Sleigh AC (1985) The extraction of boundaries using local measures driven by rules. BPRA 3rd International Conference, St. Andrew's, Scotland.

Wang H-Q, Ritchings RT and Colchester ACF. (1987) An image understanding system for carotid angiograms. Image and Vision Computing 5 79-84.

FAST TECHNIQUES FOR AUTOMATIC LOCAL PIXEL SHIFT AND RUBBER SHEET MASKING IN DIGITAL SUBTRACTION ANGIOGRAPHY

K.J. Zuiderveld, B.M. ter Haar Romeny and W. ten Hove

Department of Radiology, University Hospital of Utrecht
The Netherlands

ABSTRACT

A fast and completely automated method for reduction of movement artifacts in Digital Subtraction Angiography is presented. In the first part of the article the properties of the used matching algorithm with regard to image statistics are discussed; attention is also paid to fast implementations of the algorithm. In the second part of the article the characteristics of a practical implementation of the method are described, while the feasibility of the method is illustrated with some clinical results. We conclude that the proposed method, called Rubber Sheet Masking, compares favorably with convential interactive pixel shift methods.

1. INTRODUCTION

A major problem in clinical Digital Subtraction Angiography (DSA) is the occurrence of misregistration artifacts caused by voluntary or involuntary patient motion during the time between acquisition of images without contrast ("mask images") and acquisition of images with a contrast medium present in blood ("contrast images"). In order to alleviate this problem, techniques as remasking and pixel shift (Brody, 1985) have been developed and are usually on-line available within clinical DSA equipment.

Remasking, the selection of a new mask image (usually temporally closer to the contrast image), is routine in clinical practice. It does, however, not always solve the problem; continuous gradual motion creates artifacts in any pair of the sequence.

Pixel shift is generally implemented as a postprocessing program that allows fractional pixel shifting of the mask image in horizontal, vertical or oblique direction, thus giving the capability of removing small motion artifacts. Although clinical experience has shown that it can be of significant value in salvaging studies that otherwise would have been nondiagnostic, it is rarely applied within our department mainly for two reasons:

- The technique is not automated on the used clinical equipment. Manual registration of the mask image is done by "trial-and-error", which generally turns out to be time consuming;
- Patient motion generally is too complicated to be corrected uniformly across the whole image. The reregistration can be optimised for a certain anatomical region, but will be suboptimal in other parts of the image and may even cause or enhance local misregistration artifacts. As a result, global pixel shift is commonly applied on the same image more than once, making the procedure even more cumbersome.

In response to these observations, we developed a new method for the elimination of movement artifacts which adresses both problems sketched above: it does not require any operator interaction and automatically removes local motion artifacts across the whole subtraction image.

During the development of this method, called Rubber Sheet Masking (RSM), much attention has been paid to the properties of the used similarity measure with regard to image statistics, achievement of fast strategies for optimization of the similarity measure and the derivation of techniques for evaluating the reliability of pixel shift components. The first part of this paper summarizes the most significant observations we made in this area and is mainly focussed on methods for Automated Local Pixel Shift (ALPS).

With the RSM method the image is divided into a large number of subimages, after which ALPS is applied on each subimage. The extension of ALPS to RSM is nevertheless not straightforward; in the second part of the paper several issues regarding this extension are addressed, notably choosing the optimal subimage size, the correction of unreliable pixel shift estimates and calculating the geometrical deformation of the mask image.

The last part of this paper describes an actual RSM implementation currently under review at our department; some clinical results are briefly discussed.

2. AUTOMATED LOCAL PIXEL SHIFT (ALPS)

2.1 Suitable similarity measures

To achieve an automated registration of contrast and mask image, it is necessary to define a similarity measure which gives a quantitative indication of the success of the algorithm performing the registration. The most widely used similarity measure is probably the sum of the absolute differences (SAVD), see eg. Fitzpatrick et al. (1988) and Van Tran and Skalsky (1988). Although the SAVD measure produces excellent results in matching regions outside the blood vessels it may raise problems in areas where there is inflow of contrast medium, as was already pointed out by Venot and Leclerc (1984). Van Tran and Skalsky (1988) solve this problem by introducing a preprocessing step in which blood vessels are recognized and "masked out" with the calculation of the SAVD measure. Unfortunately this step requires a considerable amount of additional processing time.

An alternative similarity measure is the Deterministic Sign Change (DSC) proposed by Venot et al. (1984); with this method a discrete alternating pattern with depth δ is added to the contrast image $C(i,j)$ according to

$$\begin{aligned} C(i,j) &:= C(i,j) + \delta \qquad (i+j \text{ even}) \\ C(i,j) &:= C(i,j) - \delta \qquad (i+j \text{ odd}) \end{aligned} \tag{1}$$

after which the mask image $M(i,j)$ is subtracted. The DSC value can then be determined by counting the number of sign changes occurring by scanning the subtraction image $C(i,j) - M(i,j)$ line by line (or column by column). The DSC value is maximal with identical mask and contrast images and decreases as the match between mask and contrast image gets poorer.

Our observations confirmed the claim that the DSC similarity measure is generally robust, insensitive for the presence of vascular structures within the Region Of Interest (ROI) and computational cheap. We therefore decided to focus our research on ALPS algorithms using the DSC criterium as a similarity measure.

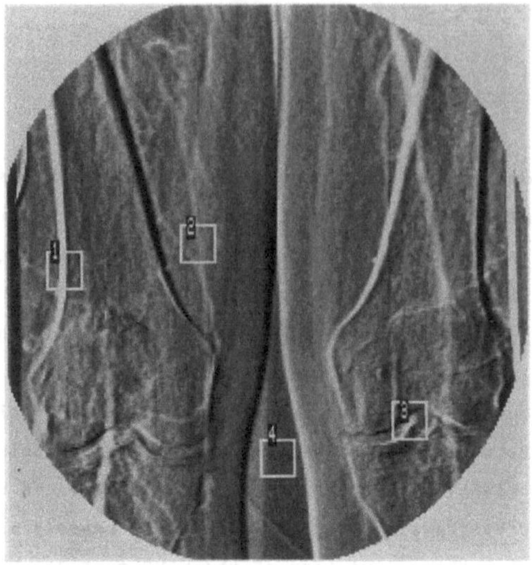

Fig. 1. DSA image of the knees (resolution 256x256). Motion artifacts were created by shifting the mask image one pixel in the x-direction; in the resulting subtraction image 4 ROI's of size 17x17 pixels containing various image structures were defined.

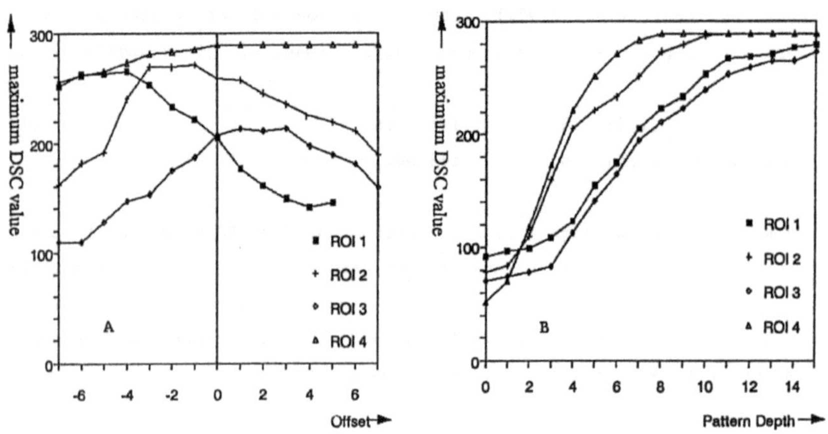

Fig. 2.

a. Maximum DSC value in each of 4 ROI's (see fig.1) as function of the greylevel offset Γ (constant DSC pattern depth δ=7, approximately twice the standard deviation of the subtraction image background noise).

b. Maximum DSC value as function of the pattern depth δ (at constant greylevel offset Γ=0).

2.2 Dependency of the DSC criterium on image statistics

A less obvious, but major drawback of the DSC criterium is the fact that it has a parameter: inappropriate values for the pattern depth δ lead to very poor results. Since image statistics largely vary on image position, examination type and especially acquisition hardware, adaptive pattern depths are virtually necessary.

In order to gain insight in the behaviour of the DSC values with varying local image characteristics, DSC values were evaluated using a clinical DSA image of the knees (see fig. 1). For each of the ROI's DSC values were calculated as function of the pixel shift (dx,dy), the pattern depth δ and a greylevel offset Γ added to the image; results are shown in fig.2.

From fig.2a we conclude that the optimal offset value Γ strongly depends on local image structure. Consequently, the DSC criterium will perform significantly better if Γ is allowed to vary across the image. This feature is an integral part of the DSC maximization algorithm as proposed by Venot and Leclerc (1984).

Fig. 2b mainly shows the obvious result that DSC values increase with δ; ROI's with severe motion artifacts tend to have considerable smaller DSC values with moderate values of δ. Note that the figure does not provide helpful clues for choosing the optimal pattern depth, i.e. the value where the calculated pixel shift is most reliable.

A suitable measure for shift vector component reliability turned out to be the normalized curvatures (2nd derivatives) of the DSC criterium, defined as

$$\begin{aligned} NCurx &= 2\ DSC(DX,DY) - DSC(DX+1,DY) - DSC(DX-1,DY) \\ NCury &= 2\ DSC(DX,DY) - DSC(DX,DY+1) - DSC(DX,DY-1) \end{aligned} \qquad (2)$$

where DSC(DX,DY) is the DSC value which is maximized at the integer displacements DX and DY.

Normalized DSC curvatures may become small or even zero in areas with structures in only one direction or without any structure at all; clinical examples are studies of the knee (in which vertical structures are dominant, eg. see ROI 1 of fig.1) and underexposed parts of the image (e.g. due to lead shielding). In such areas the calculated pixel shifts tend to get unreliable; areas with significant image structure however tend to have higher DSC curvatures as well as better pixel shift estimates.

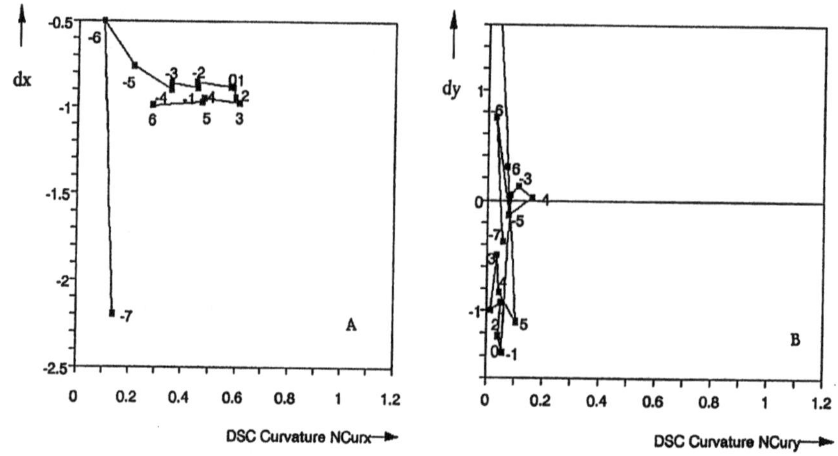

Fig. 3.

a. Normalized curvature NCurx and pixel shift dx as function of the greylevel offset Γ for ROI 3 (DSC pattern depth δ=7). This is a typical example of a ROI containing significant image structure.

b. Normalized curvature NCury and pixel shift dy as function of the greylevel offset Γ for ROI 2 (DSC pattern depth δ=7). This is a representative result for an area without much image structure.

Assuming that DSC curvature may be used to evaluate the reliability of a calculated pixel shift, and consequently providing a criterium for choosing suitable values for the DSC pattern depth δ, we investigated the dependence between normalized DSC curvatures and the calculated pixel shift (dx,dy) on both Γ and δ. Subpixel displacements were estimated using simple second order interpolation of DSC values obtained at integer displacement values:

$$dx = DX + \frac{DSC(DX+1,DY) - DSC(DX-1,DY)}{2\ NCurx} \qquad (3)$$

Results are showed in fig.3. In areas with significant image structure (fig.3a) DSC curvatures reach high values (in this example NCurx=0.6 for Γ=3), while near-optimal values for Γ also perform well. In case no significant image structures are present, maximum DSC curvatures are considerably smaller (fig.3b, where the maximum NCury=0.16 for Γ=-4), while small variations in Γ largely affect the calculated pixel shift.

Further experiments established that with reasonably large curvatures the optimum Γ value acquired from maximizing the DSC criterium approximately

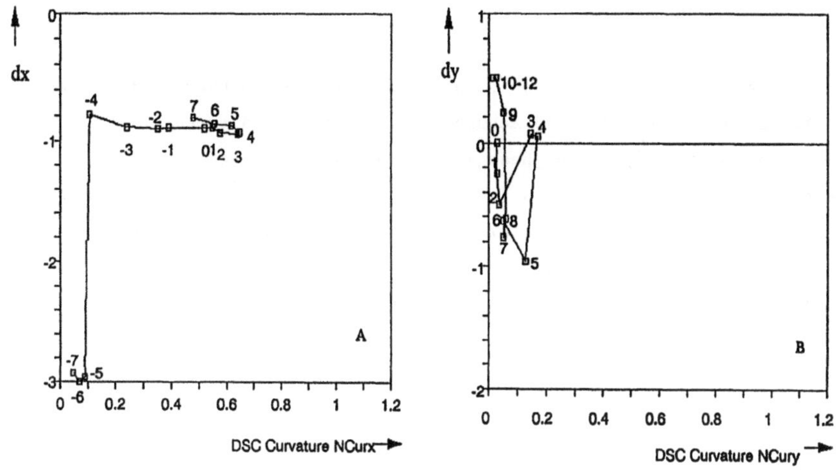

Fig. 4.

a. Normalized curvature NCurx and pixel shift dx as function of the pattern depth δ for ROI 3 (offset Γ=0).

b. Normalized curvatur NCury and pixel shift dy as function of the pattern depth δ for ROI 2 (offset Γ=0)

equals the Γ value obtained by maximizing DSC curvature (and hence reliability, see fig.2a versus fig.3a), while with small curvatures this is generally not the case (fig.2a versus fig.3b). This turns out to be helpful for improving the speed of maximization algorithms in areas with adequate image structure, since optimising the DSC value is computational a factor 3 less expensive compared with optimising DSC curvatures.

A typical relationship between DSC curvature, pixel shift and pattern depth in areas with significant image structure can be found in fig. 4a. Low values of δ lead to low curvatures and incorrect pixel shifts, while with larger δ high curvatures are obtained (maximum curvature is 0.65 for δ=11); again small fluctuations around the optimal value of δ hardly affect the estimated pixel shift.

In areas without image structure curvatures are low; in fig.4b a typical result is demonstrated. In such areas "optimal" values for δ are typically quite low.

Other experiments confirmed that regions containing significant image structure in general require larger pattern depths (with our clinical DSA images typically > 10) to obtain significant curvatures, whereas homogeneous areas should be processed using small δ values.

2.3 Optimization strategies

The choice of the optimization strategy depends on the number of parameters to be optimised, the desired accuracy and the required speed of the algorithm. Obviously the highest match quality can be achieved if all probable physical distortions (including rotation and nonisotropic scaling) are considered during the optimization of the similarity measure; this unfortunately leads to large parameter spaces and, as a result, complex optimization algorithms (as genetic search and simulated annealing; see eg. Fitzpatrick et al. (1988)) which are unrealistic for actual clinical use.

In order to accomplish clinical relevant execution times, it is necessary to introduce sequential parameter optimization (see Van Tran and Sklansky (1988)), limit the number of parameters and/or introduce specific knowledge on the properties of the similarity measure.

As the name already implies, Automatic Local Pixel Shift (ALPS) techniques should only be concerned with artifacts introduced by translation. Using the DSC criterium as similarity measure requires two additional search parameters, namely a greylevel offset (necessary to correct for fluctuations in the X-ray power) and the DSC pattern depth δ.

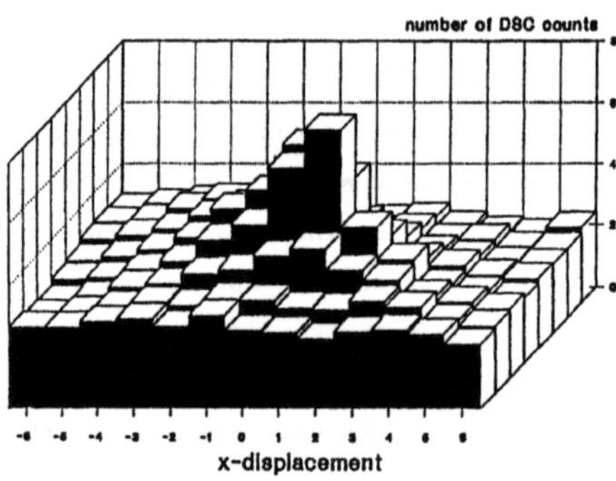

Fig. 5. Representative example of the relationship between pixel shift (dx,dy) and DSC count with optimal values for pattern depth and greylevel offset. Note the well defined maximum in the DSC count and the absence of local maxima.

The robustness of the DSC criterium, facilitating fast and very effective optimization strategies, is demonstrated in fig.5 where the DSC count at optimal greylevel offset and DSC pattern depth is shown as function of the displacement (dx,dy); this figure is fairly representative for ROI's in which significant image structure is present. In such cases simple algorithms (eg. hill-climbing, see Venot and Leclerc (1984)) turn out to be remarkably appropriate for DSC value optimization.

In areas without much image structure (where ALPS is hardly useful anyhow), using a straight-forward hill-climbing algorithm may give incorrect results due to occurrences of local DSC maxima. Since in this situation the normalized curvatures NCurx and NCury generally have small values, the hill-climbing algorithm can be modified to cope with this condition.

We observed that optimization of the offset Γ and the pixel shift (dx,dy) can be done sequentially in that order; assuming there is sufficient structure present in the image, the estimated pixel shift is hardly affected by small fluctuations in the greylevel offset Γ (see fig. 3a). As a result, intermediate DSC evaluations at different greylevel offsets (eg. at Γ-1 and Γ+1 as proposed by Venot and Leclerc (1984)) can be omitted, resulting in an almost threefold reduction of execution time without significantly affecting the quality of the registration.

Finally, the optimization strategy should provide some mechanism for adapting the pattern depth δ to local image statistics. In most areas with significant image structure virtually any reasonable pattern depth gives satisfying results (see fig.4a); an initial δ value of twice the standard deviation of the subtraction image background noise provided us with general excellent results. Only in areas with very large motion artifacts (and consequently low DSC values) larger pattern depths turned out to be beneficial; in these cases our optimization algorithm increased the pattern depth (with steps equal to the background noise standard deviation) until a maximum was found for the DSC curvatures.

2.4 ALPS: Practical implementation and an example of application

Since our goal was the achievement of a clinical useful algorithm, we implemented a very fast method which may give suboptimal results. Its main characteristics are:

- optimization of DSC values is performed by a hill-climbing procedure;
- optimization of greylevel offset Γ is followed by optimization of the pixel shift (dx,dy);
- DSC curvatures give an indication of vector component reliabilities;
- the DSC pattern depth is stepwise increased when maximized DSC value is low (typically 1/3 of the number of pixels in the ROI);
- subpixel displacements are obtained by second order interpolation of DSC values at integer pixel displacements.

The method was implemented on a Philips Analytical Processing Unit (APU); the APU is a minicomputer system dedicated for postprocessing of DSA images. DSA images were downloaded from the DSA system (Philips DVI-2 V) to the APU by a direct (IEEE-488) link. Since the APU is slow (subtraction of two 256x256 images takes approximately 20 seconds) the algorithm had to be implemented in assembly language to arrive at acceptable execution times.

A typical result of our ALPS algorithm is presented in fig.6. In this example only the horizontal pixel shift was significant. The total time required for the reregistration itself was approximately 12 seconds, which would be just clinically acceptable. Unfortunately the transfer of images, subtraction of mask and contrast image and selection of ROI's takes much longer; hence only an implementation on the DSA equipment itself (preferably in hardware) would make the ALPS method clinically feasible.

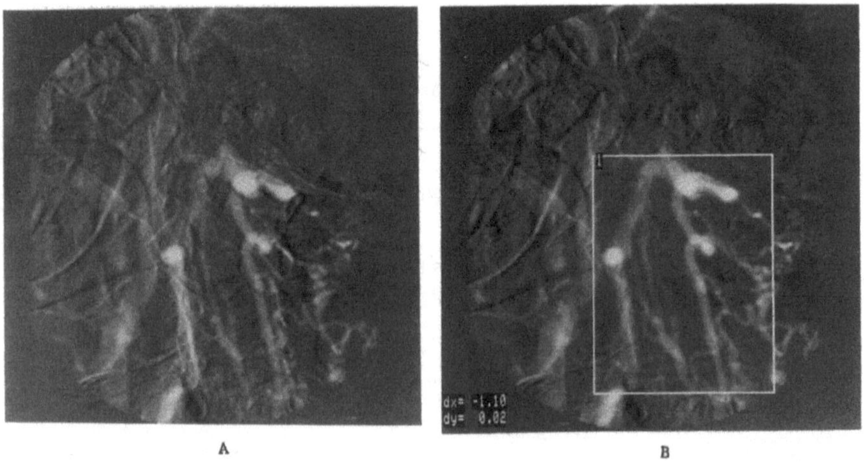

A B

Fig. 6. Typical result of the proposed ALPS algorithm with an arterial DSC of a patient with TIA's of the left cerebral hemisphere.
a. Clinical image of size 256x256 with severe misregistration artifacts.
b. Result with ALPS applied within the ROI. Note the improved visibility of small vessels.

3. RUBBER SHEET MASKING

3.1 Introduction

The Rubber Sheet Masking (RSM) technique is a logical extension of ALPS; with this technique the contrast image is divided into a large number of square subimages, to each of which the DSC maximization associated with ALPS is applied. Unreliable pixel shift estimates in the resulting displacement vector field are then corrected, after which the mask image is geometrically transformed based on the corrected vector field.

The extension of ALPS to RSM is not trivial. Several issues must be addressed, notably choosing the subimage size, methods for correction of unreliable displacement vectors, the required geometrical transformation and some implementation issues needed to achieve adequate execution speed.

3.2 Subimage size

A trade off exists between local detail ("scale") and the reliability of the calculated local pixel shift. While with smaller subimages local sensitive but less reliable vector fields are found, the more reliable displacement estimates associated with larger subimages suffer from insensitivity to local pixel shifts.

An optimal subimage size for the current RSM algorithm seems to be 32x32 pixels; in order to get more spatial information 50% overlapping subimages are used, resulting in 900 displacement vectors for 512x512 images.

A possible refinement of the RSM algorithm currently under investigation allows for automatic adjustment of the window size; it is based on the assumption that the window size can be decreased as long the corresponding normalized DSC curvature (and hence the vector component reliability) remains above a given threshold.

3.3 Correction of unreliable vectors

Since with Rubber Sheet Masking the geometrical transformation of the mask image will be based on multiple shift vectors, inaccurate pixel shift estimations may lead to severe artifacts in the final subtraction image, in opposite to ALPS where low DSC curvatures merely indicate a "don't care" situation.

We therefore developed a simple algorithm facilitating the correction of unreliable vectors, while preserving reliable shift vector estimates. A prerequisite for the proposed scheme is an estimation for the reliability of each vector component; as already was pointed out in a previous section, normalized DSC curvatures are well suited for this purpose. An additional clue on vector reliability can be obtained by taking into account that, since the human body behaves like an elastic medium, pixel shifts in adjacent subimages tend to have approximately the same value. In case a shift vector component significantly deviates from the corresponding component of adjacent subimages it is likely that the shift vector component is less reliable.

Satisfying results have been obtained using the following definitions:

$$R_x = \alpha\ \mathrm{NCurx} + (1 - \alpha)\ (\mathrm{Neighx}) \qquad (4)$$
$$R_y = \alpha\ \mathrm{NCury} + (1 - \alpha)\ (\mathrm{Neighy})$$

where R_x and R_y are the calculated component reliabilities with values between 0 (entirely unreliable) and 1 (completely reliable), Neighx and Neighy the number of adjacent shift vectors having x-components resp. y-components which differ less than 1 pixel from the calculated shift vector and α a weighing factor (typ. 0.6).

The actual correction of unreliable vectors is performed by iterative relaxation (see eg. Rosenfeld and Kak (1982)). With this algorithm initial weighing factors W proportional to differences in reliabilities are calculated, according to

$$W_x(a,b) = R_x(DX,DY) - R_x(DX+a,DY+b) \qquad (5)$$
$$W_y(a,b) = R_y(DX,DY) - R_y(DX+a,DY+b)$$

Negative weighing factors W are then set equal to zero, lest reliable shift values are affected by less reliable ones:

$$W_x(a,b) := 0 \qquad \text{if} \quad W_x(a,b) < 0 \qquad (6)$$
$$W_y(a,b) := 0 \qquad \text{if} \quad W_y(a,b) < 0$$

Finally for each vector a new value is calculated by means of a weighted average with the component values of its neighbors:

$$dx(DX,DY) := \big(dx(DX,DY) + dx(DX+1,DY)\, W_x(1,0) + dx(DX-1,DY)\, W_x(-1,0) + dx(DX,DY+1)\, W_x(0,1) + dx(DX,DY-1)\, W_x(0,-1)\big) \;/\; \big(1 + W_x(0,1) + W_x(-1,0) + W_x(0,1) + W_x(0,-1) \big) \qquad (7a)$$

and

$$dy(DX,DY) := \big(dy(DX,DY) + dy(DX+1,DY)\, W_y(1,0) + dy(DX-1,DY)\, W_y(-1,0) + dy(DX,DY+1)\, W_y(0,1) + dy(DX,DY-1)\, W_y(0,-1)\big) \;/\; \big(1 + W_y(0,1) + W_y(-1,0) + W_y(0,1) + W_y(0,-1) \big) \qquad (7b)$$

As the number of iterations increases, the vector field becomes smoother implying a reduced sensitivity to local pixel shifts. Experience showed that terminating the process after two iterations is a good compromise: all relevant unreliable pixel shift vectors are eliminated, while the amount of smoothing is still acceptable. A practical result of the algorithm can be found in fig.7.

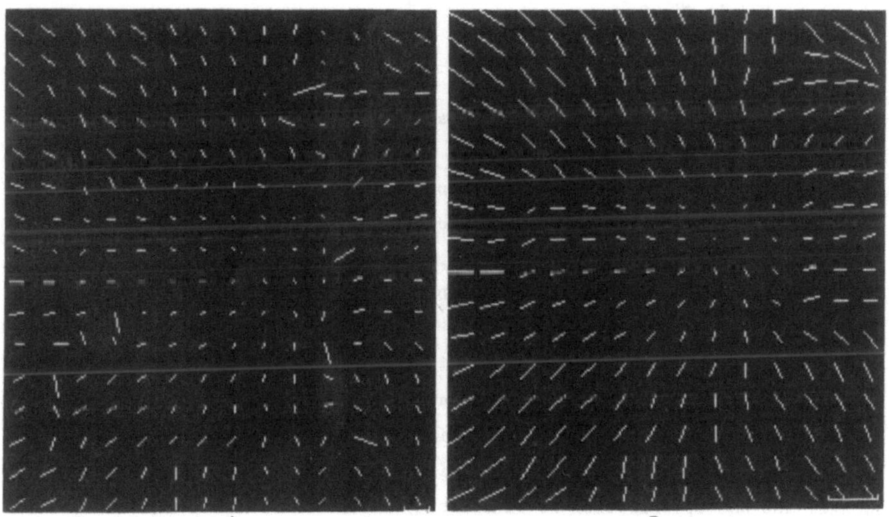

A B

Fig. 7. Example of the correction of unreliable pixel shift vectors. The original mask image (256x256) was slightly zoomed out in order to create artificial local displacements in the subtraction image.

a. First estimate of vector field representing the length and direction of local pixel shifts.

b. Vector field after correction of the unreliable vectors.

Note the differences in scale (see marker in lower right corner of the images: length is 8 pixels).

3.4 Geometrical transform of the mask image

The last step of the proposed method consists of a geometrical deformation of the mask image (warping) based on the corrected vector field. Since the shift vectors specify subpixel displacements, direct application to the contrast image is impractical: most pixels would be mapped into positions located between the pixel pattern of the original contrast image which makes the calculation of a new contrast image arduous.

These problems can be circumvented by transforming the mask image instead of the contrast image. The desired transform is obtained by inverting the vector field associated with the contrast image, after which displacement vectors associated with any pixel of the mask image can be calculated by a simple bilinear interpolation of 4 vectors associated with the obtained vector field. Since the obtained vectors specify subpixel displacements, new greyvalues for the mask image are calculated by bilinear interpolation of greyvalues corresponding with surrounding pixels.

The local deformation properties of the Rubber Sheet Masking method are demonstrated in fig.8.

3.5 Achievement of adequate execution speed

The calculation of the vector field represents the major part of the time needed for the registration process. However, execution time can be dramatically decreased by using the calculated (dx,dy) and Γ for a given subimage as initial estimates for the DSC optimization for any adjacent subimage as is demonstrated in fig.9.

Further reduction of processing time can be achieved by identifying areas without structure, since these areas will generally have an extremely high DSC count. In these regions (which typically represent 30%-70% of the total image) the DSC optimization is immediately stopped while the reliabilities R_x and R_y of vectors corresponding with these areas are set to zero.

Finally the processing time can be reduced by limiting allowed combinations of (dx,dy) values to a predefined area; if no maximum DSC value can be found in the area, the vector reliability is set to zero.

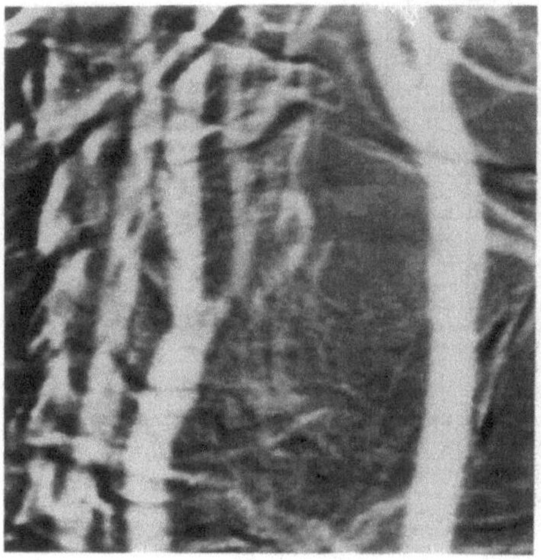

A

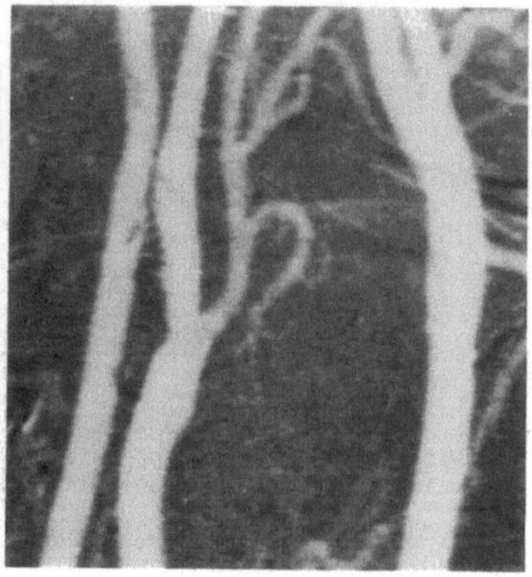

B

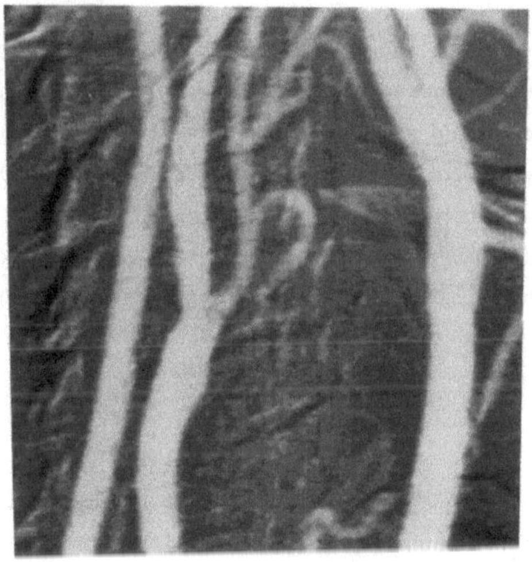

C

Figure 8.
Example of the local deformation properties of the proposed RSM method; the geometrical transform of the mask image is based on the vector field shown in fig.7b.

a. Subtraction image containing artefacts created by slightly zooming out the original mask image.
b. Rubber Sheet Masked subtraction image.
c. Subtraction image resulting from subtracting the contrast image and the original undistorted mask image of the original study.

Note that in the RSM image movement artifacts in the right common and internal carotid arteries are less visible compared with the undistorted subtraction image corresponding to the original study.

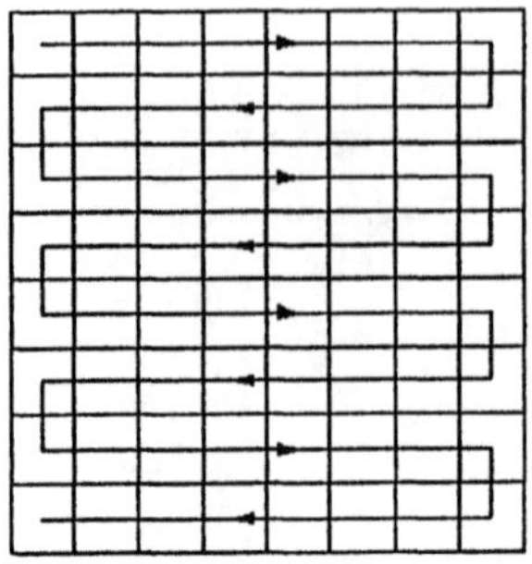

Fig. 9. Scan order associated with the RSM technique to facilitate improved processing speed by basing initial estimates for the DSC optimization on results from previous processed adjacent subregions.

3.6 Clinical results

On the current hardware the RSM method is rather time consuming, approximately 10 minutes including preprocessing of the DSA images (reconstruction from raw data, circular masking and subtraction of the original images). Therefore the method has been implemented as a program which is able to calculate multiple RSM images in batch mode.
Sets of diagnostic sequences can be processed by the technologist or radiologist in batch mode, i.e. automatically. No parameters, other than image selection, have to be set or specified.

Two examples of clinical results obtained in different anatomical regions can be found in fig.10 and fig.11, which clearly demonstrate the reduction of image artefacts caused by movement of the head and ribs. The RSM method seems to work well with DSA of the extremities. The complex motion as found with swallowing artifacts in carotid artery studies can just partially be corrected with this method. Further clinical studies including comparison with conventional pixel shift are currently performed to determine which investigations with DSA benefit most from Rubber Sheet Masking.

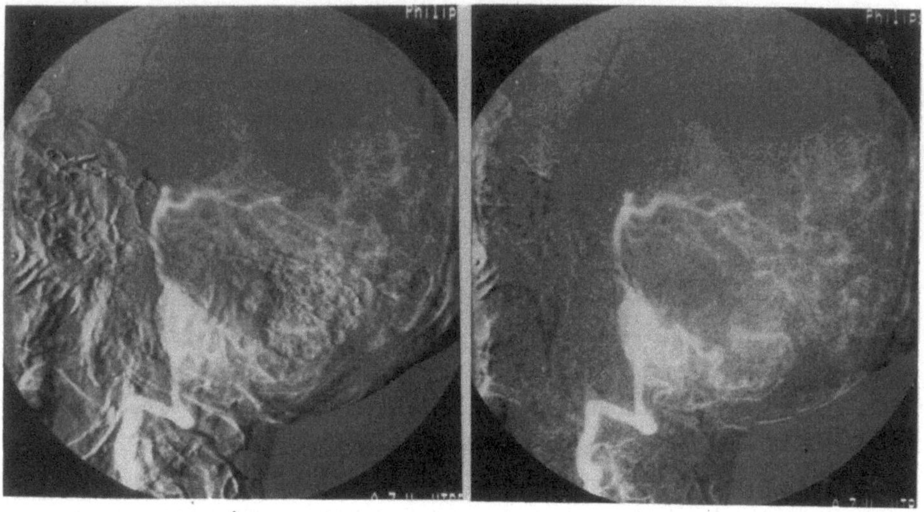

A B

Fig. 10. Example of a clinical application of the RSM method.
a. Original subtraction image (size 512x512)
b. Result obtained with RSM.
Motion artefacts have almost been completely eliminated, especially those which were introduced by movement of the mastoid cells. As a result better diagnosis of the smaller arteries at the lower side of the cerebellum is possible.

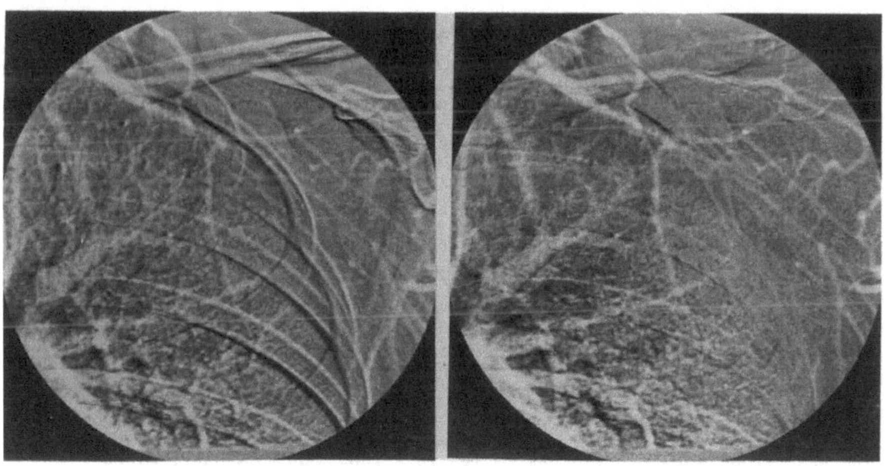

A B

Fig. 11. Example of a clinical application of the RSM method.
a. Original subtraction image (size 512x512)
b. Result obtained with RSM.
Artefacts caused by movement of the ribs are almost completely eliminated. In the RSM image the axillar artery can be much better traced and clearly shows that there is no stop in this artery; instead the vessel is hypoplastic. This can not be seen on the original image.

4. DISCUSSION

With Automatic Local Pixel Shift (ALPS) as proposed in this paper it is possible to eliminate registration artifacts within a ROI in a very effective way. The simplicity of the method makes implementation in current DSA equipment rather simple; the registration process takes only a few seconds, which is fully acceptable for clinical purposes.

The Rubber Sheet Masking method can be considered as the "ultimate" pixel shift; it is fully automatic and able to reduce the visibility of local misregistration artifacts in any part of the image. Experience in our radiology department has shown that in general RSM images seem to be superior to any global pixel shift applied on the original image.

The RSM method is simple to implement and is relatively fast; the current implementation does not use any dedicated hardware. DSA systems using faster computer technology and/or special image processing hardware compared with our APU may be able to perform Rubber Sheet Masking in only a few seconds; an implementation of RSM using transputer hardware is currently considered.

A crucial part of the proposed method is the elimination of unreliable pixel shifts; omission of this step could lead to local misregistration artifacts, which would make application of the method potentially dangerous. The algorithms developed for this purpose are simple and fast; it is assumed that only a few subimages will have unreliable shifts. If this assumption is not valid, the proposed algorithm may not be able to correct all unreliable vectors which can lead to serious misregistration artifacts. Although current work is mainly directed towards an improved detection and correction of unreliable vectors using multigrid relaxation (see eg. Terzopoulos (1986)), we did not encounter serious misregistration artifacts until now.

The RSM method does not address motion in a fundamental manner; misregistration artifacts caused by three-dimensional motion of only a part of the projected human body (eg. the human heart) are probably too complex to be completely correctable with projection images from a single direction. Secondly, the RSM method is not capable of completely correcting artifacts owing to movements which are not translational.

In general the results obtained with the RSM method are very satisfying; in our department the application of this method has become a standard clinical tool in correcting local misregistration artifacts. For wide application in routine DSA the speed should be optimised. The method is inherently suitable for single pushbutton implementation in DSA viewing equipment due to its automatic and adaptive nature.

REFERENCES

Brody, W.R. (1985). Digital subtraction angiography. IEEE Transactions on Nuclear Sciences, NS-29(3), pp. 1176-1180.

Fitzpatrick, J.M., Pickens, D.R., Mandava, V.R. and Grefenstette, J.J. (1988). The reduction of motion artifacts in digital subtraction angiography by geometrical image transformation. SPIE Medical Imaging II, Newport Beach, Vol.914, pp.379-386.

Rosenfeld, A. and Kak. A.C. (1982). Digital Picture Processing,. Vol.2, 2nd edition, pp.152-158.

Terzopoulos D. (1986). Image Analysis using multigid relaxation methods. IEEE Transactions on Pattern Analysis and Machine Intelligence, Vol. PAMI-8(2), pp. 129-139.

Van Tran, L. and Sklansky, J. (1988). Flexible mask subtraction for digital angiography. SPIE Hybrid Image and Signal Processing, Orlando - in press.

Venot, A. and Leclerc, V. (1984). Automated correction of patient motion and gray values prior to subtraction in digitized angiography. IEEE Transactions on Medical Imaging, MI-3:4:179-186.

Venot, A., Lebruchec, J.F. and Roucayrol, J.C. (1984). A new class of similarity measures for robust image registration. Computer Vision and Image Processing, 28:176-184.

List of Participants

Aarsvold, J.N.
Department of Radiology
Radiology Research Laboratory
Arizona Health Sciences Centre
Tuscon, AZ, 85724
U.S.A.

Arridge, Simon
University College Hospital
Medical Physics & Bioengineering
1st Floor, Shropshire House
1-20 Capper Street
London WC1E 6JA
England

Backer, E.
Delft University of Technology
Dept of Electrical Engineering
Mekelweg 4
Delft
The Netherlands

Barillot, C.
Neurosurgery Dept.
S.I.M. Laboratory
Ponchaillon Hospital
35000 Rennes
France

Barrett, H.H.
University of Arizona
Optical Sciences Center
Tucson AZ 85721
U.S.A.

Batista, J.
Dept. Eng Electrotecnica
Universidade de Coimbra
300 Coimbra
Portugal

Beatty, P.T.
2335 Huron Circle
Durham NC 27707
U.S.A.

Bergmann, H.
Allgemeines Krankenhaus der Stadt Wien
Abteilung fur Nuclearmedizin
Der 2 Med.Univ.-Klinik Wien
Garnisongasse 13,
A-1090 Wien
Austria

Bidaut, Luc
Hopital Erasme
Cyclotron Pet/Nuclear Med
Univ. Libre be Bruxelles
808 Route de Lennick
B1070 Brussels
Belgium

Borges, A. Rui
Dept. Electronica
Universidade De Aveiro
3800 Aveiro
Portgual

Burgess, Art
Dept. of Radiology
2211 Wesbrook Mall
Vancouver BC
Canada V5Z 1M9

Buss, Daryl D.
Department of Physiological Sciences
Box J-144 Health Science Centre
University of Florida
Gainesville, Florida 32610-0144
U.S.A.

Cawley, M.G.
Medical Physics
(Electronics 3rd floor)
Queen Elizabeth Hospital,
Edgbaston,
Birmingham, B15 2Th
England

Cochran, E.
350 N. Harrison Road. #270
Tucson, Arizona 85748
U.S.A.

Colchester, A.C.F.
Dept. of Neurology
Guy's Hospital
London Bridge SE1 9RT
U.K.

Crespi, Andrea
Presidio Ospedaliero Multizonale
"San Gerardo" USSL 64
Servizio di Fisica Sanitaria
Via Donizetti 106
20052 Monza (MI)
Italy

Dary, Christophe
DIMI
BP 1005
Hospital G. & R. Laennec
44035 Nantes Cedex 01
France

Davies, Ian
Dept. of Psychology
University of Surrey
Guildford
Surrey GU2 5HX
U.K.

Dechamps, Jos
Agfa-Gevaert
PRO/Systems Analysis
Septestraat 27
2510 Mortsel (Antwerpen)
Belgium

Defrise, Michel
Akademisch Ziekenhuis
Dienst Radioisotopen
Larbeeklaan 10
1090 Brussels
Belgium

Demeure, Roger J.J.
Service De Radiologie
Universite Catholique De Louvain
Cliniques Universitaires Saint-Luc
Avenue Hippocrate UCL 10/29.
42200 Bruxelles
Belgium

Duk, Woo Ro
Medical Image Processing Group
University of Pennsylvania
418 Service Drive
Blockley Hall 4th floor
Philadelphia PA 19104
U.S.A.

Duncan, J.S.
Yale University
Department of Diagnostic Radiology
School of Medicine
333 Cedar Street
P.O. Box 3333 New Haven,
Connecticut 06510
U.S.A.

Ericksen, James
Dept. of Computer Science
University of North Carolina
Chapel Hill NC 27599-3175
U.S.A.

Eschner, W.
Abt. Nuklearmedizin
Zentrum Radiologie der Universitat Gottingen
Robert-Koch-Str. 40
D-3400 Gottingen
FRG

Ferreira, Paulo Jorge S-G
Dept. of Electronics
Universidade Aveiro
3800 Portugal

Fidler V.
University clinic of nuclear medicine
Ljubljana,
Zolaska 7.
Ljubljana
Yugoslavia

Figueiredo, Mario
Centro De Analise e Processamento
de Sinais
I.S.T.
1096 Lisboa Codex
Portugal

Floyd Jr., C.E.
Box 3949
Duke University Medical Center
Durham,
North Carolina 27710
U.S.A.

Fonroget, J.
Service de Medecine Nucleaire
Centre Hospitalier Regional et
Universitaire d'Amiens
Hopital Sud
Ave. Rene Laennec Salouel
BP 3009
80030 Amiens Cedex
France

Fox, John
Imperial Cancer Research Fund Laboratories
P.O. Box 123
Lincoln's Inn Fields
London WC2A 3PX
UK

Friboulet, D.
Laboratoire Traitement Du Signal
et Ultrasons
LEA CNRS 1216
Batiment 502
20, Avenue Albert Einstein,
69621 Villeurbanne Cedex
France

Giakoumakis, G.E.
University of Ioannina
Department of Physics
Solid State Division
Applied Physics Laboratory
Ioannina
Greece

Gilland, David
University of North Carolina
CB # 757552 MacNider Hall 202H
Chapel Hill
NC 27514
U.S.A.

Godinho, F.
Instituto de Medicina Nuclear
Faculdade de Medicina de Lisboa
Av Prof. Egas Moniz
1600 Lisbon
Portugal

Gomez, L.
Universidad De Santiago De Compostela
Facultad De Fisica
Dept Electronica
15706-Santiago De Compostela
Spain

Green, Alan
Cancer Research Campaign Laboratories
Department of Medical Oncology
Charing Cross Hospital
Fulham Palace Road
London W6 8RF
England

Guedon, Jean-Pierre
DIMI - CHR NANTES
Hospital G & R Laennec
44035 St Herblain Cedex 01
France

Guilhem, J.B.
Service de Radiotherapie
Centre Hospitalier Regional
BP67 09
45067 Orleans Cedex 2
France

Gullberg, G.T.
Dept of Radiology
Medical Imaging Research Laboratory
Ae215 School of Medicine
Salt Lake City,
Utah 84132
U.S.A.

Hawkes, Dave
Division of Radiological Sciences
U.M.D.S.
Guy's Hospital
St Thomas Street
London SE1 9TR
U.K.

Herbst, C.P.
Department of Biophysics
Medical Faculty
University of The O.F.S.
P.O Box 339
Bloemfontein 9300
Republic of South Africa

Herman, G.T.
Medical Imaging Section
Dept. of Radiology
Hosp. of the U. of Pennsylvania
3400 Spruce St.
Philadelphia PA 19104-4283
U.S.A.

Higgins, W.E.
Biodynamics Research Unit
Mayo Clinic
Rochester, Minnesota, 55905
U.S.A.

Hoehne K.H.
Universitats-Krankenhaus Eppendorf
Int. fur Mathematik und
Datenverarbeitung in der Medizin
Martinistrasse 52
2000 Hamburg 20
F.R.G.

Houston, Alex
Dept. of Nuclear Medicine
Royal Naval Hospital Haslar
Gosport Hants PO122AA
U.K.

Jaszczak, Ron
Duke University Medical Centre
Box 3949.
Durham
North Carolina 27710
U.S.A.

Johnson, Valen
Dept. of Statistics
5734 S University
University of Chicago
Chicago Ill. 60637
U.S.A.

Jonkers, G.
Koninklijke/Shell-Laboratorium,Amsterdam
Shell Research B.V.
Postbus 3003,
1003 AA Amsterdam
The Netherlands

Jungke, Manfred
Abt. NMR
Stiftung Deutsche Klinik fur Diagnostik GmbH.
Aukammalle 33,
6200 Wiesbaden 1, Post. 2149
F.R.G.

Konig, H.A.
Siemens AG,
MR Division,
Postfach 3260,
D-8520 Erlangen
F.R.G.

Liehn, J.C.
Institut Jean Godinot
BP 171
F 51056 Reims Cedex
France

Llacer, Jorge
Lawrence Berkeley Laboratory
Cyclotron Road
Berkeley
California 94720
U.S.A.

Lokner, D.
Odsjek za fisiku
Klinicka bolnica Dr Mladen Stojanovic
Vinogradska 29
Zagreb
Yugoslavia

Luong van Tran
Dept of Electrical Engineering
Univeristy of California
Irvine CA 92717
U.S.A.

Maguire Jr., G.Q.
Columbia University
Dept of Computer Science
500 West 120 St.,
New York NY 10027-7031
U.S.A.

Malmin, R.E.
Siemens Gammasonics, Inc.
2000 Nuclear Drive
Des Plaines, IL 60018,
U.S.A.

Marques, Xavier J.A.L.
Praca Dr P Teotonio Pereira 22
4300 Porto
Portugal

Mendonca, Ana Maria
Faculdade de Engenharia
Dept.de Engenharia Electrotecnia
Rua dos Bragas
4099 Porto Codex
Portugal

Menhardt W.
Philips GmbH Forschunglab.
Postf 54 08 40
2000 Hamburg 54
F.R.G.

Metz, C.E.
Dept of Radiology
Box 225
5871 So. Maryland Ave
Chicago 11 60637
U.S.A.

Michel, C.
Universite Catholique de Louvain
Faculte de Medecine
et Centre de Recherche du Cyclotron
2, Chemin du Cyclotron
B-1348 Louvain-la-neuve
Belgium

Nahmias, Claude
Asoc. Prof. (Radiology)
Department of Nuclear Medicine
McMaster University Medical Centre
Faculty of Health Sciences200 Main Street West,
Hamilton, Ontario, L8N 3Z5
Canada

Natarajan, K.
3d Floor (South)
Medical Physics
Queen Elizabeth Hospital
Edgbaston
Birmingham, B15 2TH.
U.K.

Novario, Raffaele
Servizio di Fisica Sanitaria
Presidio Multizonale Ospedaliero
Viale Borri, 57-2100 Varese
Italy

Nunes, A.
3, chemin Champus Fusils
F-01630 Saint Genis-Pouilly
France

O, Ying Lie
Zwaardemakerlaan 23
3571 ZA Utrecht
The Netherlands

Oliver, Bill
Dept. of Computer Science
Sitterson Hall
University of North Carolina
Chapel Hill NC 27599-3175
U.S.A.

Pisani, P.
CNR Institute of Clincal Physiology
University of Pisa
Via Savi 8
56100 Pisa
Italy

Pizer, S.M.
University of North Carolina
Dept. of Computer Science
Sitterson Hall
Chapel Hill NC 27599-3175
U.S.A.

Pommert, A.
Universitat Hamburg
Universitats-Krankenhaus Eppendorf
Institut f. Mathematik u. Datenverarbeitung
in der Medizin
Martinistr 52
2000 Hamburg 20
F.R.G.

Preteux, F.
Ecole des Mines de Paris
60 Boulevard St Michel
Paris 75006
France

Robinson, Glynn Peter
Dept. of Meurology
Guy's Hospital
St Thomas St.
London SE1 9TR
U.K.

Rolland, J.P.
Department of Radiology
Radiology Research Laboratory
Arizona Health Sciences Centre
Tuscon, AZ 85724
U.S.A.

Roos, P.
Delft University of Technology
Faculty of Applied Mathematics and Informations
Room 2.312
P.O Box 356
2600 AJ Delft
The Netherlands

Rosin, Paul
Dept of Meurology
Guy's Hospital
St Thomas St.
London SE1 9TR
U.K.

Rowe, R.K.
Optical Sciences Center
University of Arizona
Tucson, Arizona 85721
U.S.A.

Sa, Luis
Dept. of Electrical and Electronic Engineering
Univ. of Coimbra
Coimbra
Portugal

Sahiner, A.V.
School of Computer Science,
PCL, 115 New Cavendish Street
London W1M 8JS
U.K.

Shepp, L.A.
A T & T Bell Labs
Murray Hill
NJ 07974
U.S.A.

Silva, Augusto
Dept. of Electronics
Universidade Aveiro
3800 Portugal

Sood, A.
Dept of Computer Science
George Mason University
4400 University Drive
Fairfax VA 22030-4444
U.S.A.

Sousa Santos, M.B.
Department of Electronics & Telecomunications
University of Aveiro
3800 Averio
Portugal

Sousa Pereira, A.M.M.
Dept of Electronics and Telecomunications
University of Aveiro
3800 Aveiro
Portugal

Staib, Lawrence
Dept. of Diagnostic Radiology
Yale University
333 Cedar St. (BML-332)
New Haven, CT 06510
U.S.A.

Tan, A.C.
Dept. of Medical Physics
10-20 Shropshire House
University College London
London WC1 6BT
U.K.

Todd-Pokropek, A.
Dept. of Medical Physics
University College London
Gower St.,
London WC1E 6BT
U.K.

Toet, A.
TNO Institute for Perception
Kampweg 5
Soesterberg
The Netherlands

Townsend, D.W.
Hopital Cantonal Universitaire
Batiment d'Appui - 1er etage
211 Geneve 4, le bilnet
Switzerland

Tsui, B.M.W.
CB#7575
152 MacNidle Hall
University of North Carolina
Chapel Hill, NC 27599-7575
U.S.A.

van Elmbt, L.
Cliniques Universitaires Saint-Luc
Service De Medecine Nucleaire
U.C.L. 10/17.80
B-1200 Bruxelles
Belgium

van Dijke, M.C.A.
Delft University of Technology
Faculty of Applied Mathematics and Informatics
Room 2. 312
P.O Box 356
2600 AJ Delft
The Netherlands

Veklerov,E.
Lawrence Berkeley Laboratory Cyclotron Road
Berkeley, California 94530
U.S.A.

Verrier, J.P.
Department of Medical Physics & Bioengineering
University College London
1st Floor Shropshire House
10-20 Capper Street
London WC1 7JA
U.K.

Viergever, M.A.
Image Processing Group
University Hospital at the Uithof
Heidelberglaan 100
3584 CX Utrecht,
The Netherlands

Vuylsteke, P.
Agfa-Gevaert N.V.
Septestraat 27 - B-2510 Mortsel (Antwerpen)
Belgium

Walker, Nick
Imperial Cancer Research Fund Laboratories
P.O. Box 123
Lincoln's Inn Fields
London WC2A 3PX
UK

Wascher, D.I.
Inst. F. Digitale Bildverarbeitung
Wastiangasse 6
8010 Graz
Austria

Wechsler, H.
Dept of Computer Science
George Mason University
4400 University Drive
Fairfax VA 22030-4444
U.S.A.

Yahyai, Abdelaziz
Universite Louis Pasteur
Laboratorire De Thermologie Biomedicale
Faculte De Medecine -
Ucr Sciences Biomedicales
1, rue Humann
67085 Strasbourg
France

Zuiderveld, K.
Department of Radiology
University Hospital Utrecht
Catharijnesingel 101
3511 GV Utrecht
The Netherlands

INDEX

NATO ASI Series F

Including Special Programmes on Sensory Systems for Robotic Control (ROB) and on Advanced Educational Technology (AET)

Vol. 64: Sensory Robotics for the Handling of Limp Materials. Edited by P. M. Taylor. IX, 343 pages. 1990. *(ROB)*

Vol. 65: Mapping and Spatial Modelling for Navigation. Edited by L. F. Pau. VIII, 357 pages. 1990. *(ROB)*

Vol. 66: Sensor-Based Robots: Algorithms and Architectures. Edited by C. S. G. Lee. X, 285 pages. 1991. *(ROB)*

Vol. 67: Designing Hypermedia for Learning. Edited by D. H. Jonassen and H. Mandl. XXV, 457 pages. 1990. *(AET)*

Vol. 68: Neurocomputing. Algorithms, Architectures and Applications. Edited by F. Fogelman Soulié and J. Hérault. XI, 455 pages. 1990.

Vol. 69: Real-Time Integration Methods for Mechanical System Simulation. Edited by E. J. Haug and R. C. Deyo. VIII, 352 pages. 1991.

Vol. 70: Numerical Linear Algebra, Digital Signal Processing and Parallel Algorithms. Edited by G. H. Golub and P. Van Dooren. XIII, 729 pages. 1991.

Vol. 71: Expert Systems and Robotics. Edited by T. Jordanides and B.Torby. XII, 744 pages. 1991.

Vol. 72: High-Capacity Local and Metropolitan Area Networks. Architecture and Performance Issues. Edited by G. Pujolle. X, 536 pages. 1991.

Vol. 73: Automation and Systems Issues in Air Traffic Control. Edited by J. A. Wise, V. D. Hopkin and M. L. Smith. XIX, 594 pages. 1991.

Vol. 74: Picture Archiving and Communication Systems (PACS) in Medicine. Edited by H. K. Huang, O. Ratib, A. R. Bakker and G. Witte. XI, 438 pages. 1991.

Vol. 75: Speech Recognition and Understanding. Recent Advances, Trends and Applications. Edited by P. Laface and Renato De Mori. XI, 559 pages. 1991.

Vol. 76: Multimedia Interface Design in Education. Edited by A. D. N. Edwards and S. Holland. XIV, 216 pages. 1992. *(AET)*

Vol. 77: Computer Algorithms for Solving Linear Algebraic Equations. The State of the Art. Edited by E. Spedicato. VIII, 352 pages. 1991.

Vol. 78: Integrating Advanced Technology into Technology Education. Edited by M. Hacker, A. Gordon and M. de Vries. VIII, 185 pages. 1991. *(AET)*

Vol. 79: Logic, Algebra, and Computation. Edited by F. L. Bauer. VII, 485 pages. 1991.

Vol. 80: Intelligent Tutoring Systems for Foreign Language Learning. Edited by M. L. Swartz and M. Yazdani. IX, 347 pages. 1992. *(AET)*

Vol. 81: Cognitive Tools for Learning. Edited by P. A. M. Kommers, D. H. Jonassen and J. T. Mayes. X, 278 pages. 1992. *(AET)*

Vol. 82: Combinatorial Optimization. New Frontiers in Theory and Practice. Edited by M. Akgül, H. W. Hamacher and S. Tüfekçi. XI, 334 pages. 1992.

Vol. 83: Active Perception and Robot Vision. Edited by A. K. Sood and H. Wechsler. IX, 756 pages. 1992.

NATO ASI Series F

Including Special Programmes on Sensory Systems for Robotic Control (ROB) and on Advanced Educational Technology (AET)

Vol. 84: Computer-Based Learning Environments and Problem Solving. Edited by E. De Corte, M. C. Linn, H. Mandl and L. Verschaffel. XVI, 488 pages. 1992. *(AET)*

Vol. 85: Adaptive Learning Environments. Foundations and Frontiers. Edited by M. Jones and P. H. Winne. VIII, 408 pages. 1992. *(AET)*

Vol. 86: Intelligent Learning Environments and Knowledge Acquisition in Physics. Edited by A. Tiberghien and H. Mandl. VIII, 285 pages. 1992. *(AET)*

Vol. 87: Cognitive Modelling and Interactive Environments. With demo diskettes (Apple and IBM compatible). Edited by F. L. Engel, D. G. Bouwhuis, T. Bösser and G. d'Ydewalle. IX, 311 pages. 1992. *(AET)*

Vol. 88: Programming and Mathematical Method. Edited by M. Broy. VIII, 428 pages. 1992.

Vol. 89: Mathematical Problem Solving and New Information Technologies. Edited by J. P. Ponte, J. F. Matos, J. M. Matos and D. Fernandes. XV, 346 pages. 1992. *(AET)*

Vol. 90: Collaborative Learning Through Computer Conferencing. Edited by A. R. Kaye. X, 260 pages. 1992. *(AET)*

Vol. 91: New Directions for Intelligent Tutoring Systems. Edited by E. Costa. X, 296 pages. 1992. *(AET)*

Vol. 92: Hypermedia Courseware: Structures of Communication and Intelligent Help. Edited by A. Oliveira. X, 241 pages. 1992. *(AET)*

Vol. 93: Interactive Multimedia Learning Environments. Human Factors and Technical Considerations on Design Issues. Edited by M. Giardina. VIII, 254 pages. 1992. *(AET)*

Vol. 95: Comprehensive Systems Design: A New Educational Technology. Edited by C. M. Reigeluth, B. H. Banathy and J. R. Olson. 1993. *(AET)*

Vol. 96: New Directions in Educational Technology. Edited by E. Scanlon and T. O'Shea. VIII, 251 pages. 1992. *(AET)*

Vol. 97: Advanced Models of Cognition for Medical Training and Practice. Edited by D. A. Evans and V. L. Patel. XI, 372 pages. 1992. *(AET)*

Vol. 98: Medical Images: Formation, Handling and Evaluation. Edited by A. E. Todd-Pokropek and M. A. Viergever. IX, 700 pages. 1992.

Vol. 99: Multisensor Fusion for Computer Vision. Edited by J. K. Aggarwal. XI, 456 pages. 1993. *(ROB)*

Vol. 100: Communication from an Artificial Intelligence Perspective. Theoretical and Applied Issues. Edited by A. Ortony, J. Slack and O. Stock. XII, 260 pages. 1992.

Vol. 101: Recent Developments in Decision Support Systems. Edited by C. W. Holsapple and A. B. Whinston. XI, 618 pages. 1993.

Vol. 102: Robots and Biological Systems: Towards a New Bionics? Edited by P. Dario, G. Sandini and P. Aebischer. XII, 796 pages. 1993.

Vol. 103: Parallel Computing on Distributed Memory Multiprocessors. Edited by F. Ozgüner. VIII, 332 pages. 1993.

MIX
Papier aus verantwortungsvollen Quellen
Paper from responsible sources
FSC® C105338

If you have any concerns about our products,
you can contact us on
ProductSafety@springernature.com

In case Publisher is established outside the EU,
the EU authorized representative is:
Springer Nature Customer Service Center GmbH
Europaplatz 3, 69115 Heidelberg, Germany

Printed by Libri Plureos GmbH
in Hamburg, Germany